3rd Bali International Seminar on Science & Technology (BISSTECH 2015)

MATEC Web of Conferences Volume 58 (2016)

Bali, Indonesia
15 - 17 October 2015

Editors:

J. Jamari
R. Handogo
E. Suryani

ISBN: 978-1-5108-2464-5

Printed from e-media with permission by:

Curran Associates, Inc.
57 Morehouse Lane
Red Hook, NY 12571

Some format issues inherent in the e-media version may also appear in this print version.

Printed by Curran Associates, Inc. (2016)

For additional information, please contact EDP Sciences – Web of Conferences at the address below.

EDP Sciences – Web of Conferences
17, Avenue du Hoggar
Parc d'Activité de Courtabœuf
BP 112
F-91944 Les Ulis Cedex A
France

Phone: +33 (0) 1 69 18 75 75
Fax: +33 (0) 1 69 28 84 91

contact-edps@webofconferences.org

Additional copies of this publication are available from:

Curran Associates, Inc.
57 Morehouse Lane
Red Hook, NY 12571 USA
Phone: 845-758-0400
Fax: 845-758-2633
Email: curran@proceedings.com
Web: www.proceedings.com

TABLE OF CONTENTS

PREFACE

These last few years, the Indonesian government continues to improve research activities in the field of food, chemical, machinery, electrical engineering, civil, management, and information in order to optimize the resources available. This policy is related with the problems faced by the Indonesian people at this time such as import of raw materials and products and high level of unemployment and poverty. In order to overcome these problems, the development of natural resource-based industries nationwide is required.

Finding solutions for the problems described above, as suggested by many, requires multiple approaches, perspectives and disciplines. And the 3rd Bali International Seminar on Technology (BISSTECH) 2015 is expected to be a forum that effectively provides opportunity to exchange of information between the government, researchers, universities, industry and society as an effort to better understand the problems which hopefully and subsequently will lead to better answers to the problems.

For this reason, the Faculty of Industrial Technology, University of Pembangunan Nasional (UPN) "Veteran" Jawa Timur (Surabaya, Indonesia), the State Polytechnic of Bali (Bali, Indonesia) and the Engineering College National Cheng Kung University (Taiwan) has agreed to jointly organize the 3rd Bali International Seminar on Science and Technology (BISSTECH) with the theme:

"TOWARDS FUNDAMENTAL AND APPLIED RESEARCH FOR SOCIO-ENVIRONMENTALLY SAFE INDUSTRIAL DEVELOPMENT"

DR. IR. NI KETUT SARI, MT

Chairman of Bisstech III
University of Pembangunan Nasional "Veteran" Jawa Timur
Surabaya, Indonesia

Om Swasti Astu, Good morning.

Minister of Research and Higher Education of Indonesia.
Governor of Bali.
Our distinguished guests, speakers, and participants of the seminar.

Welcome to "Bali International Seminar on Science and Technology" or the third BISSTECH, held by University of Pembangunan Nasional "Veteran" Jawa Timur, State Polytechnic of Bali, and National Cheng Kung University – Taiwan. Therefore, I really appreciate the topic of the seminar: "TOWARDS THE FUNDAMENTAL AND APPLIED RESEARCH FOR SOCIO-ENVIRONMENTALLY SAFE INDUSTRIAL DEVELOPMENT".

Ladies and Gentlemen,

On behalf of the committee, please allow me to deliver a short report on the seminar. The objective of this seminar is to share ideas of recent research results from industries, institutions/universities, and government related to Chemical Engineering, Food Technology, Industrial Engineering, Technology And Information Systems, Civil Engineering, Mechanical Engineering, Electrical Engineering, Accounting, Business Administration And Tourism. In this moment, we do hope that all participants will have the opportunity to share their brilliant ideas through effective communication in order to solve the problem of fundamental industrial technology sustainability.

The attendances are from 9 countries, consisting of 202 authors, 5 keynote speakers, and 48 reviewers and scientific committee. On behalf of the committee we'd like to appreciate your participation & excellent works for making this seminar possible and bringing it to success. We'd also like to express our grateful to our honorable guests from UPN "Veteran" Jawa Timur, PNB, NCKU, Sponsors (Pertamina EP, Patrajasa Logistik, Semen Indonesia, Fajar Mas Murni, Dinas Perindustrian dan Perdagangan Jawa Timur di Jember, PT. Gunung Bale, Novis Nature Vavita, Tamara Overseas Corporindo, Bapak Sony), and also Alumni of Chemical Engineering of UPN "Veteran" Jawa Timur, for the supports they have given to us. There is no better words we could say except thank you so much.

Ladies and Gentlemen,

This program will be held for 3 days. Today & tomorrow will be research results sharing sessions. Tonight, the Governor of Bali is willing to greet you at his office for gala dinner while tomorrow night, State Polytechnic of Bali will welcome you for a dinner at Jimbaran beach. The day after tomorrow, those of you who are willing to go for a sightseeing program will have the opportunity to relax and enjoy the beautiful scene and culture of Bali. Please contact the committee for registration and guidance. If you still have time, please extend your stay in Indonesia to see the other beautiful places in Indonesia, such as Surabaya, or Lombok Island, and others, which are very nice and close from Bali.

Ladies and Gentlemen,

We really hope that the seminar would give a contribution in creating a better future for mankind. We deeply apologize for all the inconveniences you might experience during the sessions of this seminar. Thank you.

Om Shanti Shanti Shanti Om.

MATEC Web of Conferences **58**, 00002 (2016) DOI: 10.1051/matecconf/20165800002
BISSTECH 2015

NYOMAN INDAH KUSUMA DEWI, SE, MBA, PH.D

Chairman of Bisstech III
Bali State Polythecnic, Indonesia

Excellencies: Minister of Research, Technology and Higher Education - Mohammad Nasir and Governor Bali - Made Mangku Pastika
Distinguished Participants,
Ladies and Gentlemen:

Good morning,
It is my privilege, as General Chair from Bali State Polytechnic, to welcome all of participants of Bali International Seminar on Science and Technology (Bisstech) III 2015 with the topic: **"Towards Fundamental and Applied Research For Socio-Environmentally Safe Industrial Development".**

This international seminar is jointly organised by Bali State Polytechnic, the Faculty of Industrial Technology, University of Pembangunan Nasional (UPN) "Veteran" Jawa Timur, and the Engineering College of National Cheng Kung University-Taiwan based on the same passion to provide a national industrial development strategy based on national natural resources. I have great confidence that Bisstech III will able to contribute ideas, strategy, theoretical and practical knowledge to achieve the underlying purpose of this seminar.

I would like to take this opportunity to express my sincere gratitude and appreciation to the honorable speakers: Professor Pao-Shan Yu from Taiwan; Professor Dr. Wolfgang W. Schmahl from Germany; Prof. Inocencio E. Buot, Jr, PhD from Philippine; and Prof. Yasushi Kiyoki from Japan for sparing their precious time to participate in this seminar. I would also like to extend my thanks to all presenters and participants for their participation. My deep appreciation to the organising committee for their hard work, without them this seminar would not be made possible.

Finally, I wish this fruitful 2 days will be beneficial for all of us as the academia and researcher by sharing and exchanging our knowledge and experiences for the better future. And enjoy your time to explore the beautiful Bali.

Thank you.
General Chair

Nyoman Indah Kusuma Dewi, SE, MBA, Ph.D

MATEC Web of Conferences **58**, 00002 (2016) DOI: 10.1051/matecconf/20165800002
BISSTECH 2015

PROF. DR. IR. H. TEGUH SOEDARTO,MP

Rector
University of Pembangunan Nasional "Veteran" Jawa Timur
Surabaya, Indonesia

Assalamualaikum Wr. Wb.
Our Distinguished Guests, Speakers, and Participants of the seminar.

Thank God for granting us with blessings, so we could gather here today in this very special moment, BALI INTERNATIONAL SEMINAR ON SCIENCE AND TECHNOLOGY (BISSTECH) 2015.
On behalf of University of Pembangunan Nasional "Veteran" Jawa Timur, please let me extend my warmest greetings and welcome our distinguished guests. I would also like to express my appreciation to each of you for attending this International Seminar, especially all of you who come from overseas.

Ladies and Gentlemen,
These last few years, the Indonesian government continues to improve research activities in the field of food, chemical, machinery, electrical engineering, civil, management, and information in order to optimize the resources available. Therefore, I really appreciate the topic of the seminar: "TOWARDS THE FUNDAMENTAL AND APPLIED RESEARCH FOR SOCIO-ENVIRONMENTALLY SAFE INDUSTRIAL DEVELOPMENT" aiming to communicate the results of research conducted by Industries, Research Institutes, Universities and Governments, both domestic and abroad based on national natural resources.

In this concern, UPN "Veteran" Jawa Timur is very pleased to join hands with Bali State Polytechnic, and National Cheng Kung University - Taiwan in holding the seminar since these institutions are on the same page in developing science and technology.
It is gratifying to note that the agenda of the seminar covering a wide range of very interesting items relating to the topic.
We believe this seminar is a great opportunity for all delegates to discuss the existing problems and find solution concerning the science and technology.

Distinguished guests, Ladies and Gentlemen,
Furthermore, may I congratulate the organizing committee for making this seminar possible. I wish this moment could not only enrich a broader knowledge, but also make new friends, and build networking in creating a better future.
Hopefully all of you are having very pleasant stay in Indonesia, especially in Bali, and if it is possible please make a visit to our campus in Surabaya, the second biggest city in Indonesia, which is only 40 minutes from Bali. I apologize for all the inconveniences you may experience during the session. Good Luck!

Wassalamualaikum Wr. Wb.
Prof. Dr. Ir. H. Teguh Soedarto, MP

Ir. MADE MUDHINA, MT

Director of Bali State Polythecnic
Indonesia

The honorable;

1. Minister of Research, Technology and Higher Education
2. Governor of Bali
3. Keynote speakers
4. Presenters and participants
5. Rector and staff of Universitas Pembangunan Nasional Veteran East Java
6. Deputy directors and head of departments of Bali State Polytechnic
7. Distinguished guests, ladies and gentlemen…..

Om Swastyastu,

First of all, let us express our great gratitude to Ida Sang Hyang Widhi Wasa, The Almighty God for His blessing today, so that we can do this International Seminar.

The honorable minister, governor, ladies and gentlemen,

As we have already known that the development and economic growth of a country is strongly influenced by some economic factors such as natural resource, human resource, capital, and managerial asset who can organize and manage the production factors. These factors should be supported by the advance of technology, as the result, some experts believe that the advance of technology plays a very important and significant role to determine the development and economic growth of that country since it tries to combine the production factors in realizing production objectives.

In general, the technology in developing countries is still left behind compared to the developed countries. It has a tendency that the gap among those countries is getting wider and in consequence it will influence the development of economy of those countries. The developing countries need different sort of technology that it is in the developed countries. If a developing country tries and uses the technology in developed countries, neglecting the existence of local culture, this may bring any problems for those developing countries due to the reason that it is neither effective nor efficient. The developing country, like Indonesia, for example, is really appropriate to apply something we call it "adaptive technology". It is a kind of technology which comes from research and development, adapted by some social considerations in order to be able to solve any real problems exist. The main indicators for the adapting process in developing technology in Indonesia should be in accordance with the consideration to employee some labor forces, use domestic material in order to add our national income.

Simple technology is strongly effective to be applied in Indonesia in order to be able to solve any problems without creating any other new and complicated problems. It is because simple technology is much easier to be understood and applied, and especially for those who are in the low culture of technology. Meanwhile, the development of mass technology which can employee big numbers of employees should be regularly continued and developed.

In relation to the development of technology, it cannot be separated from basic and applied science. Basic science will try to discover more about nature and its surrounding, then the applied one, as it is developed by Bali State Polytechnic, will try to solve any practical problems as well as to handle any problems faced by the human being itself. The collaboration between basic and applied science will finally produce any technology that can ease human to solve problems and obstacles. That is why, it is really appropriate to name this international seminar as "BISSTECH"-Bali International Seminar on Science and Technology. It can be clearly seen that science and technology will collaborate. It is a brilliant thought which will finally some media in order to help human to reach a better quality of life which will not destroy out nature.

Distinguished guests, ladies and gentlemen,

Bali State Polytechnic, as a vocational higher education, has a very strong commitment to run Three Basic Principles of Higher Education (Tri Dharma Perguruan Tinggi) by utilizing science and technology (IPTEK) in order to give direct impact to the economy of society. That is why; Bali State Polytechnic has a strong willing to cooperate with UPN Veteran, Surabaya, East Java to run this international seminar with the theme "TOWARDS FUNDAMENTAL AND APPLIED RESEARCH FOR SOCIO-ENVIRONMENTALLY SAFE INDUSTRIAL DEVELOPMENT BASED ON ENGINEEERING AND TECHNOLOGY". This seminar aims at facilitating researchers to exchange their ideas and thoughts as well as to build any network among them in this country, which later on trying to find any right solution of our nation' problems through the development of technology that you have done in your research.

At last, on the behalf of the committee, I would like to extend my special thank to Minister of Research, Technology and Higher Education, Governor of Bali, for your excellent support to this seminar. Furthermore, I also would like to thank to keynote speakers, presenters, and all participants for your contribution in this prestigious international seminar. The last but not least, we would like to thank to UPN Veteran East Java which has cooperated with Bali State Polytechnic to run this international seminar.

Finally, I wish you a very good luck and all the best for the seminar. I do hope this seminar will give us any positive contribution. Thank you

Om Shanti, Shanti, Shanti, Om

edp sciences

MATEC Web of Conferences

Statement of Peer review

In submitting conference proceedings to MATEC Web of Conferences, I certify to the Publisher that I adhere to the **Policy on Publishing Integrity** of the journal in order to safeguard good scientific practice in publishing.

1. All articles have been subjected to peer review administered by the proceedings editors.
2. Reviews have been conducted by expert referees, who have been requested to provide unbiased and constructive comments aimed, whenever possible, at improving the work.
3. Proceedings editors have taken all reasonable steps to ensure the quality of the materials they publish and their decision to accept or reject a paper for publication has been based only on the merits of the work and the relevance to the journal.

Title, date and place of the conference

Title: The 3rd Bali International Seminar On Science & Technology (BISSTECH 2015)

Date: October 15-17, 2015

Place: Grand Inna Kuta Hotel, Bali, Indonesia

Proceedings editor(s):

J. Jamari, Dr. (University of Diponegoro, Semarang, Indonesia) *

Renanto Handogo, Ph.D. (Sepuluh Nopember Institute of Technology, Surabaya, Indonesia)

Erma Suryani, Ph.D (Sepuluh Nopember Institute of Technology, Surabaya, Indonesia)

Date and editor's signature

March 31, 2016

J. Jamari, Dr.

ACTIVATION OF TRASS ROCK AS BLEACHING PALM OIL (CPO)

Laurentius Urip Widodo[1], Sukirmiyadi, Siswanto and Ely Kurniati
University of Pembangunan Nasional "Veteran" Jawa Timur
Surabaya, East Java, Indonesia
E-mail : gerak_samodro3@yahoo.com[1]

ABSTRACT

Trass rock is a rock originating from volcanic eruption materials which has experienced weathering. It has the similar composition as bentonite. Therefore, trass rock can also be used as raw material as a bleaching earth to substitute bentonite. In this study, a two-stage process was conducted: to do the activation process of trass rock and stage bleaching process of palm oil (CPO). In activation process, trass rock was performed by using hydrochloric acid, then trass rock was crushed until it could pass of the 200 mesh. After that, the soft rock was weighed and added to the acid solution with the ratio of: 1 : 10 with the concentration of hydrochloric acid solution 5 N and its activation time was 4 hours at the temperature of 105^{0}C. Meanwhile, for the bleaching process, at first, oil was treated with degumming and netralization process. Furthermore, oil bleaching process was carried out by adding the activated trass . Then trass was put into the oil when its temperature was reaching up to 186^{0}C. Heating was continued until its temperature was reaching up to 198°C for 45 minutes. In this activation process, it was obtained that the content of SiO_2 was 37.8%, Al_2O_3 of 4.4% and the ratio of SiO_2/Al_2O_3 is 8.6. In the bleaching process of palm oil (CPO), the best condition of trass addition was 4% of the oil weight, and its bleaching time was 45 minutes, the red color intensity obtained was 6.9 and yellow was 6.1, with FFA as big as 2.13% and peroxide value was 5.45 meq/kg.

Keywords: activation,trass rock, bleaching, CPO.

INTRODUCTION

Trass rock (or tuff) is a kind of rock originated taken from eruption material of volcano or volcanic sediment. Trass is formed when it becomes decayed especially that of containing andesitic composition (Kurum, 2009). As Indonesia has so many volcanoes in all parts of it, trass can be found everywhere in a very great quantity. Up to now, trass is only known as a natural cement and it is limited as a basic/raw material and mixture of cement. Therefore, we have to think of it as another alternative of use that seems to have higher economic value. Trass is actually the blend of active aluminia and silicate from which it contains of silica amorf (Edwin, 2007). The chemical composition of trass is almost the same as that of bentonite from which contains the blend of aluminia and silicate. Therefore, trass is possible to be used as an alternative of bleaching earth basic material.

Nowadays in Indonesia, the plantation of palm is getting more and more developed. Therefore, oil making process needs bleaching earth as a bleaching media when the palm oil (CPO) is manufactured into frying oil. Up to now, bentonite is the only one thing used as a raw material in bleaching earth making. However, we have to know that later on bentonite will be used up because it is mine goods. Bentonite can be classified into two, Na-bentonite and Ca-bentonite. The characteristics of Na-bentonite is that it has a big content of Na^+, it can expand and suspend if it is dispersed into the water. This material can not be used as a raw material of bleaching earth. Meanwhile, in Ca-bentonite, it has more Ca^{2+} and Mg^{2+} than Na^+. The characteristics of Ca-bentonite is that its absorption power is only a little if it is dispersed into the water, it will precipitate and will not be suspended (Fu, 2011). As a matter of fact that trass has the similar composition as bentonite, it can also be used as bleaching earth. Due to the fact that trass has so many impurities covering its surface, trass needs to be activated at first before being used as a bleaching earth. Some impurity components are such CaO, MgO and Fe2O3. These impurities can be reduced or even removed by acid activation. Finally, it can increase the absorption power from the trass. The mass ratio of bentonite and the volume of acid solution is 1:10 (w/v) (Foleto, 2011). Acid can create new pore so that it will increase the acidity of its surface by cation substitution such as Al^{3+}, Fe^{3+} and Ca^{2+} with H^+ (Ejikeme, 2013). The absorption power of color substance/essence of bleaching earth is because the existence of Al ion available in absorbent surface. Acid activation is aimed at increasing both the degree of Al_2O_3 and SiO_2. This activation will then be obtained the ratio of Al_2O_3 and SiO_2. The increase of SiO_2 is very important because it can increase the content of Si-OH (Cylanol) on the adsorbent surface, that is trass (Makhoukhi, 2009). During the bleaching process of the dye, peroxide and other impurities are removed from the raw material of oil. Through this process, the appearance of oil product can be improved. Its color becomes clearer and its production stability might be improved (Falaras, 2000). Bleaching earth can also remove chlorophyll, carotenoids, phosphorlipids, metals. This removal can be done through adsorption only (Makhoukhi, 2009). The new pore is created from the removal of some ion such as Al^{3+}, Fe^{2+}, Fe^{3+}, Mg^{2+} from the open octahedral layer at the edge of smecteet. More over, silicon (Si) is relieved from the tetrahedral layer of precipitate in the form of silica amorf with H^+ion. In fact, activation process might increase the surface area and its diameter of pore from the clay. This process can absorb more pigment (Khan, 2014). Here is the chemical reaction in activation process with trass aid as follow:

Reaction of dissolving substance.

$$CaO + H_2O \longrightarrow Ca(OH)_2 \quad (1)$$

$$Na_2O + H_2O \longrightarrow 2NaOH \quad (2)$$

Reaction rock trass with chloride acid as the activator.

$Al_2O_3 + 6HCl \longrightarrow 2AlCl_3 + 3H_2O$ (3)
$Fe_2O_3 + 6HCl \longrightarrow 2FeCl_3 + 3H_2O$ (4)
$MgO + 2HCl \longrightarrow MgCl_2 + H_2O$ (5)
$CaO + 2HCl \longrightarrow CaCl_2 + H_2O$ (6)
$Na_2O + 2HCl \longrightarrow 2NaCl + H_2O$ (7)

In general, the color of palm oil (CPO) before being processed, is dark red. Therefore, bleaching process needs to be done to eliminate it. To do bleaching, palm oil must be treated through pretreatment process, that is degumming and netralization. The purpose of degumming process is to remove the sap without reducing the existence of fat acid in the oil and to precipitate fosfatide which can not be soluble in the water. To do degumming process, palm oil (CPO) must be heated up to 80^{0}C, then added with phosphate acid of 85% as much as 0.15% from oil weight, stir within 15 minutes. The purpose of neutral-lization process of palm oil (CPO) is to reduce or remove Free Fat Acid (FFA) in the palm oil. The existence of neutralization process mostly employed is using the solution of NaOH. For netralization process, the concentration NaOH of solution employed was 11.1% of the weight (16^{0}Be) and stirred up continually within 25 minutes while being heated at the temperature of ±59^{0}C.

The experiment was aimed at activating the trass rock so that it could be used as a bleaching earth that had much higher economic value. So far, trass rock was only used as a raw material of cement making and a mixture of concrete. The activation process of chloride acid (HCl) was expected to obtain the bleaching earth from trass as a raw material that had a high absorption power like bentonite. Therefore, trass was used as a bleaching of palm oil (CPO) by varying the weight of trass in oil.

METHODOLOGY

The materials required in this research were trass rock, HCl, aquades and palm oil (CPO). Meanwhile, some instruments/equipments required were iron mortar, stove/ heater, stiring tank, filter paper, funnel and oven/ dryer.

The research was carried out in two process stages, the activation process of trass rock and bleaching process of palm oil (CPO). In activation process, trass rock was crushed until smooth, then it was screened by using a screener of 200 mesh. The trass that could pass down a screener of 200 mesh was washed by using water to remove the dirt or soil attaching on the trass, then screened again and then dried up in the oven. Having been washed and dried up, activation of trass and some solution of chloride acid (with ratio 1:10) was conducted (150 gr of trass rock in 1500 cc of acid solution). The solution concenteration of chloride acid 5 N with time allocation of activation process within 4 hours was heated until its temperature was reaching up to 105^{0}C. After that filtrating was done to separate the solid matter from the liquid. The solid matter obtained was then washed by using aquades until the pH of water was up to ±5. Then, it (solid matter) was dried up in oven with the temperature 110^{0}C. The dried solid matter was then analyzed to see the content of SiO_2, Al_2O_3,CaO, MgO and Fe_2O_3 by using X-Ray Fluorescene Mini Pal 4. Meanwhile, for the stage of bleaching process, at first oil was treated by degumming and neutralization process. In these two process, oil was heated up to 80^{0}C then added with 85% of phosphate acid as much as 0.15% of the weight of oil and stirred within 15 minutes. After that the temperature was lowered up to 60^{0}C and added with NaOH 11.1% (16^0 Be) as much as 2% of the volume of oil and stirred within 25 minutes. Next, oil was cooled down and refined to separate oil from the soap. The netralized oil was then bleached by adding the activated trass taken from the best activation process as much as 2%; 4%; 6% and 8% of oil weight. The trass was put into the oil when its temperature reached up to 186^{0}C and continued up to 198^{0}C as long as 45 minutes. Then this bleaching oil was analyzed its color intensity by using konicaminolta CR-10. Meanwhile, the content of FFA and peroxide was analyzed through titration.

RESULT

Based on the analysis result, it was known that the quality of trass rock before activating process as shown in the (Table-1) below:

Table- 1. Result Data of Former Trass Rock Analysis Before Activating Process

Former Material Original	Level % Weight					Ratio $\frac{SiO_2}{Al_2O_3}$
	SiO_2	Al_2O_3	Fe_2O_3	CaO	MgO	
Before Washing	23.4	3.7	23.8	27.1	0.5	6.32
After Washing	19.3	3	26.9	26.4	0.5	6.43

Meanwhile, the analysis result of trass rock after activation process using solution activator of chloride acid (HCl) could be seen at (Table- 2) below:

Table- 2. Result Data of Trass Rock Analysis After Activating Process with Chloride Acid (HCl)

Trass Activation		Level % Weight					Ratio $\frac{SiO_2}{Al_2O_3}$
Time (Hour)	Concentration of HCl	SiO_2	Al_2O_3	Fe_2O_3	CaO	MgO	
4	5 N	37.8	4.4	9.38	20.6	0.6	8.6

The following was the quantity analysis result of palm oil (CPO) before and after treatment process as shown at (Table- 3).

Table- 3. Analysis Data of Palm Oil (CPO) Before Treatment Process

Raw material of oil	FFA (%)	Peroxide value (Meq/ Kg)
Former oil	3.51	19.90
Netralization oil	2.45	13.95

The next was the analysis result of oil after bleaching process as shown at (Table- 4).

Table- 4. Analysis Data of Palm Oil (CPO) as The Result of Bleaching Trass by HCl Activation

Oil after being neutralized		Oil Color		FFA (%)	PV Meq /Kg
		Red	Yellow		
		7.5	5.9	2.45	13.95
Bleaching Time	**Trass Add %Weight**	**Oil Color**		**FFA (%)**	**PV Meq /Kg**
		Red	**Yellow**		
45 minutes	2	7.0	5.9	2.16	5.9
	4	6.9	6.1	2.13	5.45
	6	7.1	6.0	2.19	11.05
	8	7.3	5.8	2.37	12.75

DISCUSSION

Natural trass rock was perhaps still mixed with soil. Therefore, it needed to be cleaned or washed before being activated. However, the washing had to be done when the size of trass particle had not been smooth or its size was about 8-10 mesh. This condition was meant to protect some required particles oftrass from removing. In this research the washing of trass rock was done after the trass rock had been shattered smoothly until it could pass the screen of 200 mesh. This might reduced the level of some required particles. (Table- 1) showed that some important particles such as SiO_2 and Al_2O_3 required as a bleaching earth were removed. Before being cleaned/ washed, the level of SiO_2 was 23.4% and it decreased into 19.3% after being washed. The decrease also happened in Al_2O_3. Before being cleaned, it was 3.7% and it decreased into 3% after being cleaned. Meanwhile, the element of impurities were getting increased, Fe_2O_3 = 23.8% became 26.9%. Therefore the washing had to be done carefully and correctly, so that SiO_2 and Al_2O_3 would not decreased.

Chloride acid was used as an activator in trass rock activation to removed impurities of Fe_2O_3, CaO and MgO available in the trass rock. The activation result showed that Fe_2O_3 element got reduced more than the others. These was due to Fe_2O_3 was very reacted to chloride acid so that during activation process, reaction mostly accured between chloride acid and Fe_2O_3. This could be seen in (Table- 2). It was known that using chloride acid as an activator could reduced impurities much available in the trass rock especially impurities of Fe_2O_3. In fact, chloride acid could decrease Fe_2O_3 up to 9.38% from the former Fe_2O_3 in the trass rock of 26.9%. In this activation process it was known that the other impurities in the trass rock did not get reduced signify-cantly. This could happened because consentration of acid used as an activator was only little a before it reacted to Fe_2O_3. Therefore it could no longer reacted to CaO and MgO available in the trass rock. More over, reaction would happen to the reactive element and less reactive and then continued up to the element which was not reactive at all. The decrease of oxides was due to reaction occured inside the acid (Ejikeme, 2013).The result obtained true activation of chloride acid could be seen at (Table- 2). It was known that the level of SiO_2 was 37.8%, Al_2O_3 = 4.4% and ratio of SiO_2 and Al_2O_3 was 8.6.This might cause the quantity of four was increasing If its surface was enlarged, the level of acidity would increase as an effect of structure changes (Alendaroglu, 2003). Therefore the activation of trass rock and chloride acid was considered to be good enough. The activation result of trass rock and chloride acid was then used as a bleaching of palm oil (CPO).

However before bleaching, palm oil was pre-treated at first. In fact the result of this pretreatment could decrease its FFA and the quantity of peroxide available in the oil could be seen at (Table- 3). The decrease of FFA was due to its reaction with NaOH added during netralization process. Formerly the FFA was 3.51% and it decreased in to 2.45%. This decline was only a little because the addition of NaOH was only 2% of the oil volume. If the quantity of NaOH was too much, it might cause the quantity of oil reduced. Furthermore, the result of oil bleaching employing trass as a result of activation with chloride acid could be seen at (Table- 4). It was shown that the color adsorption, FFA and the quantity decrease of peroxideoccured during the trass addition of 4% of the oil weight within 45 minutes. Moreover, before bleaching it was known that the intensity of red color was 7.5 and the yellow one was 5.9. This caused the color texture of oil tended to look red. But after bleaching process by adding the trass of 4% of the oil weight, the intensity of red color became 6.9 and the yellow became 6.1. The intensity decrease of red color and the intensity increase of yellow one made the oil texture became yellower. (Table- 4) showed that the more percentage of trass addition it did not mean that the intensity decrease of red color became higher. This was due to the adsorption balance that had already been achieved between a mixture of adsorbent and oil. This condition restrained the pigment to be adsorbed from the over dosage of adsorbent (Usman , 2013). Therefore, the more percentage of trass addition, the more increase the level FFA would be. Furthermore,

(Table- 4) also showed the former FFA was 2.45% if it was added with trass of 4%, its FFA decreased in to 2.13%. Meanwhile if it was added with trass of 8%, the level of FFA was 2.37%. (Table- 4) also showed that the more percentage of trass addition the higher number the peroxide would be (Wannahari, 2012). He said that in general the used of more adsorbent did not mean to reduce the value of PV (Peroxide Value). This was due to the saturation in use of several adsorbent from which level had to be back to the limit point of adsorbent to adsorb wholly. The former peroxide value was 13.95 meq/kg and the trass addition of 4%, the peroxide value decreased in to 5.45 meq/kg. Meanwhile with the trass addition of 8%, the peroxide value was 12.75 meq/kg. Therefore in bleaching process the higher addition of trass, it even decreased its adsorption power of trass as a bleaching earth.

CONCLUSION

The research result concluded that the washing of trass rock as the basic/ raw material should not be in smooth condition, but should be in bigger particles to maintain SiO_2 and Al_2O_3 from removing. Meanwhile the activator of HCl with its acid concentration of 5 N and activation time 4 hours, it could be found that the level of SiO_2 was 37.8%, Al_2O_3 = 4.4% and ratio of SiO_2/ Al_2O_3 was 8.6. In bleaching process, the best condition of trass addition was 4% of the oil weight with the bleaching time allocation was 45 minutes. This condition was obtained that the intensity of red color was 6.9 and the yellow one was 6.1, the level of FFA was 2.13% and its peroxide value was 5.45 meq/kg.

REFERENCES

Alemdaroglu T.,2003. Investigation of the Surface acidity of a Bentonite Modified by Acid Activation and Thermal Treatment, Turk J Chem Vol.2, p.675 – 681

Ejikeme E.M. , Egbuna S.O. , Ejikeme P.C.N. , 2013. Optimal Bleaching Performane of Acid Activated Ngwu – langwu Clay, International Journal of Engineering and Innovative Technology (IJEIT) Vol.3,Issue 5, p.13-19

Falaras P., Lezou F.,Seiragakis G. and Dimitrios P., 2000. Bleaching Properties Of Alumina–Pillared Acid - ctivated Montmorillonite, Clays and Clay Minerals,Vol.48, No.5, p.549-556

Foletto E. L., Colazzo G. C., Volzone C. and Porto L. M. ,2011. Sunflower Oil Bleaching By Adsor tion Onto Acid-Activated Bentonite , Brazilian Journal of Chemical Engineering, Vol. 28, No. 01, p. 169 -174

Fu Y., Chung D.D.L., 2011. Coagulation of oil in water using sawdust, bentonite and calcium hydroxide to form floating sheets, Applied Clay Science 53,p. 634–641

Khan I.H., *Tabasum F., Iqbal S.*, 2014. Utilization Of Locally Available Bentonites For Water Purification By Activation, A scientific journal of Comsats – Science Vision, Vol.20, No.1, p. 67-73

Kurum S.,Akkoca D.B.,2009. The Petrographic Properties of the Epiclastites in Neogen Volcanic Succession and Usage as Admixture in Pozzolanic Cement Production (Cemisgezek/Tunceli,Turkey), Ozean Journa l of Applied Sciences 2(1), p.91-101

Makhoukhi B.,Didi M. A., Villeminb D. and Azzouzc A., 2009. Acid Activation of Bentonite For Use as a Vegetable Oil Bleaching Agent, Grasas Y Aceites, 60 (4), Julio-Septiembre, p. 343-349

Usman M.A., Oribayo O., Adebayo A. A.,2013. Bleaching of Palm Oil by Activated Local Bentonite and Kaolin Clay from Afashio Edo - Nigeria , Chemical and Process Engineering Research, Vol.10, p.1-11

Wannahari. R.,Nordin.M.F.N.,2012.Reduction of Peroxide Value in Used Palm Cooking Oil Using Bagasse Adsor – bent, American International Journal of Contemporary Research, Vol.2, No.1, p.185-191

ANTIOXIDANT ACTIVITY AND SENSORY PROPERTIES CARROT (*Daucus carrota*) SOYGHURT

Enny Karti Basuki Susiloningsih, Ulya Sarofa, Fauziah Imroatus Sholihah
University of Pembangunan Nasional "Veteran" East Java, Surabaya, Indonesia
E-Mail : ennykartibasuki@gmail.com

ABSTRACT

Carrot (*Dascus carota L.*) is one of the more commonly used vegetables of human nutrition. It is rich in beta carotene, ascorbic acid, tocopherol and classified as vitaminized food. Combination of carrot juice and yoghurt produce a nutritionally balanced food. The aim of this study is to determine the best combination of carrot juice concentration and was fermentation duration. soyghurt was prepared in the laboratory scale production from soymilk and carrot juice was fermented with *Streptococcus thermophillus* and *Lactobacillus bulgaris*. Carrot soyghurt was prepared by blending soymilk with 10, 15 and 20% carrot juice before fermentation 14, 16 and 18 hours. The carrot soyghurt produced from 15% carrot juice and 16 hours fermentation were accepted of panelists, having 140 of taste score, 104 of flavour score, 118 of color score and 94.5 of viscocity, 6.52% of protein content, 1.51% of fat content and 52.49% of activity antioxidant.

Keywords: antioxidant activity, sensory properties, fermentation, carrot soyghurt

INTRODUCTION

The oxidative damage caused by free radicals and other reactive oxygen species plays a significant pathological role in human diseases (Wang *et al.*, 2006). Among these diseases cancer has an important place. According to observational epidemiological studies, diets containing mainly fruits and vegetables are related to a lower cancer incidence, especially cancers from the gastrointestinal tract. This is due in part to the dietary antioxidant content of fruit and vegetables (Serrano *et al.*, 2007). So it is clear that natural antioxidant from foods may reduce the oxidative damage on human body (Lin and Yen, 1999).

Soymilk has a high amount of protein, iron, unsaturated fatty acids and niacin, but it is low in fat, carbohydrates, calcium. It has little saturated fat and no cholesterol, hence it is known to reduce the risk of heart disease (Kim *et al.*, 2006, Jinapong *et al.*, 2008). It is safe for people with lactose intolerance or milk allergy and for children with galactosemia. However, the beany flavor due to the presence of aldehydes and alcohols and flatulence caused by indigestible oligosaccharides limit the widespread consumption of soymilk (Blagden and Gilliland, 2005, Yang and Zhang, 2009). Soy-based foods are promising supplements to overcome existing protein-calorie-malnutrition problems (Khare *et al.*, 1994). However, its consumption may lead to digestive problems associated with the presence of raffinose and stachyose. One method of overcoming such problem and also to improve the acceptability is by fermentation (Hati *et al.*, 2013).

Fermentation improves the bioavailability of isoflavones, assists in digestion of protein, provides more soluble calcium, enhances intestinal health, and supports immune system. Soybeans contain the highest concentration of isoflavones, among all foods (Ismail and Hayes, 2005, Kuo *et al.*, 2006). The range of health benefits such as cancer prevention, reduced risk of osteoporosis, valuable role in chronic renal disease, and protection against cardiovascular disorders are claimed to be due to the presence of these isoflavones (Orhan *et al.*, 2007). However, soymilk is the aqueous extract of whole soybeans (Glycine max). Soymilk is considered as a suitable economical substitute for cow's milk and an ideal nutritional supplement for lactose- intolerant population (Dhananjay *et al.*, 2006). Soybeans is a rich source of isoflavones, which are reported to have beneficial estrogenic effects (Adlerereutz, 2002, Brouns, 2002, Corn *et al.*, 2004) with potential bioactive antioxidant properties. Probiotic lactic acid bacteria, when grown in soy milk with whey protein concentrate have the ability to utilize phenolic components producing phytase enzyme and reducing the polyphenol content after fermentation. Free radicals and other reactive oxygen species are generated by exogenous chemicals or endogenous metabolic processes in food systems or human body. The radicals may cause oxidative damage by oxidizing biomolecules and results in cell death and tissue damage (Kehrer, 1993).

Carrot (*Dascus carota L.*) is one of the more commonly used vegetables of human nutrition. It is rich in beta carotene, ascorbic acid, tocopherol and classified as vitaminized food (Hashimoto and Nagayama, 2004). Combination of carrot juice and yoghurt produce a nutritionally balanced food. Carrots are good source of carbohydrate, calcium, phosphorous, iron, potassium, copper, manganese and sulphur. It is an excellent source of vitamin A, B1, B2, C, E, thiamin, folic acid and riboflavin but lack in protein and fat. Yoghurt is rich in protein and fat but is deficient in iron and vitamin C. Yoghurt is an important dairy product, particularily for consumers with lactose intolerance (Mohamed *et al*, 2014) Supplementation of yoghurt with selected vegetables will provide additional health properties, especially concerning with antioxidant properties and will result in the development of novel functional dairy products (Lejuo *et al*,2014). Blending of yoghurt with carrot juice would produce a nutritionally rich food (Ikken *et al.*, 1999, Raum, 2003). The intake of carrot as potent antioxidants, appear to be associated with better health. It is not only preventing vitamin A deficiency but also cancer and other diet related human diseases. It has greater cytotoxic effect against cancer cell and reducing the enzymes that promote the conversion of precarcinogens to carcinogens. It may also enhance the immune system, protect against stroke, high blood pressure, osteoporosis, cataracts, arthritis, heart disease, bronchial ashma and urinary tract infections (Beom *et al.*, 1998, Sun *et al.*, 2001, Seo and Yu, 2003).

Soybeans are the most important food source of isoflavones that exhibit antioxidant activity. Carrots are food source of vitamin C (ascorbic acid), vitamin E (alfa tocopherol) and carotenoids as non enzymatic antioxidants. Combination of carrot juice and soyghurt produced a nutritionaly balanced food and an ideal nutritional supplement for lactose intolerant population. This study was therefore, aimed to find the best combination of carrot juice concentration and its fermentation duration in soyghurt.

METHODOLOGY

Materials used in this research were soybeans, starters (*Lactobacillus bulgaricus* and *Streptococcus thermophillus*), carrot, low fat skimmed milk, sucrose, and aquadest. The equipment used in this research were magnetic stirrer, autoclave, digital balance, centrifuge, spectrophotometer, micro Kjeldahl and several glass tools to support the research.

Preparation of carrot juice used carrot roots were washed thoroughly, skins were removed, peeled by sharp knife and cut longitudinally into halves. These halves were stem blanched for ten minutes to inactivate pectinase and peroxidase enzymes, in order to tenderize the carrot tissues. The juice was obtained by mixer these in blender with sieves. The obtained juice was analyzed for its antioxidant activity using DPPH.

Preparation of soy milk used manufacture of soy milk was started from soybeans sorting and cleaned under running water. Then, they were immersed in water for 8 hours and washed thoroughly. Soybeans were boiled in water with a ratio of 1:8 until it reached temperature of 90°C. Boiled soybeans were blended until it turned into a mush, then they were filtered to get the soy milk. Furthermore, soy milk as raw material was analyzed for its antioxidant activity using DPPH.

Preparation of inoculum used stock cultures were prepared by mixing nutrient broth grown cultures with sterile, incubation at 37°C for 24 hours. As much as 1% cultures were transfered into 10% (w/v) of skim milk, and put into incubation at37°C for 48 hours.

Preparation of carrot soyghurt used soy milk and various concentrate of carrot juice (10%, 15% and 20% v/v) with a total volume 300 ml and added sucrose of 4% (w/v), and skim milk of 20% (w/v) were homogenated. The mixtures was then pasteurized at 85°C for 15 minutes, and cooled to 40-45°C. Starter was added at the rate of 5% (*L. bulgaricus:S. thermophillus* = 1:1) was homogenized and put in various of incubation (14, 16 and 18 hours) at room temperature. The obtained mixture was analyzed for antioxidant activity using DPPH, whereas fat by AOAC (1990) and protein by AOAC (1990).

Sensory evaluation used twenty panelists student of Food Technology of Faculty of Industrial Technology Indusrty University of Pembangunan Nasional "Veteran" Surabaya Indonesia experience in the use and evaluation of carrot soyghurt. They evaluated 20 g portions of each carror soyghurt samples using an unstructured hedonic scale 5 with ends anchored, 1 dislike very much and I like very much. The samples were presented at random and individually evaluated by Nelson and Trout (1981).

Statistical analysis was performed using two ways analysis of variance (Anova) and the mean were compared across groups by DMRT and the significant differences were determined at $p < 0.05$.

RESULTS AND DISCUSSION.

Activity Antioxidant of Raw Materials

The antioxidant activity of soy milk, carrot juice and soyghurt were 13.72%, 8.35% and 32.83%, respectively. Bioactive compounds presented in soybeans varied greatly with the cultivar, weather and geographical sowing location (Hoeck *et al*., 2000, Seguin *et al*., 2004), which was similar to that of bioactive of carrot.

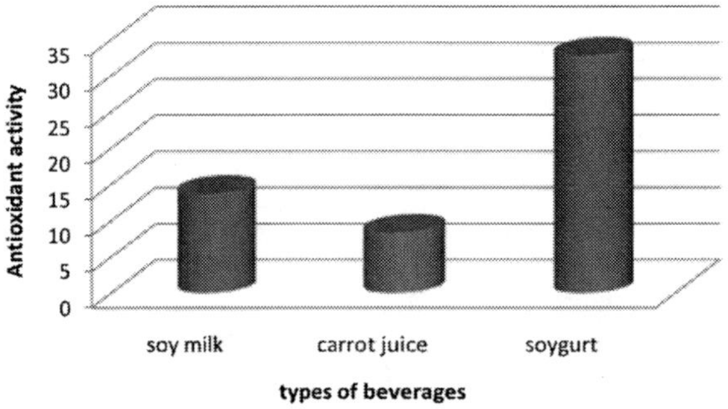

Figure-1. antioxidant activity of soymilk, carrot juice and soygurt

Protein content

The total of protein content in carrot soyghurt ranged from 6.745 to 6.34% for increased fermentation time and from 6.63% to 6.45% for increased concentration of added carrot juice. Protein content of carrot soyghurt had no significant different ($p>0.05$), were presented in Table 1.. Protein decreased during fermentation time and increased concentration of added carrot juice in carrot soyghurt. It was found that the curd tention rate decreased with the increased of the concentration of added carrot juice, on contrary, the protein decreased with the increased fermentation time and increased of added concentration juice carrot. A nearly similar finding was reported by Galal *et al.* (2003). Probiotic organisms are rich in proteolytic activity, and increase free amino acid content in carrot soyghurt.

Table-1. Protein carrot soyghurt.

Fermention time (h)	Total protein (%)	Carrot juice (%)	Total protein (%)
14	6.74	10	6.63
16	6.47	15	6.49
18	6.34	20	6.45

Reactive oxygen species can cause fragmentation of the peptide chain, alteration of electrical charge of proteins, cross-linking of protein, and oxidation of specific amino acids and therefore lead to increased susceptibility to proteolysis by degradation by specific proteases (Kelly and Mudway, 2003). There are many proposed mechanisms for the therapeutic effects of isoflavones, including inhibition of protein tyrosine kinase, binding to estrogen receptors, inhibition of production of reactive oxygen species (Wei *et al.*, 1995).

Total fat and activity antioxidant

Total fat content in carrot soyghurt ranged from 1.19% to 1.92% and antioxidant activity from 30.07% to 60.72% for increased concentration of added carrot juice and time fermentation. Total fat of carrot soyghurt decreased for added concentration of carrot juice and increased fermentation time which has a significant different ($p<0.05$) in carrot soyghurt (Table 2). Probiotic organisms are rich in lipolytic activity, increased free fatty acid content in carrot soyghurt. Lipid soluble vitamin E is concentrated in the hydrophobic interior site of cell membrane and is the principal defence against oxidant induced membrane injury. Vitamin E donates electron to peroxyl radical, which is produced during lipid peroxidation. A-Tocopherol is the most active form of vitamin E and major membrane bound antioxidant in cell. Vitamin E triggers apopsis of cancer cells and inhibits free radical formations (White *et al.*, 1997). Reactive oxygen species can induce lipid peroxidation and disrupt the membrane lipid bilayer arrangement that may inactivate membrane-bound receptors and enzymes and increase tissue permeability (Girotti, 1985). Products of lipid peroxidation,such as malondialdehyde and unsaturated aldehydes, are capable of inactivating many cellular proteins by forming protein cross linkages (Montuschi, *et al.*,2000, Morrison *et al.*, 1999, Nowak *et al.*, 1999). Activity antioxidant of carrot soyghurt increased for concentration of added carrot juice, also fermentation time increased and was statistically significant ($p<0.05$) in carrot soyghurt. Flavonoids are considered as phenol compounds with highest antioxidant activity due to their chemical structure. Among dietary flavonoids, isoflavones, especially genistein, shown one of the highest antioxidant activities (Heim *et al.*, 2002).

Table-2. Fat and antioxidant activity of carrot soyghurt.

Carrot juice (%)	Fermentation time (h)	Total fat (%)	Antioxidant activity (%)
10	14	1.92^{f}	30.07^{a}
	16	1.85^{f}	42.97^{b}
	18	1.70^{e}	46.65^{c}
15	14	1,70^{f}	50.32^{d}
	16	1.51^{cd}	52.49^{e}
	18	1.42^{c}	55.20^{f}
20	14	1.57^{d}	57.03^{g}
	16	1.32^{b}	59.09^{h}
	18	1.19^{g}	60.72^{i}

The same letter in the same column indicates no significant differences.

The inhibition of ascorbate autoxidation observed with soymilk may be attributed to the action of isoflavones and tocopherols, the main phenols found in soybean (Persky and van Horn, 1995). On the other hand, liberation of aglycone genistein and diadzein through the catalytic action of beta glucosidase during fermentation (Chien, 2004) and the intracellular antioxidants of starter organism (Lin and Yen, 1999) may account for the increased ascorbate-autoxidation inhibition found with the fermented soymilk. To prepare the probiotic fermented soy curd, previously found that the bifidobacteria and lactic acid bacteria were beneficial to each other when grown simultaneously (Wang *et al.*, 2002). Water soluble vitamin C (ascorbic acid) provides intracellular and extracellular aqueous phase antioxidant capacity primarily by scavenging oxygen free radicals. It converts vitamin E free radicals back to vitamin E. Its plama levels have shown to decrease with age (Bunker, 1992, Mezzetti *et al.*, 2006). Carotenoids are pigments found in plants. Primarily, ß-carotene has been found to rect with peroxyl, hydroxyl and superoxide radicals. Carotenoids show their antioxidant effects in low oxygen partial pressure but may have pro-oxidant effects at higher oxygen concentration (Rice *et al.*,1997). Both carotenoids and retinoic acid are capable of regulating transcription factors (Niles, 2004).

Organoleptic of carrot soyghurt

Sensory properties of carrot soyghurt were evaluated for the flavor, color, viscosity and taste. Flavor ranged from 88 to 109, color from 46 to 150, viscosity from 66 to 132.5 and taste was from 74 to 155.5 for increased concentration of added carrot juice and time fermentation in carrot soyghurt. The effect of fermentation time and carrot juice level on these total score flavor, color, viscosity and taste were significant different ($p <0.05$) (Table 3).

Table-3. Organoleptic examination of carrot soyghurt.

Carrot juice (%)	Fermentation time (h)	Total score of flavor	Total score of color	Total score of viscosity	Total score of taste
`10	14	95	48	132.5	155.5
	16	97	46	132.5	129.5
	18	92	55	137	129.5
15	14	88	81	128.5	108.5
	16	104	118	94.5	140
	18	109	124.5	75	79
20	14	108.5	132.8	66	79
	16	102	143	67	74
	18	104	150	67	74

Product of *Lactobacillus bulgaricus* such as flavor and *Streptococcus thermophillus* such as taste carrot soyghurt. Taste was cause by lactic acid, acetaldehyde and volatile compound producing bacteria. The addition of 15% carrot juice to soymilk showed to be the best concentration as it got the highest evaluation mark during the 18 hours fermentation. It implies no unpleasant odor, a pleasant level of acidity and pleasing balance of flavor during the fermentation period. Color score increased for concentration of added carrot juice and time fermentation carrot soyghurt. The color of carrot soyghurt was caused by ß-caroten. With increased concentration of added carrot juice and along the fermentation time in carrot soyghurt, color of carrot soyghurt became a thick orange. Viscosity score decreased for concentration of added carrot juice and increased time fermentation. The addition of 10% carrot juice to soymilk showed to be the best concentration as it got the highest evaluation mark during the 14 hours fermentation.

CONCLUSIONS

The results of the present investigation are of pratical value.The increase in the concentration carrot juice and fermentation time gave significant effect on reducing fat content and viscosity while increasing antioxidant activity, flavor, color and taste increased and does not have significant effect on protein content. With the carrot juice's concentration of 15% and fermentation time of 16 hours, carrot soyghurt can be considered acceptable.

REFERENCES

Adlercreutz H. 2002. Phyto-oestrogens and cancer, Lancer Oncology 3: 364-373.

Beom J., Young S. and Myung H. 1998. Antioxidant activity of vegetables and blends in iron catalyzed model system, J. Fppd Sci. and Nutr. 3: 309-314.

Blagden T. and Gilliland S. 2005. Reduction of levels of volatile components associated with the beany flavor in soymilk by *Lactobacilli and* Streptococci, Journal of Food Science: Food Microbiologyand Safety 70(3): 70-76.

Brouns F. 2002. Soya isoflavones: a new and promising ingredient for the health foods sector, Food Research International 35: 187-193.

Bunker V.W. 1992. Free radicals, antioxidants and ageing, Medical Laboratory Science 101: 3780-3785.

Chien H.I. 2004. Change of isoflavones contents in cultured soymilk fermented with lactic acid bacteria and bifidobacteria, M.S. Thesis, National Taiwan University, Taipei, Taiwan.

Corn Well T., Cohick W. and Raskin I. 2004. Phytoestrogen and human health effects, Phytochemistry 65: 995-1016.

Davies K.J. 1987. Protein damage and degradation by oxygen radicals. I General aspects, Journal of Biological Chemistry 262: 9895-9901.

Dhananjay S., Kulkarni, S. S., Kapanoor K. G., Naganagouda V. K. and Veerappa H. M. 2006. Reduction of flatus-inducing factors in soyilk by immobilized a-galactosidase, Biotechnology and Applied Biochemistry 45: 51-57.

Galal E.A., Mahmoun A., El-Fakhany A. and Moawad A. 2003. Effect of adding carrot puree on organoleptic, chemical and microbiological quality, Proc. The 1st International content Food for Better Health, NRC, 18-20 October (2003), Cairo, Egypt.

Girotti A.W.1985. Mechanisms of lipid peroxidation in vivo and in vitro, Journal Free Radicical Biology and Medicine 1: 87-95.

Hahimoto, T. and Nagayama T. 2004. Chemical composition of ready to eat fresh carrot, Journal of The Food Hygienic Society of Japan 39: 324-328.

Hati, S., Vij, S., Mandal, S., Malik, R.K., Vandna, 2013, α-Galactosidase activity and oligasacharides utilization by probiotic *lactobacilli* during fermentation of Soy Milk, Journal of Food Processing and Preservation (doi. 10.1111/jfpp.12063).

Hoeck J.A., Fehr W.R. Murphy P.A. Welke G.A. 2000. Influence of genotype and environment on isoflavone contents of soybean, Journal Crop Science 40: 48-51.

Heim K.E., Tagliaferro A.R. and Bobilya, D.J. 2002. Flavonoid antioxidants: Chemistry, metabolism and structure activity relationships, The Journal of Nutritional Biochemistry 13: 572-584.

Ikken Y., Cambero I., Marin M., Martner A., Hars I. and Morales P. 1998. Antimutagenic effect of fruit and vegetable aqueous extracts against N-nitrosamine evaluated by the Amestest, Journal of Agricultural and Food Chemistry 1: 353-376.

Ismail B. and Hayes K. 2005. ß-glucosidase activity toward different glycosidic forms of isoflavones, Journal of Agricultural and Food Chemistry 53: 4918-4924.

Jinapong N., Suphanthaarika M. and Jamnong P. 2008 Production of instant soymilk powders by ultrafiltration, spray drying and fluidized bed agglomeration, Journal of Food Engineering 84: 194-205.

Kehrer J.P. 1993. Free radicals as mediators of tissue injury and disease, Critical Reviews Toxicology 23: 21-48.

Kelly F.J. and Mudway I.S. 2003. Protein oxidation at the air-lung interface, Amino Acids, 25: 375-396.

Khare S.K., Krishna J. and Manishwara M.N. 1994. Hydrolysis of flatulence-causing oligosaccharides by agarose-entrapped Aspergillus oryzae cells, Biotechnology and Applied Biochemistry 19: 193-198.

Kim S., Berhow M., Kim J.,Chi H., Lee S. and Cheng I. 2006. Evaluation of soyasaponin, isoflavone, protein, lipid and free sugar accumulation in developing soybean seeds, Journal of Agricultural and Food Chemistry 54: 10003-10010.

Kuo I., Cheng W., Wu R., Huang C. and Lee K. 2006. Hydrolysis of black soybean isoflavone glycosides by Bacillus subtilis natto, Biotechnological Products and Process Engineering 73: 314-320.

Lejuo D.N., Tadecesz G. and Matgorxata T. 2014. Yoghurts with addition of selected vegetables : Acidity, antioxidant properties andsensory quality, Acta Scientiarum Polonorum Technologia Alimentaria 13(1) 35-42.

Lin M.Y. and Yen C.L. 1999. Antioxidative ability of lactic acid bacteria, Journal of Agricultural and Food Chemistry 47: 1460-1466.

Menzzetti A., Lapenna D., Romano F., Costantini F. and Piendomenico S.D. 1996. Systemic oxidative stress and its relationship with age and illness, Journal of The American Geriatries Society 44: 823-827.

Mohamed A.G., Abeer F.,Zayan, Nadia and Shalulin M. 2014. Physiochemical and Sensory evaluation of youghurt fortified with dietary fiber and phenolic compounds, Journal of Life Science 11(9) 816-822.

Montuschi P., Collins J.V., Ciabattoni G., Lazzeri N., Corradi M., Kharitonov S.A. and Bames P.J. 2000. Exhaled8-isoprostane as an in vivo biomarker of lung oxidatiye stress in patients with COPD and healthy smoker, American Journal of Respiratory Critical Care Medicine162: 1175-1177.

Morrison D., Rahman L., Lannan, S. and Max Nee W. 1999. Epithelial permeability, inflammation, and oxidant stress in the air spaces of smokers, American Journal of Respirratory and Critical Care Medicine159: 473-479.

Niles R.M. 2004. Signaling pathways in retinoid chemoprevention and treatment of cancer, Mutation Research 555: 81-96.

Nowak D., Kasielski M., Pietras T. and Bialasiewicz P. 1999. Increased content of thiobarbituric acid reactive substances and hydrogen peroxide in the expired breath condensate of patients with stable chronic obstructive pulmonary disease no significant effect of cigarette smoking, Journal Respiratory Medicine 93: 389-396.

Orhan I., Ozcelik B., Kartal M., Aslan S., Sener B. and Ozguven M. 2007. Quantification of daidzein, genistein and fatty acids in soybeans and soy sprouts,and some bioactive studies. Acta Biological Cracoviensia Series Botanica 49: 61-68.

Persky V. and Van Horn. 1995. Epidemiology of soy and cancer perspectives and directions, Journal of Nutrition 125: 709-712

Raum R.,2003, Microbiological quality of health food and organic foods, Journal Netherland Milk and Dairy 14: 130-134.

Rice-Evans C.A., Sampson J., Bramley P.M. and Holloway D.E.1997. Why dowe expect carotenoids to be antioxidants in vivo?, Free Radical Research 65: 381-398.

Seguin P., Zheng W., Smith D.L.and Deng W.,2004. Isoflavone content of soybean cultivar grown in castern Canada, Journal Science Food and Agricultural 84: 1327-1332.

Seo A. and M Yu. 2003. Toxigenic fungi and mycotoxins,In: hand book of industrial mycology (Andrea, Z ed.) Academic Press London, p: 233-246.

Serrano J., Goni I. and Sauro-Calixto F. 2007. Food antioxidant capacity determined by chemical methods may underestimate the physiological antioxidant capacity, Food Research International 40: 15-21.

Sun M.S., Mihyang K. and Song J.B. 2001. Cytotoxicity and quinine reductase induced effects of Daucas carrot leaf extracts on human cancer cells, Kor. Journal of Food Science 30: 86-91.

Wang Y.C., Yu R.C. and Chou C.C. 2002. Growth and survival of bividobacteria and lactic acid bacteria during the fermentation by ß-galactosidase from Mortierella vinacea. Entrapment of ß-galactosidase within polyacryamide gel, Journal of Food Science 41: 173-175.

Wei H., Bowen R., Cai Q. 1995. Antioxidant and antipromotional effects of the soybean isoflavone genistein, Proceeding of the Society for Experimental Biology and Medicine 208: 124-130.

White E., Shannon J.S. and Patterson R.F. 1997. Relationship between vitamin and calcium supplement use

DOI: 10.1051/matecconf/20165801002

and colon cancer. Cancer Epidemiol Biomarkers and Prevention 6: 769-774.

Yang H. and Zhang I. 2009. Changes in some components of soymilk during fermentation with the basidiomycete Ganoderma lucidum, Food Chemistry 112: 1-5.

BIOETHANOL PRODUCTION FROM LIQUID WASTE OF RICE FLOUR WITH BATCH PROCESS

Sari Ni Ketut*, Sutiyono S, Luluk E, Ernawati Dira, Wesen Putu, Tatik Sri Hajati
University of Pembangunan Nasional "Veteran" Jawa Timur, Surabaya, East Java, Indonesia
E-Mail: sari_ketut@yahoo.com

ABSTRACT

Liquid waste rice flour is abundant liquid wastes but it is still underutilized. So far, it is only used for process water in factory production, wastewater, and even regarded as environment pollution. Rice flour liquid waste has higher levels of glucose, starch and protein which can be used as one of ethanol producers. This study aims to assess the process of hydrolysis, fermentation, and batch distillation process, as well as to search for alternative raw materials products of bioethanol. There are three processes of making bioethanol, namely biological hydrolysis process done by using bacillus; the process of fermentation by using Saccharomyces cerevisiae (SC); and batch distillation. After the third process was done, the results were: glucose is 5% - 10% in the process of hydrolysis; ethanol content is 11% - 16% in the fermentation process; and the levels are high enough for bio-ethanol, which is 95% - 96% in the batch distillation process. So it can be concluded that the liquid waste of rice flour can be used as raw materials for the manufacture of alternative bioethanol.

Keywords: batch distillation, fermentation, hydrolysis, liquid waste of rise flour

INTRODUCTION

Bioethanol is obtained from biomass and bio-energy crops proposed as one viable alternative to gasoline (Demirbas, 2011). Lignocellulosic biomass is one of the main potential sources for the production of bioethanol global economy. Agriculture, forestry (soft wood and hard wood) and industrial waste are major lignocellulosic biomasses (Limayem *et al.*, 2012). Balat *et al.*, (2008) investigated the production of bioethanol from lignocellulosic biomass using process pretreatment, hydrolysis, fermentation and recovery of ethanol, thus obtained ethanol to levels below 16%, with a further distillation process obtained a 95-96% ethanol content.

The research by Nibedita *et al.*, (2012) on production of bioethanol from agricultural waste using PROFER pretreatment method reported an increase of the concentration of fermentable sugar after saccharification enzyme process. The conversion of glucose and xylose fermentation to ethanol requires new technologies, in order to obtain ethanol levels below 16%.

Research of Saravana *et al.*, (2014) on production of bioethanol from sago pith waste (SPW) using microwave hydrothermal catalytic hydrolysis with carbon dioxide gave a result of higher energy savings compared to previous techniques without enzyme, acid or alkaline catalyst, deriving 15.6% ethanol content.

Starch is a component that is more complex than a disaccharide. Before undergoing a fermentation process, the starch must be broken by using the amylase enzyme and form a disaccharide maltose component. Using enzymes, maltase and maltose, starch is hydrolysed into glucose (Karimi *et al.*, 2006). The reaction formula of glucose from starch is as follows:

$$(C_6H_{10}O_5)_n + \tfrac{1}{2}nH_2O \xrightarrow[\text{Amylase}]{\text{Enzyme}} \tfrac{1}{2}nC_{12}H_{22}O_{11} \qquad (1)$$

$$\tfrac{1}{2}nC_{12}H_{22}O_{11} + \tfrac{1}{2}nH_2O \xrightarrow[\text{Maltase}]{\text{Enzyme}} \tfrac{1}{2}nC_6H_{12}O_6 \qquad (2)$$

Processes performed in this research were pre-treatment, chemical or enzymatic hydrolysis, fermentation and product separation or distillation. With a proper pre-treatment strategy on hydrolysis enzyme, it can increase the efficiency of lignocellulosic biomass and lignin inhibiting the saccharification process. Various pre-treatment approach has been used in the past such as using acid or alkaline, hydrogen peroxide, steam explosion, hot liquid, ammonia fiber expansion pretreatment (Teymouri *et al.*, 2005), sodium chorite pretreatment (Kumar *et al.*, 2009) and biological pretreatment. The purpose of using dilute acid pretreatment is to remove the hemicellulose and sugar recovery component. Among all the methods of pretreatment, acid pretreatment method of biomass with dilute sulfuric acid has long been recognized as an important step to eliminate the hemicellulosic fraction of lignocellulosic substrates conserve biological conversion of cellulosic biomass into ethanol (Kuhad *et al.*, 2010).

Research of Alok *et al.*, (2012) suggested the acid pretreatment of waste paper hydrolized as raw material for production of bioethanol can optimize sulfuric acid hydrolysis and hydrolized acid fermentation paper waste process with Pichia stipites, thus obtained ethanol content of 77.54%. With an additional distillation process, it obtained 95-96% ethanol content.

The fermentation process is affected by microorganisms and requires good nutrition to get a good fermentation. Proper nutrition for microorganisms is nitrogen which can be obtained from the addition of NH3, ammonium salts, peptone, amino acids and urea, liquid Nitrogen is needed to 400-1000 g / 1000 L, Phosphate is needed for 400 g / 1000 L (Kuhad *et al.*, 2010).
In the fermentation process, glucose is converted to ethanol with the following reaction:

$$\underset{\text{Glucose}}{C_6H_{12}O_6} \longrightarrow \underset{\text{Ethanol}}{2C_2H_5OH} + 2CO_2 \qquad (3)$$

Equation (1) is a Differential-algebraic-model equation (Daes) for batch distillation of multi-component system, assuming no form of two liquid phases. The above equation is a forward-finite-difference, which will be acquired under the liquid composition (xi, j + 1) as a function of Δξ (dimensionless time), which is obtained as follows (Henley *et al.*, 1998):

$$x_{i,j+1} = x_{i,j} + (y_{i,j} - x_{i,j})\,\Delta\xi \qquad (4)$$

Where the composition of the initial liquid in the bottom ($x_{i,j}$) and Δξ were determined, while the vapor composition ($y_{i,j}$) is calculated using the equation bubble T (Henley et al., 1998).

Predecessor researchers for bioethanol made from starch have obtained quite good cellulose and bioethanol results. The purpose of this study was to look for alternative raw materials, reviewing the process of hydrolysis, fermentation, and batch distillation process to produce bioethanol with high ethanol content. In this research, we located raw materials from wastewater, using three processes (hydrolysis, fermentation and distillation batch) simultaneously and attained bioethanol production levels of 95-96% ethanol as substitute materials.

METHODOLOGY

From the results of laboratory analysis, it is known that ethanol-forming elements are glucose and starch. The average concentration of glucose is about 6%, starch in the rice is about 79%, while the starch in the rice flour wastewater is quite high, about 30%.

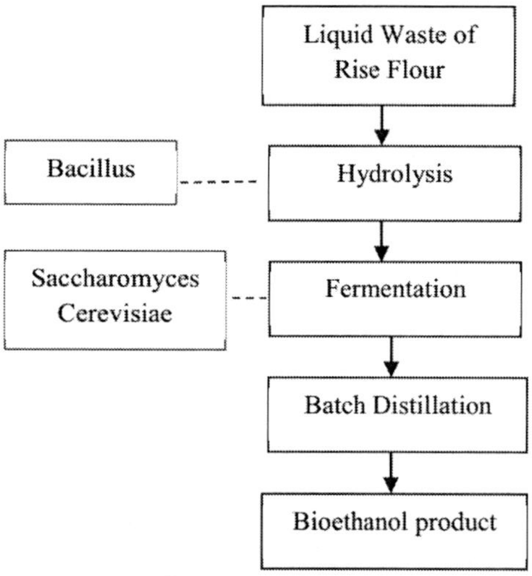

Figure-1. Chronology of bioethanol production using hydrolysis, fermentation and distillation batch

Pre-treatment was performed using waste water and filtration of plant rice flour to get a high glucose and starch and the hydrolysis process with the bacillus. A good quality bioethanol product is determined from several parameters, such as the degree of acidity (pH), Bacillus ratio for rice flour mill effluent, SC starter concentration, time of fermentation, and batch distillation time.

Quality analysis for raw materials and products of bioethanol was performed through laboratory work. The analysis was carried out by instrumentation and gravimetry, using Gas Chromatography (GC) and Spectrophotometer, which analyzed the concentration of starch, glucose, ethanol, crude protein and N, P, K, Ca, Mg, S.

Hydrolysis process procedure

Hydrolysis process at figure 1 was performed under the following conditions: a temperature of 30 °C, hydrolysis time of 1 hour with 200 rotations per minute (rpm). The varied conditions were: the ratios of the volume of liquid waste bacillus were 1:2; 5:4; 10:7 and volume of: 5, 7, 9, 11, and 13 (% v/v). Once the process is completed, we obtained filtrate and solids. The filtrate would be processed using fermentation to obtain ethanol and solids levels can then be used as composts.

Fermentation process procedure

The result of glucose hydrolysis process has not qualified yet, so the next step is adding citric acid or NaOH. Citric acid is added to the filtrate until it reached a fermentation pH of 4.5, approximately. Next is adding the starter to the fermented solution in an anaerobic condition. In the fermentation process (Figure 1), the permanent conditions were a temperature of 30 °C, pH of 4.5; and the fermentation volume is equal to the amount of filtrate volume of hydrolysis process results. The changed conditions were varying the ratios of the volume of Bacillus waste which were: 1:2; 5:4; and 10:7 with the addition of Saccharomyces cereviceae of: 5, 7, 9, 11, and 13 (% v / v) and fermentation time of 4, 6, 8, 10, and 12 days, then continued with performing ethanol level analysis.

Distillation process procedure

Results obtained from the fermentation were inserted into the distillation flasks to obtain ethanol from glucose. Batch distillation process as shown in Figure 1 was run at a temperature of 78°C, equipped with a total condenser and heater. After the volume of the solution reached 10% of the volume of bottom feed, the batch distillation was stopped, and then the solution was analyzed for its level of ethanol.

RESULTS AND DISCUSSION

Quality Raw Materials

Liquid waste used as study materials was derived from rice flour waste from PT. Boga Sari mill. Carbohydrates are the main source of calories for humans, apart from protein and fat. Carbohydrates having the chemical formula of $(CH_2O)_n$, are formed from the reaction of CO2 and H2O with the help of sunlight through the process of photosynthesis in plant cells that contain chlorophyll. Ingredients that are a source of carbohydrates come from roots and stems of plants such as cassava, sago, and sugar cane; whereas from grains are rice, corn and soybeans. Rice flour mill effluent is the raw material came from carbohydrates types of grains, such as rice.

According to the results of laboratory analysis, starch content in rice flour mill effluent is of 9.282%, averagely. This means that if all elements are completely hydrolyzed, a large amount of glucose will be obtained. From 100 liters of liquid waste of rice flour mill, we can produce a maximum of 9.282 liters of glucose. In addition to starch, glucose is also present in the flour wastewater, increased in bulk. In 100 grams of flour liquid waste, glucose can be produced to a maximum of 3,786 liters. Noting the high composition of glucose and starch in the flour mill waste water, the hydrolysis process is expected to run perfectly, so that the amount of glucose and starch was actually relegated to 12.568 liter glucose. The process of filtering waste flour mill is done by using gauze and was done repeatedly to get a clean result before going to hydrolysis process. In addition to the natural filtration, sedimentation was also done by keeping it for a while so it would form naturally, and then the filtrate was taken and sedimentation were used as compost sediment.

The effect of pH on adding H-Cl can be determined. With additional volume greater than the H-Cl, a low pH would be obtained. Due to the required pH of the fermentation is 4.5, the addition of H-Cl volume by 7% v / v is the closest, for various volume of wastewater. Before performing the process of hydrolysis, the pH of the filtrate was measured in accordance with the provisions of the fermentation process which is approximately 4.5. To obtain a pH of 4.5, Na-OH was added if the pH of the filtrate was below 4.5, while citric acid was added if the pH of the filtrate was above 4.5.

Hydrolysis Process

Starch is a component that is more complex disaccharide; since starch must be broken using amylase into disaccharide components, such as maltose. By using enzyme or bacillus, maltose is hydrolyzed into glucose.

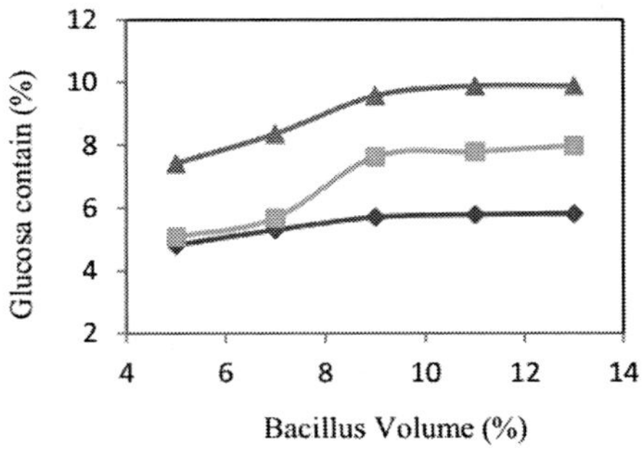

Figure-2. Changes on glucose contains with an addition of bacillus in liquid waste of rice flour (ratio bacillus/volume waste = 1:2: ♦, ratio bacillus /volume waste = 5:4: ■, ratio bacillus/volume waste = 10:7: ▲)

Figure 2 shows the effect of glucose levels for bacillus volume addition. The greater the volume of bacillus, the higher the level of glucose. On the addition of Bacillus volume above 9%, the graph shows a constant profile, because the performance of Bacillus began to decline and die from time to time, so it would be needed to add more Bacillus. The addition of Bacillus volume between (3-9) % v/v showed an increase in glucose levels, therefore showing the growth phase of Bacillus, so it can decompose starch into glucose optimally. Before the fermentation process is done, the filtrate optimum glucose level was measured at about 16%; if the glucose level is more than 16%, dilution is done. If the glucose level is less than 16%, the addition of glucose is made.

Fermentation Process

The fermentation process is a fermentation process that does not use oxygen (anaerobic process). To control the production of ethanol from sugar is quite complex. Substrate concentration, oxygen, and ethanol products, all of them affect yeast metabolism, cell survival, cell growth, cell division, and the production of ethanol. Appropriate selection of yeast strains and a high tolerance for concentration substrate or alcohol is important to improve results. The first fermentation is basic maintenance of seeds fermentation or starter preparation. Starter was inoculated until it was ready to be a fermenter, and then put into the fermentation substrate, which used SC seed fermentation.

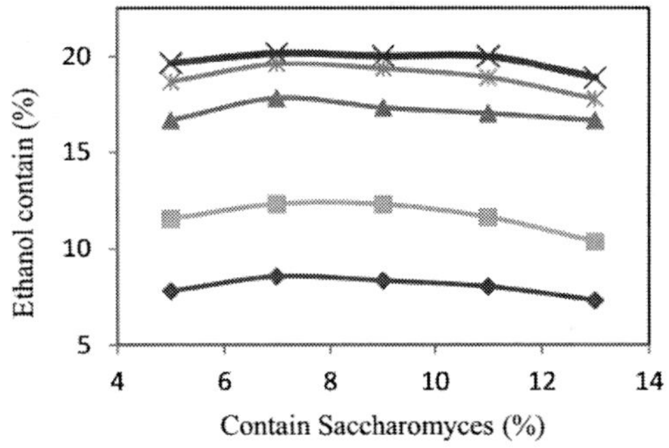

Figure-3. Changes on ethanol contains with an addition of Saccharomyces in liquid waste of rice flour (fermentation time = 4 days: ♦, fermentation time = 6 days: ■, fermentation time = 8 days: ▲, fermentation time = 10 days: x, fermentation time = 12 days: *)

Figure 3 shows the level of concentration of the filtrate which affects glucose residue. For a number of 6, 9, 11 and 13% v/v SC starters, the maximum residual glucose levels will be obtained in (1.3 - 3). The percentage, as in the reactor tank, of filtrate amount and SC starter were low hydrolyzed, so that the fermentation process is not optimal. By increasing the number of hydrolysis filtrate and SC beginner, residual glucose became smaller, because it has been fermented into ethanol.

After analyzing the residual glucose concentration of the fermentation process, the percentage of SC of the liquid (filtrate) was examined. In addition of 7%, it showed a smaller residual glucose levels compared to the addition of 6, 9, 11 and 13% starter. In initial research and Journal, an addition of 7% of the volume of fluid was admitted. The influence of the level of residual filtrate glucose levels, for 6, 9, 11 and 13% SC starters, a maximum residual glucose level was obtained in (1.5 to 5)%. It is because in the reactor tank, the filtrate is hydrolyzed and the SC starter is still low, so the fermentation process is not optimal. With the increasing number of hydrolyzed filtrate and SC starter, the residual glucose becomes smaller, since it was fermented into ethanol. According to preliminary research and the Journal, the best fermentation time is six days. Six days of fermentation is the best because it is the optimum phase of the SC. Below six days, there will be an adjustment to SC growth and after six days there will be a regeneration phase or change phase of SC.

Batch Distillation Process
Ethanol is obtained from the fermentation and distillation batch process of mini-scale plant. By choosing the best condition of fermentation, the fermentation time is 10 days. The concentration of ethanol from distillation batch shows the range of 95% to 96% and yield 31.66% to 33.3%. Economically, it meets the necessary technical requirements of ethanol, which is 95%, and the resulting yield is also high.

Table-1. Ethanol contain and yield on distillation batch

RatioHCl to raw materials	SC contain [%v/v]	Glucose contain [%v/v]	Ethanol contain [%v/v]	Ethanol contain [%v/v]	Yield [%]
1 : 2	5	4.82	19.63	91.5	30.5
	7	5.30	20.15	93	31,8
	9	5.70	20.01	94	32.3
	11	5.78	19.99	92.5	31.6
	13	5.81	18.87	91	32.3
5 : 4	5	5.08	18.63	92	30.2
	7	5.68	20.46	94.5	31.8
	9	7.63	20.88	96	33.3
	11	7.78	19.33	95	32.6
	13	7.98	18.27	94	32.1
10 : 7	5	7.41	19.63	94	30.4
	7	8.35	21.9	94.5	32.8
	9	9.56	20.44	95	33.1
	11	9.87	19.66	94.5	32.9
	13	9.88	18.77	93	32.6

CONCLUSIONS

Glucose levels of 9.98% were obtained in the process of hydrolysis of rice flour liquid waste, while ethanol in the fermentation process is 16% to 20.88%. Raw materials used to produce bioethanol indicate that the levels of ethanol 95% to 96% with ethanol yield 31.69% to 33.3% can be used to design a batch distillation column. By identifying the concentration and yield of ethanol, the liquid waste of rice flour raw material is more profitable than other raw materials.

ACKNOWLEDGEMENT

The authors would like to acknowledge the financial support of the Ministry of National Education of the Republic of Indonesia with the National Strategic Competitive Grant, Contract Number: 180/SP2H/PL/DIT.LITABMAS/V/2013.

REFERENCES

Alok Kumar Dubey, P.K. Gupta, Neelam Garg, Sanjay Naithani. 2012. Bioethanol Production from Waste Paper Acid Pretreated Hydrolyzate with Xylose Fermenting Pichia Stipitis. Carbohydrate Polymers, 88: 825-829.

Balat M, Balat H, Cahide O. 2008. Progress in Bioethanol Processing. Proggres Energy Combust Sci, 34: 551-73.

Demirbas A. 2011. Competitive Liquid Biofuels from Biomass. Applied Energy, 88: 17-28.

Karimi K. Kheradmandinia S, Taherzadeh MJ. 2006. Conversion of Rise Straw to Sugars by Dilute Acid Hydrolysis. Biomass and Bioenergy, 30: 247 – 253.

Kuhad, R. C., Gupta, R., Khasa, Y. P., & Singh, A. 2010. Bioethanol Production from Lantana Camara (Red Sage) Pretreatment, Saccharification and Fermentation. Bioresource Technology, 101: 8348-8354.

Kumar, A., Singh, L., K., & Ghose, S. 2009. Bioconversion of Lignocellulosic Fraction of Water-Hyacinth (Eichhornia Crassipes) Hemicellulose Acid Hydrolysate to Ethanol by Pichia Stipitis. Bioresource Technology, 100: 3293-3297.

Limayem A. Ricke SC. 2012. Lignocellulosic Biomass for Bioethanol Production; Current Perspectives, Potential Issues and Future Prospects. Progres Energy Combust Sci, 38: 449-67.

Nibedita Sarkar, Sumanta Kumar Ghosh, Satarupa Bannerjee, Kaustav Aikat. 2012. Bioethanol Production from Agricultural Wastes: An Overview. Renewable Energy, 37: 19-27.

Teymouri, F., Laureano-Peres, L., Alizadeh, H., & Dale, B, E. 2005. Optimization of the Ammonia Fiber Explosion (AFEX) Treatment Parameters for Enzymatic Hydrolysis of Corn Stover. Bioresource Technology, 96: 2014-2018.

Saravana Kannan Thangavelu, Abu Saleh Ahmed, Farid Nasir Ani. 2014. Bioethanol Production from Sago Pith Waste Using Microwave Hydrothermal Hydrolysis Accelerated by Carbon Dioxide. Applied Energy, 128: 277-283.

MATEC Web of Conferences **58**, 01004 (2016) DOI: 10.1051/matecconf/20165801004
BISSTECH 2015

BIOETHANOL QUALITY IMPROVEMENT OF COFFEE FRUIT LEATHER

Luluk Edahwati[1*], Dyah Suci P.[1], Nana Dyah S. [1],Tri Widjaya[2], Ali Altway [2], Domas Cahyo [1], Atika Nandini [1].

Department of Chemical Engineering
[1]University of Pembangunan Nasional "Veteran", East Java, Surabaya 60294
[2]Sepuluh Nopember Institute of Technology, Surabaya 60111
[1]*Corresponding Author's E-mail: edahwatiluluk@yahoo.co.id

ABSTRACT

Recently, Indonesia's dependence on petroleum is to be reduced and even eliminated. To overcome the problem of finding the needed alternative materials that can produce ethanol, in this case as a substitute material or a transport fuel mix, boosting the octane number, and gasoline ethanol (gasohol) can be conducted. In the red coffee processing (cooking) that will produce 65% and 35% of coffee beans, coffee leather waste is a source of organic material with fairly high cellulose content of 46.82%, 3.01% of pectin and 7.68% of lignin. In this case, its existence is abundant in Indonesia and optimally utilized. During the coffee fruit peeling, the peel waste is only used as a mixture of animal feed or simply left to rot. The purpose of this study was to produce and improve the quality of the fruit skin of bioethanol from coffee cellulose. However, to improve the quality of bioethanol, the production of the lignin content in the skin of the coffee fruit should be eliminated or reduced. Hydrolysis process using organosolve method is expected to improve the quality of bioethanol produced. In particular, the use of enzyme *Saccharomyces* and *Zymmomonas* will change the resulting sugar into bioethanol. On one hand, by using batch distillation process for 8 hours with *Saccharomyces*, bioethanol obtains high purity which is 39.79%; on the other hand, by using the same batch distillation process with *Zymmomonas,* the bioethanol obtains 38.78%.

Keywords: delignification, organosolve method, Saccharomyces C, Zymomonas M, bach distillation.

INTRODUCTION

Bioethanol has a great potential as an alternative fuel in various energy sectors, especially in the transport sector. In 2007, the United States of America became the largest bioethanol producer in the world with a capacity of fuel alcohol production of 51.5 billion liters from 180 plantations with bio-refinery (Walker, 2010).

The production and the use of bio-ethanol has attracted more and more attention as a strategy for reducing greenhouse gas (GHG) emissions and improving global energy security. In general, the commercial production of bio-ethanol is mainly dependent on the fermentation of sucrose from sugar cane and molasses, or glucose from starch-based crops such as corn, wheat and cassava (Davis, RogerSB, Pearcec, & Peirisa, 2006). Brazil together with the US uses ethanol approximately 60.0% of the ethanol world production by utilizing sugar cane and corn (Chandel et al., 2007). However, using food crops to ethanol production could increase food safety concerns (MR Schmer, 2008). Bioethanol is obtained from biomass and bioenergy crops which have been declared as one viable alternative to gasoline (Demirbas, 2011).

Lignocellulosic biomass is one of the main potential sources for the production of bioethanol global economy. Agricultural, forest (soft wood and hard wood) and industrial wastes are some major lignocellulosic biomass (Limayem et al., 2012). Balat et al. (2008) studied the production of bioethanol from lignocellulosic biomass by using several subsequent processes such as pretreatment, hydrolysis, fermentation and recovery of ethanol, thus it obtained ethanol to the levels below 16%, then, with a further distillation process, it will obtain 95-96% ethanol content.

Saravana et al. (2014) conducted a study about the production of bioethanol from sago pith waste (SPW) by using microwave hydrothermal catalytic hydrolysis with carbon dioxide, resulting in higher energy savings compared to previous techniques without the enzyme, acid or alkaline catalyst, in which the process derived ethanol with the content of 15.6 %.

The technology for lignocellulosic ethanol production mainly relies on pre-treatment, chemical or enzymatic hydrolysis, fermentation, and product separation or distillation. With proper pretreatment strategy on hydrolysis enzyme, it can increase the efficiency of lignocellulosic biomass and lignin to inhibit the saccharification process. Various pre-treatment approaches have been used in the past such as acid or alkaline pretreatment, pretreatment hydrogen peroxide, steam explosion, hot liquid, ammonia fiber expansion pretreatment (Teymouri et al., 2005), sodium chlorite pretreatment (Kumar et al., 2009) and biological pretreatment. The purpose of using dilute acid pretreatment is to remove the hemicellulose and sugar recovery component. Among all the methods of pretreatment, acid pretreatment method of biomass with dilute sulfuric acid has long been recognized as an important step to eliminate the hemi-cellulosic fraction of lignocellulosic substrates to conserve the biological conversion of cellulosic biomass into ethanol (Kuhad et al., 2010). Reaction formation of glucose from cellulose can generally be written as follows:

$$\underset{\text{Cellulose}}{(C_6H_{10}O_5)_n} + n\,H_2O \longrightarrow \underset{\text{Glucosa}}{nC_6H_{12}O_6}$$

The fermentation process is affected by microorganism, and it requires good nutrition to get a good fermentation. Proper nutrition for microorganisms is nitrogen which can be obtained from the addition of NH_3, ammonium salts, peptone, amino acids and urea, liquid Nitrogen needed approximately to 400-1000g/1000L, Phosphate needed approximately to 400g/1000L (Kuhad et al., 2010).

In the fermentation process, glucose is converted to ethanol with the following reaction:

$$\underset{\text{Glucosa}}{C_6H_{12}O_6} \xrightarrow{SC\ /\ Zymmomonas} \underset{\text{Ethanol}}{2C_2H_5OH} + 2CO_2$$

According to the researcher predecessor, bioethanol made of cellulose obtained results which were quite good bioethanol. The purpose of this study is to look for alternative raw materials, and to review the process of hydrolysis, fermentation, and batch distillation process to produce bioethanol with high ethanol content. Original research which is located on the second generation of alternative materials is elephant grass, by simultaneously using three processes (hydrolysis, fermentation and distillation batch) and technical bioethanol production to the levels that are 95-96% ethanol substitute material.

METHODOLOGY

From the results of laboratory analysis, it is discovered that the cellulose content of coffee skin is 46.82%, 3.01% of pectin and 7.68% of lignin. Coffee skin is dried and ground into powder. To get a high glucose, cellulose hydrolysis process is done by employing organosolve method which uses an acid solution that is environmentally friendly citric acid. Product quality bioethanol is determined from multiple parameters that affect as the concentration of acid additions, additions % ethanol, fermentation time, % mo and time addition of batch distillation. The analysis of the quality of raw materials and products bioethanol is made by laboratory analysis. Instrumentation analysis was performed by using Gas Chromatography (GC) and Spectrophotometer, in which it will particularly analyze the ethanol content.

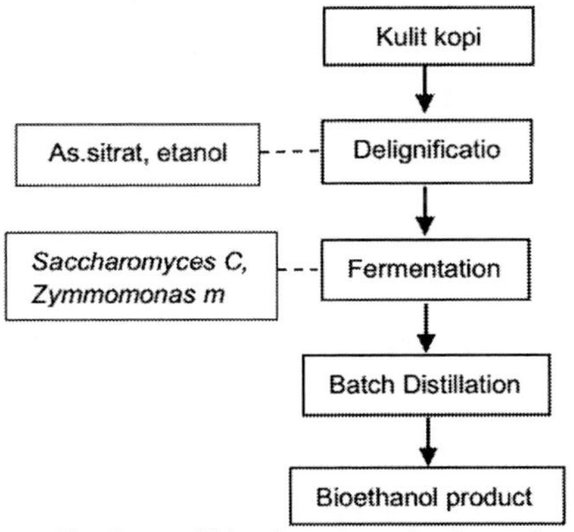

Figure-1. The flow of bioethanol production using hydrolysis, fermentation and distillation batch.

The procedure of delignification process

The condition is still being done at the temperature of 80 °C and the stirring speed of 600 rpm. Changes on conditions are made on the addition of citric acid (1:10 ; 1:12 ; 1:14 ; 1:16 ; 1:18), %addition of ethanol (10, 20, 30, 40, 50) as well as the length of stirring (1 ; 1.5 ; 2 ; 2.5 ; 3 hours). The solids are separated from the solution, and then, made into the delignification process. The solution (filtrate analyzed as pectin). Analysis is performed on the concentration of lignin separated both in the filtrate and in the sediment.

The procedure of fermentation process

The results of glucose hydrolysis process that has not qualified to do the process of adding citric acid or NaOH. Citric acid is then added to the filtrate as a result of hydrolysis in which it will be fermented until it reaches the fermentation pH approaching 4.5. Next, the starter is added to the solution, then, it is fermented in anaerobic conditions by closing the bottle tightly and watching it for a certain time. The changes of fermentation time conditions are 3, 4, 5, 6, 7 days, and Zym starter SC are 9, 10, 11 %. Then, the ethanol levels are analyzed from those data.

The procedure of distillation process

Results obtained from the fermentation are inserted into the distillation flask to obtain ethanol from glucose. Batch distillation process is run at the temperature of 78 °C. After the volume of the solution bottom stays 10% of the feed volume, batch distillation is stopped, and then analyzed for the ethanol levels.

RESULTS

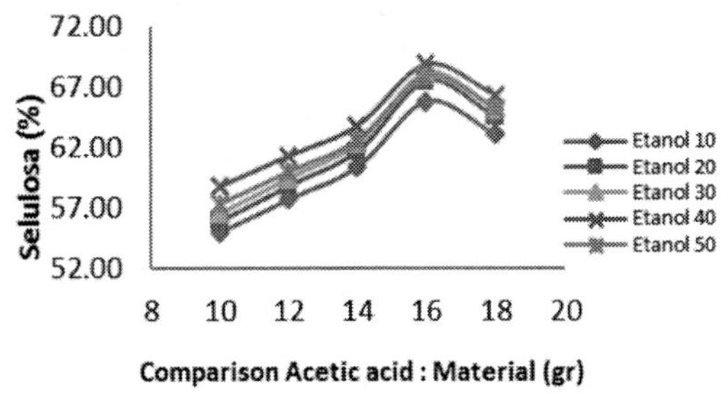

Figure-2. Cellulose content of the sludge after the delignification.

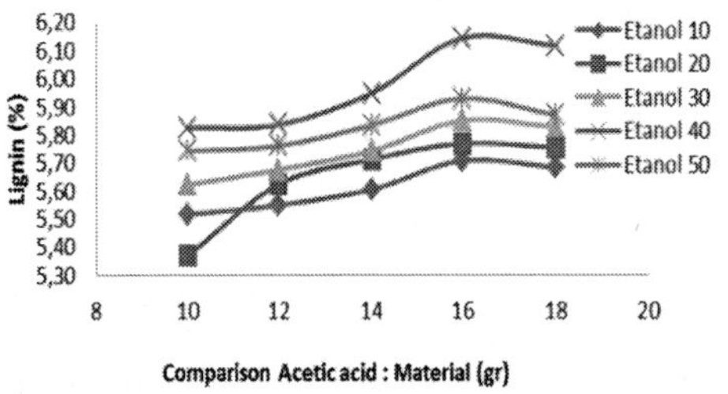

Figure-3. Lignin content is obtained after a process of delignification

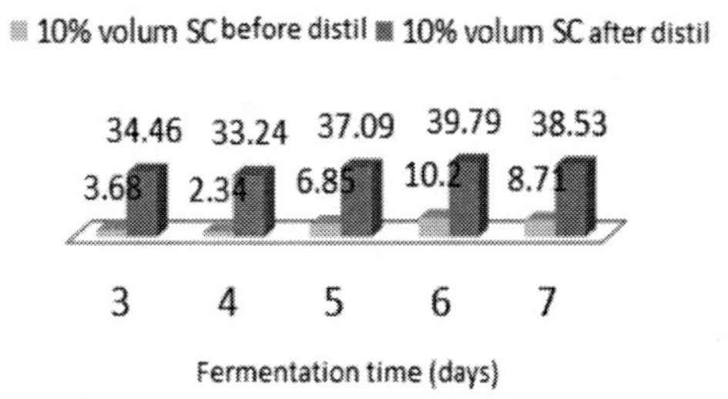

Figure-4. Levels of bioethanol using *Saccharomyces*

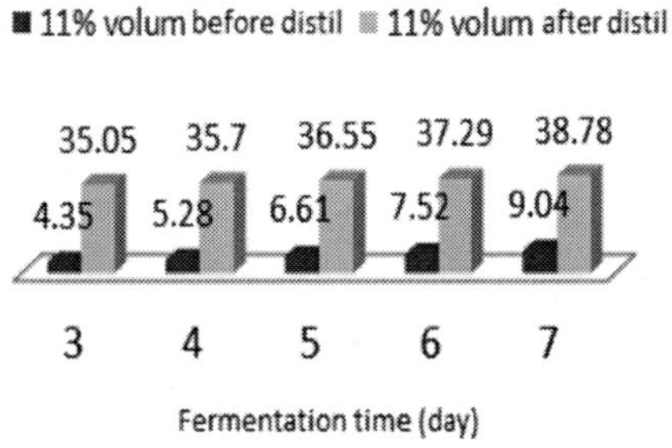

Figure-5. Levels of bioethanol using *Zymmomonas*

DISCUSSIONS

From Figure 2 and 3, the greater the solution of cellulose obtained, the greater the cooker is. However, at certain times it shows a decrease in the amount of cellulose. It happens the same to lignin. That is because the greater the concentration of the solution cooker, the greater the lignin content dissolves. Nevertheless, when the concentration is too high, it will cause damage to the cooker cellulose, and cellulose esterification reaction occurs in which in this case, reaction between the alcohol formed cellulose ether. Thus, it causes a decrease in cellulose and lignin produced

In the fermentation process (Figure 4 and 5) that uses Saccharomyces, best conditions are obtained at a concentration of 10 % with a time starter for 6-day fermentation, producing ethanol content amounted to 10.20 %. After the fermentation process is followed by batch distillation process for 8 hours, it produces ethanol with a concentration of 39.79 %. Whereas, the fermentation process using mobilis Zymmomonas obtained the best conditions at 7 days and the starter concentration of 11 %. In the beginning, the ethanol content is 9.04 %, then, after the distillation process the ethanol content amounts to 38.78 % .

CONCLUSION

Based on the research goal which is to assess the delignification process, fermentation, and batch distillation process, as well as finding alternative raw materials, the product of bioethanol coffee fruit peel can be used as an alternative raw material for the production of bioethanol. Still, it is possible to proceed the conduct of further distillation to obtain ethanol content of 96%. Thus, it is very possible for a commercial-scale pilot plan.

ACKNOWLEDGEMENT

The authors would like to acknowledge the financial support of the Ministry of National Education of the Republic of Indonesia through the PEKERTI Grant, Contract Number: SPP/0074/Lit/LPPM/II/2015.

REFERENCES

Balat M.H. Balat and Cahide. 2008. Progress in Bioethanol Processing. Prog. Energy. Combust. Sci., 34, 551-73.

Demirbas A. 2011. Competitive Liquid Biofuels from Biomass. Appl. Energy, 88, 17-28.

Henley E.J.J. D. Seader and D. K. Roper. 2011. Separation Process Principles, 3rd Edition, New York: John Wiley & Sons Inc.

Kuhad R.C.R. Gupta Y.P. Khasa and A. Singh. 2010. Bioethanol Production from Lantana Camara (Red Sage): Pretreatment, Saccharification and Fermentation. Biores Tech, 101, 8348-8354.

Kumar A. L.K. Singh and S. Ghose. 2009. Bioconversion of Lignocellulosic Fraction of Water-Hyacinth (Eichhornia Crassipes) HemicelluloseAcid Hydrolysate to Ethanol by Pichia Stipitis. Biores Tech, 100, 3293-3297.

Limayem A. S.C. Ricke. 2012. Lignocellulosic Biomass for Bioethanol Production: Current Perspectives, Potential Issues and Future Prospects. Prog. Energ. Combust. Sci., 38, 449-67.

Teymouri F.L. Laureano-Peres H. Alizadeh and B.E. Dale. 2005. Optimization of the Ammonia Fiber Explosion (AFEX) Treatment Parameters for Enzymatic Hydrolysis of Corn Stover. Biores Tech, 96, 2014-2018.

Thangavelu S.K. A.S. Ahmed and F.N. Ani. 2014. Bioethanol Production from Sago Pith Waste Using Microwave Hydrothermal Hydrolysis Accelerated by Carbon Dioxide. Appl. Energ., 128, 277-283.

Walker G.M. 2010. Bioethanol Science and Technology of Fuel Alcohol. Global Production of Bioethanol. Graeme M. Walker & Ventus Publishing Aps.

BIOGAS PRODUCTION AND REMOVAL COD – BOD AND TSS FROM WASTEWATER INDUSTRIAL AlCOHOL (VINASSE) BY MODIFIED UASB BIOREACTOR

Isni Utami , Sri Redjeki, Dwi Hery Astuti and Sani
University of Pembangunan Nasional "Veteran" Jawa Timur,
Surabaya, Indonesia
Email: isniutami@yahoo.com

ABSTRACT

Biogas production and decreased organic loading of vinasse using a modified UASB bioreactor has been done successfully. Vinasse is waste from the ethanol industry which contains COD: 9.360 mg / L , BOD : 4.013 mg/L, and TSS: 317.5 mg/L. The purpose of this research was to study the performance of bioreactors Upflow Anaerobic Sludge Blanket (UASB) to decompose the vinasse into biogas or methane. UASB operating principle is to distribute wastewater in the bioreactor to flow upward through the sludge blanket by setting the hidrolic retention time (HRT). Four UASB bioreactor columns were used in this experiment wherein each with a capacity of 50 L in volume; 23 cm inside diameter, and 120 cm. The variations of hydraulic capacity followed the variations of HRT in the range of 72-36 hours. Modifications were carried out on the top of column UASB with the aim of preventing gas losses and increasing the flowrate of gas out from the top of the column. The results showed that HRT increased from 36 h to 72 h followed by an increase in COD removal efficiency of 55.64% to 66.81%; BOD_5 from 67.85% to 74.58%; and TSS from 66.69% to 84.19%. The maximum volume of biogas produced was in the range of 5.826 L / day (42.89% methane) to 7.930 L / day (methane 58.06%).

Keywords: Biogas, Methane, UASB, Vinasse,

INTRODUCTION

Vinasse is wastewater from the ethanol fermentation industry which still contains high enough organic matter, so that it can be used as raw material for biogas production. The waste water can be decomposed by microorganism to degrade organic material in a bioreactor. The organic matter parameter, measured as BOD or COD, can be used as a substrate for bacteria (Ashish et al., 2014). Substrate utilization will be able to produce the removal of chemical contaminants (Okpokwasili and Nweke, 2005), forming a new microorganism cells and decomposition product (Schuner and Jarvis, 2009; Vadivel et al., 2014).

Decomposition of organic wastes can be done with the anaerobic system (Guerrero et al., 1999; Driessen and Yspeert, 1999; Liu et al., 2002). Anaerobic wastewater treatment can generate methane gas (CH_4) and carbon dioxide (CO_2), which is known as biogas. Biogas is used as an alternative fuel that is environmentally friendly, thus the liquid waste vinasse is utilized commercially. Methane fraction varies depending on the substrate contained in wastewater (Marchaim, 1992) .

Anaerobic degradation of organic compounds can be efficient and stable when there is a balance metabolism rate beetween the acid-forming bacteria and bacteria that form CH_4 (Yu and Fang, 2003). Imbalance can occur due to excessive organic loading or changes in operating conditions (Tabatabaei et al., 2011) which would trigger the accumulation of acidic intermediate products that are inhibiting bacteria methanogens (Marchaim, 1991).

According to (Mulligan et al., 1993), the loading of 36.5 kg COD / m^3 / day can produce gas by 12 m^3 / m^3 / day.

UASB (Upflow Sludge Blanket Anaerobic)

UASB is one way of anaerobic wastewater treatment which operated continuously. In the UASB bioreactor, wastewater flowed vertically from the bottom upward through the sludge blanket towards which there is a bacterial decomposition of waste. UASB performance was strongly influenced by Hydrolic Retention Time (HRT); Hydrolic Loading Rate (HLR); Velocity Up (Vup); influent flow; the concentration of organic waste; Organic Load Rate (OLR); and variable operating temperature and degree of acidity (pH) (Alphenar et al., 1993; Aslan and Sekerdag, 2008; Farajzadehha et al., 2012; Ramakrishna et al., 2013).

The anaerobic process using conventional UASB have limitations in processing capacity, due to mass transfer limitations caused by the mixing process, conditions are less than optimal in the reactor (shin et al., 2003).

This process is mostly applied in industrial waste with an organic load of the medium, which has a concentration of COD (Chemical Oxygen Demand) in the range of 3000-7000 mg / L (Lettinga and Hulshoff, 1991).

Long enough HRT will allow longer contact between the anaerobic sludge wastewater so that the process of decomposition of organic substances to be better (Farajzadehha et al., 2012). However, according to Van Haandel (2006), a very long HRT may adversely affect granulation process sludge in UASB bioreactor, while a very short HRT can lead sludge to drift out of its biomass reactor. UASB technology for wastewater treatment has been widely studied with a variety of performance parameters (Van Haandel, 2006; Ramasamy et al., 2004; Supriyanto et al., 2010)). Most of the study were prioritized on the removal of COD, BOD_5 and TSS; while not much has been done for the production of biogas using UASB.

This study aims to assess the performance of a modified UASB reactor for treating vinasse waste into biogas. Modifications was done at the top of the column UASB so that the volume of space above the columns are not too big to prevent gas losing so that gas products formed may soon be out of the reactor.

METHODOLOGY

Vinasse waste was obtained from PT Energi Nusantara Agro, Mojokerto, Indonesia. Vinnase analysis showed that the COD: 9,630 mg / L; BOD: 4,013 mg / L; TSS: 317.5 mg / L and a pH of 6.5. Activated sludge was taken from the slurry manure (as methanogenic anaerobic bacteria).

Acclimatization of microbes was performed for five days. To remove oxygen, Nitrogen gas was used. After the acclimatization stage is completed, it was followed by the stage of running for 18 days. Influent pH is maintained at around 7 by adding CaO. Column-shaped reactor served as a tank anaerob with dimensions of 23 cm in diameter; 120 cm in height with a volume of 50L. The volume of sludge was 30% of the volume of waste. Figure 1 shows a circuit of UASB.

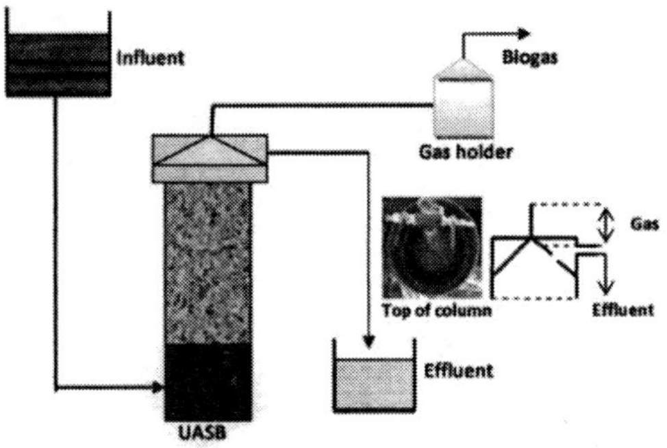

Figure-1. Diagram of UASB Process

Waste water flows into the influent tank and then flowed into the UASB reactor at a predetermined flow rate. Liquid waste flowed from the bottom of the reactor then flows upward through the mud bed anaerobic in which microorganisms sludge in contact with the substrate in wastewater. On the other hand, sludge bed that consists of microorganisms that grow naturally in the form of granules with a diameter of 0.5 to 2 mm has a high deposition speed so that it can fight the wash out of the system at high hydraulic loads. Anaerobic degradation process will produce gas (biogas containing CH_4 and CO_2), and the movement of gas bubbles being released upwards causing upheaval hydraulic reactor blend naturally without using a mixer. At the top of the reactor, there is a separation of solids, liquids and gases by a three-stage separator (also known separator gas-liquid-solid). Biogas will be captured through the hose, while the liquid effluent will flow into the effluent tank. Discharge arrangement adapted to Hydrolic retention time (detention time) of each reactor is 72 hours, 60 hours, 48 hours and 36 hours with a reactor volume of 50 liters. Table 1 shows the variation of HRT relationship with flowrate, HRL and Vup.

Table-1. Corellations of HRT, flowrate, HRL, and Vup

HRT h	**Flow rate** L/h	**HRL** $m^3/m^2/h$	**Vup** m/h
72	0.6944	0.0167	0.0167
60	0.8333	0.02	0.02
48	1.0416	0.025	0.025
36	1.3888	0.033	0.033

The analysis of characteristics of the waste was carried out by COD analysis using SNI 6989.73-2009 and BOD_5 6989.72-2009. While GC 7900 gas analyzer used the Biogas Porapak Q method (Area ASTD). Gas volume measurements were performed daily, while gas composition analysis was done every 2 days and TSS analysis every 3 days.

RESULT AND DISCUSSION

Figure 2 shows the correlation of HRT and UASB operation time to the reduction of COD. COD decreased high enough up to twelveth day and then decreased slowly up to the eighteenth days. This trend did not occured on a decrease of BOD in which up to eighteen days it tends to be stable (Figure 3). HRT increase will be followed by a decrease in the concentration of COD and BOD_5, it shows that there has been a decomposition of organic material by microorganisms activity.

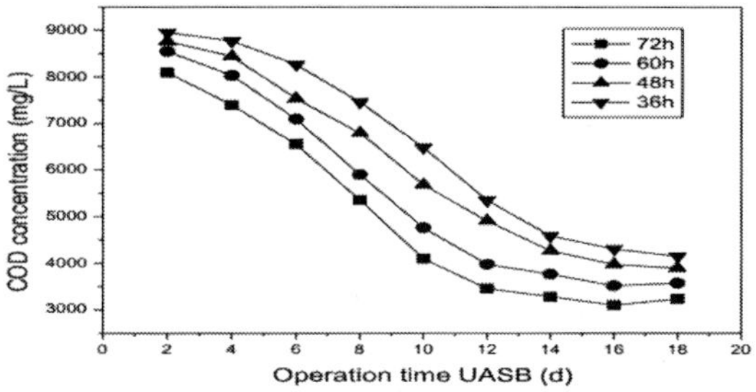

Figure-2. Correlation of HRT and UASB operation time to reduction of COD

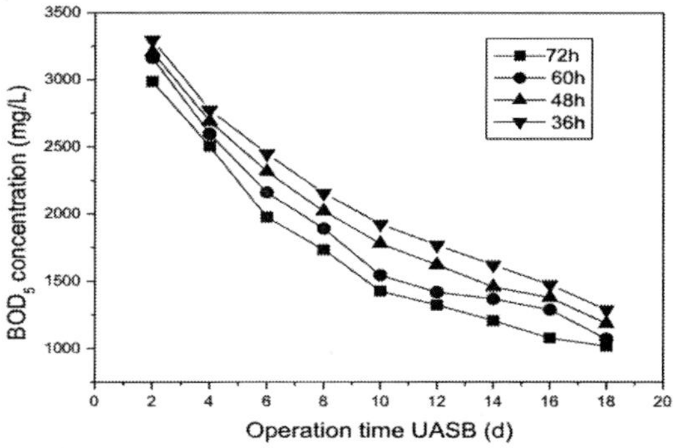

Figure-3. Correlation of HRT and UASB operation time to reduction of BOD_5

Increasing the residence time in the reactor means that the longer the waste is in the system. As a result, the contact between the biomass with the substrate is also getting longer so that the process of biological degradation takes place the better, and consequently a reduction in total COD and BOD_5 increased. Organic materials contained in the vinasse are described as complex compounds into the simple compounds. The maximum reduction in COD and BOD_5 obtained for HRT 72 h, on 16th day of COD concentration reached 3,106 mg / L while on 18th day match BOD_5 concentration reached 1,020 mg / L.

Figure 4 showed the effect of HRT and operation time UASB on the percentage removal of BOD_5 and COD. The removal of BOD_5 is higher than COD removal both for HRT 72 h and for HRT 36 h. The removal of

DOI: 10.1051/matecconf/20165801005

BOD_5 on HRT 72 h is not much different from the removal of BOD_5 at 36 h, but for COD removal is big enough difference between HRT 72 h and HRT 36h.

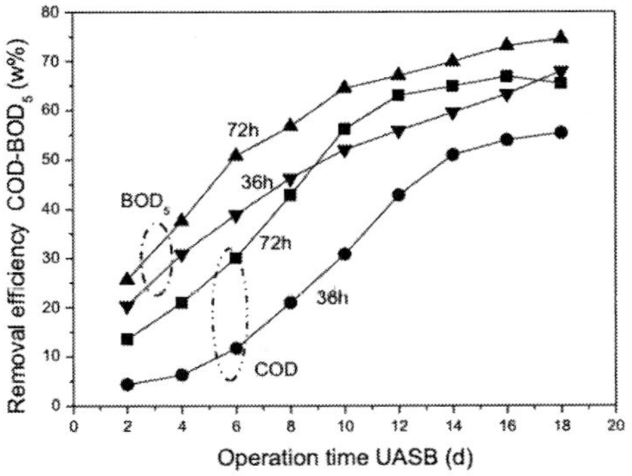

Figure-4. Effect of HRT and UASB operation time on removal efficiency of COD dan BOD_5

The maximum COD removal efficiency ranges from 66.81% - 55.64% and BOD_5 ranges from 74.58% - 68.85%. The largest removal HRT obtained in 72 h with a percentage removal of COD of 66.81% and BOD_5 74.58%. Long enough HRT will provide opportunities for anaerobic sludge and wastewater to contact longer so that the process of decomposition of organic matter becomes better. A longer HRT may adversely affect sludge granulation process in UASB reactor, while a very short HRT can lead biomass to drift out from reactor.

Setting the flow rate of vinnase which flows through the bottom of the reactor is necessary to study its effect on the removal of the organic load (COD and BOD_5). Variation of the flow rate will also cause the changes on the Hydrolic Loading Rate and upflow velocity (Vup). Table 1 shows the corellations of rate, HRL, and Vup to the removal efficiency of COD and BOD_5. The results showed that the greater the flow rate, the greater the HRL and Vup.

Hydrolic Loading Rate (HLR) is a velocity of flow in the reactor, is directly proportional to flow rate and inversely proportional to the cross-sectional area of the reactor. Vup is the flow velocity to pass through the cross section of the UASB bioreactor, is proportional to the flow rate and inversely proportional to the surface area of bioreactor. Operational of Vup on the UASB bioreactor varies according to the flow rate of the bioreactor. Upflow velocity has two opposite effects which can increase the contact time between biomass and substrate, making it possible to eliminate organic matter better. Whereas reducing upflow velocity can interfere with the contact between biomass substrates.

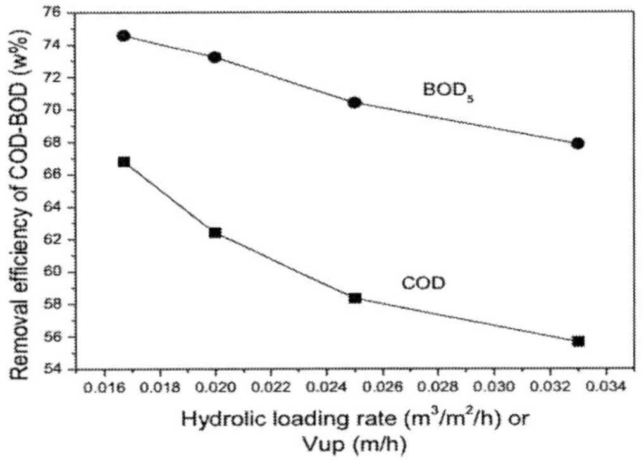

Figure-5. Effect of HLR to the efficiency removal of COD dan BOD_5

The results showed that the smaller the HLR (or Vup) at the same concentration load, COD and BOD_5 removal efficiency was increased. The level of efficiency of COD and BOD_5 decreased when the hydraulic and organic loading is increased. Quantitatively, the results obtained by the efficiency of COD and BOD_5 largest in HLR 0.0167 m^3 / m^2 / h with a percentage of 66.81% removal of COD and 74.58% removal of BOD_5.

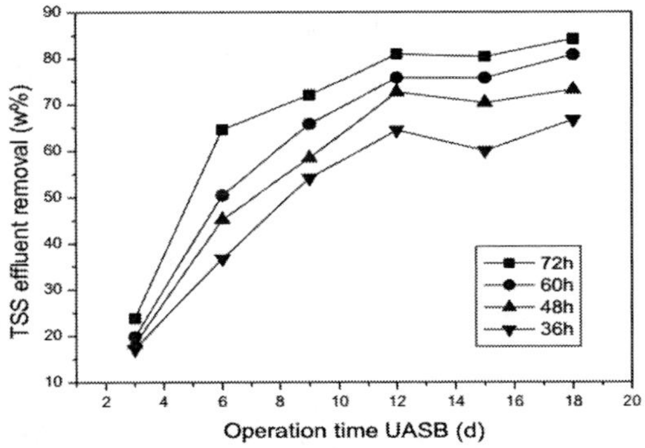

Figure-6. Correlation of HRT and operation time UASB to the efficiency of reduction of TSS consentration

Figure 6 shows the correlation of HRT and operation time UASB to the efficiency of reduction of TSS consentration. Effective time to decompose TSS is 12 days. The results showed that the greater the HRT, the greater the reduction of TSS efficiency and it will be followed by an increase in sludge bed with increasing time. The biggest reduction efficiency of TSS is in the range of 66.69 to 84.19 %, and up to a maximum reduction in HRT of 72 h with a percentage of 84.19% decreased TSS.

Biogas production

Figure 7 shows the correlation of COD and BOD_5 reduction on methane volume on 72 h HRT. COD decreased from about 8,000 to 3,200 mg / L, while the BOD_5 decreased from 2,985 to about 1,000 mg / L. On the other hand the volume of methane increased from 0.67 to 4.68 L / d. It can be indicated that the amount of methane gas that formed more influenced by COD reduction rather than by BOD reduction. However, the percentage of BOD_5 removal (25.62 - 74.58 %) is higher than the percentage of COD removal (13.5 - 66.81 %).

DOI: 10.1051/matecconf/20165801005

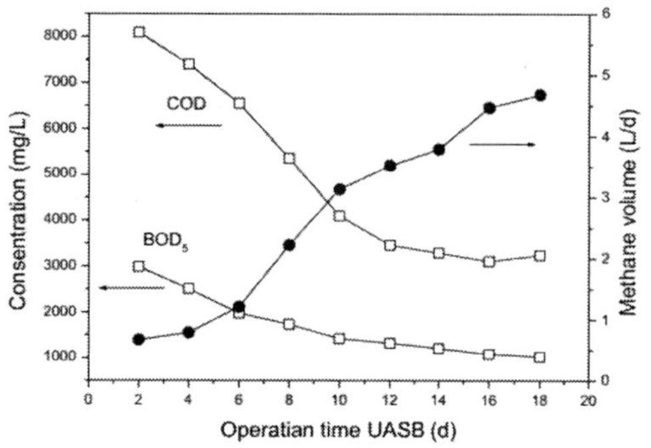

Figure-7. Correlation of COD and BOD_5 reduction on methane volume at 72 h HRT

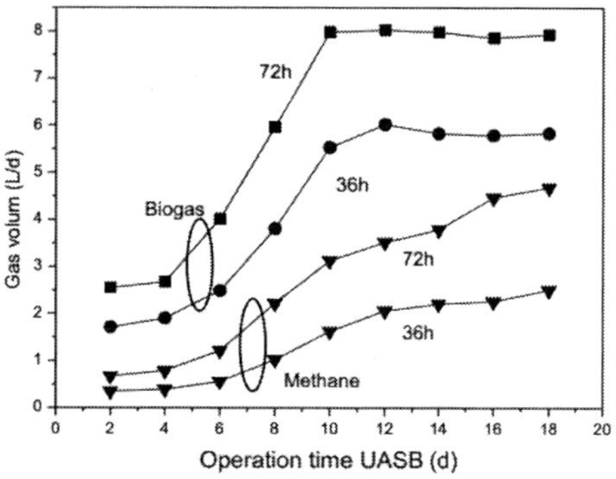

Figure-8. Correlation of HRT and operation time UASB to the production of biogas

Anaerobic degradation process involves a number of different bacteria but the process is driven mainly by two types of reactions, namely acidogenesis and methanogenesis. In the first stage of acidogenic, organic matter is decomposed into VFA (volatile fatty acids), then metabolysis at a later stage by methanogenic bacteria to produce methane gas (biogas). Figure 8 shows the biogas and methane production during fermentation. Biogas production is almost stable after the 10th day at 7.93 L/d (72 h HRT) and 5.826 (36 h HRT) since the process has reached steady conditions. However, the methane gas production is still slightly increased despite stable biogas condition.

Figure 9 shows the effect of HRT on production of gas on the 18th day. Methane product reached a concentration in the range of 42.89 to 58.06%. Increasing concentrations of methane from HRT 48 h to 60 h is considerably higher than of HRT 36 to 48 and from HRT 60 h to 72h.

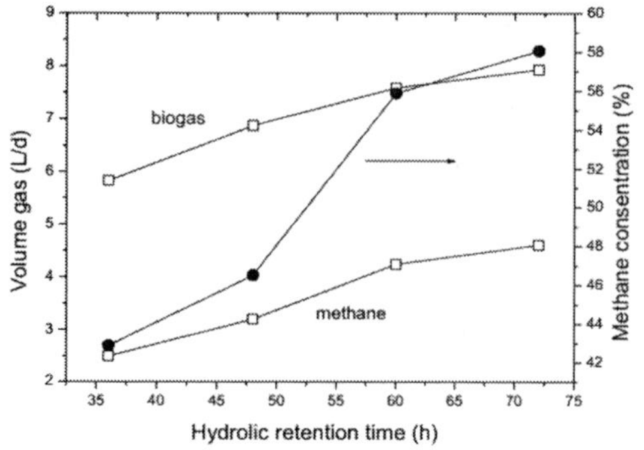

Figure-9. Effect HRT on gas product and methane concentration obtained at the 18th day.

CONCLUSION

Vinasse can be used as raw material for biogas production. The maximum biogas production at 72 hours obtained biogas volume of 7,930 L / day with the composition of the gas 58.06% (4.6042 L CH_4 / day).

COD; BOD_5; TSS has a relationship with a hydraulic capacity (HRT, debit, VUP, HLR). The smaller the discharge, the smaller Vup / HLR is getting and the higher efficiency, and vice versa. The greater the HRT, the removal of COD; BOD_5 and TSS increased so that the efficiency of separation is increasing. COD removal efficiency value; BOD_5 and TSS optimum occurred in HRT of 72 hours with an organic load of 9360 mg / l on day 18 which gained 67% COD removal; BOD_5 and TSS 75% 84.19%.

REFERENCES

Ashish,T., Oprakash S., 2014, Study of Characteristics and Treatment of Dairy Industry Wastewater ,J.App &Environ. Micro, 2 (1),16-22

Aslan, Sibel, Sekerdag , Nusret., 2008, The Performance of UASB Reactors Treating High-Strength Wastewaters. Journal of Environmental Healt,Find Article. Com

Alphenaar, P. A., Visser, A. A., and Lettinga., G., 1993, The Effect of Liquid Upward Velocity and Hydrolic Retention Time on Granulation in UASB Reactors Treating Wastewater with High Sulphate Cotent Biosource Technology, 43, 249.

Driessen, W., and Yspeert, P.,1999. Anaerobic Treatment of Low, Medium and High Strength Effluent in The Agro-Industry. Wat. Sci. Tech., 8, 221-228.

Farajzadehha, Soheil. 2012. Lab Scale Study of HRT and OLR Optimization in UASB Reactor for Pretreating Fortified Wastewater in Various Operational Temperatures. APCBEE Procedia 1, 90 –95.

Guerrero, L., Omil, F., Mendez, R., and Lema, J.M., 1999, Anaerobic Hydrolysis and Acidogenesis of Wastewaters from Food Indsutries with High Content of Organic Solids and Protein. Wat. Res., 33, 15, 3281-3290.

Haandel, A.C., 2006, Anaerobic Reactor Design Concepts for the Treatment of Domestic Wastewater, RevEnviron. Sci.BioTechnol, 5-21.

Liu, H.W., Walter, H.K., Vogt, G.M., Vogt, H.S., Holbein, B.E., 2002, Steam Pressure Disruption of Municipal Solid Waste Enhances Anaerobic Digestion Kinetics and Biogas Yield, Biotechnol. Bioeng. **77**, 121-129.

Lettinga, G., and Hulshoff pol, L.W., 1991,UASB Process Design for Various Types of Wastewaters. *Wat. Sci. Tech.*, 24, 87-107

Mulligan,C., Bechara,F.S., Meunier,J.et Chebib,J., 1993, Anaerobic Treatment of Cheese Dairy Wastewater Using the SNS Bioreactor. Water Pullutan Reaserch Jaurnal of Canada, 28, 597-619.

Marchaim, U., 1991, Biogas Processes for Sutainable Development.
http://www.citeseerx.ist.tsu.edu/viewdoc/download?doi.
Accessed 17 Juny 2011

Marchaim, U., 1992, Biogas Processes for Sutainable Development.Shmona: MIGAL., Galilee Technological Centre Kiryat Shmona, Israel.

Okpokwasili, G.C., and Nweke, C.O., 2005, Microbial Growth and SubstrateUtilization Kinetic. African Journal of of Biotecnology, 5(4), 305-317.

Ramakrishna Rao, R,S., Bala Prasad,S,B., Sekhar P,S,R., S.A. Rahiman, S,A., 2013, Performance Studies on Wastewater Treatment Efficiency of an Artificial Wetland, J.Environ Eng & Techno., 2 (3), 40 – 35

Ramasamy, E.V., Gajalakshmi, S., Sanjeevi, R., Jithes, M.N., Abbasi, S.A., 2004, Feassibility Studies on the Treatment of Daiary Wastewater with Upflow Anaerobic Sludge Blanket Reaktors. Bioresource Technology, 93, 2009-212.

Soepriyanto, Tantowi,I., Murtina, D. L. and Bernadeta, N., 2010, Vinase Processing of Alcohol Industry Wastewater into Biogas Using UASB Bioreaktor, Journal Purifikasi, 11 (1), 11-20

Schnurer, A., and Jarvis, A., 2009, Microbiological Handbook for Biogas Plants
http://www.ocfarmbiogas.ca/Microbiological%20
Handbook%20for%Biogas%20Plants.pdf.

Shin, H.S., Han, S.K., Song, Y.C., dan Lee, C.Y., 2001, Performance of UASB Reactor Treating Leachate from Acidogenic Fermenter in the Two-phase Anaerobic Digestion of Food Waste. *Wat. Res.*, **35**, 3441-3447.

Tabatabaei, M., Sulaiman, A.M., Nikbakht, N., Yusof, Najafpour, 2011, Influential parameter on Biomethane Generation in Aerobic Wastewater Treatments Plants. Accessed 4 October 2011

Vadivel, R Minhas, P.S, Kumar, P. Singh, Y., Nageshwar, R., Nirmale, A., 2014, Significance of vinasse waste management , Afr.J.Agric.Res., 9 (38), 2862-2873.

Yu, H.Q., and Fang, H.H.P., 2003, Acidogenesis og Gelatin – Rich Wastewater in an Upflow Anaerobic Reaktor:Influence of pH and Temperature.Water Research 37(1), 55-56.

SYNTHESIS AND CHARACTERISATION OF STRUVITE FAMILY CRYSTALS BY AN AQUEOUS PRECIPITATION METHOD

S.Sutiyono[1], Luluk Edahwati[1,2], Dyah Suci Perwitasari[1,2] ,
Stefanus Muryanto[2,3], J. Jamari[2] and Anatasius P.Bayuseno[2]

[1]Departement of Chemical Engineering, University of Pembangunan Nasional "Veteran" Jatim.
[2]Center for Waste Management, Mechanical Engineering Graduate Program,
Diponegoro University, Tembalang Campus, Semarang 50275, Indonesia.
[2,3]Office of Research and Department of Chemical Engineering, UNTAG, Bendhan Dhuwur Campus, Semarang 50233, Indonesia.
E-mail: apbayuseno@gmail.com

ABSTRACT
Struvite (NH_4MgPO_4•$6H_2O$) is a mineral often found as a scale deposit in an industrial process and piping. The common occurrence of struvite formation is a wide variety of environments. pH solution and environmental temperature are important parameters that control precipitation and subsequent decomposition of struvite. The study aims to characterise the struvite precipitated out the solution through chemical and mineralogical analysis. In this work, struvite solution was prepared by mixing NH_4OH , $MgCl_2$ and H_3PO_4 with the different purity grade (A and B) in the mol ratio of 1 : 1 : 1 in the stirred glass vessel tank with volume of 500 ml, and then stirred at 200 rpm for 70 minutes. Temperature of 30 ^{O}C and an initial pH of 10 were selected. The precipitated deposit was dried at room temperature for 48 hours of being analyzed. Material characterization of the deposits was conducted using XRPD Rietveld method of qualitative and quantitative mineralogical phase analysis. XRF was employed for bulk chemical elemental analysis of the scale obtained. The sample A using analytical grade powder provided the major phases of struvite-(K) and struvite. In constrast, the major phase of newberyite and minor phases of lepidocrocite (FeOOH) and niter (KNO_3) was observed by following changes in the lowest purity of powder (Sample B). Analysis of the experimental data suggests a solution mediated transformation process as a possible mechanism of struvite transformation.
Keywords: *Bulk elemental analysis; Characterization; Struvite scalling, XRPD Rietveld method*

INTRODUCTION

Struvite (NH_4MgPO_4•$6H_2O$) or magnesium ammonium phosphate (MAP) is a mineral often found as a precipitate in an industrial process and pipping system. Struvite precipitated in wastewater treatment plants make some problems causing deposits in pipe walls [Gaterell *et al.*, 2000; Priestley et.al., 1997]. It is also known as a bio-mineral in urinary tracts and kidney. However, precipitation of struvite has been addressed as a recovery method of ammonia or phosphate from industrial and domestic wastewaters [Tünay et,al., 1997; Shin and Lee, 1997; Schulze-Rettmer, 1991; Snoeying and Jenkins, 1980; Loewenthal et.al., 1994]. It has attracted considerable attention because of its occurrence of different environments and its utility as a resource. Struvite has been reported to be a highly effective source of nitrogen, magnesium and phosphorus for plants and can be used as a slow release fertilizer at high application rates without damaging plant roots [Whitaker and Jeffery, 1970; Battistoni et.al., 1997; 2000].

Struvite precipitation may be controlled by an alkali pH levels and temperatures. Moreover, struvite decomposition occurs to a slow low heating, while this decomposition is accompanied by developing both ammonia and water at temperatures below 40°C. However, more rapid heating enables the decomposition to occur at around 80°C. Thus synthesis route of minerals from struvite-saturated solutions may be carried at high temperature due to the increasing solubility of struvite.

Further, in the air, struvite may transform to form dittmarite ($MgNH_4PO_4H_2O$) at 103 ^{O}C [Abdelrazig and Sharp, 1988], but in water, it decomposes slightly at lower temperature of 60 ^{O}C [Sarkar, 1991]. Additionally, the remaining water of crystallisation and NH_4 are liberated at 235 ^{O}C and followed by formation of the pyrophosphate at 575 C [Abdelrazig and Sharp, 1988]. Here, magnesium pyrophosphate is considered as the most stable phase at high temperature. However, the different minerals may be produced in-between temperature. The stable temperature range of struvite decomposition may be related to the occurring process in the presence of nitrogen, air, or water [Frost et al., 2004]. In turn, the presence of water enables the stable temperature range of the generated minerals to lower values. Thus, this conditional process can provide important

DOI: 10.1051/matecconf/20165801006

information about the possible resulting minerals from struvite-saturated solutions to elevated

Struvite precipitation occurs to the presence of Mg^{+2} (M), NH_4^+ (N) and PO_4^{-3} (P) and the thermodynamic solubility product, K_s, is surpassed according to following reaction:

$$Mg^{2+} + NH_4^+ + PO_4^{3-} + 6H_2O \rightarrow MgNH_4PO_4.6H_2O \quad (1)$$

$$K_s = \left[Mg^{2+}\right]\left[NH_4^+\right]\left[PO_4^{3-}\right] \quad (2)$$

However, the precipitation of struvite has been reported to reduce with the pH, in which HPO_4^{-2} is predominantly precipitated in the reaction instead of PO_4^{-3} according to following reaction:

$$Mg^{2+} + NH_4^+ + HPO_4^{2-} + 6H_2O \rightarrow MgNH_4PO_4 \cdot 6H_2O + H^+ \quad (3)$$

Struvite crystals have been reported to be formed in the pH range of 7-11. In this range pH, a few crystals may be formed including trimagnesium phosphate [$Mg_3(PO_4)_2 \cdot 6H_2O$], newberyite [$MgHPO_4 \cdot 3H_2O$], and bobierrite [$Mg_3(PO_4)_2 \cdot 8H_2O$] [Musvoto et al., 2000]. Here, the solution pH and concentration of magnesium and phosphate controls the formation of these crystals.

The present study was undertaken to examine temperature and pH on struvite crystallization. The emphasis of this study was to synthesize struvite crystals and discuss the impact on the quality of the recovered product. The degree of crystallization for struvite products was analyzed by the Rietveld refinement method of x-ray powder diffraction (XRPD). Total elemental composition of solid crystals was conducted by wavelength dispersive X-ray fluorescence (XRF).

METHODOLOGY

Sample preparation

Struvite precipitation experiments were carried out in a batch reactor with a volume of 500 ml mixed with magnetic stirrer. The pH measurements are made with pH meter E588 Metrohm Herisau and pH is adjusted by KOH solutions. The chemicals used for synthesis were technical and analytical grade and hereby designated as sample A and B. In this work, struvite solution was prepared by mixing NH_4OH, $MgCl_2$ and H_3PO_4 in the mol ratio of 1:1:1

temperature. But it cannot be used to predict the mineral formation of the solution.

and then stirred at 200 rpm for 70 minutes. Temperature of 30 ^OC and an initial pH of 10 were adjusted. The precipitated deposit was dried at room temperature for 48 hours of being analysed.

Bulk chemical analysis

X-ray fluorescence (XRF) was employed for measurement of total elemental chemistry (mass of element/dry mass of solid crystal) with lower detection limits (LDL)- down to a few ppm (order of less than 10 ppm) for all size fractions (<100 μm) of the solid crytals. Further, batches of 8 grams of the sample were mixed with 2 g of acetone as a binder. The acetone was employed to increase sample densification. The mixture was finely homogenised in an agate bowl. The powder pellets were subsequently pressed on 20 Kg/cm^2 for 2 minutes and finally analyzed by XRF for all elements present in the solid, except for those lighter than fluorine (atomic mass=19).

X-ray powder diffraction (XRPD) and Rieveld analysis

XRPD measurements were conducted on A Philips PW 1710 Diffractometer equipped with a Cu tube and measurement parameters: 40 kV, 40 mA, 0.02^O step, 2 s per step, 5 – 90° 2θ. Phase identification was made with (Powder Diffraction File)-PDF-2 Phillips software. Rietveld refinement of the XRPD data was carried out with Program Fullprof-2k, version 3.30 [Rodriguez-Carvajal, 2005], with the crystal structure database from the published data in Mineralogical Society of America (see Table 1). Results presented here were obtained with refining fundamental parameters such as: the 2θ scale zero position, the polynomial fitting for the background with six coefficients, the phase scale factors, the cell parameters, the peak asymmetry and the peak shape functions, the atomic coordinates, anisotropic temperature factors, and the preferred orientation for struvite family crystal with March-model. Site occupation factors were not refined during Rietveld refinement, because of all struvite peaks are overlapped by peaks of other minerals. The error of the Rietveld quantifications can be estimated to about 2 wt. % (1ρ) per phase.

Table-1. Crystal structure models used in Rietveld refinements

Structure models	PDF number	Authors
Struvite [$MgNH_4PO_4 \cdot 6H_2O$]	PDF#71-2089	Whitaker and Jeffery (1970)
Stuvite-(K) [$KMgPO4 \cdot 6H_2O$]	PDF#35-0812	Graeser et al. (2008)
Newberyite [$MgHPO_4 \cdot 3H_2O$]	PDF# 76-3368	Abbona et al. (1979)
Lepidocrocite [FeOOH]	PDF#89-6096	Pernet, Joubert, Berthet-Colominas. (1975) 17, 1505
Niter [KNO_3]	PDF#81-0070	Lu, H.M., Hardy.(1991). 44, 7215

The final Rietveld scale factors were converted to PCs-by-weight using the "ZMV" expressions below:

$$\text{wt.\%(analyte phase)} = \frac{[s(ZMV)]_{analyte\ phase}}{\Sigma[s(ZMV)]_{all\ phase}} \quad (4)$$

where s, Z, M and V signify the phase scale factors; number of formula units/cell; formula weights and unit cell volumes, respectively.

RESULTS

Bulk chemical compositions

Chemical compositions of the crystal solid were analysed for major and trace elements. The chemical compositions are presented in the form of element weight percents, independent of the actual form of chemical binding in the residues. The light elements oxygen and carbon could not be analysed. However, the cations may be mainly bound in oxide compounds.

Table 2. Chemical elements of precipitating solids

Elements	P	Mg	K	Ca	Ni	Fe	Cu	Ti	Cr	Zn
Wt.%	61.9	12.0	9.8	8.5	4.3	1.2	0.7	0.6	0.3	0.2
Esd	1.4	2.0	0.1	0.4	0.3	0.1	0.1	0.1	0.1	0.0

Powder XRD analysis

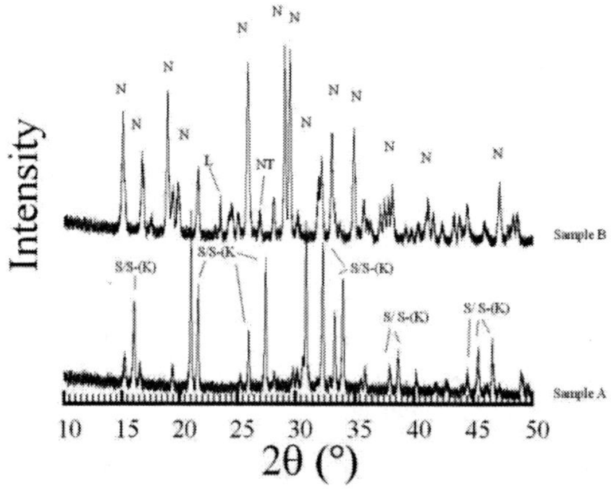

Figure-1. XRPD pattern of the samples A and B obtained at temperature of 30 °C and pH 10. The principal diffraction peaks of the predominant minerals are shown for reference purposes. The peaks are labelled L (lepidocrocite); N (newberyite); NT (niter); S (struvite), and S-(K) (struvite K) .

Table 3 Mineralogical phase composition of the synthetic materials using samples of A and B

Sample A			Sample B		
	wt.%	eds		wt.%	eds
Struvite-(K)	87.64	1.08	Newberyite	96.91	0.61
Struvite	12.36	0.88	Lepidocrocite	1.40	0.01
			Niter	1.69	0.04
Total	100.00		Total	100.00	

DISCUSSIONS

Table 2 presents the chemical compositions of the crystal solid. The sum of the major components and trace element concentrations for these materials are notably less than 100 %, because the remainder of the materials may comprise largely water and carbon dioxide (which were not analysed for).

The mineralogy phases of crystal solid controlling the magnesium, ammonium, phosphate levels in the solution were initially identified by XRPD search-match method. In this method, all XRPD patterns of sample A agree well with the PDF#71-2089 value for struvite, and the DF#35-0812 value for struvite-(K), while PDF#70-2345 value for newberyite [$MgHPO_4 \cdot 3H_2O$], PDF# 76-2301 value for lepidocrocite [FeOOH] and PDF#81-0070 value for niter [KNO_3] were found in the sample B. The '*candidate* and/or *probable*' minerals found in the search-match program were subsequently proved by the full profile Rietveld refinement, as peaks of phases which have been missed in the search match or mistakenly assigned phases clearly stand out in the difference plot of the calculated and the measured diffraction profile (Prince, 1993; Rietveld, 1969;Winburn *et al.*, 2000) (fig.1).

The quantitative results are given in weight percent normalised to 100 %. The major mineralogical phases in the sample A are struvite-(K) in addition to struvite. The significant amount of newberyite is obtained in the sample B with some minor of lepidocrite and nitrite produced. This indicates that the use of different chemical grade in solution provided the different minerals, although the molar ratio is same. The highest purity of chemical powder has influenced the formation of struvite family crystals, but the lowest purity may be favoring in formation of newberyite crystals.

Both the struvite and struvite-(K) has been known to be iso-structural minerals [Graeser et.al, 2008]. The struvite-(K) precipitation may occur in the solution according to following reaction:

$$Mg^{2+} + K^+ + PO_4^{3-} + 6H_2O \rightarrow KMgPO_4 \bullet 6(H_2O)_{(solid)} \quad (5)$$

The presence of KOH in the solution may react to ions of Mg^{2+} and PO_4^{3-} to form struvite-(K). Here Mg^{2+} ion was partly consumed for this mineral, while the rest of Mg ion may be used for struvite formation. Moreover, solution pH controls the distribution ammonia nitrogen between NH_4^+ and NH_3 according to following equilibrium reaction:

$$NH_4^+ + H_2O \Leftrightarrow NH_3 + H_3O^+ \quad (6)$$

where the equilibrium constant (K_a), is defined as [Stumm and Morgan, 1996].

$$K_a = \frac{[NH_3][H_3O^+]}{[NH_4^+]} = 5.7 * 10^{-10} \quad (7)$$

The high pH values have been reported to favor NH_3 formation which is in equilibrium with air according to Henry Law and promotes air stripping of ammonia [Stumm and Morgan, 1996]. At pH 10 examined during the study, much ammonia could be lost at 30 ^{o}C, resulting to the minor struvite formation. In contrast, newberyite was found in quite abundance in the sample B. Newberyite is known as a rare mineral and can be formed in the absence of ammonium. The absence of ammonium may be lost at the temperature studied.This mineral was subsequently precipitated out the solution according to the following reaction [Abbona et al., 1979]:

$$Mg^{+2} + HPO_2^{-4} + 3H_2O \rightarrow MgHPO_4 \bullet 3H_2O_{(solid)} \quad (8)$$

The formation of this mineral is commonly accompanied by other phosphates (*e.g.* struvite and hannayite) [Abbona et al 1988]. Here pH and molar ratio of MAP concentration are particularly an important parameter in the crystal formation and aggregation. The significant amount of newberyite was produced in the study, because this mineral can be developed significantly with MAP of 1:1:1 molar ratio [Abbona et al., 1979].

CONCLUSIONS

The study proved that the use of different purity of material in the synthesis of struvite provided the different minerals. Use of the technical grade material enables major struvite to form in the condition pH selected. In contrast, newberyite is the major mineral in the synthesis of material using the analytical grade chemical. As shown in Table 2 where certain hazardous substances such as Cu, Cr and Zn can also be shipped in the process of formation of struvite. It is also reinforced by the results shown in Table 3 where the purity of the chemical will affect the results stuvite formed though the same molar ratios are used.The pH for ammonia recovery as struvite precipitation in the technical grade sample is 10. However, the same pH used in the analytical grade sample has experienced the loss of ammonia to air due to stripping since high pH favors NH3. Newberyite precipitation has also been observed as low as pH 10.

Acknowledgements

The work was supported by the University of Pembangunan National, Surabaya Indonesia under the PhD research grant program.

REFERENCES

Abbona F., Boistelle R., Haser R. (1979). Hydrogen bonding in $MgHPO_4 \cdot 3H_2O$ (newberyite). Acta Crystallographica, B35, 2514-2518.

Abdelrazig B.E.I., Sharp J.H. (1988). Phase changes on heating ammonium magnesium phosphate hydrates. Thermochimica Acta 129,197-215.

Abbona F., Lundager Madsen H. E., Boistelle R. (1988). The final phases of calcium and magnesium phosphates precipitated from solutions of high to medium concentration J. Cryst. Growth. 89, 592-602.

Battistoni P., Fava G., Pavan P., Musacco A., Cecchi F. (1997). Phosphate Removal in Anaerobic Liquors by Struvite Crystallization Without Addition of Chemicals: Preliminary Results, *Wat. Res.*, 31, 2925-2929.

Battistoni P., Pavan P., Prisciandaro M., Cecchi F. (2000). Struvite Crystallization: A feasible and reliable way to fix phosphorus in anaerobic supernatants, *Wat. Res.*, 34, 3033-3041.

Frost R.L., Weier M.L., Erickson K.L. (2004). Thermal decomposition of struvite, J. Therm. Anal. Calorim. 76,1025-1033.

Gaterell M.R., Gay R., Wilson R., Gochin R.J., Lester J.N. (2000). An economic and environmental evaluation of the opportunities for substituting phosphorus recovered from wastewater treatment works in existing UK fertiliser markets, *Env. Technol.*, 21, 1067-1084.

Graeser S., Wostl W., Bojar H.P., Berlepsch P., Armbruster T., Raber T., Ettinger K., Walter F. (2008). Struvite-(K), $KMgPO4 \cdot 6H_2O$, the potassium equivalent of struvite a new mineral. European Journal of Mineralogy 20, 629–633.

Lu H. M. and Hardy J. R. (1991). First-principles study of phase transitions in KNO_3. Physical Review. B 44, 7215.

Loewenthal R.E., Kornmuller U.R.C., Van Heerden E.P. (1994). Struvite Precipitation in Anaerobic Treatment Systems, *In 17th International Symposium on Anaerobic Digestion, Capetown, South Africa,*498-507.

Musvoto E. V., Wentzel M. C., Ekama G.A. (2000). Integrated Chemical-Physical Process Modeling-Ii. Simulation Aeration Treatment For Anaerobic Digester Supernatant. *Water Research,* 34, 1868 - 1880.

Pernet M., Joubert J.C., Berthet-Colominas C. (1975). Etude par diffraction neutronique de la forme haute pression de FeOOH, Solid State Communications, Volume 17, Issue 12, 15, Pages 1505–1510.

Priestley A.J., Cooney E., Booker N.A., Fraser I.H. (1997). Nutrients in wastewaters-ecological problem or commercial opportunity. Proceedings of the 17th Federal Convention of the Australian Water and Wastewater Association, Melbourne, **1**, 340-346.

Rodriguez-Carvajal J., Program Fullprof.2k, version 3.30, Laboratoire Leon Brillouin, France, June 2005.

Snoeying V.L., Jenkins D. (1980). *Water Chemistry*, John Wiley and Sons Inc., New York.

Sarkar A.K. (1991). Hydration/dehydration characteristics of struvite and dittmarite pertaining to magnesium ammonium phosphate cement systems. J. Mater. Sci., 26, 2514–2518.

Schulze-Rettmer R. (1991). The Simultaneous Chemical Precipitation of Ammonium and Phosphate in the Form of Magnesium Ammonium Phosphate, *Water Sci. Technol.,* 23, 659-667

Stumm W., Morgan J.J. (1996). *Aquatic Chemistry,* Wiley-Interscience, New York .

Shin H.S., Lee S.M. (1997). Removal of Nutrients in Wastewater by Using Magnesium Salts, *Env. Technol.,* 19, 283-290.

Tünay O., Kabdasli I., Orhon D., Kolçak S. (1997). Ammonia removal by magnesium ammonium phosphate precipitation in industrial wastewaters, *Water Sci. Technol.,* 36, 225-228.

Whitaker A., Jeffery J.W. (1970) . The crystal structure of struvite, $MgNH_4PO_4{\cdot}6H_2O$. Acta Crystallographica B26,1429-1440.

EFFECT OF CALCIUM ADDITIVE ON THE CRYSTALLIZATION OF STRUVITE

Dyah Suci Perwitasari [1], Luluk Edahwati [2], S.Sutiyono [3], Novel Karaman [4], J. Jamari [5], Stefanus Muryanto [6], Athanasius.P.Bayuseno [7]
[1,2,3,4] University of Pembangunan National "Veteran" Jawa Timur, Surabaya, East Java, Indonesia
[1,2,4,5,6,7] Mechanical Engineering Graduate Program, Diponegoro University, Tembalang Campus, Semarang 50275, Indonesia
[6] Department of Chemical Engineering, UNTAG University in Semarang, Bendhan Dhuwur Campus, Semarang 50233, Indonesia
E-Mail: [1] saridyah05@gmail.com, [7)] apbayuseno@gmail.com

ABSTRACT

Crystallization of struvite [$MgNH_4PO_4.6H_2O$] may lead to the deposition of scale which may create significant problems in the process pipes, pumps and other industrial equipment. However, struvite precipitation can be benefited for phosphate recovery for use of fertilizer. The aim at the present work was to investigate calcium additive on struvite precipitation. The experiment was carried out in a batch mode using a 1-liter Pyrex glass vessel mechanically agitated for 200 rpm. The scale-forming solution was prepared for mixing solutions to $MgCl_2$ and $NH_4H_2PO_4$ with Mg^{+2} , NH_4^+ and PO_4^{-3} in a molar ratio of 1: 1:1. The crystallization temperature of 30 and 40 ^oC was selected. Ca was added into the crystallizing solution to chloride dihydrate 0.4M ($CaCl_2$•$2H_2O$). Then each solution was pH adjusted to 9 by addition of KOH. The crystals obtained were characterized using SEM for morphology, EDS for elemental analysis as well as XRPD Rietveld analysis for crystaline phases. The induction periods varied from 10 to 90 min, which means that the struvite crystals began forming 10 to 90 min after mixing of the solution. It was observed that the Ca additive may inhibit the struvite crystallization. SEM analysis revealed that the struvite crystals obtained were predominantly of irregular prismatic morphology. Furthermore, the EDS pattern revealed that the elemental composition of the crystals consisted of Ca, Cl, S Mg, N, and P, providing that many crystalline phase found in the crystals such as Gypsum, $CaCl_2$, struvite, struvite-(K) and sylvite. It was observed that the Ca additive appeared to inhibit the struvite crystallization.

Keywords: Ca-additives, SEM-EDS analysis, struvite and XRPD Rietveld analysis

INTRODUCTION

Struvite precipitation is now considered as a prospective technology for phosphate recovery from waste solutions, municipal sewage, industrial wastewater, liquid manure (Doyle et al., 2003; Doyle and Parsons, 2002). Recently, the struvite precipitation is becoming interesting (Doyle et al., 2003), because the final product has a potential market for fertilizer industries. This can be only achieved by controlling the properties of product, which reflect to its composition [nitrogen (N), phosphorus (P) and magnesium (Mg) ions in equal molar concentrations (Booker et al., 1999). While the chemical purity and crystal size distribution of the crystal product can be obtained from the precipitation process by controlling values of process parameters and also designing of equipment and apparatus (Rawn et al., 1937, Snoeyink and Jenkins, 1980). Indeed for the struvite precipitation, the properties and quantity of the final product formed has become a considerable concern in order to be successful and economically valuable at full scale. Consequently, the purity and size of crystal recovered has to be handled for a reuse as fertiliser, whereas the granule formed has to be strong enough for practical reason (collection, transport and land filling). The purity is sometimes controlled for environmental reasons.

Furthermore, the significant factor influencing the process and eventually the crystal quality is the presence of contaminant in the solution such as calcium, chloride, carbonates, sulphates, nitrates, fluorides, fluorosilicates (Doyle and Parsons, 2002, Mohajit et al., 1989). These impurities may inhibit or promote the precipitation reaction, and influence the rate of crystal nucleation and growth, and finally control the crystal shape and possible particle agglomeration (Booker et al., 1999; Ohlinger et al., 1999). In an alkaline solution and the presence of calcium ions, calcium phosphates (in wastewater systems commonly a poorly crystallised hydroxylapatite) or calcium carbonate, which can react effectively with phosphate or carbonate ions according to the following equations:

$$5Ca^{2+}+3PO4^{3-}+H_2O \rightarrow Ca_5(PO_4)_3OH+H^+$$

$$Ca^{2+}+CO_3^{2-} \rightarrow CaCO_3$$

The occurrence of phosphates or carbonate mineral may worsen the chemical composition of the product and limits the possibilities of its further use (Stratful et al., 2001; Kofina and Koutsoukos, 2003). Furthermore, the Ca ions present in the solution may control in the struvite formation, in which it will either compete for phosphate ions or interfere with the crystallisation of struvite. As a consequence of obstruction for crystallisation of struvite, the change in the characteristics of struvite crystals may occur, however this area of research has not been examined in detail in terms of qualitative and quantitative. Thus only limited study has been conducted in this area of research (Le Corre et al., 2005). Here, an improved understanding on struvite precipitation out a solution due to the presence of ion additives is required.

The aim of the present research was to examine how Ca-ion may affect struvite crystals nucleation and growth, and discuss the impact of this ion on the quality of the crystals. In the present study, struvite crystals were grown by the prepared solution containing MAP composition and then characterized by X-ray powder diffraction (XRPD) for mineralogical phase composition and SEM analysis with EDS for elemental analysis and morphology.

METHODOLOGY

Materials and crystallization experiments

A batch of crystallization experiments was conducted to examine the effect of Ca ions on struvite crystal growth. Here struvite crystallization was developed from mixed solutions of magnesium chloride hexahydrate ($MgCl_2{\cdot}6H_2O$) and ammonium phosphate ($NH_4H_2PO_4$), which were supplied with analytical grade of chemical product by Merck, Germany. The solution was prepared in a 1-L beaker from equal volumes (400 mL) of $MgCl_2{\cdot}6H_2O$ and NH_4H_2PO4 so that the Mg concentration in the final solution was that the ratio of Mg:N:P was 1:1:1. Then each solution was pH adjusted to 9 by addition of KOH. Ratios Mg:Ca:Cl calculated were 2:1:1. Each stock solution was then filtered through a 0.22 μm paper filter to eliminate the presence of impurities, and dried in a desiccator at room temperature. In the present study, all experiments were conducted at room temperature.

Analytical methods

The dried sample of crystals was then characterized by X-ray powder difraction (XRPD) (Philips 1830/40) for mineralogical phase composition, while elemental composition and morphology of the precipitates were examined by scanning electron microscopy (SEM; JEOL JSM 5200) equipped by EDS.

Thermodynamic chemical modeling

The Visual MINTEQ software program version 3.0 was employed in the study to estimate the solution equilibrium using input of the activities of the various ions present in the solution. This program provided the calculation results on the solubility of solids, simulating equilibrium and speciation of inorganic solutes in the laboratory solution. The Visual MINTEQ can also estimate every precipitated solid phases in oversaturated condition equilibrium. Type of calculation selected for Visual MINTEQ evaluation in this study was the possible solids. The species used in the model were calculated by MINTEQ program by entering input parameters at fixed pH value of 9 and temperatures of 30 and 40 ^{o}C. By using these values, the program calculated Mg^{+2}, NH_3, NH_4 , H_3PO_4, $H_2PO_4^-$, HPO_4^{-2}, $PO4^{-3}$, and $MgNH_4PO_4$. Moreover estimated outputs of minerals were compared to the analysis of XRPD results. The chemical composition of the synthetic waste water for the input of the Program is given in Table-1.

Table-1. The approximate composition of the synthetic wastewater.

No	Parameter	Concentration (molal)
1.	Mg	0.20403
2.	NH_4	0.34182
3.	PO_4	0.18297
4.	K	0.21158
5.	Ca	0.22520
6.	Cl	0.12555

RESULTS

Table-2. Model prediction of species for chemical composition of the synthetic waste water.

	pH 9 and temperature of 30 ^{o}C		pH 9 and temperature of 40 ^{o}C	
Component	**% of total concentration**	**Species name**	**% of total concentration**	**Species name**
Mg^{+2}	67.366	Mg_2^+	67.331	Mg^{+2}
	0.194	$MgOH^+$	0.433	MgOH+
	7.293	$MgCl^+$	7.496	MgCl+
	1.119	$Mg(NH_3)_2^{+2}$	2.486	$Mg(NH_3)_2^{+2}$
	1.39	$MgPO_4$-	1.55	$MgPO_4$-
	22.639	$MgHPO_4$ (aq)	20.703	$MgHPO_4$ (aq)
$NH4^{+1}$	70.526	NH_4^{+1}	55.938	NH_4^{+1}
	1.336	$Mg(NH_3)_2^{+2}$	2.968	$Mg(NH_3)_2^{+2}$
	25.278	NH_3 (aq)	38.913	NH3 (aq)
	2.665	$CaNH_3^{+2}$	2.061	$CaNH_3^{+2}$
	0.195	$Ca(NH_3)_2^{+2}$	0.119	$Ca(NH_3)2^{+2}$
PO_4^{-3}	0.022	$PO4^{-3}$	0.022	$PO4^{-3}$
	5.429	HPO_4^{-2}	4.409	$HPO4^{-2}$
	0.022	$H2PO_4^-$	0.017	H2PO4-
	1.55	$MgPO_4^-$	1.729	MgPO4-
	25.244	$MgHPO_4$ (aq)	23.086	$MgHPO_4$ (aq)
	10.292	$CaHPO_4$ (aq)	9.082	$CaHPO_4$ (aq)
	55.794	$CaPO_4^-$	60.06	CaPO4-
	0.014	$CaH2PO_4^+$	0.012	CaH2PO4+
	1.456	$KHPO_4^-$	1.411	KHPO4-
	0.167	K_2HPO_4 (aq)	0.162	K_2HPO4 (aq)

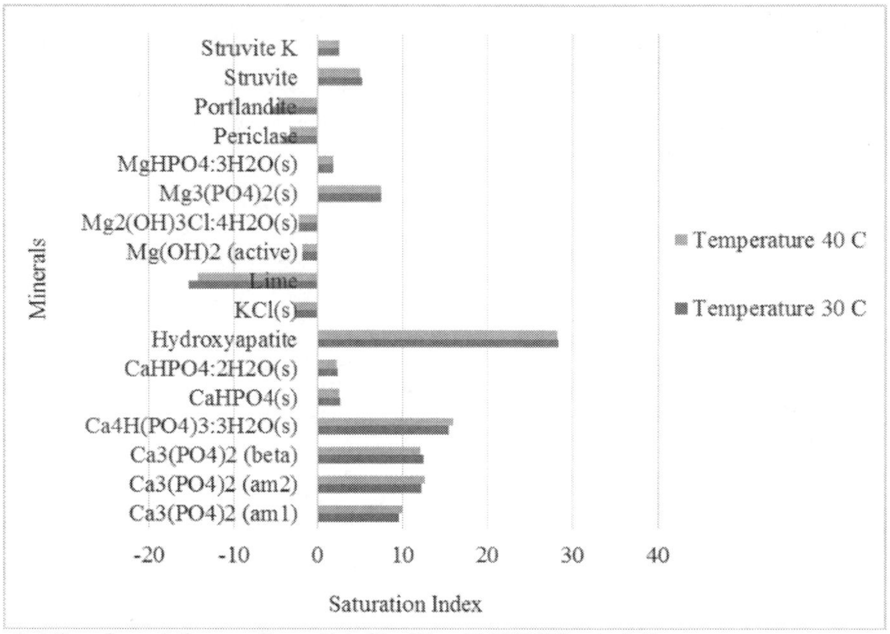

Figure-1. Mineral speciation results predicted by Visual MINTEQ calculation.

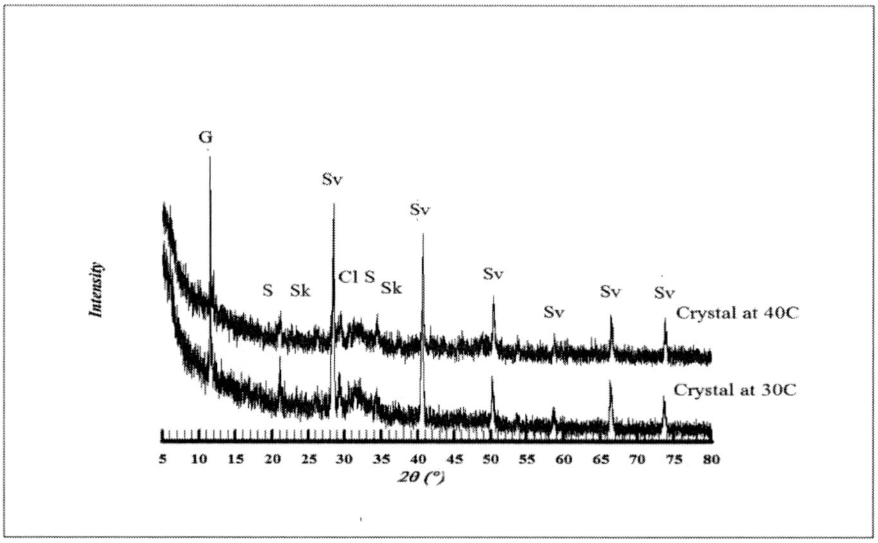

Figure-2. XRPD patterns of the precipitates obtained at temperature of 30 and 40 ^{o}C and pH 9. The principal diffraction peaks of the predominant minerals are shown for reference purposes. The peaks are labelled Cl ($CaCl_2$), G (gypsum), S (struvite), SK (struvite-K) and Sv (sylvite).

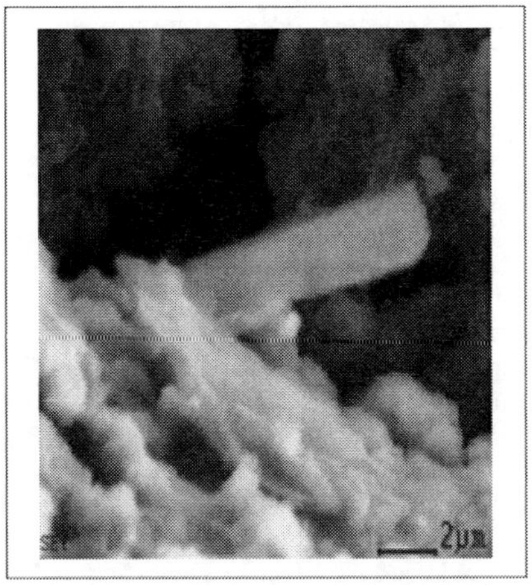

Figure-3. a) Typical SEM morphology for the crystals

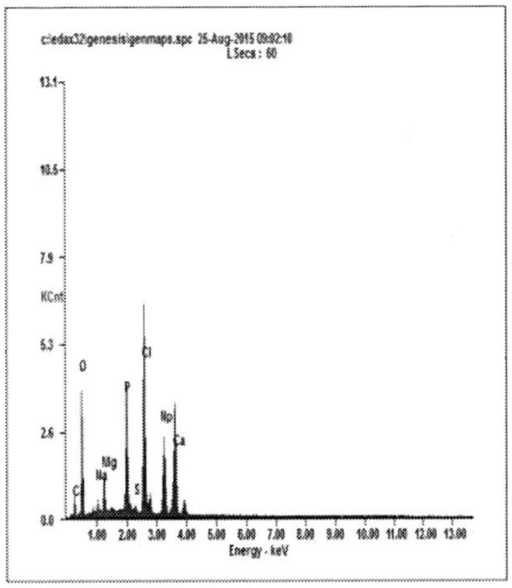

b) EDS analysis of the deposits formed at pH 9 and temperature of 30 ^{O}C.

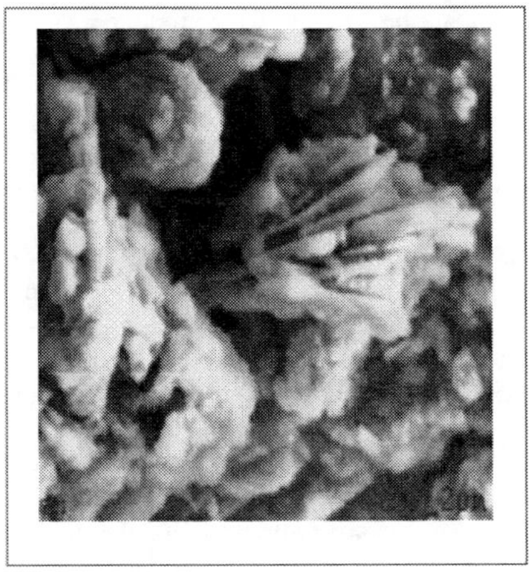

Figure-4. a) Typical SEM morphology for the crystals

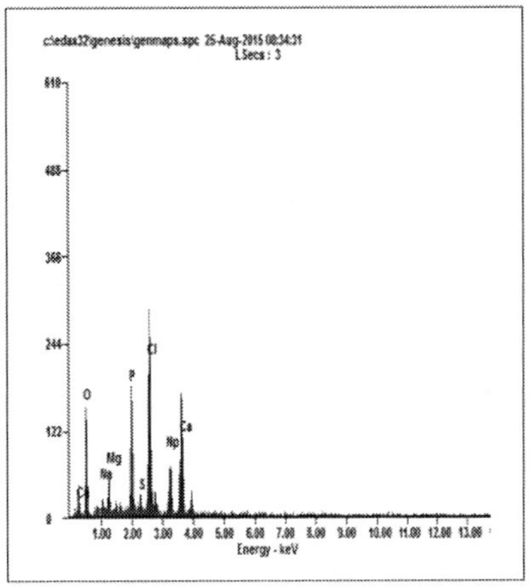

b) EDS analysis of the deposits formed at pH 9 and temperature of 40 ^{O}C.

DISCUSSIONS

Chemical equilibrium model and mineral speciation results

The precipitation potential of minerals was predicted by the Visual MINTEQ using the values in Tables-1. The chemical equilibrium model was calculated in the program provided in Table-2 . At temperature of 30 ^{O}C, when the Mg_2: NH_4 :PO_4 molar ratio was 1:1:1 and pH 9, the highest NH_4 removal was obtained as 70.526%. However, at temperature of 40 ^{O}C, when the Mg^{+2} :NH^{+4} :PO_4^{-3} molar ratio was 1:1:1 and pH 9, the lowest NH_3 removal was obtained as 55.938%. At the equilibrium pH 9 and in the presence of Ca ions, the element Mg^{2+}, NH^{4+}, and PO_4^{3-} may form ion complexes, such as MgOH+, MgCl, $Mg(NH_3)_2{}^{+}2$, $MgPO_4$-, MgHPO4+, $NH_4{}^{1-}$, $Mg(NH_3)_2{}^{+2}$, NH_3, $CaNH_3{}^{+2}$, $Ca(NH_3)_2{}^{+2}$, $Ca(NH_3)_2{}^{+2}$, $HPO_4{}^{-2}$, H_2PO_4-, $MgPO_4$-, $MgHPO_4$, $CaHPO_4$, $CaPO_4$-, CaH_2PO_4+, $KHPO_4$-, K_2HPO_4 in the system. These ion complexes agree with the finding results of previous works (Bouropoulos and Koutsoukos, 2000; Ohlinger et al., 2000).

Furthermore the predicted minerals were precipitated out the solution shown in Figure 1. The saturation indexes (SI) value is presented in the study for estimating the possibility for the formation of mineral speciation. A system can be oversaturated when $S > 1$, in which mineral precipitation potential exists. If $S \leq 1$, the system can be at equilibrium or undersaturated for minerals and the precipitation may not occur. Therefore for the mineral growth and accumulation, a positive SI values must be usually required. As can be seen, some minerals at temperature conditions had a positive SI value. In this pH 9, a few crystals can possibly form as hydroxyapatite, newberyite, struvite-(K), struvite, and $Mg_3(PO4)_2$. These crystals formed was influenced by solution pH and concentration of magnesium and phosphate On the other hand, minerals of sylvite were undersaturated. These estimated minerals found in the system were subsequently verified in the analysis of XRPD data.

Figure 2 shows the results of X-ray powder diffraction analysis conducted on the formed precipitates at temperatures of 30 and 40 ^{O}C. The presence of minerals was identified by the location of the intensity peaks, corresponding to the standard database lines (PDF-Powder Diffraction File) such as $CaCl_2$ (PDF#740992), gypsum (PDF#761746), struvite (PDF#35-0812), struvite-K (PDF#70-2345). These results also confirm that peaks at Struvite and Struvite K at low intensity were found in the pH 9 and at temperatures of 30 and $40^{O}C$. It indicates that those minerals may be formed into the low concentrations. They were further confirmed by comparison between solid analysis of phosphate/magnesium concentrations on the Visual Minteq program (Table 2). It has been shown that the precipitation of Struvite reduces the pH which suggests that $HPO4^{-2}$ would precipitate in the reaction rather than $PO4^{-3}$ according to following reaction:

$$Mg^{2+} + NH4^{+} + HPO4^{2-} + 6H^{2}O \longrightarrow MgNH_4PO_46H_2O + H^{+}$$

It must be yet pointed out that other precipitates may occur for pH values above 9 (Bouropoulos and Koutsoukos, 2000; Ohlinger et al., 2000).

Undesirable amounts of precipitated Struvite may be also contributed to the presence of Ca ions. For example, calcium ions (Ca^{2+}) with the typical concentrations of 30-60 mg/L in municipal wastewater plants has been suggested as representative ions for hampering struvite crystal nucleation and growth. Calcium ions can actively react with phosphate to form calcium phosphates. Previous works on the effects of calcium on struvite crystallization reported that calcium ions with struvite co-precipitation may retard the nucleation induction and inhibit the growth of struvite crystal formation (Le Corre et al., 2005). Moreover, calcium, as an impurity, could be a negative factor of struvite formation. The presence of calcium at high levels in the synthesized wastewater would inhibit struvite formation, because of the possible formation for calcium-phosphorus precipitates. Basically, the presence of impurities in the solution affects the growth rate of crystals by blocking of active growth sites and inhibiting the increase of crystal size. (Prywer and Olszynski, 2013).

Solid precipitates were characterized as the field scanning electron microscopes (SEM) coupled with energy dispersive X-ray analysis (SEM-EDS) . The SEM investigation into pH 9 is shown in Figure 3. The SEM images reveal that the deposits formed have typical irregular prismatic shaped crystals, while the most common elements identified with EDS are O, Na , Mg , P , S , Cl , N , and Ca . These elemental analyses further confirm that the deposits formed into this condition are gypsum, $CaCl_2$ and Struvite, in agreement with XRPD . In some of literature, the formation of struvite crystals can be found in the pH range of 7-11 (Bouropoulos and Koutsoukos, 2000; Ohlinger et al., 2000). In this range pH, a few crystals may form as trimagnesium phosphate $[Mg_3(PO_4)_2.6H_2O]$, magnesium hydrogen phosphate $[MgHPO_4.3H_2O]$, and bobierrite $[Mg_3(PO_4)_2.8H_2O]$ (Musvoto et al., 2000). However, in this study these three of kind the crystals formed were not identified by the XRPD Rietveld method, because these minerals may be precipitated in the order of days and are influenced by solution pH and concentration of magnesium and phosphate (Musvoto et al., 2000).

The SEM images of the precipitates formed into condition 40 ^{O}C and initial pH 9 are shown in Figure 4. The needle-shaped crystals morphology is clearly found. The crystals may be Struvite as indicated elements of Mg, N, P and O in the EDS spectrum . Other minerals of gypsum and $CaCl_2$ are possible and elements of Ca , Cl , S and O are shown in the EDS spectrum . However , minerals of Struvite-(K) and sylvite (KCl) are not shown in the SEM-EDS analysis , though both minerals can be identified by the XRPD method. These minerals may be identified with other spots. Here successful P recovery should need an effective nucleation and growth of Struvite crystals so that desirable amounts of precipitated Struvite can be recovered typically from the gravitational settling process.

CONCLUSIONS

The impact on Mg and Ca solution concentrations on struvite precipitation has been investigated using the analytical tools including XRPD and SEM-EDS. It has been shown that the presence of Ca ions in solution has a significant impact on struvite crystallisation in terms of the shape and purity of the product recovered. Gypsum, $CaCl_2$, struvite, struvite-(K) and sylvite can be identified as mineral species developed in the solution. In the present study, it has been found that the calcium concentration present inhibits the struvite growth, or affects struvite crystallisation and leads to the formation of many minerals rather than crystalline struvite. Ca impact on struvite crystal morphology will have to be considered in the future in order to improve the recovery and reuse as a fertiliser. Further works needs to be done on the effect of Ca concentration on identification of phases development in the synthetic waste water system.

ACKNOWLEDGEMENTS

The work was supported by the University of Pembangunan National "Veteran" Surabaya, East Java, Indonesia and PhD research projects.

REFERENCES

Bouropoulos, N. C. & Koutsoukos, P. G. 2000. Spontaneous Precipitation Of Struvite From Aqueous Solutions. Journal Of Crystal Growth, 213, 381 – 388.

Booker N.A., Priestley A.J. and Fraser I.H. 1999. Struvite formation in wastewater treatment plants: opportunities for nutrient recovery. Environmental Technology, 20, 777-782.

Doyle J.D., Parsons S.A.2002. Struvite formation, control and recovery. Water Res., 36, 3925- 3940.

Doyle J.D.,Oldring K., Churchley J., Price C., Parsons S.A. 2003. Chemical control of struvite precipitation.

Journal of Environmental Engineering-ASCE, 129 (5), 419-426.

Kofina A.N. and Koutsoukos P.G. 2003. Nucleation and crystal growth of struvite in aqueous media. New prospectives in phosphorus recovery. Wasic Workshop, Istanbul (Turkey).

Le Corre, K.S. E. Valsami-Jones, P. Hobbs, S. A. Parsons. 2005. Impact of calcium on struvite crystal size, shape and purity Journal of Crystal Growth, 283, 514-522.

Mohajit X., Bhattarai K.K., Taiganides E.P. and Yap B.C. 1989. Struvite deposits in pipes and aerators. Biological Wastes, 30, 133-147.

Musvoto, E. V., Wentzel, M. C. and Ekama, G. A. 2000. Integrated Chemical- Physical Process Modeling-Ii. Simulation Aeration Treatment For Anaerobic Digester Supernatant. Water Research, 34, 1868 - 1880.

Ohlinger K.N., Young T.M. and Schroeder E.D. 1999. Kinetics effects on preferential struvite accumulation in wastewater. Journal of Environmental Engineering, 125, 730-737.

oung, T. M. and Schroeder, E. D. 2000. Postdigestion Struvite Precipitation Using A Fluidized Bed Reactor. Journal Of Environmental Engineering, 126, 361 - 368.

Prywer J., Olszynski M. 2013. Influence of disodium EDTA on the nucleation and growth of struvite and carbonat apatite, Journal Of Crystal Growth, 375, 108-114.

Rawn A.M., Perry Banta A. and Pomeroy R. 1937. Multiple-stage sewage sludge digestion. American Society of Civil Engineers, 93-132.

Snoeyink V.L and Jenkins D. 1980. Water Chemistry . John Wiley and sons, New York.

Stratful I., Scrimshaw M.D. and Lester J.N. 2001. Conditions influencing the precipitation of magnesium ammmonium phosphate. Water Res., 35, 4191-4199.

EFFECTIVENESS OF ENVIRONMENTAL MANAGEMENT BASED ON TRASH IN THE CITY OF DEPOK

Ismiyati[1], Irfan Purnawan[1] and Muh. Kadarisman[2]
[1]Department of Chemical Engineering, Faculty of Engineering, University of Muhammadiyah Jakarta, Indonesia
[2]Department of Public Administration Science, Faculty of Social and Politic Sciences, University of Muhammadiyah Jakarta, Indonesia
E-mail: ismiyati.umj@gmail.com

ABSTRAK

Indonesian government efforts, particularly in the city of Depok, regarding decent and healthy environmental management in the waste sector have to be enhanced. This is a complex and diverse problems, even though the government has set into the form of Law (UU) Number 18 Year 2008 on Waste Management to build a strategic partnership of stakeholders in waste management. The Government has also issued Government Regulation of the Republic of Indonesia Number 81 Year 2012 on "Management of household waste and household-like waste". Further research is required on what extent that the implementation of waste management had been undertaken, as well as the necessity to study the methods of environmental insight in waste management. Waste management needs to be comprehensive and integrated from upstream to downstream, thus providing economic benefits, people healthiness and environmental safety. The purpose of this study is to analyze the effectiveness of environmental management based in Depok. This study uses qualitative design, which is to observe, understand and expose reality or empirical phenomena; hence the proper method used is descriptive method. Instrument to gather the data is completed by observation and in-depth interviews. The type of data required includes primary and secondary data. The informant is selected by purposive sampling, and data analysis is treated with triangulation technique. The results present that in the framework of the effectiveness of environmental management based trash in Depok showed that 1) Institutional Factor for Urban Waste Management of Depok has run properly, 2) Private partnership (commercial solution) to be involved in urban waste management in Depok is being explored, 3) Applicable technology in fostering a culture of 3R (reduce, reuse and recycle) on the entire community of Depok, has been done with the declaration of Depok Trash Care towards Trash Clean 2020 in March 2014.

Keywords: community, environment, waste management

INTRODUCTION

Waste is one of those words that are so widely used we often forget to question what it really means (Urbanbandit, 2012). Wastes are materials that are not prime products (that is products produced for the market) for which the initial user has no further use in terms of his/her own purposes of production, transformation or consumption, and of which he/she wants to dispose (United Nations Environment Program (UNEP), 2005). According to (Williams, 2005), waste(s) (also known as rubbish, trash, refuse, garbage, junk and litter), is very subjective; one person may deem an item to be waste whilst another might see it as a resource.

Waste products arise from our ways of life and it is generated at every stage of production and development. It spans all stages of human activities, from manufacturing to consumption (Oyelola, et.al, 2011). It is especially a serious problem in developing countries where generation of waste per unit of output is much higher than that in the developed countries because of inefficiency in manufacturing processes (Cointreau-Levin, 2006).

Depok city is a city with a relatively high population growth with more than 2 million people which occupy 11 districts and 63 villages. The growth itself brings problems of waste that increase linearly. Although the Government has issued the Law (UU) No. 18 of 2008 on Waste Management and Government Regulation of the Republic of Indonesia No. 81 Year 2012 on Management of household waste and household-like waste, but people of Depok was not considering the environmental conditions yet, even Depok ever received the predicate of dirtiest city across Indonesia in 2010. Depok City local Government along with the communities were very anxious and concerned over the predicate. Following up on the Indonesian Government Regulation No. 81 of 2012, particularly Chapter II regarding the waste management strategy, and Article 4 (3) which states that district / city government formulate and establish policies and strategies for district / city waste management. Depok City Government has issued a decree for Depok area No. 5 Year 2014 on Waste Management in an effort to overcome the problems of waste, mainly reduces the capacity of waste dumped in Final Disposal (TPA) which increase significantly. A number of policies have launched a movement of Clean Depok with priorities in some sectors and locations, such as the opening of the waste bank program in every neighborhood community which trigger residents to sort inorganic waste and then deposit it in the waste bank.

Another one is construction program of Waste Processing Unit, which processes organic waste into compost. In addition, cooperation with the institutions or technical ministries such as the Ministry of Public Works and the Ministry of Environment, as well as cooperation with Cell in order to achieve community empowerment, is starting to intensify information delivery. It is not merely managing physical trash only, but also involves the community through effective campaigns. Since 2012, Depok has launched Depok Waste Concern. In March 2014, Depok has also been declared a Depok Waste Concern towards Clean Waste at 2020.

The purpose of this study was to analyze the effectiveness of local government policies, community empowerment, environmental management and waste management technologies in Depok. The scope of this study includes the effectiveness of local government policies, community empowerment, environmental management and waste management technologies aspects.

LITERATURE REVIEW

In every aspect of human life, several unwanted materials (newspapers, broken bottles, aluminum cans, flower trimmings, etc) are generated. These materials are discarded simply because they are considered waste to that effect (Abagale, et.al, 2012). (CTMA, 2004) stated that the total stream of waste generated within a community is often categorized into municipal waste, industrial waste, constructional and demolition waste. While (Alexander, 2015) cited that household waste streams in the county identified nine main types of materials: paper/card, plastic film, dense plastic, textiles, miscellaneous combustibles, glass, ferrous metal, garden waste and food waste.

Solid wastes are all the wastes arising from human and animal activities that are normally solid and are discarded or intend to discard as useless or unwanted. The term solid waste is all inclusive, encompassing the heterogeneous mass of throwaways from the urban community as well as the more homogeneous accumulation of agricultural, industrial, and mineral wastes (Tadesse, 2004; Agwu, 2012).

Solid waste management is the process of collecting, storing, treatment and disposal of solid wastes in such a way that they are harmless to humans, plants, animals, the ecology and the environment generally. The unhealthy disposal of solid waste is one of the greatest challenges facing developing countries (Kofoworola, 2007). While Magutu and Onsongo (2011) have said that Waste management practices differ for developed and developing nations, for urban and rural areas, and for residential and industrial producers.

Hasan (2004) reported that public awareness, community participation, appropriate legislation, and strong technical support are key components to successful municipal solid waste management. Subramanian (2005) argued environmentally sound management of MSW has received a prime focus by both the international community and most national governments. This is evidenced by previous papers about municipal solid waste in some less-developed Countries on various continents or island nations (Goorah, Esmyot, & Boojhawon, 2009; Nasrabadi, Hoveidi, Bidhendi, Yavari, & Mohammadnejad, 2008). Thus, a comprehensive MSW management regime should include administrative, financial, legal, planning, and engineering functions (Ramachandra & Varghese, 2003).

METHODOLOGY

The method used in this research is descriptive method. The method is used for trying to find the fact about the effectiveness of Government Policies and local Government Policies of Depok West Java related with waste based environmental management, with the proper interpretation and accurately describe the nature and actions of the phenomenon of groups and individuals at the empirical level. Research design used in this study is a qualitative approach, which observe and capture reality and examine the behavior of individuals and groups as well as the experience of informants.

The qualitative approach is to introspect, retrospect, describing as it is, to experience and discover verstehen, uniqueness deeply, research a phenomenon, observe empirical causality and form a theory of data (Ndraha, 2003). The emphasis of the qualitative approach in this study is based on the consideration that the focus in this study is to reveal the process and find the meaning of the local Government policy of Depok, West Java related to sustainable development and environmental insight. Qualitative approach is more relevant for disclosure process and interpretation of the meaning in a study.

Thus, the qualitative approach here is to understand the process of investigating the social or human problem, based on the creation of a holistic picture formed by the words, and reporting detailed views of informants. Furthermore, through triangulation techniques, researchers conducted a crosscheck data obtained from one informant and comparing data from interviews with observations with regard to local Government policy of Depok, West Java related to sustainable development and environmental insight. Overall data are obtained in the form of primary and secondary data. The main data source research using qualitative approach are words and actions, the rest is additional data, documents and others.

Data Requirements, Source Data and Data Types

In this study, the data required is related to sustainable development policies and environmental insight in Depok, West Java. The main data source is a qualitative research design with words and actions, the rest is additional data. The type of data needed is primary data which obtained directly from the informant, while secondary data is obtained from documents, literature as well as various product regulations relevant to the focus of research.

Informant

Informant, is a source of primary data which very important in this study. Therefore, the manner and requirement to determine the informant is crucial in this research in order to address concerns and research purposes. Informant is a source of primary data and determined by purposive sampling with direct appointment by researchers based on consideration that the informant actually understand the substance of the research (Moleong, 2006).

In this study, the informants are: (1) The City Secretary of Depok (Informant I); (2) The Secretary of the Department of Cleansing and Plantation (Informant II); (3) The Secretary of the Environment Agency Depok (Informant III); (4) and (5) Waste Bank Field Coordinator and Waste Management Unit (Informant IV & V).

Data Collection Techniques / Instruments of Research

a. Interviews. Interviews were conducted in order to obtain sufficient data about the object of research. In this context, the researchers used in-depth interviews, with specified outline.
b. Observation (participant observation). This study using participative observation, which puts researchers as an integral part of the research object. Phenomena associated with this research object are direct observation on the activities at the sites. Data obtained through subsequent observations are recorded systematically, photographed or recorded.

c. Visual images. Researchers obtain information and informant's opinions including the state of the research location recorded in the form of visual images, namely tape cassette and camcorder.
d. Primary and secondary data as well as examine various documents and regulation product relevant to the object of research.

Validity Assessment and Data Processing Techniques

Assessment of the validity of data in this study used triangulation techniques. This technique is achieved by checking and cross-checking comparison of data from observation and interview, as well as comparing the results of interviews with the contents of the document. It also compared the interview results of one informant with another (Cresswell, 2002).

Based on this technique, data processing procedure is accomplished as follows.

a. Collecting data by determining the data source and differentiates it in the category of primary and secondary data. Sorting or classifying the data and then performed in a structured and systematic recording of relevant data.
b. Criticizing the available data. Critics of the data are prepared to control and conduct recheck on the data whether relevant or not to be presented and interpreted. Furthermore, through triangulation techniques, researchers conducted a crosscheck data obtained from one informant with another and comparing data from interviews with observations and the existing formal data.

RESULTS AND DISCUSSIONS

Government Policies Effectiveness

According to Indonesia's National Encyclopedia, effectiveness is something that shows the level of achievement of a goal. An attempt could be said effective if it ideally achieves the goal of business.

Effectiveness of Government Policies concerning waste management refers to the (Act) No. 18 Year 2008 on the Waste Management, the Indonesian Government Regulation Number 81 Year 2012 as well as the latest Regulation No. 5 of 2014 on Waste Management. Regulation No. 5 of 2014, which in Chapter IV focuses on the planning, waste reduction and handling. Part one is "Planning and Master Plan" shows: 1) the City Government arrange reduction and waste management plan as outline in the strategic plan and annual work plans.

Next 2) Plan for waste reduction and handling as number 1) shall at least contain: a) waste reduction target b) providing waste reduction and handling facilities through landfill. c) The pattern of regional cooperation, partnership and community participation, d) financing requirements which provided by the local government and the community, and e) Plan for the development and utilization of environmentally friendly technology in meeting the needs of reuse, recycle and waste handlers.

Furthermore, in the Master Plan includes: 1) The local government prepared a master plan of waste management for a period of 10 (ten) years as outlined in the annual work plan. Master Plan as mentioned at 1) shall contain at least: a) restrictions on landfill waste (reduce); b) recycling bins (recycle); c) the reuse of waste (reuse); d) waste segregation; e) waste collection; f) waste transporting; g) waste treatment; h) final waste processing; and i) funding.

The results of interviews with informants I, II, III and the results of direct observation on the effectiveness of the implementation of government policies adopted by the central government and the local government as set out in Regulation Depok No. 5 of 2014 resulted that basically the local Government through the relevant Department, namely the Department of Cleansing and Plantation, has been implement to develop "Waste Bank" contained almost all neighborhood communities with the amount of approximately 500 waste banks, as well as the construction of Waste Treatment Unit located in districts, in order to overcome the accumulation of waste and reduce the capacity of the waste which brought to the final disposal of waste. The Waste Management Units which have been built and spread across districts up to present are stated in **Table-1**.

Referring to Waste Processing Unit, which spreads about 30 units across 11 districts and 63 villages, it means that Waste Processing Units are available almost every two villages. The construction of Waste Composting Unit was initially started with many protests from the communities which refuse its existence. With persuasive approach and the use of environmentally friendly technologies, until now the local government of Depok superbly managed to build Waste Processing Unit and planned to build it in every area of the village.

Table-1. Construction of Waste Processing Unit (UPS) up to 2014

No	Location	Land Status	Surface Area
1	TPA Cipayung (Hanggar 1)	L	± 540 m^2
2	TPA Cipayung (Hanggar 2)	L	± 540 m^2
3	TPA Cipayung (Hanggar 3)	L	± 540 m^2
4	TPA Cipayung (Hanggar 4)	L	± 540 m^2
5	Jl. Merdeka (Hanggar 1), Abadijaya, Sukmajaya	SP/ PF	± 466.24 m^2
6	Jl. Merdeka (Hanggar 2), Abadijaya, Sukmajaya	SP/ PF	± 430 m^2
7	Jl. Jawa, Beji, Depok Utara	SP/ PF	± 540 m^2
8	Jl. Sadewa Raya Depok Tengah	SP/ PF	± 316.9 m^2
9	Kampung Lio, RT 03/07, Bojong, Pondok Terong	O	± 700 m^2
10	Jl. Pulo Mangga RT 02/05, Grogol, Limo	O	± 568 m^2
11	Bukut Rivaria, RW 014, Bedahan, Sawangan	SP/ PF	± 1000 m^2
12	Cimpaeun RT 03/16, Cimanggis	SP/ PF	± 500 m^2
13	Cilangkap RT 01/011, Cimanggis	SP/ PF	± 400 m^2
14	Kampung Sasak, RT 03/06, Limo	O	± 700 m^2
15	Jl. Mesjid Nurul Huda, RT01/06, Bojongsari Sawangan	O	± 1000 m^2
16	Jl. Cilobak Raya V, RT 02 /01, Pangkalan, Jati, Limo	O	± 500 m^2

No	Location	Land Status	Surface Area
17	Jl. Bulak Timur RT 01/06, Cipayung, Pancoran Mas	O	± 600 m^2
18	Komplek Wika Realty Unit II, Tamansari Puri Bali, Curug, Sawangan	SP/ PF	± 500 m^2
19	Permata Regency, Jl. Raya Citayam, Pondok Jaya, Pancoran Mas	SP/ PF	± 500 m^2
20	Jl. Al-Hidayah, Pangkalan Jati Baru, Limo	SP/ PF	± 500 m^2
And there are 10 new Waste Processing Units in the development process. Note: L = Landfill SP = Specific Facilities PF = Public Facilities O = Ownership			

Source: DKP Administrative Information System, Depok 2014

Aspects of Community Empowerment
Household Scale Community approach

The very effective program in waste management is community empowerment which involving the community, especially at the level of home neighbors for sorting waste. Directing public awareness is not an easy effort. It is associated with culture changes and perspective. By involving all potential group that exists in society such as Cadre, Dasawisma, Youth, Prayer Groups, Religious and Community Leaders, as well as the universities that work in a coordinated, well-planned and sustainable, through continuous socialization, is expected to managed the culture changes and perspective. According to (Abduli, 1997), the development of permanent education and training programs for personnel involved in medical, hazardous, and municipal solid waste management is highly recommended. Hence, another program which also important related to public awareness is the socialization of materials about waste management in Depok. Through education in elementary and junior high school, children are also expected to be able to have perspective and culture that is more environmentally friendly.

The Government's main program for addressing the problem of waste is the 3R (reduce, recycle and reuse), by establishing Waste Bank and the construction of Waste Processing Unit (UPS). The program has been promoted since 2014 along with the publication of Regulation No. 5 Year 2014 concerning waste management. Through the Waste sorting socialization movement, continuously in every neighborhood, the waste banks are expected to be available in every neighborhood, while the residents should active and focus on waste management independently. Waste Bank is a concept to sort waste collection and has a management similar with banks, with waste saved instead of money. Residents who save also called customers and owned saving book.

The savings could be used for payment of electricity and water. Even residents might also borrow money that could be paid by dry waste that collected every week. Waste is saved in the form of dry waste or inorganic waste such as cardboard, glass, plastic bottle, can, iron, glass bottle, aluminum and mixed. The price of each kilogram varies according to its kind, as shown in **Table-2** below.

Table-2. Dry Waste Prices in Waste Bank Depok

Dry Waste types	Price per kg (Rp)
Cardboard	1.100
Plastic cup (clean)	5.500
Plastic bottle	3.000
Glass bottle	300
Can	1.800
Iron	2.300
Aluminium	9000
Mix	1.400

Source: DKP Administrative Information System, Depok 2014

Observations and interviews result with informants with the present of waste bank in each community neighborhood, citizens are very enthusiastic, participation of them to save in waste banks is quite high, which means that there is an increasing awareness of the public to take care of the environment. Almost in every community neighborhood in the area of Depok has opened waste bank they open every week in average. Citizens carrying dry waste that has been collected and sorted then brought it to the bank, weighed and write it in the saving book. Dry waste from the waste banks are collected and sent to the waste bank owned by Department of Cleansing and Plantation which also distributed in the districts. From the district waste banks, the dry waste then brought to industrial waste collector. The average dry waste collected in every district at waste collector is about 600 kg to 1000 kg per week; with funds returned to the communities are about 6 million to 10 million every week. With total of 11 districts, means that collected dry waste as much as 6,600 kg to 11,000 kg. This is greatly helps the government of Depok, particularly in reducing the volume of the final disposal bins located in Cipayung, which lately has been full.

Waste Processing Unit Development that has been opened in several districts, that currently has reached 30 units; also expect the participation of the community in sorting organic or wet waste start from home. Through continuous socialization, residents in home neighborhood expected to sort organic / wet waste and place it in a small bucket, then by community officers is taken and collected in collecting point, then officers from the Department of Cleansing and Plantation subsequently collect it every 2 days and processed into compost at the Waste Treatment Unit of each district. The program is supposed to be effective in dealing with the waste problem if all residents are able to cooperate.

The results of direct observation and interviews with the informant, during the two years program of Depok Waste Care, until now the community has been implementing the program Depok Waste Care and environmental awareness towards Depok Waste Free, particularly the sorting of organic waste, recently reached 15% but from time to time through continuous socialization, started moving towards improvement, public awareness is also moving to increase. Some residents have also applied sorting starting from community neighborhood.

Regional Scale Approach

Program conducted by regional-scale approach is an effort to change the paradigm of waste management, namely collect-transport-dispose become collect-process-benefits. The programs are to build the Waste Processing Unit in the lower region scale (sub-district). Land required for 1 Waste Processing Unit is approximately 500 m^2. In a four-year period, it is expected that waste processing units will dominate waste treatment in the city of Depok which took over the role of waste landfill.

With the achievement of performance indicators, integrated waste management is waste management at Waste Management Unit before being transported to the Final Waste Disposal. One (1) waste processing unit can handle as many as 30 m^3 / day that will produce 2.4 m^3 of recycled materials such as metal, paper and plastic that still have economic value, compost that equivalent with 1 ton of compost per day and absorb as much as 14 workers. The compost produced is marketed and used by farmers in the region of Depok, particularly on agriculture carambola, guava, vegetables and ornamental plants which estimated to use compost as much as 15,200 tons per year.

Depok Waste Concern Program is quite helpful in reducing the waste problem in local government of Depok. The capacity of waste dumped into the final disposal, is now decrease to 20%. Household waste has been divided into inorganic waste and be brought to the Waste Bank, while organic waste processed in Waste Processing Unit, and the rest of residual waste dumped into the final disposal site (TPA). In accordance with Law No. 18 of 2008 on waste management stated that waste should not be disposed in the landfill, unless already be sorted out from the house and handled through UPS, while TPA can only hold the residual, dangerous and toxic wastes.

Aspects of Environmental Management

Clean environment is the desire of everyone. It is not easy to create such a clean and tidy environment that looks great on view. Along with the advanced level of education, people's minds and technologies, environmental awareness to have clean conditions must be improved. Clean and healthy environment conditions will make occupants feel comfort while healthy body is also maintained. Many benefits would be obtained by people when their environment is clean and tidy. Clean environment will diminish certain source of disease. This is certainly related to health matter. Moreover, with a clean environment, we will feel comfortable being at home.

Therefore, it is each individual responsibility to maintain cleanliness and healthiness of our environment. Although other external factors have influences to the health and clean conditions, but a clean home environment is the most important thing and should be prioritized in maintaining. Clean environments, which are free from waste, can prevent from disasters such as floods during monsoon. One of the causes of flooding in various regions is due to the large trash scattered thereby inhibiting the flow of water. This is one of the very bad behaviors that seemed to have become habit of most Indonesian society, especially in urban areas.

Observing and interviewing the respondents, it is then concluded that the habit of clean living in Depok still low yet has improved. Especially as the Depok City Government has built several waste treatment units, which is capable of reducing the waste problem about only 20%.

The habitude of disposing waste to rivers and ditches means it is difficult to generate cleaner environment, however, to change the habit was not a simple thing to do. Limited land to be used as waste treatment site is one of the reasons of why people in Depok choose to dispose waste carelessly.

To generate a clean environment is the responsibility of every single human being, not only the government though its regulations. There are several steps to create a clean environment such as:

1. Provide awareness about the importance of a clean environment to public, especially to children so that this awareness will continue to grow by age. Familiarize clean life at early age is proven to be more effective than at mature.
2. Apply the habit of disposing waste to its place as early as possible. Introducing the habit to children will be more beneficial to create the pattern of good behavior in the subconscious. Waste is always a classic matter yet can be solved with simple things. Disposing waste into the trash bin for example, is the simple beginning to address the problem.

Aspects of Technology

Waste is often seen as a phenomenon that cannot be solved effectively. Various strategies have been carried out both by the Government of Depok and environment community. Public awareness is still low which roles only 15% but the gradually increased through continuous information and dissemination by Depok City Government through the Department of Sanitation, which is optimistic to process waste, organic communities increasingly Waste Processing Unit.

Likewise, the manufacture of "Waste Bank" is run by people in their environment. Public awareness to separate inorganic waste in the form of plastic, tin, rubber, tin, metal, iron and other materials has been noble, as it can be exchanged into savings. Inorganic waste can be recycled to reproduce recycle stuffs. Moreover, some areas even produce a good quality handicraft from inorganic materials.

As for organic waste such as leaves fall, trash from agricultural residue, waste from the kitchen vegetables, and other types of organic waste, they can processed to make compost that can be used by the community itself as fertilizers. Compost also has economic value and business opportunities that can be further developed. An added value can be obtained from the processing of organic waste into compost. Compost made from organic waste material can also be used for the treatment of plants in organic farming systems.

Manufacturing process of compost from organic waste

Producing compost from organic waste is quite simple and using simple tools such as buckets, chopping machines, sieving machines, spades, forks, levers, scales, and thermometer.

The compost material is organic waste such as leaves, vegetable scraps, grass, or even trash from agricultural residues and household waste. Depok has its own method for composting process which is not involving active materials such as sugar, EM4. However, the bacteria obtained from the leaf litter leaves were allowed to stand for 3 days. From its process, the leaves' decomposition will grow bacteria, which is the active ingredient of fermentation process.

How to make

1. Organic waste collected from households, leftover vegetables from the market divided and separated from the rest of the inorganic such as plastic, etc., while the leaves of leaf litter from the felling of trees chopped and allowed to stand for 3 days until the bacteria generates. The leaves must have sufficient moisture content by giving splashes of water to grow the bacteria.
2. Combine ingredients strips of household waste and leaf litter in the ratio 1: 1. Then pile them for 3 months. Each pile is then weighed, coded, and marked the start date. Observe the temperature during the process every two weeks, as the temperature will continue to rise to approximately 80° C due to the fermentation process. After reaching the maximum temperature, typically the pile will have reduced in temperature. As the temperature drops, the pile is being conversed, to give the opportunity to the fermentation process by keeping the humidity. The processes are applied several times until a stable temperature reaches room temperature.
3. Next, the piles are sieved to obtain compost and allowed to stand for 1 month.
4. Fertilizer compost is ready to be harvested, put in pouch bags and weighed with a capacity of 10 kg per bag.

For the record, if the water in the trash is excessive, it will cause a foul odor causing maggots grow, conversely, if the temperature is hot, then cooled or inverted, then the degrading microorganism will die causing poor quality organic compost.

Waste processing technology in Depok is quite inefficient. It is necessary to add the cost of EM4 and other supporting chemicals. Improvement need to be conducted as there are many visits by the governments and various other areas for a comparative study on waste management in Depok.

CONCLUSIONS

In the realization of the policy and regulation of Government of Depok in environmental management based on effective waste, it can be concluded that:

1. Institutional Regulations for Waste Management in Depok has been effective. This was evidenced by the good cooperation between the government and the public of Depok especially in socializing the regulation. It has been proven that the society is enthusiastically joining the socialization event in many areas in Depok.
2. Public awareness to create a clean environment of Depok has been growing rapidly. This is evidenced by the presence of active Waste Bank in many areas in Depok. However, in creating public awareness of clean living from organic waste is only about 15%, but it is continued to rise in line with the amount of construction waste Processing Unit growing and declining capacity of waste in landfill.
3. Waste treatment technologies in Depok Waste Processing Unit, is simple in technology yet efficient and effective, as it need no chemicals such as EM4 and sugar solution, but only bacteria as the result of decomposition. The resulting compost is a standardized compost.

REFERENCES

Abagale, K. F., Mensah, A., and Osei, A. R. 2012. Urban Solid Waste Sorting in A Growing City of Ghana. International Journal of Environment and Sustainability, Vol.1, No. 4, Pp 18-25, available online at www.sciencetarget.com, accessed on November 20th, 2015

Abduli, M.A. 1997. Solid Waste Management in Guilan Province, Iran. Journal of environmental health, vol. 59, no. 10, pp. 19-24.

Agwu, M. O. 2012. Issues and Challenges of Solid Waste Management Practices in Port-Harcourt, Nigeria - a Behavioral Perspective. American Journal of social and management of sciences, available online at http://www.scihub.org/AJSMS/PDF/2012/2/AJSMS-3-2-83-92.pdf, accessed on November 19th, 2015.

Alexander, F.O., Anan, J. & Matilda, O.B. 2015. An Assessment of Solid Waste Management System in the Kumasi Metropolis. Journal of Arts and Humanities, vol. 4, no. 3, pp. 27-39, available online at http://www.theartsjournal.org/index.php/site/article/view File/663/354, accessed on November 19th, 2015.

Cape Town Metropolitan Assembly (CTMA). 2004. Integrated waste management. Republic of South Africa

City Regulation of Depok Number 5 Year 2014 about Waste Management

Cointreau-Levin, S. 2006. Occupational and Environmental Health Issues of Solid Waste Management: Special Emphasis on Middle- and Lower–income Countries. World Bank Report, and Pp 1-41, available online at http://siteresources.worldbank.org/INTUSWM/Resources/up-2.pdf, accessed on November 20th, 2015

Cresswell, John W. 2002. Research Design. Quantitative & Qualitative Approaches, New York: Sage Publication, Inc.

Goorah, S.S.D., Esmyot, M.L.I., & Boojhawon, R. 2009. The Health Impact of Nonhazardous Solid Waste Disposal in a Community: The Case of the Mare Chicose Landfill in Mauritius. Journal of Environmental Health, 72(1), 48–54.

Government Regulation of the Republic of Indonesia Number 81 Year 2012 about Management of household waste and household-like waste

Hasan, S.E. 2004. Public Awareness is Key to Successful Waste Management. Journal of Environmental Science and Health, A39(2), 483–492.

Kofoworola, O. F. 2007. Recovery and Recycling Practices in Municipal Solid Waste Management in Lagos, Nigeria. Waste Management Journal, Vol. 27, No. 9, Pp 1139-1143.

Law (UU) Number 18 Year 2008 about Waste Management.

Maguto, P.O., and Onsongo, C.O. 2011. Operationalizing Municipal Solid Waste Management. Integrated Waste Management Journal, Vol. II, Pp 1-9, available online at http://www.intechopen.com/books/ integrated-waste-management-volume-ii/operationalizing -municipal-solid-waste-management, accessed on November 19th, 2015

Moleong, Lexy J. 2006. Qualitative Research Methodology, 22nd edition, Bandung: PT Remaja Rosdakarya

Nasrabadi, T., Hoveidi, H., Bidhendi, G.N., Yavari, A.R., & Mohammadnejad, S. 2008. Evaluating Citizen Attitudes and Participation in Solid Waste Management in Tehran, Iran. Journal of Environmental Health, 71(5), 30–33.

Ndraha, Taliziduhu. 2003. Methodology of Indonesian Government, 3rd edition. Jakarta: Bina Aksara, p. 57.

Oyelola, O. T., Babatunde, A. I., and Abiodun, A. A. 2011. Appraisal of Municipal Solid Waste Management in Lagos Metropolis. Wilolud Journals, Vol. 2, No. 2, Pp 1-7, (available online at https://www.researchgate.net/ publication/265349558_APPRAISAL_OF_MUNICIPAL _SOLID_WASTE_MANAGEMENT_IN_LAGOS_MET ROPOLIS), accessed on November 20th, 2015.

Ramachandra, T.V., & Varghese, S.K. 2003. Exploring Possibilities of Achieving Sustainability in Solid Waste Management. Indian Journal of Environmental Health, 45(4), 255–264

Subramanian, K. 2005. Solid Waste Management Issues in Indian Cities. Chennai, India: The Hindu, available online at http://www.hindu.com/ accessed on November 21st, 2015.

Tadesse, T. 2004. Solid Waste Management. Lecture Notes for Environmental and Occupational Health Students, Ethiopia Public Health Initiative, Pp 1-199.

United Nations Environment Programme. 2005. Solid Waste Management, Chapter III: Waste Quantities and Characteristics, Pp 31-38, available online at hhtp://unep.or.jp, accessed on November 21st, 2015.

Urbanbandit Website. 2012. What is waste? Zero Waste Advocate and Urban Environmentalist", available online on http://urbanbandit.ca/2012/11/05/what-is-waste/, accessed on November 19th, 2015.

Williams, P. 2005. Waste Treatment and Disposal, Second edition, John Wiley and Sons Ltd, Chichester, Pp 380.

EFFECTS OF GIVING PEDADA FRUIT FLOUR (PFF) ON DIGESTA PROFILE AND SCFAs HYPERCHOLESTEROLEMIC RATS

Jariyah*[1], Simon Bambang Widjanarko[2], Yunianta[2], Teti Estiasih[2]
[1] University of Pembangunan Nasional "Veteran"Jawa Timur, Surabaya, Indonesia
[2]University of Brawijaya, Malang, East Java, Indonesia
E-mail : jariyahupn65@gmail.com

ABSTRACT

Pedada Fruit Flour (PFF) contains 8.9% of pectin, it includes one type of soluble dietary fiber. It has hypocholesterelemic property. The effect of PFF to the digesta cholesterol profile, and production of SCFA (short chain fatty acids) has not been done before. The aim of this study is to determine the impact of giving PFF on digesta profile and short chain fatty acids (SCFAs). Twenty-five Wistar rats were divided into 5 treatment groups consist of two normal control, hypercholesterolemic group, and 3 treatment of hypercholesterolemic groups feed were supplemented with 3, 6 and 9% PFF. Feeding was given for 28 days, on the 30th day surgery was done to take the digesta, then performed an analysis of digesta profile includes weight, water content, cholesterol and SCFAs levels. The results showed that supplementation of 6 and 9% PFF had significant effect on weight, water contents and cholesterol of digesta with the highest propionic: acetic acid ratio in the group treated with supplementation of 9% PFF (0.46) followed by supplementation group 6% and 3% i.e., 0.41 and 0.38, respectively. The ratio of propionic: acetic acid can reduce of serum cholesterol levels.

Keywords : PFF, digesta profile, SCFAs, hypercholesterolemic.

INTRODUCTION

Pedada Fruit Flour is a flour made from the pulp of mangrove spesies Pedada (*Sonneratia caseolaris*). The PFF soluble fiber content is higher than of seaweed species (Jariyah *et al.,* 2013). Soluble fiber such as pectin in the digestive system has hypocholesterolemic effect and has the effect of resistance against coronary heart disease through a decrease in cholesterol (Suido *et al.,*2002). The PFF has been reported that it significantly decreases the total cholesterol, LDL cholesterol, triglycerides (Jariyah *et al.,* 2013) and blood glucose (Jariyah *et al.,* 2014).

Mechanism of dietary fiber lowers cholesterol, namely: inhibiting the absorption of cholesterol, lowering cholesterol transfer availability to the blood flow is reduced, preventing the synthesis of cholesterol, lowering the energy density of foods that reduce cholesterol synthesis and increasing the excretion of bile (Tim and Slavin, 2008). Maryanto *et al.,* (2013), suggested that the mechanism of cholesterol reduction can be assessed by analyzing the SCFA (short chain fatty acids) and cholesterol digesta. SCFA is the result of fermentation of dietary fiber in the colon such as acetate, propionate and butyrate (Henningsson *et al.,* 2002; Hijova and Chmelarova, 2007; Fotschki *et al.,* 2014). Propionic acid after entering the bloodstream and liver can inhibit the action of the enzyme HMG-CoA reductase and reduce cholesterol synthesis (Lupton and Turner, 2000; Saravanan and Ignacimuthu, 2015). Meanwhile butyric acid is widely used as an energy source in colon and have anti-inflammatory properties that are important for maintaining the health and healing of colon cells. Acetic acid plays a role in the synthesis of long-chain fatty acids, including cholesterol, glutamine, glutamate and beta-hydroxybutyrate, and can be used as an energy source (Syamsir, 2007; Topping, 1996). This study is a part of a previous research by Jariyah *et al.* (2013) and aims to determine the effect of PFF on digesta profile and SCFAs in hypercholesterolemic rats.

METHODOLOGY

Pedada fruit flour was mixed with AIN-93M standard feed as many as 3, 6 and 9% by Reeves *et al.* (1993) modified. Meanwhile high cholesterol feed has been made by previous researchers Jariyah *et al.* (2013). As measured research subjects as much as 25 male Wistar rats, cholesterol kit No. 10 130 021, ether, acetic acid, propionate and butyrate standard.

Twenty-five male Wistar rats weighing ± 200g were adapted for one week to be fed and watered adlibitum. Furthermore, mice were divided randomly into 5 groups, each group consisting of five rats. Group 1 was the positive control group (normal control), the other groups were conditioned hypercholesterolemia with high cholesterol fed. Group 2 was a group of hypercholesterolemic rats (negative control). Group 3, 4 and 5 were groups of hypercholesterolemic rats dosed with PFF in a standard feed by 3%, 6% and 9%. Feeding was done for 28 days and on day 30 a surgery was done to take digesta. To be analyzed for water content, weight, cholesterol and SCFAs. Analysis of SCFAs included acetic acid, propionic and butyric using gas chromatography, whereas cholesterol analysis was performed using Liebermann-Burchard method with a spectrophotometer.The research was carried out after obtaining the Ethical Clearence from University of Brawijaya No. KEP-101-UB / 2012.

Statistical Analysis

Data were analyzed by One Way Anova SPSS V 16 with a confidence level of 5%, followed by LSD test 5% to determine the differences between the treatment groups.

RESULTS AND DISCUSSIONS

The weight, water content and cholesterol leves in ceacum digesta of rats

Feeding containing PFF showed a significant difference to the weight of the digesta rats ($p < 0.05$), as presented in Table 1. The weight of the digesta increased with increased dietary addition of PFF in the feed. The weight of digesta gained due to the increased PFF added, digesta weight of groups 5 was 6.3% (2.66±0.07), higher than that of group 3 and 4. Likewise, the water content of the digesta showed that there was an increase in the group of rats with PFF 6% and 9%, while the addition of 3% PFF gave no significantly difference with the standard feed group, the rising of water content was follwed by the increasing PFF added to the feed (Table 1).

PFF addition of 3%, 6% and 9% in the feed can increase cholesterol levels in digesta of rats significantly when compared to the group fed with the AIN-93M standard. The highest of cholesterol levels in digesta were found in the group with the addition of 9% PFF, followed by 6%, 3%, hypocholesterolemic and normal control group i.e., 2.19% ; 1.91% ; 1.53%; 1.05%; and 0.94% respectively.

The increase in weight of the digesta prove that dietary fiber contained in the PFF affect the increased weight of the digesta. Insoluble fiber has a great ability to bind water so the digesta become heavier, while the soluble dietary fiber will increase the weight of digesta to bind fat and bile acids in the intestine to be issued together with feces. These results are similar with the results of research conducted by Jenkins *et al.,* (1998) that the feeding resistant starch type 2 and 3 in rats may increase stool weight by 23%.

Water content of the digesta incresed as added PFF was increased, it indicated that the effect of PFF fiber increased stool water in the gastrointestinal tract. Meier (2008) stated that fiber is necessary to maintain the normal funtioning of gastrointestinal tract in large bowel as prebiotic affect, dilution of colonic content, reduction in availability of toxins and bile acids, increase of flatulence and increase in stool water. Liong and Shah (2006) reported that the levels of faecal water increased along with the increase of feeding high fiber probiotics and sinbiotics.

The increased cholesterol levels in the digesta due to the addition of PFF dietary fiber contained in PFF able to bind cholesterol in the intestine, causing the accumulation of cholesterol, especially in the digesta which were removed together with feces. The more cholesterol is tied, the more cholesterol is also present in the digesta. The high levels of cholesterol showed the low absorption of cholesterol in the intestine (Lupton and Tumer, 2000). Zhou *et al.* (2015), repoted that dose response relationship between increased dietary fiber intake and dreased total cholesterol to HDL cholesterol

Table- 1. Weight, water content, and cholesterol levels in the digesta of rats during treatment

Treatment	Weight of ceacum digesta /day (g)	Water Content of ceacum digesta (%)	Cholesterol level of ceacum digesta (%)
Normal control (Group. 1)	0,92±0,06 b	60,51±2,96b	0,94±0,06d
Hyperchol. control (Group 2)	0,97±0,07 b	59,37±4,32b	1,05±0,16d
Hyperchol.+ 3% PFF (Group 3)	1,21±0,11 b	62,81±3,27ab	1,53±0,05c
Hyperchol.+ 6% PFF (Group 4)	1,73±0,10 a	64,74±3,34 a	1,91±0,03b
Hyperchol. + 9% PFF (Group 5)	2,66±0,07 a	67,26± 5,59a	2,19±0,16a

Different letters indicate differences in one column
Hyperchol.= Hypercholesterolemia

Short Chain Fatty Acids (SCFAs) in the ceacum digesta of rats

The results of the short chain fatty acids (SCFAs) composition in the digesta of rats on hypocholesterolemic effect are presented in Table 2. It showed that the administration of PFF has significant effect on levels of acetic, propionic and butyric acid ($p < 0.05$). The acetic and propionic acid content given of 3% PFF treatment (group 3) were lower than group 4 and 5.

SCFA which was obtained from dietary fiber fermentation in the gut of the cholesterol synthesis depends on the ratio of acetic acid and propionic acid in the blood vessels of the portal. Acetic and propionic acid first reaches the liver that affects the metabolism of carbohydrates and fats (Henningsson *et al.,*2002; Lupton and Turner, 2000).

The highest ratio of propionic : acetic acid in the cecum digesta SCFA components contained in the treatment of PFF addition of 9 % (0.46) followed by groups of PFF 6 % (0.41), PFF 3 % (0.38), hypercholesterolemia (0.30), and normal group (0.26). Cummings *et al.* (2001) reported that molar ratio of acetate : propionate : butyrate acid from fermentation of wheat and oat bran were 64:16:20, from the fermentation of pectin was 80 : 12 : 8, whereas for starch at 62:15:23, while the non- starch polysaccharide type has a molar ratio of 63 : 22 : 8. In general terms, acetate appears to contribute 50-60% of total SCFA while propionate and butyrate gave 20-24% and 15-20%, respectively (Topping, 1996).

Short-chain fatty acids (SCFAs), especially propionate acid can decrease of serum cholesterol levels (Chen and Anderson, 1984; Wright *et al.,* 1990; Bridges *et al.,*1990; Chen *et al.,* 2003) The mechanism of cholesterol reduction by SCFAs fermented dietary fiber, especially propionic acid, was inhibited by the enzyme HMG-CoA reductase. This enzyme plays a role in the formation of mevalonate which is the main product in the formation of cholesterol, with the decreasing of the activity of the enzyme HMG-CoA, mevalonate would not be formed, along with cholesterol (Chen and Anderson, 1984; Saravanan and Ignacimuthu, 2015). Vuyst and Leroy

(2011) stated that the butyrate acid has an important role against colon health, it has two functions i.e., as an energy source for the colonic epithelium and provides important effects on gene expression of colon cells. So it has known to have ability to inhibit the promotion phase of carcinogenesis (Bird *et al.*, 2006; Kapkac *et al.*, 2003).

Table-2. Composition of acetic , propionic and butyric acid in the digesta of rats after treated

Treatment	Acetic Acid (mM) (A)	Propionic Acid (mM) (B)	Butyric Acid (mM)	B/A
Normal control (Group1)	76.32±11.34a	18.78±2.10ab	24.88±2.15a	0.26 ± 0.07
Hyperchol.(Group 2)	20.01±8,77b	6.48±3.74c	5.63±3.42c	0.30 ± 0.06
Hyperchol.+3% PFF (Group 3)	11.06±1.34bc	4.28±0.89c	5.24±1.23c	0.38 ± 0.03
Hyperchol.+ 6% PFF (Group 4)	36.20±8.77c	14.27±1.75b	14.81±2.11b	0.41 ± 0.05
Hyperchol. + 9% PFF (Group 5)	51.31±5.72c	23.52±1.45a	12.33±1.74b	0.46 ± 0.02

Differences letter stating the difference in one column .
Hyperchol.= Hypercholesterolemia

CONCLUSIONS

In conclusion, this study on rats fed with PFF on feed during 4 weeks treatment was able to give effect on increased of weight, water content and cholesterol in feses together with the propinic acid ratio of SCFAs composition ceacum digesta of rats, so that it can be used as an alternative to decrease cholesterol.

REFERENCES

Bird A.R., Brown I.L. and Topping D.L. 2006. Low and high amylose maize starches acetylated by a commercial or a laboratory process both deliver acetate to the large bowel of rats. Food Hydrocolloids, 20 (8): 1135-1140.

Bridges S.R., Anderson J.W., Deakins D.A. and Wood C.A. 1990. Oat bran increases serum acetate of hypocholesterolemic men. American Journal of Clinical Nutrition, 56: 455 – 459.

Chen H.L., Wang C.H., Chang C.T. and Wang T.C. 2003. Effects of Taiwanese Yam (*Dioscorea japonica* Thunb. Var. *pseudojaponica* Yamamoto) on Upper Gut Function and Liipid Metabolism in Balb/cMice. Basic Nutritional Investigation. Nutrition, 19:646-651.

Chen W.J. and Anderson J.W. 1984. Propionate may mediated the hypocholesterolemic effects of certain soluble plant fibers in cholesterol-fed rats. Proceedings of the Society for Experimental Biology and Medicine, 175: 225-218.

Cummings J.H., Mcfarlane G.T. and Englyst H.N. 2001. Prebiotic digestion and fermentation. American Journal of Nutrition, 415-420.

Henningsson A.M., Bjorck I.M.E and Nyman M.G.L. 2002. Combinations of indigestible carbohydrates affect short-chain fatty acid formation in the hindgut of rat. Journal of Nutrition, 132: 3098-3104.

Hijova E. and Chmelarova A. 2007. Short chain fatty acids and colonic health. Bratisi Lek listy, 108 (80) : 354-358.

Fotschki B., Jurgoński A., Juśkiewicz J., Kołodziejczyk K. and Sójka M. 2014. Effects of Dietary Addition of a Low-Pectin Apple Fibre Preparation on Rats. Polish Journal of Food Nutrition Sciences, 64(3): 193-199.

Jariyah, Azkiyah L., Widjanarko S.B., Estiasih T., Yuwono S.S and Yunianta. 2013. Hypocholesterolemic effect of Pedada (*Sonneratia caseolaris*) fruit flour in wistar rats. International Journal of PharmTech Research, 5(4): 1619-1627.

Jariyah, Widjanarko S.B., Yunianta and Estiasih T. 2014. Hypoglycemic effect of Pedada (*Sonneratia caseolaris*) Fruit Flour (PFF) in alloxan-induced diabetic rats International Journal of PharmTech Research, 7(1) : 31-40.

Jenkins D.J.A., Vuksan C ., Kendall W.C., Wursch P., Jeffcoat R., Waring S. and Wong E.1998. Physiological effects uf resistant starches on fecal bulk, short chain fatty acids, blood lipids and glycemic index. Journal of American College of Nutrition, 17(6): 609-616.

Kapkac M., Erikoglu M., Tuncyurek P., Ersin S., Esassolak M., Alkanat M. and O. Sipahioglu. 2003. Fiber enriched diets and radiation induced injury of the gut. Nutrition Research, 23: 77-83.

Liong M.T. and Shah N.P .2006. Effects of a Lactobacillus casei synbiotic on serum lipoprotein, intestinal microflora, and organic acids in rats. Journal of Dairy Science, 81:1390-1399.

Lupton J.R. and Turner D. 2000. Dietary Fiber, In Biochemical and Physiological Aspects of Human Nutrition. London: WB Sounders Company.

Maryanto S., Fatimah S., Sugiri and Marsono Y. 2013. The effect of red guava fruits supplementation on the SCFA and cholesterol production in caecum of hypercholesterolemic rats. AGRITECH, 33 (3):334-339.

Meier R.F. 2008. Basic in clinical nutrition: Fiber and short chain fatty acids. The European e-Journal of Clinical Nutrion and Metabolism, xx : 1-3.

Reeves P.G., Nielsen F.H. and Fahey G.C.1993. AIN-93 Purified diets for laboratory rodents: final report of the American Institute of Nutrition Ad Hoc Writing Committee on the formulation of the AIN-76A rodent diet. American Institute of Nutrition. Journal of Nutrition, 123: 1939-1951.

Saravanan M and Ignacimuthu S. 2015. Hypocholesterolemic Effect of Indian Medicinal Plants - A Review. Medical Chemistry , 5(1):40-49.

Suido H., Tanaka T ., Tabei, Takeuchi A., Okita M., Kishimoto T., Kasayama S. and Higashino K. 2002. A mixed green vegetable and fruit bererage decreased the serum level of low- density lipoprotein cholesterol in hypercholesterolemic patients. Journal Agricultural and Food Chemistry, 50 (11): 3346-3350.

Syamsir E. 2007. *Function to diet of SCFAs for health.* Elvira_itp's blog. It was accessed on August 1, 2014.

Tim D.A. and Slavin J.L. 2008. Dietray fiber and the relationship to chonic diseases. American Journal of Lifstyle Medicine, 2:233.

Topping D.L. 1996. Short chain fatty acids produced by intestinal bacteria. Asia Pacific Journal Clinical Nutrition, 5:15-19.

Vuyst L.D. and Leroy F. 2011. Cross-feeding between Bifidobacteria and butyrate producing, colon bacteria expalins bifidobacterial competitiveness, butyrate production, and gas production. International Journal of Food Microbiology, 149 (1): 73-80.

Wright R.S., Anderson J.W. and Bridges S.R. 1990. Propionate inhibits hepatocyte lipid synthesis. Proceedings of the Society for Experimental Biology and Medicine., 195: 29.

Zhou Q., Jiang Wu , Jie Tang, Jia-Ji Wang, Chu-Hong Lu and Pei-Xi Wang .2015. Beneficial Effect of Higher Dietary Fiber Intake on Plasma HDL-C and TC/HDL-C Ratio among Chinese Rural-to-Urban Migrant Workers. International Journal of Environmental Research Public Health , 12:4726-4738.

EMULSION COPOLYMERIZATION OF STYRENE-CO-MALEIC ACID AND INCORPORATION OF CU_2O INTO ITS MATRIX

Yenny Meliana*, Sri Fahmiati, Agus Haryono
Polymer Chemistry Group, Research Center for Chemistry, Indonesian Institute of Sciences
Kawasan Puspiptek Serpong, Tangerang Selatan, Banten, Indonesia
*)E-mail: yenn001@lipi.go.id

ABSTRACT

Incorporation of inorganic materials into polymer matrix using emulsion polymerization technique provides many advantages. In our work, we synthesized poly(styrene-co-maleic acid) with Cu_2O incorporation into the polymer matrix. The process was optimized by varying monomers weight ratio and Cu_2O weight percentage. The products were characterized by Fourier Transform Infra Red spectroccopy and Particle Size Analyzer. FTIR result showed that Cu was attached to carboxylic group indicated by the shift of C=O band. From particle size analysis, the variation of Cu2O % weight did not show any significant difference, but for 15%wt of Cu_2O, the particle distribution was dispersed more uniformly.

Keywords: copolymer, emulsion copolymerization, Cu_2O.

INTRODUCTION

Incorporation of inorganic materials into polymer matrix by polymerization has been done in many research (Mahdavian a et al., 2015). Many tipes of inorganic particles, such as calcium carbonate, titanium dioxide, and silica were used in the research. By using polymerization technique, the agglomeration can be avoided and the mechanical properties can be improved.

Many polymerization technique has been done to achieve the purpose such as free radical polymerization, suspension polymerization, emulsion polymerization etc. Emulsion polymerization provides advantages such as high molecular weight colloidal polymer, more environmentally friendly and safe process (Yamak, 2013). Besides, narrow particle size distribution can be obtained from this type of polymerization (Rivas, 2000).

Shao-Hai et al has made styrene-maleic acid via free radical copolymerization and encapsulated Pigment Red 122 into it. The result showed that with increasing molar content of maleic acid, the particle size decreased first and then increased. Otherwise, the stability of dispersion was increased first and then decreased.

Deb et al studied that the metal complex clusters fairly uniformly distributed in the polymer beads with local concentrations of metal higher than stoichiometric amounts , they investigated the crosslinked styrene-maleic acid copolymer complexes of some transition metals and their adsorption behavior. The copolymerization of styrene-maleic acid was carried out suspension polymerization.

In our study, we incorporated Cu_2O into poly(styrene-co-maleic acid) via emulsion polymerization. The effect of monomer ratio, Cu_2O percent weight on structure and particle size distribution were investigated.

METHODOLOGY

Materials that carried out for this study were Maleic Acid and Sodium Dodecyl Sulfate (SDS) were purchased from Merck. a,a'-Azobisisobutyronitrile (AIBN) as initiator was purchased from Aldrich. Styrene monomer was purchased from PT. Brataco. Distilled water was used in this study. Cu_2O was supplied from PT. Sigma Utama, Cibinong, Indonesia.

Polystyrene-co-maleic acid-Cu copolymer was synthesized by emulsion polymerization. The reaction was carried out in 300 ml three-neck flask equipped with a magnetic stirrer, silicon oil bath, a condensor, and a thermometer). Styrene, AIBN, and Cu_2O were prepared as oil phase solution. Styrene and AIBN was stirred at 600 rpm until well mixed then Cu_2O was added to the solution. The water phase solution was prepared by stirring the SDS, maleic acid, and distilled water. Then, oil phase solution was added to water phase solution and mixed well for 30 minutes. Deoxygenation was performed by nitrogen gas flushing for 5 minutes at atmospheric pressure. The reaction temperature was elevated until it reach 75°C. Then reaction was carried out for 3 hours. The product consisted of three layers, the viscid white forth at the top, the green solution at the middle and yellow precipitate in paste form at the bottom. The paste was separated from the solution and then dried in oven for 72 hours at 40°C, crushed, and then characterized by Fourier Transform InfraRed Spectrophotometer (FTIR) and Particle Size Analyzer (PSA).

IR spectra of the copolymers were recorded with a FTIR spectrum analyzer (IR Prestige-21, Shimadzu) with ATR 800 method. The average diameter of the particle was measured by Particle Size Analyzer (Nanoplus, 660 nm).

RESULTS AND DISCUSSIONS

The synthesis of polystyrene-co-maleic-acid)-Cu copolymer has been investigated with variation of monomer weight ratio and weight percent (wt.%) of Cu_2O. Table 1 and 2 showed the formulations of these variations.

Copolymerization reaction between styrene and maleic acid is supposed as below:

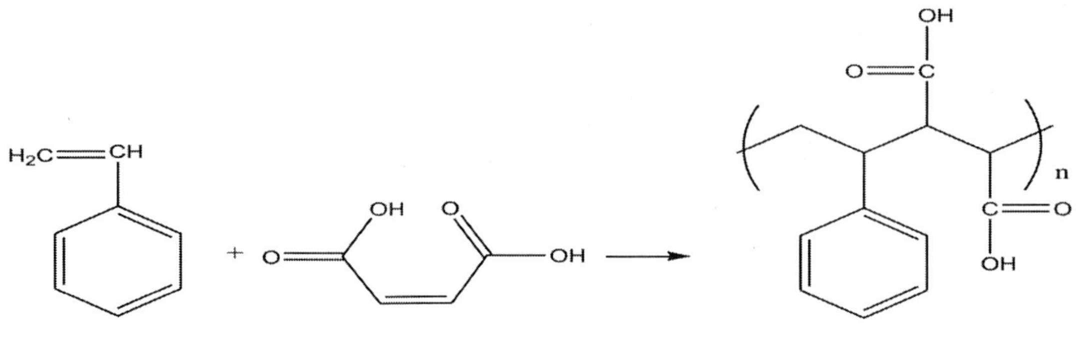

FIGURE-1. The proposed reaction of styrene and maleic acid to form copolymer poly(styrene-co-maleic acid)

TABLE-1. Monomer weight ratio variation

Code	Monomer weight ratio (w/w)		AIBN (ml)	SDS (g)	Aquadest (ml)	Cu_2O (wt.%)
	Styrene	Maleic Acid				
PE 1	1	1	1	1,4	250	0
PE 2	1	1				5
PE 3	2	1				
PE 4	3	1				
PE 5	4	1				

TABLE-2. Cu_2O weight percent variation

Code	Monomer weight ratio (w/w)		AIBN (ml)	SDS (g)	Aquadest (ml)	Cu_2O (wt.%)
	Styrene	Maleic Acid				
PE 1	1	1	1	1,4	250	0
PE 2						5
PE 6						15

1. FTIR result

The IR spectrum is used to confirm the formation of this copolymer. Figure 2 and 3 shows the spectra of variation used in this research.

Based on FTIR spectra, it can be seen that all the copolymers showed the characteristic stretching vibration of –CH at 3500-3300 cm^{-1}, stretching vibration of C=O at 1687– 1705 cm^{-1}, stretching vibration of C-OH from carboxylic acid at 1261-1381 cm^{-1}, C=C aromatic ring at 1506-1595 cm^{-1}.

We also investigated the weight percentage (wt.%) of copper (Cu) to know the optimal Cu adsorbed into the polymer matrix. Table 2 shows Cu_2O weight percentage variation used in our research.

Based on FTIR spectrum, it can be seen that all the copolymers in this variation showed the characteristic stretching vibration of CH aromatic stretching at 3500-3300 cm^{-1}, C=C aromatic stretching vibration at 1577-1625 cm^{-1}, stretching vibration of C=O at 1699–1705 cm^{-1}, C-OH at 1257-1382 cm^{-1}. The IR Spectra for Cu_2O at 15%wt showed the shift of C=O band to lower wave number. Intensity reduction of the carboxylate peak indicated acid consumption from carboxylic and showed that metal carboxylates were formed. From FTIR spectra, it was suggested the Cu attachment to the carboxylic group.

2. Particle Size Analysis

Table 3 showed mean particle diameter for each variation. There were no significant difference in particle size for each variation, but the 15%wt weight of Cu_2O showed smaller particle diameter than the 5%wt. This phenomenon was supposed as that amount of Cu_2O that entrapped into polymer matrix was same for all of monomer ratio and % weight of Cu_2O. Or in other word, the amount of styrene and Cu2O %weight did not give much significant effect on the particle size of this composite.

TABLE-3. Mean diameter of poly(styrene-co-maleic acid)-Cu_2O particle with varied %weight of Cu_2O

Code	Styrene (mol)	Maleic acid (mol)	% weight Cu2O	Mean diameter (d) μm
PE 2	1	1	5	1,8
PE 4	3	1	5	1,7
PE 6	1	1	15	1,1

Figure 5 and 6 describes the volume distribution for 5%wt and 15% wt of Cu in monomer ratio 1:1. The more Cu_2O percent weight showed narrower particle volume distribution. This result indicated that with 15%wt of Cu_2O, the particle distribution was dispersed more uniformly in aqueous media than that of 5%wt (6).

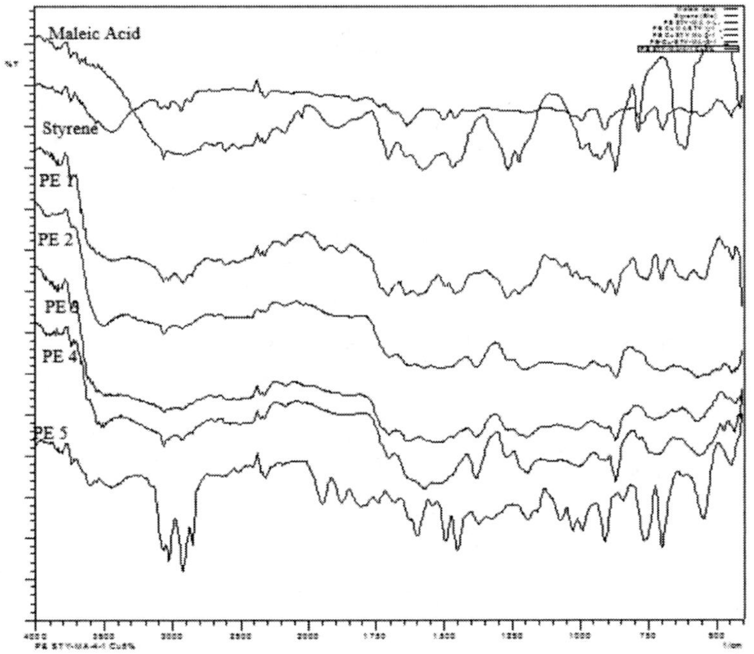

FIGURE-2. FTIR spectra of Poly(styrene-co-maleic acid)-Cu with monomer weight ratio variation.

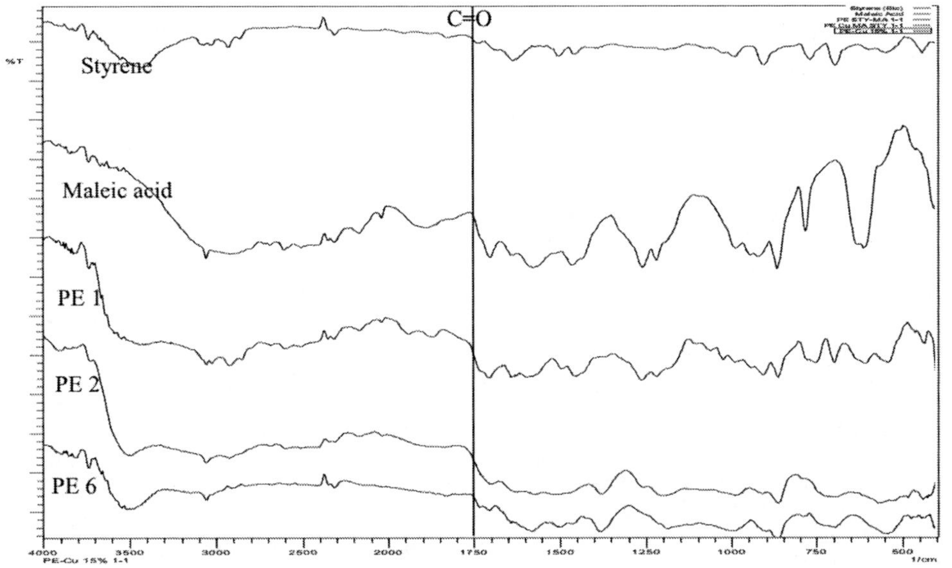

FIGURE-3. FTIR spectra of Poly(styrene-co-maleic acid)-Cu with Cu_2Oweight percent variation

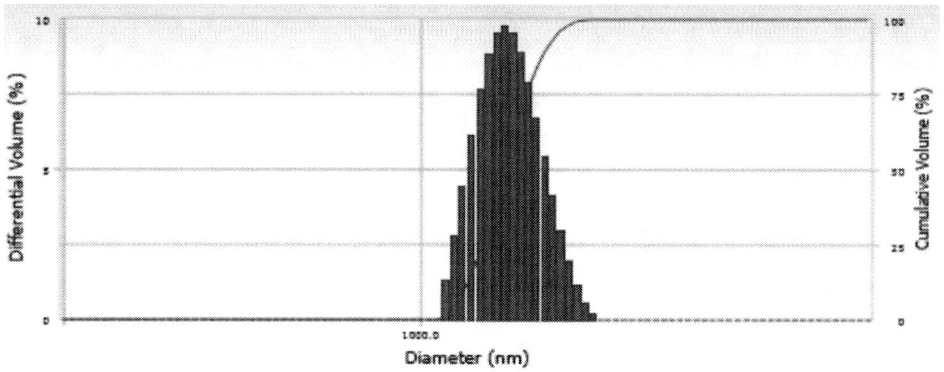

FIGURE-4. Volume distribution Poly(Sty-co-MA) 1:1 Cu 5%

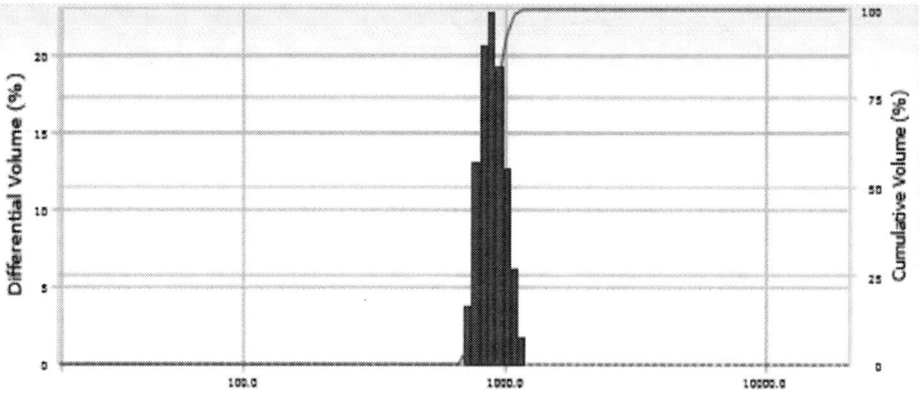

FIGURE-5. Volume disribution Sty-Ma 1:1 Cu 15%

CONCLUSIONS

The copolymer poly (styrene-co-maleic acid)-Cu was synthesized via emulsion copolymerization. From FTIR spectra, it was confirmed that the Cu was attached to carboxylic group indicated by the shift of C=O band. From particle size analysis, the amount of styrene and Cu2O %weight did not give much significant effect on the particle size of this composite. For 15%wt of Cu2O, the particle distribution was dispersed more uniformly in aqueous media than that 5%wt.

ACKNOWLEDGEMENT

The authors thanked Ani Chusnaeni for her helpful contribution during this work, Mrs. Dieni Mansur for the discussion and Mrs. Yulianti Sampora for FTIR measurement. The financial support from Indonesian Institute of Sciences through Competitive Programme is fully acknowledged.

REFERENCES

Rahim-Abadi M. M., Mahdavian A. R.,Gharieh A., Salehi-Mobarakeh H. 2015. Chemical modification of TiO_2 nanoparticles as an effective way for encapsulation in polyacrylic shell via emulsion polymerization. Progress in Organic Coatings. 88 : 310–315.

Xiaokun Ma, Bing Zhou, Yanhui Deng, Ye Sheng, Chengyu Wang, Yan Pan, Zichen Wang. 2008. Study on CaCO3/PMMA nanocomposite microspheres by soapless emulsion polymerization. Colloids and Surfaces A: Physicochem. Eng. Aspects. 312 : 190–194.

Mahdavian A. R., Ashjari M. , Makoo, A. B. 2007. Preparation of poly (styrene–methyl methacrylate)/SiO_2composite nanoparticles via emulsion polymerization : An investigation into the compatiblization. European Polymer Journal. 43 : 336–344

Hale Berber Yamak. Emulsion Polymerization: Effects of Polymerization Variables on the Properties of Vinyl Acetate Based Emulsion Polymers. 2013.

L. Rivas Bernabel., Pereira Eduardo., Martinez Esteban., Ignacio Moreno-Villoslada. 2000. Metal Ion Interaction with Poly (2-Acrylamide-2-Methyl-1-Propanesulfonic Acid-co-Methacrylic Acid), Bol. Soc. Chil. Quím 45.

Shao-Hai Fu, Kuan-Jun Fang. 2007. Preparation of Styrene-Maleic Acid Copolymers and Its Application in Encapsulated Pigment Red 122 Dispersion. Applied Polymer Science.105 : 317–321.

Deb P. C., Rajput L. D., Agrawal A. K., Singh P. K., Hande V., and Sasane S.2005. Crosslinked styrene-maleic acid copolymer complexes of some transition metals and their adsorption behavior. Polym. Adv. Technol. 16: 681–687.

A. Matei., I. Cernica., O. Cadar., C. Roman., V. Schiopu. 2008. Synthesis and characterization of ZnO – polymer nanocomposites. Article in International Journal Of Material Forming.

Galindo R.B. 2012. Encapsulation of Silver Nanoparticles in a Polystyrene Matrix by Miniemulsion Polymerization and Its Antimicrobial Activity. Research Article ISRN Nanotechnology.

Weiss C. K., Landfester K. Miniemulsion Polymerization as a Means to Encapsulate Organic and Inorganic Materials. 2010. Advance Polymer Science. 233: 185–236.

ESTIMATING GREENHOUSE GAS EMISSIONS LEVEL OF A NATURAL GAS PIPELINE – CASE STUDY FROM A TO B POINT IN WEST JAVA-INDONESIA

Cindy Dianita, Asep Handaya Saputra
Department of Chemical Engineering, Faculty of Engineering, University of Indonesia
E-mail : cindydianita@yahoo.com

ABSTRACT

Indonesia is one of the highest greenhouse emitters in the world. As a response of this problem, Indonesia declared the national action plan to focus on national greenhouse gas (GHG) reduction by 26 % by 2020. To achieve this target, Government puts energy sector as one of the top priorities since it is the second strongest contributor to national GHG emissions. The main purpose of this paper is to apply the method of fugitive emissions calculation to the existing natural gas pipeline in Indonesia. Fugitive emissions are the major component of GHG emissions from natural gas systems and methane (CH_4), the primary component of natural gas pipeline, is a potent GHG. Tiered approaches from Interstate Natural Gas Association of America (INGAA) are implemented in this paper as the estimation guidelines. A case study of a natural gas pipeline system in Indonesia is analyzed to compare the GHG emissions level resulted from Tier 1 and Tier 2 methods. In these methods, the input data are pipeline length, the number of compressor stations, and the number of meter and pressure regulation stations. In this case, the GHG emissions level of Tier 2 is significantly different from Tier 1. The variation of pipeline length shows that for the length under 479.2 miles, Tier 1 gives lower amount of CO_2 equivalent than Tier 2. The differences of these estimation methods and results can be furtherly developed to provide relevant information and recommendation for the Companies and Government to record the emissions level from natural gas transmission pipeline according to their needs and purposes.

Keywords: emissions, GHG, pipeline, Indonesia

INTRODUCTION

The demand for energy in Indonesia will continue to grow in the coming years and the domestic consumption for natural gas is increasing fast. The government of Indonesia through The Ministry of Energy and Mineral Resources launched the National Gas Policy Roadmap 2014-2030. This book records the strong demand for gas is expected to be significantly high in 2015- 2025 because the government sets the target to optimize the consumption of natural gas in domestic market as it is environmentally friendly fuel and cheaper than other fossil fuels. This target also in line with the national target of a 26 % reduction in Greenhouse Gas (GHG) emissions by 2020.

Currently, the transportation of natural gas in Indonesia is dominated by pipeline and the government of Indonesia plans to build transmission and distribution pipelines of natural gas to supply a range of customers across the country. Compared with other forms of transport, pipelines provide more continuous, stable, and high-capacity [1].

Although natural gas is considered as clean fuel, the issue of indirect GHG emissions from gas production and transport has come up in the discussions about the energy supply [2]. It is a potent GHG when it is released into the atmosphere as uncombusted methane. The U.S. Environmental Protection Agency (EPA) reported the fugitive emissions of methane from oil and gas extraction and pipeline transmission are the main anthropogenic source of methane in the United States and the second largest source globally [3] . Interstate Natural Gas Association of America (INGAA) stated that fugitive emissions are the biggest methane emission source account for transmission sector [4]. Methane is a far more potent greenhouse gas GHG than CO_2 with an estimated global warming potential 86 times greater than CO2 on a 20 year basis and 34 times greater on a 100 year basis [5,6,7,8,9]. GHG emissions from natural gas delivery must be quantified to evaluate the environmental impacts of natural gas transportation and to develop the emission reduction strategies.

The main purpose of this paper is to apply the emission estimation methodologies INGAA to calculate the fugitive emissions of the existing natural gas pipeline in Indonesia using Tier 1 and Tier 2 methods.

INGAA is a trade organization that advocates regulatory and legislative positions of importance to the natural gas pipeline industry in North America. The guidelines published by INGAA are not only implemented in the US but also have become the references for natural gas industry in other countries.

In Indonesia, the guidelines for GHG emissions refers to the Intergovernmental Panel on Climate Change (IPCC) guidelines for national GHG inventories. Thus, most of GHG emissions of gas pipeline in Indonesia is reported on the basis of IPCC. This paper is intended to characterize the GHG emissions of a natural gas pipeline system in West Java, Indonesia by referring to INGAA methodologies and procedures. Moreover, the differences of these estimation methods and results from Tier 1 and Tier 2 methods can be furtherly developed to provide relevant information and recommendation for the Companies and Government to record the emissions level from natural gas transmission pipeline according to their needs and purposes.

METHODOLOGY

Indonesia Ministry of Environment defines fugitive emissions as unintentional leaks from any activities of energy production and distribution such as flaring, venting, leaks from pipe connection and valves and methane release from coal mining activities [10]. In general, fugitive emissions from oil and gas activities may be attributed to the several sources: fugitive equipment leaks, process venting, evaporation losses, disposal of waste gas streams by venting or flaring, accidents and equipment failures [11].

There are 3 tiers in estimating the emission of gas pipeline system by referring to INGAA guidelines. Due to the availability of the data, GHG emissions are calculated on the basis of Tier 1 and Tier 2.Tier 1 method only requires pipeline length while Tier 2 calculates the fugitive emission of transmission pipeline based on pipeline length as well as the quantity of compressor stations and meter/regulator stations. Table 1 and Table 2 summarize the emission factors for fugitive emissions based on Tier 1 and Tier 2 methods. GHG emissions are calculated by multiplying the activity data by the emission factor. The methane fugitive emissions is then converted using Global Warming Potential (GWP) of gas to get the number of CO_2 equivalent.

Table-1. Emission Factors for Fugitive Emissions of Tier 1

Activity Data	GHG	Emission Factor	Emission Factor Units
Pipeline length	CH_4	7923	lb CH_4/mile-year
	CO_2^1	7.59	lb CO_2/mile-year
	CO_2^2	466.7	lb CO_2/mile-year

* $^1CO_2$ from CH_4 oxidation
$^2CO_2$ from pipeline leaks

The calculations of GHG emissions for a natural gas pipeline system from A to B Point in West Java (Indonesia) have been performed. The total length of this gas transmission pipeline system counts 220 km and contains 3 compressor stations and 3 meter/regulator stations. The fractions of CH_4 and CO_2 in the natural gas are assumed as 93.4% and 2% respectively.

Table-2. Emission Factors for Fugitive Emissions of Tier 2

Activity Data	GHG	Emission Factor	Emission Factor Units
Pipeline length	CH_4	23.08	lb CH_4/mile-year
	CO_2^1	7.59	lb CO_2/mile-year
	CO_2^2	1.52	lb CO_2/mile-year
Number of compressor station	CH_4	1259400	lb CH_4/station-year
	CO_2	72747	lb CO_2/station-year
Number of meter/regulator station	CH_4	2533	lb CH_4/station-year
	CO_2	146.34	lb CO_2/station-year

* $^1CO_2$ from CH_4 oxidation
$^2CO_2$ from pipeline leaks

RESULTS AND DISCUSSIONS

The calculation results of fugitive emissions by referring to Tier 1 and Tier 2 methods of INGAA are presented by Table 3 and Table 4. From the two tables, it is obvious that the Tier 2 result is 3.5 times higher than Tier 1. For this case, as the Tier 2 parameter of GHG consists of 3 factors, the highest emissions comes from compressor stations (more than 99%).A simulation is also performed to find the condition that gives the similar number of fugitive emission for both Tier 1 and Tier 2. A similar result is shown when the pipeline length is 771.2 km. A change in number of compressor stations or meter/regulator stations cannot give the same result for the two Tiers with the same length of pipeline (220 km).

Table-3. Result of Tier 1 Fugitive Emissions

Activity Data	GHG	Tonnes GHG/mile-year	Tonnes CO_2 equivalent
Pipeline length	CH_4	491.28	10316.94
	CO_2 from oxidation	0.47	0.47
	CO_2 fugitive leaks	28.94	28.94
Tonnes CO2 equivalent from pipeline length			10346.35

Table-4. Result of Tier 2 Fugitive Emissions

Activity Data	GHG	Tonnes GHG/mile-year	Tonnes CO_2 equivalent
Pipeline length	CH_4	1.43	30.05
	CO_2 from oxidation	0.47	0.47
	CO_2 fugitive leaks	0.09	0.09
Tonnes CO2 equivalent from pipeline length			30.62
Number of compressor stations	CH_4	1713.78	35989.39
	CO_2	98.99	98.99
Tonnes CO2 equivalent from compressor stations			36088.38
Number of meter/regulator station	CH_4	3.45	72.38
	CO_2	0.20	0.20
Tonnes CO2 equivalent from meter/regulator stations			72.58
Tonnes CO2 equivalent from transmission line			36191.58

From Tier 1 and Tier 2 methods, the pipeline length is considered as the parameter of fugitive emissions calculation. Referring to the definition of fugitive emission in the Guidelines of National Greenhouse Gas Inventory of Indonesia, fugitive emission is unintentional leaks from any objects in energy production and distribution. For pipeline, the connectors beteween two pipeline segments as well as valves and components attached to the pipelines are source for leaks. For this reason, the pipeline length data can be considered as the information to develop the fugitive emissions estimation.

In this work, the effects of operating pressure and diameter of gas pipeline are not calculated although they are considered as the main operating parameters for gas pipeline system. These two parameters may contribute to the pipeline leaks and identify as activity factor of emission source. However, the leaked volumes do not significantly affect the total volume of pipeline leaks [12].

The overall fugitive emissions calculation result of Tier 1 is lower than Tier 2. However, it is much higher compared to Tier 2 for pipeline length variable. It is due to the difference in emission factor. Tier 1 has higher emission factor for CH_4 and CO_2 from pipeline leaks to cover other variables that are not included in the calculation but considered as potential leak sources. On the other hand, Tier 2 puts smaller emission factor for pipeline length variable but complete the calculation with two other parameters i.e number of compressor stations and meter/regulator stations. From the calculation, the emissions calculation result from compressor station is the highest than pipeline length and meter/ regulator stations.

Principally, compressor is one of the main equipment in transporting natural gas through pipeline to provide adequate pressure of the gas to reach the end users. As the distance between the source and end users is far, it needs some compressor stations at certain points to maintain the operating pressure. Even in a compressor station consists of more than one compressor to gradually increase the operating pressure as a compressor has technical limitation to increase the pressure to a certain value. The compressor is worked as the result of the driver, typically a gas or diesel engine or gas turbine. The driver of compressor is a potential leak source that release GHG emissions, primarily CH_4 and CO_2. The higher the length of pipeline , the higher possibility of natural gas pressure drop. Consequently, it needs more compressor stations and resulting more GHG emissions. The high methane emissions as the result of compressor activity was also recorded in Russian long distance gas transport system [13]. There are some recommended options to improve the level emissions from compressor activity such as the replacement of centrifugal compressor seal oil systems and the installation of low bleed pneumatic devices [13].

Another variable of Tier 2 calculation is the number of meter/regulator stations. Meter/regulator stations are installed to measure the flow of gas along the pipeline. These stations are important to monitor, control, and acccount for the natural gas flow in the pipeline. The regulating equipment has vital roles as the regulator that reduce the delivery pressure to customer as well as to protect a section of a pipeline with a lower maximum operating pressure. These meter/regulator stations basically consist of meters, valves, fittings, instrumentation, and controls which contribute to the release of GHG to atmosphere. To position the pressure regulators, it is common to use gas operated pneumatic devices. These type of penumatic devices can release the gas to atmosphere when the regulator is activated. However, as the effect of leaks from meter/regulator stations is not as high as compressor stations, Tier 2 method applies lower emission factors for this activity factor.

The results of Tier 1 and Tier 2 is totally different. Principally, Tier 2 is an improvement method of Tier 1. This approach has additional activity factors but the errors from this method may be higher than Tier 1 if the facililty of the gas transmission pipeline system is different from typical industry averages. Although the accuracy of Tier 1 is less than Tier 2, but it can be used as a estimation tool to calculate emissions when the availability of the data is limited such as when the prelimanary design phase or when the facility has not been constructed yet. The quality of the data, the purpose of the estimation activity, as well as cost factor should be noticed in selecting the proper approach information. The initial estimation can be used as reference and recommendation for environmental assessment as well as preparing the GHG inventories in order to obtain government permission.

In certain conditions, the result of Tier 1 can be similar as Tier 2. For the case in this work, a similar result is obtained when the pipeline length is 771.2 km while the number of compressor stations and meter/regulator stations are maintained. A change in number of the stations (compressor stations or meter/regulator stations) will not give the same emission result for the length of 220 km. The result of Tier 2 can be lower than Tier 1 when the pipeline length is more than 771.2 km with the same number of compressor stations and meter/regulator stations. However, to maintain the same number of these stations is technically not possible, as the increase of the distance (pipeline length) will need more compressor stations and meter/regulator stations to control the pressure and flow rate of the natural gas along the pipeline.

CONCLUSIONS

The Tier 1 and Tier 2 methods from INGAA provide different required input data and result of gas pipeline fugitive emissions. In this work, Tier 2 shows a 3.5 times higher emission level than Tier 1 for a pipeline system in West Java, Indonesia. Although Tier 2 is developed with more detail factors of emission sources, but it does not mean that the lower accuracy method (Tier 1) can not be applied to estimate the emission level in natural gas industry. The cost for developing and documenting average facility level component counts for the estimation also need to be considered whether it is appropriate with the purpose of inventory.

REFERENCES

[1] Mohitpour, M., Botros, K, and Van Hardeveld, T. 2008. *Pipeline Pumping and Compression Systems: A Practical Approach*. New York: ASME, p.1.

[2] Lechtenboehmer, S., Fischedick, M., Dienst, C. and Hanke, T. 2003. *GHG emissions of the natural gas life cycle compared to other fossil fuels (in Europe)*. Beijing: 3 rd International Methane and Nitrous Oxide Mitigation Conference, pp.790-798.

[3] U. S. Environmental Protection Agency, 2012. *Methane Emissions*. [online] Washington DC. Available at: http://epa.gov/climatechange [Accessed 5 Sep. 2015].

[4] Interstate Natural Gas Association of America, 2005. *Greenhouse gas emission estimation guidelines for natural gas transmission and storage*. Washington DC: INGAA.

[5] Shindell, D, Kuylenstierna, JCI, Faluvegi, G, Milly, G, Emberson, L, Hicks, K, Vignati, E, Van Dingenen, R, Janssens-Maenhout, G, Raes, F, Pozzoli, L, Amann, M, Klimont, Z, Kupiainen, K, Höglund-Isaksson, L, Anenberg, SC, Muller, N, Schwartz, J, Streets, D, Ramanathan, V, Oanh, NTK, Williams, M, Demkine, V, Fowler, D. 2012. Simultaneously Mitigating Near-Term Climate Change and Improving Human Health and Food Security. *Science*, 335(6065), pp.183-189.

[6] Howarth, R., Shindell, D., Santoro, R., Ingraffea, A., Phillips, N. and Townsend-Small, A. 2012. Methane Emissions from Natural Gas Systems. *National Climate Assessment*, (Ref. no. 2011–0003).

[7] Alvarez, R.A, Pacala, S., Winebrake, J., Chameides, W. and Hamburg, S. 2012. Greater focus needed on methane leakage from natural gas infrastructure. Proceedings of The National Academy of Sciences of The United States of America, p.Vol 109, no.17.

[8] Townsend-Small, A., Tyler, S., Pataki, D., Xu, X. and Christensen, L. 2012. Isotopic measurements of atmospheric methane in Los Angeles, California, USA: Influence of fugitive fossil fuel emissions. *Journal of Geophysical Research*, 117(D7).

[9] Schlesinger, W. and Bernhardt, E. 2013. *Biogeochemistry: An Analysis of Global Change*. 3rd ed. San Diego CA: Academic Press.

[10] Indonesia Ministry of Environment, 2012. *Pedoman penyelenggaraan inventarisasi gas rumah kaca nasional*. Jakarta: Indonesia Ministry of Environment.

[11] Picard, D. 2006. *Fugitive Emissions from Oil and Natural Gas Activities, Good Practice Guidance and Uncertainty Management in National Greenhouse Gas*. [online] IPCC. Available at: http://www.ipcc-nggip.iges.or.jp/public/gp/bgp/2_6_Fugitive_Emissions_from_Oil_and_Natural_Gas.pdf [Accessed 22 Aug. 2015].

[12] Campbell, L. and Stapper, B. 1996. *Methane Emissions From The Natural Gas Industry*. U.S. Environmental Protection Agency.

[13] Lechtenböhmer, S., Dienst, C., Fischedick, M., Hanke, T., Fernandez, R., Robinson, D., Kantamaneni, R. and Gillis, B. 2007. Tapping the leakages: Methane losses, mitigation options and policy issues for Russian long distance gas transmission pipelines. *International Journal of Greenhouse Gas Control*, 1(4), pp.387-395.

FUNCTIONAL PROPERTIES OF LEUCAENA LEUCOCEPHALA PROTEIN CONCENTRATES RESULTED SEPARATION OF ULTRAFILTRATION MEMBRANE

Dedin Finatsiyatull Rosida*) , Nur Hapsari, Taufik Hidayah

Departement of Food Technology
University of Pembangunan Nasional "Veteran" Jawa Timur, Surabaya, Indonesia
Email: *) dedin.tp@upnjatim.ac.id

ABSTRACT

Protein is a polypeptide, which is composed of a series of amino acids, with a relatively very large molecular weight, which ranges from 8,000 to 10,000 Da. Separation of protein can be done by chemical, enzymatic or using an ultrafiltration membrane. Applications of protein concentrates product depends on its functional properties such as water absorption, oil absorption capacity and power emulsions, and foaming power which will affect the quality of product. Food product that use protein concentrates in the form of sausages, meatballs, bakery and beverages. Techniques of membrane separation are generally based on particle size and molecular weight of the thrust in the form of pressure differences, and different electric field concentration. This study aims to determine the role of membrane ultrafiltration to separate proteins and produce the characteristic functional properties. This study used two main factors, ultrafiltration membrane pressure (2.33 atm, 3.66 atm) and the separation time (15 minutes, 30 minutes, 45 minutes). The resulted showed leucaena seed protein concentrates obtained from the best treatment of 3.66 atm pressure and separation time 45 minutes. Concentrates of leucaena seed had the functional properties of the water absorption capacity of 3.3 g water/g of solid, oil absorption capacity of 1.43 g of oil/g solid, Bulk density of 0.55 g/ml, power emulsion 51.11% and power foam 9.8%.

Keywords: Protein, leucaena, ultrafiltration membrane, functional properties

INTRODUCTION

In several processing industries, separation of different components from a mixture is an important unit operation. Sometimes the separated component is an important product. There is a variety of technologies available for use in separations, each operating based on physical and chemical properties of the mixture. One of the fundamental separation processes that brought about a major change in processing of protein concentrate-product streams is membrane separation technology. Membrane separations work on the basis of differences in size and shape of the molecules. Today dairy industry accounts for major share in the total membrane area installed (Marella et al., 2013).
Leucaena leucocephala has been used by the community as shade trees, erosion control, firewood and fodder. Leucaena seed utilization is very limited whereas seeds contain high protein (30-40%) (Suprihatin, 2009). Rahayu et al., (2005) shows that the use of leucaena on soysauce containes protein 20.86%. The use of protein concentrates more desirable than flour bean because protein concentrates have functional properties that are large and relatively free of toxic factors and carbohydrates that can not be digested (Neto et al., 2001 cited in Adebowale and Lawal, 2003). The previous research has produced leucaena seed protein concentrate that had the protein 56.30 % (db) on the concentration of enzyme treatment of sheep rumen fluid (10 mL/100g) and incubation time of 48 h (Dedin F.Rosida et al., 2015). Applications of this protein concentrates depend on its functional properties of proteins such as water absorption, oil absorption capacity, power emulsions, and foaming. The functional properties are characterized by the physicochemical characteristics of food proteins that can determine behavior in food during processing, storage, and consumption.

The production of protein concentrates can be obtained using membrane fractionation process triggered by the pressure difference between the feed stream and the product stream (permeate). In this study, protein concentrates was made from separation method using ultrafiltration membranes. The principle of the use of ultrafiltration membranes are fractionated by molecular weight ultrafiltration membrane pore qualify. Factors that affect the membrane include molecular size, molecular shape, membrane materials, the characteristics of the solution, and operational parameters such as pressure, temperature, concentration, pH, ionic strength, polarization (Eden, 2012).

Membrane technology has several advantages compared to other processes, such as: Separation can be carried out continuously, energy consumption is generally relatively low, membranes process can easily be combined with other separation processes (hybrid processing), separation can be performed in a condition that is easily created, easy to scale-up, no need for additional material and the membrane material is so easily used

Reverse Osmosis (RO), Nanofiltration (NF), Ultrafiltration (UF) and Microfiltration (MF) are four commonly used membrane separation process. These processes have evolved from consistent research and development in the area of new membrane material development as well as applications (Marella et al 2009).

Chemical properties of the membrane which is important in membrane separation processes, among others: (1) the nature of the hydrophilic or hydrophobic; (2) charge on the ion; (3) resistance to high temperatures and certain chemical substances; and (4) the appeal against the particles in the feed (Osada and Nagawa, 1992). Membrane processes are grouped by size separation and applications that can be applied (Gutman, 1987).

Membrane microfiltration and ultrafiltration have been applied in the food industries for the purpose of concentrating, purifying biopolymer to a solution of macromolecules such as proteins and polysaccharides. The industrial thermal processing of foods may have a severe impact on the sensorial and nutritional properties of the final product. Membrane technologies have been extensively studied as alternative processes. Forward osmosis (FO) is a promising membrane technology to be used in food industries. The only driving force of the process is the osmotic pressure difference between the two solutions that flow in counter-current mode on opposite sides of a permeable membrane. Thus, the main advantages of FO, compared to both thermal and conventional membrane processing, include low hydraulic pressure, low treatment temperature, low fouling tendency, high solids content processing capability and easy to scale-up (Sant'Anna et al., 2012). The separation membrane performance is influenced by the characteristics of the membrane used. Assessment of the characteristics of such a membrane that covers the structure and pore size and other physical properties of the chemical. The main parameters used in the assessment of the performance of the membrane filtration is the flux and rejection (Osada and Nakagawa, 1992). This study aims to determine the role of membrane ultrafiltration to separate proteins and produce the specific characteristic functional properties

METHODOLOGY

Raw materials was Leucaena leucocephala seeds. The study began with soaking seeds for 24 hours, then stripping the skin and let dried. The next steps were grinding and sieving to 100 mesh.

Leucaena flour was added 0.01 N HCl 100 mL to stirr for 1 hour and neutralized with 0.01 N NaOH solution of 100 mL. The starting of the protein separation used ultrafiltration membranes with a pore size of the membrane 0.45 µm. The pressure levels used during the separation were: 2,333 atm and A2: 3.667 atm, while the fractionation time were: 15 minutes, 30 minutes and 45 minutes. The last step was washing with 200 ml of distilled water and dried. The protein concentrate parameters was observed to include: Protein yield, Bulk density (Okezie and Bello, 1988), the absorption of oil (Soluski and Fleming, 1977), permeability (Lin et al., 1974), power foam (Widowati et al., 1998), capacity and emulsion stability (modification Franzen and Kinsella, 1979).

Analysis of Water Absorption was done by taking a sample of 1 g and 10 mL of distilled water for 2 minutes vortex, chilled for 1 hour at room temperature and then centrifuged at 3000 rpm for 25 minutes. Filtrate volume was measured. Water absorption (water mL /g sample) was calculated by the following equation: (10 mL - vol filtrate mL)/ W sample (g) (Lin et al, 1974)

Analysis of Oil Absorption was done by taking a sample of 0.5 g plus 3 mL of soybean oil vortex for 2 minutes and then allowed to stand at room temperature. The mixture was centrifuged at 3000 rpm for 25 minutes. Supernatant was poured into 10 mL measuring cup and free oil was observed. Volume oil absorption (oil mL/g sample) measurements performed in two repetitions with the following equation: (3 mL of oil – filtrate)/W sample (g) (Soluski and Fleming, 1977)

RESULT AND DISCUSSION

Protein Value

Nutrient content in leucaena with skin is protein 19.75%, fat 5.58%, moisture 14.31% and ash 5.66%. The results of the analysis flour leucaena seed without seed skin is obtained moisture 12.26%, ash 6.73, fat 3.19% and protein 55.38% (Dedin F.Rosida et al 2012). It is suspected that leucaena was processed into flour without involving the skin, where the skin contains cellulose and flouring processes can reduce the moisture content so that the protein concentration in leucaena seed flour was higher. Leucaena seed flour without skin contains approximately 50% protein (Budianto et al., 1987)

The average levels of protein concentrates in leucaena seed with pressure treatment were between 47.79% - 49.42% (Table 1), the greater the pressure inside the membrane, larger the protein content. This occured because the fluid passing through the membrane more filtered with increased pressure. The size of the protein ranged from 0.03 to 0.3 μm, while the pore size microfiltration membrane ranged from 0.1 to 1.0 μm. so leucaena seed proteins can pass through the membrane pore size of 0.45 μm.

Table-1. Protein value of Leucaena protein concentrate

Treatment	Protein value (%)
Pressure (atm)	
2.33	47,79 ± 2,266[a]
3.66	49,42 ± 2,031[a]
Time (minutes)	
15	47,77 ± 2,732[a]
30	48,76 ± 1,422[a]
45	49,28 ± 2,516[a]

The average levels of leucaena seed protein concentrates with a separation time of 15 minutes was 47.77%, 48.76% at 30 minutes, 49.28% at 45 minutes (Table 1). Time is one thing that plays a role in improving the protein content outcomes. The longer time given the higher ability of material to go through the membrane by the pressure exerted, there are deficiencies which may be generated from a given length of time that a build up of material in the pores of the membrane making the material can not be passed through membrane. Membrane performance can be decreased by the presence of fouling and polarization concentration. Polarization concentration and fouling can limit the membrane separation process since both cause a decrease in the permeate flux (Wenten, 1999).

Membrane serves to separate material based on the size and shape of the molecule to hold the components of the feed which has a larger size than the pores of the membrane and to skip parts that have a smaller size. Using membrane filtration in addition to function as a means of separation also serves as a means of concentration and purification of a solution passed on the membrane.

According to the desired separation process, the membrane can be divided into microfiltration, ultrafiltration and reverse osmosis with a sequence of increasingly smaller pore size. Among the three types of membrane, ultrafiltration has the most extensive applications in industries such as fermentation, food and beverage industry, semi-conductors and others (Osada and Nakagawa, 1992).

Functional Properties

The water absorption of leucaena concentrates with pressure was 3.12 - 3,22 g water/g solid (Table 2). The water absorption of leucaena concentrates with a time separation was 3.03 - 3.28 g water/g solid. Water absorption or water holding capacity is the ability of proteins to bind water during force use, pressure, centrifugation and heating (Zayas, 1997). The higher pressure showed an increase in the protein content of leucaena concentrates that affected the increased absorption of water.

Table -2. Functional Properties of Leucaena protein concentrate

Treatment	Water absorption (g air/ g solid)	Oil absorption (g oil/ g solid)	Bulk density (%)	Power emulsion (%)	Power Foam (%)
Pressure (atm)					
2.33	3,12 ± 0,249[a]	1,2 ± 0,186[a]	0,58 ± 0,013[a]	47,67 ± 2,422[a]	7,55 ± 1,648[a]
3.66	3,22 ± 0,192[a]	1,3 ± 0,228[a]	0,56 ± 0,015[a]	50,20 ± 2,465[a]	8,61 ± 1,746[a]
Time (minutes)					
15	3,03 ± 0,234[a]	1,25 ± 0,238[a]	0,58 ± 0,017[a]	47,84 ± 3,037[a]	7,0 ± 1,789[a]
30	3,2 ± 0,141[a]	1,23 ± 0,115[a]	0,58 ± 0,012[a]	48,78 ± 1,990[a]	8,1 ± 1,033[a]
45	3,28 ± 0,232[a]	1,38 ± 0,294[a]	0,56 ± 0,015[a]	50,17 ± 3,208[a]	9 ,0 ± 1,828[a]

The water absorption is also affected by the concentration of the protein and the presence of other components such as hydrophilic polysaccharides, fats, salt, and is also influenced by the length of the heating and storage conditions. The higher the concentration of protein in a food, the water absorption capacity is increasing (Zayas, 1997 in Gunawan, 2009).

The oil absorption of leucaena concentrates with pressure was 1.2 - 1.3 g water/g solid (Table 3). The oil absorption of leucaena concentrates with a time separation was 1.25 - 1.38 g oil/g solid. Characteristics of food products involves many interactions between proteins and fats, such as the formation of emulsions, the emulsification of fat in meat, absorption flavor and texture of the dough. Kinsella (1979) describe the mechanism of oil absorption as a physical phenomenon of trapped oil. The absorption of oil is determined by the binding of fat by non polar groups on the protein (Kinsella 1979; Sathe et al. 1982).

In a food product in the form of powder, fat binding is affected by particle size. Protein in the form of powder with low density and small particle size can absorb and trap more oil than high density protein (Zayas, 1997). Fat binding ability of the protein is essential for the application of protein as meat replacement and extension with respect to mouthfeel and flavor retention involving the holding capacity of fat in the product. The value of oil absorption capacity of leucaena protein concentrates compare to oil absorbency of commercial soy protein concentrates (ISOPRO) was 1.1704 g oil / g solid (Lin et al., 1974).

The bulk density of leucaena concentrates with pressure treatment was 0.58% - 0.56% and treated-separation time was 0.58% - 0.56%. Bulk density is the mass of particles that occupy a certain volume unit (Wirakartakusumah, et al., 1992). Bulk density stated the brevity of a material in occupied volume. The greater the bulk density, the more compact the material because it can occupy the same volume in a greater weight. Bulk density is also the ratio between the weight of the volume of material that is expressed in g/mL. Peleg and Bagley (1983) states that the Bulk density depends on several factors, namely the attraction between the particles, particle size and surface area of the particles touch. Bulk density of protein concentrates are influenced by increased cohesiveness of particles and moisture. The statement was reinforced by the opinions Wirakartakusumah et al., (1992), which states that in general, water absorption associated with increased cohesion, mainly caused by the liquid bridges between the particles. Therefore, food powder which is hygroscopic at high water levels caused a decrease in bulk density.

Bulk density is influenced by the particle size, the properties of materials, material composition, and may also be influenced by the degradation of the molecules in the material due to processing. So the bulk density values can be caused by the degradation of starch molecules, proteins, fats, and others when treated early ripening so that the molecules occupy a smaller space (Winata 2001).

The emulsion power of leucaena concentrates with pressure treatment was 47.67% - 50.20%. The mean power emulsion of leucaena concentrates with a treatment time of separation was 47.84% - 50.17%. The emulsion is defined as a mixture of two or more liquids that do not dissolve each other, which is a liquid dispersed in the form of globules or small grains in other liquids. Dispersed liquid called discontinuous phase while dispersing a fluid which is called the continuous phase (Fardiaz, et al., 1992). Protein emulsion properties is the ability of a protein to form an emulsion. Protein emulsion activity is the ability of a protein in the formation of the emulsion and its stabilize (Zayas, 1997). The protein solubility, surface hydrophobicity and flexibility emulsion protein molecules influence the properties of globular proteins. Therefore, the ability of the membrane to produce optimal protein concentrates power is associated with an emulsion of a protein concentrates (Nakai et al. 1980),.

The higher the solubility of proteins, the higher the ability to form a layer that envelops oil emulsion doplet so that the activity increases (Suwarno et al., 2003). Emulsion formation by proteins is influenced by internal and external factors. Internal factors include pH, ionic strength, temperature, molecular weight surfactants, and the type of protein. External factors include the type of equipment used, and the speed of stirring. Until now, there has not been established any standard methods for measuring the properties of these emulsifiers, therefore, the observations of the various laboratories vary widely with each other and can not be compared (Damodaran, 1996).

The foam power of leucaena concentrates with pressure treatment was 7.55% - 8.61%. The average foam power of leucaena concentrates with a treatment time of separation was 7% - 9% (Table 2). The ability of the protein to form a stable foam are indispensable in the production of various foods, such as ice cream, cake, bread, whipped cream and mousse. Protein concentrates is often used in products creamer, milkshake, meat analogues, and so on. Foam is a dispersed structure that combines two colloidal liquids, such as a protein solution and gas dispersing medium as a dispersed phase. The nature of the formation of foam is a protein in which the surface properties of the surface properties is influenced by surface tension. Foam formation associated with reduced surface tension at the air-water system due to the adsorption of protein molecules (Okezie and Bello, 1988). Foam power increases as concentrations of protein increases due to an increase in the thickness of the interfacial layer, the foam stability is obtained when the maximum elasticity of the surface is also maximum.The capacity and stability of the foam is better done at high concentrations because of the increased viscosity of the protein and facilitates the formation of multilayer film that is cohesive and the interfacial surface (Damodaran, 1996).

The previous research has produced leucaena seed protein concentrate that had functional properties of oil absorption capacity of 3.42 mL/g, the bulk density 0.56 g/mL, power emulsion 52.75%, 6.39% moisture content on the concentration of enzyme treatment of sheep rumen fluid (10 mL/100g) and incubation time of 48 h (Dedin F.Rosida et al., 2015).

CONCLUSION

Preparation of leucaena protein concentrates used ultrafiltration membrane of 0:45 μm and obtained greater pressure and longer time to produce to increase in protein content, water absorption, power and emulsion foam power. The optimal functional properties of leucaena protein concentrates was produced at 3.66 atm pressure and within 45 minutes found the ability of the foam by 9.8%, 51.11% emulsion power, water absorption 3.3 g water/g solid, 1.43% oil absorption and bulk density 0.59 g mL

REFERENCES

Adebowale K.O and Lawal O.S. 2003. foaming, gelation and Electrophoretic Characteristics of Mucuna Bean (Mucuna pruriens) Protein Concentrates. Journal of Food Chem. 83: 237-246.

Budijanto S, Sitanggang AB and Murdiati W. 2011. Characteristic physico chemical properties and functional winged seed protein isolate (Psophacarpus tetragonolobus L). Journal of Food Technology and Industry. XXII. (2): 130-136

Damodaran S. 1996. Amino Acids, Peptides, and Proteins. Inside: Fennema, O. R (Ed). 1996. Food Chemistry. New York: Marcel Dekker Inc.

Dedin F. Rosida, Constantia F. 2012. The Study on role of red rice in (soybean-Leucaena leuococephala) Tempe Quality. Journal of Rekapangan ISSN 1978-4163. Vol. 6 No.1 : 64-72

Dedin F. Rosida, Yulistiani R, Ardiani W, Dedid Cahya Happyanto. 2015. Production of leucaena seed protein concentrate by using hydrolase enzymes from the rumen of sheep. Wulfenia Journal. Vol.22.issue.9 (September 2015): 190-204

Eden J.M., Widmann M., Grawe D, Rast S. 2012. Skill, Correction, and Downscaling of GCM-Simulated Precipitation. American Meteorological Society. AMS Journals online. Volume 25, Issue 11 (June 2012)

Fardiaz D., Andarwulan N., Wijaya H and Puspitasari N. L. 1992. Technical Analysis and Chemical Properties of Functional Food Components. PAU Food and Nutrition, IPB, Bogor.

Gutman R. G. 1987. Membrane Filtration. The rheological of Pressure Driven Crossflow Procces. IOP Publishing Ltd. England.

Hanson L. P. 1974. Vegetables protein processing. London: Hayes Data Corp. Inside: Ogunwolu S.O., Henshaw F.O., Mock H.P., Santros A., Awonorin S.O. 2009. Functional Properties of Protein Concentrates and isolates Produced From Cashew (Anarcaduim occidentale L.) Nut. Journal of Food Sci. 115: 852- 58.

Hapsari A.W. 2009. Study of physicochemical properties, functional protein, and antioxidant capacity in protein concentrate bean sprouts lablab (Lablab Purpureus (L.) Sweet). Thesis. Faculty of Agricultural Technology. Bogor Agricultural Institute.

Kartika Y. D. 2009. Characterization of functional properties winged seed protein concentrate (Psophocarpus tetragonolobus L). Thesis. Faculty of Agricultural Technology. Bogor Agricultural Institute.

Kinsella J. E 1979. Functional properties of soybean protein. Clock. Oil. Chem. Soc. 56: 242-258.

Lin M. Y., Humbert G. S., and Solunski F. W. 1974. Certain functional properties of sunflower meal products. Journal. Food Sci. 39: 368.

Marella C . 2009. Whey protein fractionation using membrane separation technology. dissertation. South Dakota State University, Brookings.

Marella C., Muthukumarappan K and Metzger L. E. 2013. Application of membrane separation technology for developing novel dairy food ingredients. Journal Food Process Technol 2013, 4:9 http://dx.doi.org/10.4172/2157-7110.1000269

Okezie B. O and Bello A. B. 1988. Physicochemical and functional properties of winged bean flour and isolate compared with soy isolate. Journal Food Sci. 53 (2): 450-454.

Osada Y and Nakagawa T. 1992. Membrane Science and Technology. Marcel Dekker, Inc., New York

Peleg M and Bagley E. B. 1983. Physical Properties of Foods. Westport: AVI Publishing Co. Inside: Ogunwolu, SO, FO Henshaw, HP Mock, A. Santros, SO Awonorin. 2009. Functional properties of protein concentrates and isolates produced from cashew (Anarcaduim occidentale, L.) Nut. Journal Food Sci. 115: 852-858.

Sant'Anna Voltaire, Ligia Damasceno Ferreira Marczak, Isabel Cristina Tessar. 2012. Membrane concentration of liquid foods by forward osmosis: Process and quality view. Journal of Food Engineering. Volume 111, Issue 3, August 2012, Pages 483–489

Sathe S.K., Deshpande S.S and Salunkhe D. K. 1982. Fuctional properties of winged bean (Psophocarpus tetragonolobus (L.) DC) Proteins. Journal Food Sci. 47: 503-509.

Solunski F and Fleming S. E. 1977. Chemical functional and nutritional properties of sunflower protein products. Journal Am. Oil Chem. Soc 54: 100-105

Suwarno and Maryani. 2003. Potential hyacinth bean (Lablab purpureus (L) Sweet) as raw materials of protein isolates. Thesis. Faculty of Agricultural Technology. Bogor Agricultural Institute.

Suprihatin. 2009. Protein Hydrolysis of Leucaena seeds. Surabaya: UNESA University Press.

Wenten I. G., 1999. Industrial Membrane Technology. Department of Chemical Engineering. Industrial Technology Faculty. Bandung Institute of Technology Duo

Widowati S., Susi Wijaya S. K and Yulianti R. 1998. Globulin Fraction and Nature Functional Protein Isolates of Soybean Varieties Ten Indonesia. Food Crops Agricultural Research 17: 52-58

Winata A. Y. 2001. Characterization of wheat breadfruit (Artocarpus altilis) drying drum precooking results. Essay. Faculty of Agriculture, IPB, Bogor.

Wirakartakusumah M.A., Abdullah K and Syarief A. M. 1992. Physical properties of food materials. Bogor: Inter-University Center of Food and Nutrition, Bogor Agricultural University.

Zayas J. F. 1997. Functionality of Proteins in Food. Berlin: Springer-Verlag.

INHIBITION OF STRUVITE CRYSTAL GROWTH IN THE PRESENCE OF HERBAL EXTRACT *ORTHOSIPHON ARISTATUS* BL.MIQ

Stefanus Muryanto[1*], Sri Sutanti[2], Mega Kasmiyatun[1]

[1]Department of Chemical Engineering and Office of Research, Universitas 17 Agustus 1945 Semarang, Semarang, Indonesia

[2]Akademi Kimia Industri Santo Paulus, Semarang, Indonesia

E-mail: stefanusmuryanto@yahoo.com.au

ABSTRACT

The formation of urinary stones is a serious problem worldwide. Struvite ($MgNH_4PO_4.6H_2O$) is known as one of the main components of the stones. Traditionally, herbal extracts treatment is recommended. This paper discusses the effects of Orthosiphon aristatus Bl.Miq leaves extract (OA) on struvite crystals grown by diffusion gel method. After three weeks the crystals were separated from the gel, air dried and stored for characterisation. The crystals exhibit different morphologies depending upon the location of growth. It was found that OA caused the crystals to have noticeable surface defects, discolorisation and fragmentation. Also, OA (up to 5.0%) decreased the growth rate. XRD analysis confirms struvite crystallinity. Pure crystals and those grown with OA showed identical XRD pattern, indicating that the crystalline nature may be unaffected. For crystals with OA, FTIR spectrum shifted to higher wave numbers, implying that the various organic compounds in OA might be absorbed on the crystal surface. TG-DTA studies showed similar thermal behaviour for both pure crystals and those with OA. However, the later revealed lower endothermic temperature, indicating lower thermal stability. Fragmentation and lower thermal stability of struvite indicated that OA was a potent inhibitor of struvite and could be effective for urinary stone treatment.

Keywords: crystal growth, Orthosiphon aristatus, SEM, struvite, XRD

INTRODUCTION

The formation of urinary calculi is a serious clinical problem worldwide (Suguna *et al.*, 2012). Phosphate minerals, such as magnesium ammonium phosphate hexahydrate ($MgNH_4PO_4.6H_2O$), mineralogically known as struvite, is one of the main components of the calculi. It was also reported that recurrence after treatment is relatively high (ElMahdy and Persad, 2014; Suguna *et al.*, 2012; Chauhan, 2011; Miyaoka and Monga, 2009; Grohe *et al.*, 2006).

The urinary calculi (= *urinary stones, renal stones* or *kidney stones*) may arise due to chemical activities within the body (Suguna *et al.*, 2012; Chauhan, 2011). The condition favourable for the formation of the stones includes high concentrations of calcium salts (Ca-oxalate and Ca-phosphate), infections of urinary tracts, pH, and a decrease in the body's natural ability to inhibit the formation of the crystals (Suguna *et al.*, 2012; Rajendran and Keefe, 2010).

Traditionally, urolithiasis or kidney stone disease can be prevented or treated with herbs (*herbal medicines* or *herbal extracts*). These herbs include *Origanum vulgare*, *Moringa oleifera*, *Asparagus racemosus*, *Rotula aquatica*, *Mimosa pudica*, *Trigonella foenum graecum*, *Nigella sativa*, and *Punica granatum* (Khan *et al.*, 2011; Butterweck and Khan, 2009). These studies indicate that herbal extracts could inhibit the growth of the crystals forming the kidney stones. As opposed to treatment using antibiotics, these herbs were claimed to cause no bacterial resistance and other adverse effects (Miyaoka and Monga, 2009). As such, the herbal medicines have gained significant popularity.

A number of research has also been done to explore the effects of the leave extract of *Orthosiphon aristatus* (OA) – commonly known as cat whiskers in Malay, or Java tea, on urolithiasis (Yam *et al.*, 2013; Abdelwahab *et al.*, 2011; Adam *et al.*, 2009; Yam *et al.*, 2007). These studies focused on the medicinal properties of OA.

However, only a few studies have been conducted on the effects of herbs on the crystal growth of struvite as one of the main components of renal calculi. These herbs include *Boerhaavia diffusa* Linn, *Commiphora wighti*, *Rotula aquatica* Lour, and *Citrus medica* Linn (Chauhan, 2011). As far as the present authors know, the effects of OA leaves extract on the crystal growth and properties of struvite has not been examined. Reported in this paper are the results of such studies discussed in terms of morphology, crystallinity, elemental composition, thermal and mechanical properties.

METHODOLOGY

The crystallisation experiment was done using a single diffusion gel method, which is a standard method as a substitute for an *in vivo* model. This method enables a continuous observation of the crystal growth (Chauhan and Joshi, 2013; Chauhan, 2011; Rajendran and Keefe, 2010).

Preparation of the gel

The gel selected for the growth medium was silica gel, since it has a number of favourable characteristics: it requires no other chemicals, irreversible, elastic, has a stable form, and does not react with the growing crystals (Chauhan, 2011).

To grow the struvite crystals within the gel, two solutions containing the crystal-forming components were mixed. The two solutions would diffuse through the gel, so that ions and molecules of the crystal-forming components would react slowly and in a controlled mode. This contact would yield an insoluble phase in the form of crystals. The crystallisation process can be easily conducted using ordinary lab test tubes.

The silica gel (Na_2SiO_3) was diluted with distilled water as necessary and would react as follows (Chauhan, 2011):

$$Na_2SiO_3 + 3\ H_2O \longrightarrow H_4SiO_4 + 2\ NaOH$$

Further, H_4SiO_4 would polymerise and release the water, to form covalent bonds. The bonds repeat continuously to form a three dimensional structure, which can be a perfect medium for crystal growth (Suguna *et al.*, 2012).

Preparation of the leaves extract of the Orthosiphon aristatus Bl.Miq (OA)

The leaves were purchased locally, washed using flowing tap water and rinsed with distilled water. Then, the cleaned leaves were sun-dried, crushed into powder using a household blender, and stored in a tight container. The powder was subsequently mixed with distilled water and boiled (50 g powder/500 mL water) in a beaker until the volume reduced to half the initial volume. The liquid obtained was then vaccuum-evaporated using a modified Soxhlet extractor until the liquid turned into a paste. The paste, which is the OA extract, was stored in a refrigerated place and ready for the experiment.

Crystallisation of struvite

1.0 M $NH_4H_2PO_4$ solution (reactant 1) was mixed with Na_2SiO_3 solution in a test tube until a gel was formed. Then, 1.0 M $MgCl_2$ solution (reactant 2) was poured on top of the gel. Next, the test tube was sealed with a silver film and placed in a closed cupboard to prevent contamination. The reaction occurred in the gel as follows:

$$NH_4H_2PO_4.2H_2O + Mg(CH_3COO)_2.4H_2O \longrightarrow MgNH_4PO_4.6H_2O + 2\ CH_3COOH$$

In the case of OA trials, the magnesium acetate solution was mixed with the extract (0.1% to 5.0%) prior to the experiment. Crystals started appearing at the interface between the gel and the supernatant one day after the initiation of the run. After three weeks the crystals were separated from the gel, air dried and stored for characterisation, which included morphology, crystallinity, elemental composition, thermal and mechanical properties.

The growth of the crystals in the gel was observed visually and recorded everyday during the growth period, which lasted up to three weeks. Beyond that period the crystals started dissolving which could be due to the release of acetic acid (see the reaction equation).

Once the growth ended, the gel containing the crystals were extracted from the test tube and the crystals separated. The crystals were washed cautiously with distilled water and subsequently dried in an oven (40°C) for 24 hours. The dry crystals were then stored for subsequent analysis.

Characterisation of the struvite crystals

Characterisation of the crystals included morphology, crystallinity, elemental composition and thermal properties. Thus, it required scanning electron microscope (SEM-EDX, XRD), Fourier transform infra red (FTIR), and TG-DTA.

RESULTS AND DISCUSSION

Crystal Growth

Single gel diffusion method was used to grow struvite crystals because the method imitates the gel-like medium in human body wherein the urinary stones develop (Suguna *et al.*, 2012).

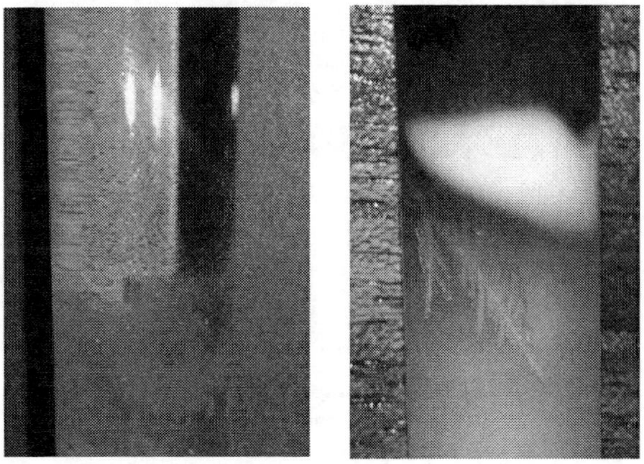

Figure-1. Dendritic crystals appeared near the gel-liquid interface: pure (left); with OA (right)

In the current work, struvite crystals were grown in the absence and in the presence of OA at room temperature. Tiny crystals, which were predominantly dendritic, appear at the gel-liquid interface one day after the experimental run (Fig 1). Other forms, e.g. needle-like and plate-like can also be seen. It is also seen that there is a change in morphology of the crystals along the test tubes from top to bottom. It can be postulated that the change was due to the availability of the struvite-forming components as well as their diffusivity in the gel (Suguna *et al.*,-2012).

OA was found to affect the morphology and size of crystals. It can be argued that the organic components of OA could have been affecting the growth rate of the crystals by adsorption on the crystal surface and blocking the active growth sites resulting in the inhibition of the growth. This current work indicates that OA may affect the nucleation and growth of struvite resulting in poor yield.

Morphological Analysis

Figure 2 compares the morphological characteristics of the struvite crystals obtained. In the absence of OA the surface of the crystal appears smooth (left). On the other hand, in the presence of OA (5.0%), the surface of the crystals are rougher and fragmented.

This findings is similar to that of Suguna *et al.* (2012) on the crystallisation of struvite doped with fluoride. It is postulated that the change in surface characteristics of the crystals with OA could be caused by the adsorption of OA on the crystal surface.

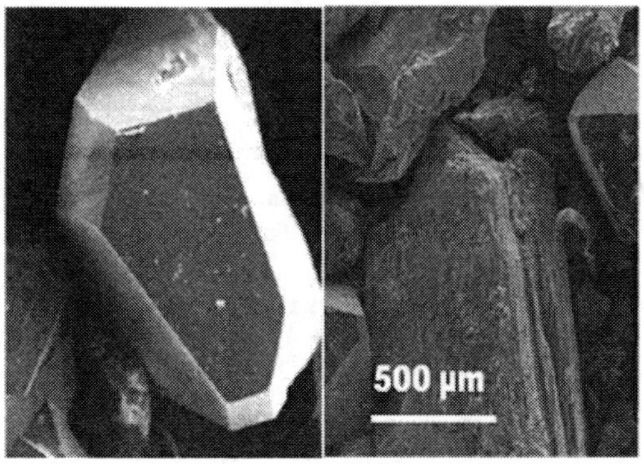

Figure-2. Morphology of the crystals: without OA (left); with OA (right)

X-Ray Diffraction Analysis

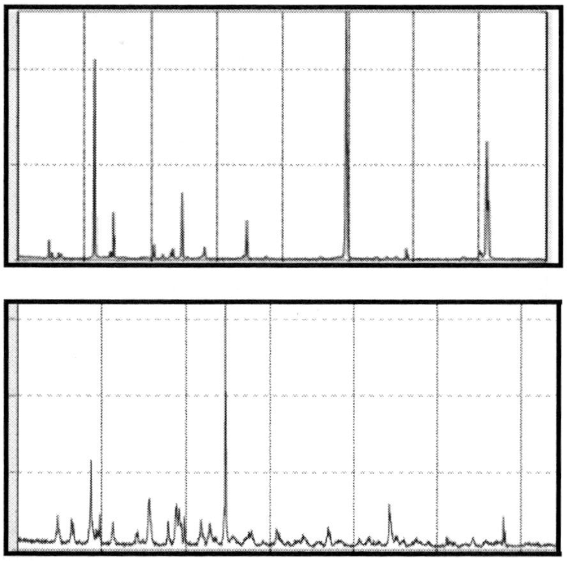

Figure-3. XRD pattern of struvite in the current work: without OA (top); with OA (bottom)

X-ray diffraction (XRD) analysis is believed to be a reliable method for analysis of crystalline materials, including the crystalline phases of struvite. X-ray identification was performed using Cu-Kα monochromated radiation in a conventional Bragg-Brentano (BB) parafocusing geometry. Results of the XRD analysis (the peak positions and peak heights of the spectrum) were checked against the data for the spectrum of struvite standard (JCPDS – #015762).

Figure 3 depicts the XRD analysis for the crystals. The diffraction peak positions and intensities of the crystals obtained mostly match up with the standard data. Overall, it can be stated that the crystals produced were struvite.

FTIR Analysis

The FTIR analysis was carried out to examine the chemical bondings which subsequently provide the information of the molecular structure of the compounds. Figure 4 depicts the recorded spectrum obtained for the struvite crystals.

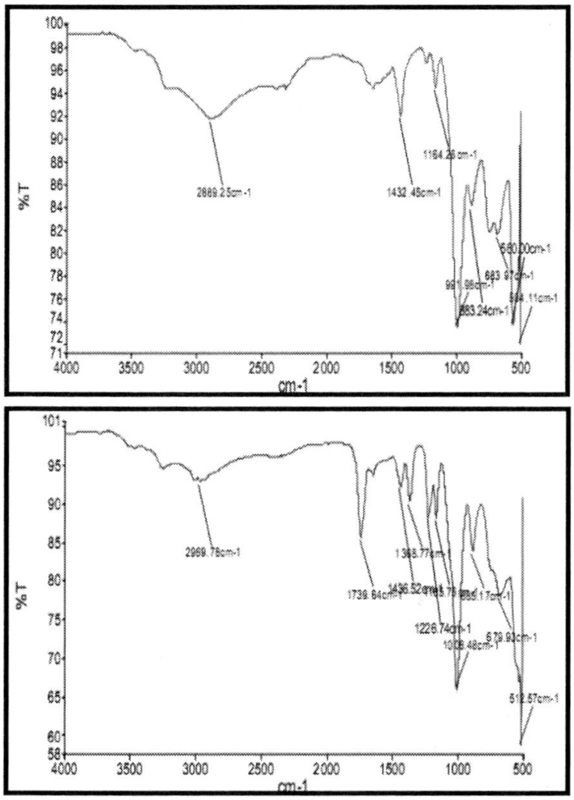

Figure-4. FTIR spectrum struvite in the current work: without OA (top); with OA (bottom)

The spectrum clearly shows the four absorption regions as expected for the struvite: metal-oxygen bonds, NH_4^+ units, PO_4^{3-} units, and water of crystallization.

A broad asymmetric band between 2100 and 3600 cm^{-1} can be attributed to O-H and N-H stretching vibrations. A medium broad band at 2889.25 cm^{-1} indicates the ν_1 symmetric stretching vibration of N-H in NH_4^+. Bending modes of water of crystallisation (H-O-H) is seen at absorption of around 1700 cm^{-1}. There is a strong vibration taking place at 1432.45 cm^{-1} which suggests ν_4 asymmetric bending vibration of N-H in NH_4^+. Another strong vibration at 1164.26 cm^{-1} indicates ν_4 asymetric bending vibration of N-H. A very sharp absorption peak at 991.98 cm^{-1} indicates the NH_4^+ rocking. The metal-oxygen bond is shown as the weak absorption band at 683.97 cm^{-1}. The weak bands between 500 and 600 cm^{-1} can be attributed to PO_4^{3-} units. Overall, the FTIR spectrum recorded confirms the presence of the four compound species of struvite: metal-oxygen bond (due to Mg), NH_4^+, PO_4^- and water of crystallisation. As can be seen in Figure 4, for the experiments with OA (bottom) all absorption peaks shifted to higher wave numbers which indicate the influence of OA.

TG-DTA Analysis

For the crystals without OA, there is a major weight loss of about 75% between 80°C and 150°C. This loss denotes the elimination of six water molecules and one ammonia molecule (Suguna *et. al.*, 2012). In addition, this coincides well with the DTA curve which shows a strong endothermic peak at 140°C. In the higher temperature region between 150 and 710°C, two molecules of $MgHPO_4$ combine to form magnesium pyrophosphate ($Mg_2P_2O_7$) with the loss of one water molecule, and resulting in nearly 53% of the sample remains stable. The following reactions were postulated to occur during the heating of the crystals (Suguna *et al.*, 2012).

$$MgNH_4PO_4.6H_2O \longrightarrow MgH_4PO_4 + NH_3 + 6\ H_2O$$

$$MgH_4PO_4 \longrightarrow Mg_2P_2O_7 + H_2O$$

The thermal behaviour of struvite with OA is similar in terms of weight loss (Fig 5 – bottom). However, there is a reduction in endothermic peak, which was recorded at as low as 110°C. This reduction indicates the lowering of the thermal stability of the crystals formed under the influence of OA. Thermal properties of the crystals obtained were examined at temperature range between 30°C and 1000°C at the rate of 20°C per minute under nitrogen atmosphere (Fig. 5). The exothermic peak of the struvite without OA is slightly higher than 700°C, whereas for the crystals with OA is slightly lower than 700°C. This condition further confirms the reduced thermal stability of the crystals with OA.

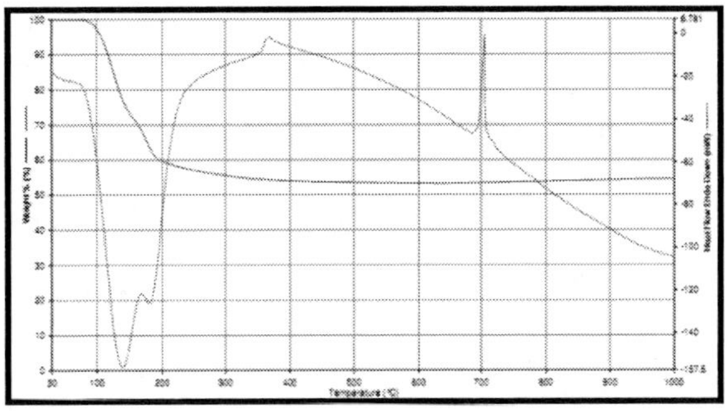

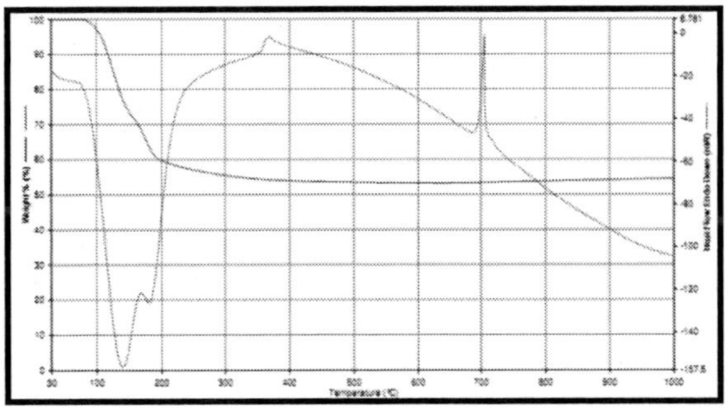

Figure-5. TG-TGA analysis of the crystals in the current work: without OA (top); with OA (bottom)

Mechanical Properties Analysis

The mechanical properties of the crystals were examined with the intention to contribute to the development of the techniques for crystal fragmentation. This could be useful for the design of medical instrumentation aimed at destructing urinary stones. The mechanical properties of the crystals were tested using Vickers hardness tester, and it was found that the struvite crystals with OA are softer than those without OA (results of the analysis were not shown).

CONCLUSIONS

Struvite ($MgNH_4PO_4.6H_2O$) is known as one of the main components of the urinary stones. Traditionally, herbal extracts are recomended for the urinary treatment. This paper presents an investigation on the effects of *Orthosiphon aristatus Bl.Miq* leaves extract (OA), commonly known as Java tea, on struvite crystals grown by a single diffusion gel method.

Under SEM the crystals exhibit different morphologies: dendritic, prismatic, platelet and needle-like, depending upon the location of growth. It was also found that for the crystals with OA, the crystal surface was rough which could indicate the adsorption of OA on the surface.

The XRD of the crystals confirms the crystalline phase of struvite. Both pure crystals and those grown with OA showed identical XRD pattern, indicating that the crystalline nature seems to be unaffected by OA.

The FTIR spectrum proves four absorption peaks: water of crystallisation, NH_4^+, PO_4^{3-}, and Mg-oxygen bonds, which agrees with previous studies. For the crystals with OA the spectrum indicates shifting to higher wave numbers, implying that components of OA might be absorbed on the crystal surface.

TG-DTA studies (30^{0}C to 1000^{0}C; 20^{0}C/min; nitrogen atmosphere) showed similar thermal behaviour for both pure crystals and those with OA. However, the later revealed a decrease in endothermic peak temperature, indicating lower thermal stability.

Vickers microhardness test showed that crystals with OA showed lower hardness.

Properties of the struvite crystals grown with OA seemed to be affected by the various organic compounds contained in OA. Fragmentation, lower thermal stability and reduced hardness of the crystals proved that OA was a potent inhibitor of struvite and could be effective for urinary stone treatment.

ACKNOWLEDGMENTS

The authors are grateful to DIKTI – the Indonesian Ministry of Education and Culture for the research fund granted under Hibah Fundamental scheme: 003/K6/KM/SP2H/PENELITIAN_BATCH_I/2015.

REFERENCES

Abdelwahab S.I., Mohan S., Elhassan M.M., Al-Mekhlafi N., Mariod A.A., Abdul A.B., Abdulla M.A., Alkharfy K.M. 2011. Antiapoptotic and antioxidant properties of Orthosiphon stamineus Benth (cat's whiskers): intervention in the Bcl-2-mediated apoptotic pathway. Evidence-Based Complementary and Alternative Medicine 2011. doi: 10.1155/2011/156765. (accessed 25 June 2015).

Adam Y., Somchit M.N. and Sulaiman M.R. 2009. Diuretic properties of Orthosiphon stamineus Benth. Journal of Ethnopharmacology 124(1): 154-159.

Butterweck V. and Khan S.R. 2009. Herbal medicines in the management of urolithiasis: alternative or complementary?. Planta Medica 75:1095-1103.

Chauhan C.K. and Joshi M. J. 2013. In vitro crystallization, characterization and growth-inhibition study of urinary type struvite crystals. Journal of Crystal Growth 362: 330-337.

Chauhan C.K. 2011. Growth and characterization of struvite and related crystals. PhD thesis, Saurashtra University, India.

Elmahdy M. and Persad R. 2014. Aetiology and prevention of recurrent renal calculi. Trends in Urology and Men's Health 5: 36-38. doi:10.1002/tre.408

Grohe B., Rogers K. A., Goldberg H. A. and Hunter G. K. 2006. Crystallization kinetics of calcium oxalate hydrates studied by scanning confocal interference microscopy. Journal of Crystal Growth 295: 148-157.

Khan A., Bashir S., Khan S.R. and Gilani A.H. 2011. Antiurolithic activity of Origanum vulgare is mediated through multiple pathways. BMC Complementary and Alternative Medicine 11: 96-111.

Miyaoka R. and Monga M. 2009. Use of traditional Chinese medicine in the management of urinary stone disease. International Brazilian Journal of Urology 35(4): 396-405.

Rajendran K. and Keefe C.D. 2010. Growth and characterization of calcium hydrogen phosphate dehydrate crystals from single diffusion gel technique. Crystal Research and Technology 45(9): 939-945.

Suguna K., Thenmozi M. and Sekar C. 2012. Growth, spectral, structural and mechanical properties of struvite crystals grown in presence of sodium fluoride. Bulletin of Materials Science 35(4): 701-706.

Yam M.F., Lim C.P., Ang L.F., Por L.Y., Wong S.T., Asmawi M.Z., Basir R. and Ahmad M. 2013. Antioxidant and toxicity studies of 50% methanolic extract of Orthosiphon stamineus Benth. BioMed Research International, article ID 351602. http://dx.doi.org/10.1155/2013/3516022. (accessed on 26 June 2015).

Yam M.F., Basir R., Asmawi M.Z. and Ismail Z. 2007. Antioxidant and hepatoprotective effects of Orthosiphon stamineus Benth.standardized extract. American Journal of Chinese Medicine 35(1): 115-121.

MATEC Web of Conferences **58**, 01014 (2016) DOI: 10.1051/matecconf/20165801014
BISSTECH 2015

NUMERICAL OF BIOETHANOL PRODUCTION FROM LIQUID WASTE OF RISE FLOUR BY DISTILLATION PROCESS

Sari Ni Ketut*, I Nyoman Abdi, Wesen Putu, Retno Dewati
University of Pembangunan Nasional "Veteran" Jawa Timur, East Java, Indonesia
E-Mail: sari_ketut@yahoo.com

ABSTRACT

The availability of liquid waste of rise flour can be maintained in continuously and abundant, and is one of waste water that is underutilized, but liquid waste of rise flour has a starch content which can be used as one of the producers of ethanol. This study has the objective to produce bioethanol and a prototype bio-ethanol industry, and perform simulations before conducting experiments, as well as a reference in developing the bio-ethanol industry in Indonesia.
Based on preliminary analysis obtained starch content of the rice is about 79% b/b, while the starch content of the liquid waste of rise flour is quite high about 30% v/v. The process of preliminary research that has been done is the hydrolysis with Bacillus then followed by bacterial fermentation of glucose by Saccharomyces cerevisiae. The quality of ethanol produced is influenced by the levels of starch, glucose, acidity (pH), the volume ratio of waste with Bacillus volume, the number of bacteria Saccharomyces cerevisiae, fermentation time that has been done in the preliminary study. The filtrate resulting from the hydrolysis and fermentation processes of contains ethanol about 16% v/v carried out a batch distillation process to obtain bioethanol, variables in a batch distillation process is ethanol composition and temperature. Before experimental batch distillation process simulation ternary system Ethanol-Water-Impurities (EWI) with a model of Differential-Algebraic-Equations (DAEs), using the programming language Mathlab version 7 in advance, so that in the determination of the variables can be more effective and more research costs cheap.
The results obtained experimentally study the composition of bioethanol a maximum of 95% to 96%, the results of experiments and simulations EWI ternary system form the temperature profile, the profile of the composition of liquid and vapor composition profile dimensionless time functions both at the bottom and in the distillate shows the results of the same approach. The simulation results before used reference in experiments performed the validation beforehand, so that the ternary system simulation EWI after validation of reference can be used in experiments.

Keywords: batch distillation, bioethanol, DAEs, liquid waste of rise flour

INTRODUCTION

Conventional crops such as corn and sugar cane cannot meet global demand for bio-ethanol production, because their main value on food and animal feed. Therefore, agricultural waste is an attractive feedstock for bioethanol production. Agricultural wastes are cost-effective, renewable and abundant. Bioethanol from agricultural waste could be a promising although the process has several challenges and limitations such as biomass transportation and handling. Efficient methods of pretreatment with high in sugar, fermented using enzym, as well as improve the efficiency of the whole process. The conversion of glucose and xylose to ethanol requires new fermentation technology, to make the entire process cost effective (Nibedita, *et al.*, 2012).

The production of bioethanol from sago pith waste (SPW) using microwave hydrothermal accelerated hydrolysis by carbon dioxide. Structural changes in the SPW examined after hydrolysis, the purity of ethanol after fermentation, and distillation. Energy consumption is evaluated for microwave hydrothermal, maximum 43.8% and 40.5% glucose theoretical ethanol yield is obtained. Fermentation ethanol yield coefficient was 0.47 (g ethanol per g glucose) were 15.6 SPW ethanol per 100 g of dry (Saravana, *et al.*, 2014). Starch is a component that is more complex than a disaccharide, before fermentation, starch must be broken by using amylase in maltose, a disaccharide component. By using enzymes, maltase and maltose is hydrolyzed to glucose (Karimi, *et al.*, 2006).

In the process of separation, thermodynamic data of the most dominant influence on the performance of the process is the equilibrium phase. One of the modern thermodynamics correlation in behavior mempersentasikan mix is not ideal UNIQUAC equation, the approximate equilibrium ternary and quaternary systems can be done only based on experimental data of binary systems. Models UNIQUAC activity coefficient equations developed from a binary mixture, and has the advantage for application in a mixture of multi-component system because it only requires a binary parameter (does not require additional parameters). But the loss of the model does not always succeed in estimating equilibrium multi-component system that shows a mixture which is not ideal especially for couples with a mix that has limited solubility such as butanol-water. To overcome this necessary equilibrium data measurement accurately binary system and the estimated model parameters of the model so that the activity coefficient of these parameters can be used to estimate the vapor-liquid equilibrium multi-component system accurately (Handogo, *et al.*, 1997).

In the chemical industry, the fermentation process is one way to get chemical compounds with the aid of microorganisms, fermentation products entering the next stage of separation. At this stage it is very important to produce a product with a certain purity, one of the tools commonly used in the separation process is a batch distillation column. Separation processes in the industry generally separation of multicomponent and binary separation rare, therefore it is very important to review the multicomponent batch distillation. Design of multi-component batch distillation is generally obtained by performing a simulation, in order to obtain simulation results closer to the actual state of the thermodynamic accurate data is needed.

Theory of Thermodynamic and Separation

In the chemical industry, the fermentation process is one way to get chemical compounds with the aid of microorganisms, fermentation products entering the next stage of separation. At this stage it is very important to produce a product with certain purity, one of the tools commonly used in the separation process is a batch distillation column. The separation process in the separation of multicomponent general industry and rarely binary separation, therefore it is very important to review the multicomponent batch distillation. Design of multi-component batch distillation is generally obtained by simulation, in order to obtain simulation results closer to the actual state of the accurate thermodynamic data is needed.

In the separation process, the data thermodynamics of the most dominant influence on the performance of the process is the equilibrium phase. One of the modern thermodynamic correlations in presentation behavior is not ideal mix UNIQUAC equation, the approximate equilibrium ternary and quaternary systems can be done only by the experimental data of binary systems. Activity coefficient models with UNIQUAC equation was developed from a mixture of binary, and has advantages for applications in multi-component mixtures because the system requires only binary parameters (does not require additional parameters). But the loss of the model does not always succeed in estimating equilibrium multi-component system that shows the mix is not ideal especially for couples with a mix that has limited solubility such as butanol-water. To overcome this necessary equilibrium data measurement systems and models of binary accurately estimate the parameters of the model so that the activity coefficient of these parameters can be used to estimate the equilibrium vapor-liquid multi-component system accurately (Handogo, *et al.*, 1997).

Iterative procedure to find the temperature of the bubble that is looking for a saturation temperature of the pure components temperature saturated (Ti) on pressure (P) (Prausnitz, *et al.*, 2001). T price as the starting price will be used to determine the saturated vapor pressure of a substance to be estimated with the Antoine equation. Activity coefficient γi obtained from: UNIQUAC equation. Model-Differential-Algebraic Equations (DAES) for batch distillation ternary system, assuming it does not form a two-phase liquid by Doherty and Perkins (1978) as follows:

$$d\xi = \frac{-dx_i}{(x_i - y_i)} \qquad (1)$$

$$\xi = \ln\left(\frac{W_o}{W}\right) \qquad (2)$$

With forward-finite-difference, Equation (2) becomes:

$$x_{i,j+1} = x_{i,j} + (y_{i,j} - x_{i,j})\,\Delta\xi \qquad (3)$$

The liquid composition initially at bottom (xi, j) and $\Delta\xi$ determined, while the vapor composition (y_i, j) is calculated using Equation bubble-T (Prausnitz, 2001).

METHODOLOGY

System simulation is studied for EWI ternary system with the composition shown in Table 1.

Table-1. The feed composition EWI

Feed composition (fraction mole)			
Run	Water	Butanol	Impurities
1	0,8	0,1	0,1
2	0,7	0,2	0,1
3	0,7	0,1	0,2
4	0,6	0,3	0,1
5	0,6	0,1	0,3
6	0,5	0,1	0,4
7	0,4	0,1	0,5

To calculate the saturated vapor pressure is used equations Antoine, Antoine parameter data is shown in Table 2. (Prausnitz, 2001), in which the temperature (T) in units of K and the saturated vapor pressure (PSAT) in units of Bar. To calculate the activity coefficient (γ) using equation UNIQUAC, where the price of binary interaction parameters UNIQUAC (uij), molecular volume data (r) and molecular surface area (q) shown in Table 3, assuming the price of z = 10.

Table-2. Parameters *Antoine* EWI

	Parameter Antoine		
Component	A	B	C
Impurities	4.23	4.90	5.37
Water	207.01	1475.14	1758.22
Ethanol	228.06	182.74	230.92

Source: Prausnitz, 2001

Table-3. EWI ternary system parameters UNIQUAC

Component	r	q
Impurities	2.3635	2.3279
Water	3.4812	3.0533
Ethanol	2.1054	1.9720

Source: Prausnitz, 2001

Feed composition shown in Table 1 to calculate the saturated vapor pressure is used equations Antoine,

Antoine parameter data is shown in Table 2 (Prausnitz, 2001), in which the temperature (T) in units of K and the saturated vapor pressure (P^{SAT}) in units of Bar.

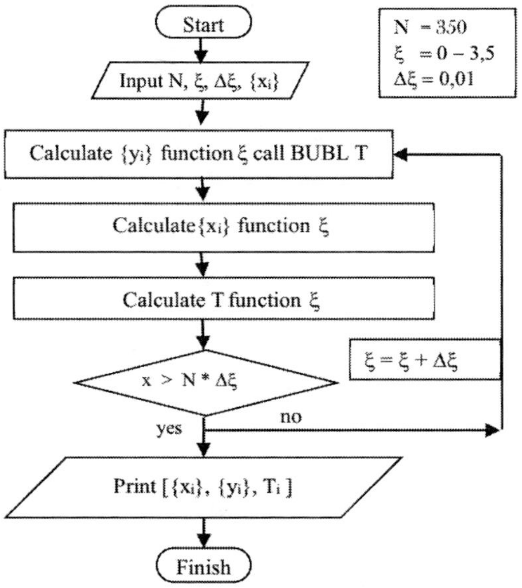

Figure-1. Algorithm batch distillation

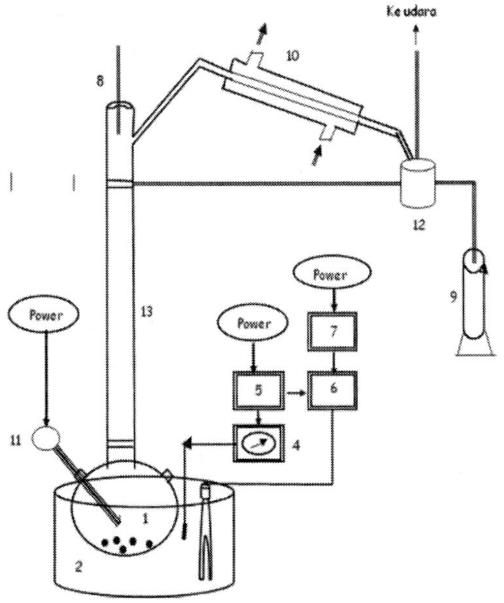

Figure-2. A set of simple batch distillation apparatus

Caption:

1. Still Pot	8. Thermometer
2. Oil batch	9. Measuring cup
3. Heater	10. Condensor
4. Thermocouple	11. Thermocouple RTD
5. Controller	12. Separator column
6. Contactor	13. Packed column
7. Regulator	

Experimental procedures EWI ternary system using a simple batch distillation, as follows:

a. The simple batch distillation equipment used as in Figure 2.

b. EWI determined volume of 3500 ml mixture with EWI feed composition (mole fraction) are shown in Table 1.

c. The simple batch distillation process is operated in a state of total reflux.

d. Once the temperature at the bottom with temperatures approaching the same distillate, taken a few drops of volume in the bottom of each ξ = 0, 1, 2, 3

e. EWI Then distillate composition analyzed using Gas Chromatography

f. Once the volume in the still-pot approximately 10%, then a simple batch distillation process is stopped.

RESULT AND DISCUSSION

Temperature profile at the bottom shows the results closer to the temperature in the distillate, because the batch distillation process is simple to operate in conditions of total reflux. Therefore, the temperature profile simulation results as shown in Figure 3.

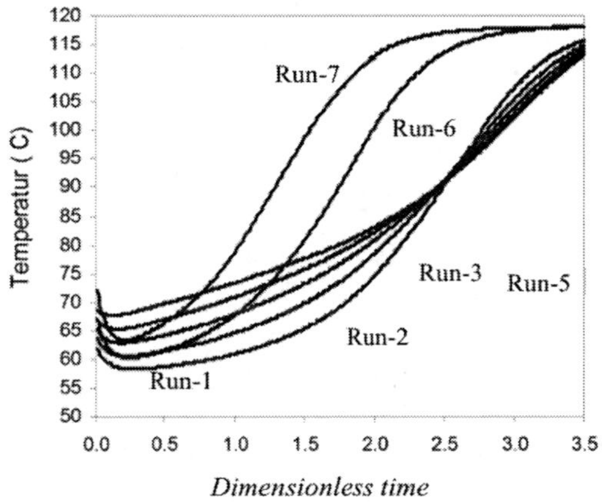

Figure-3. The temperature profile EWI ternary system for Run-1 to Run-7

From Figure 3 shows the temperature profile EWI for Run 1 to Run-7 as a whole rose against the dimensionless time. This is because the components that evaporated with the larger portion components of ethanol, so it takes a greater temperature to vaporize the components butanol which has not evaporated.

DOI: 10.1051/matecconf/20165801014

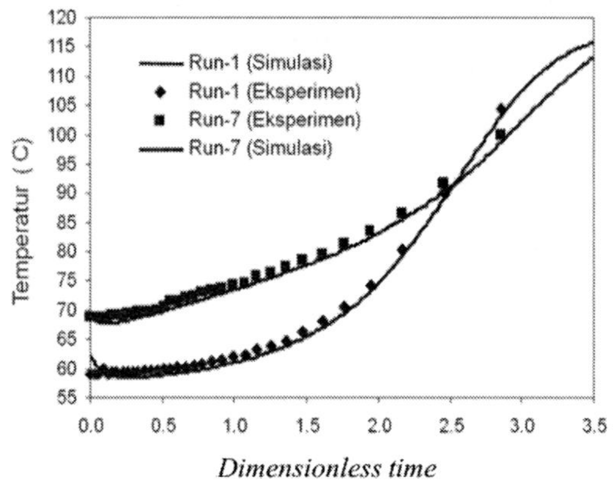

Figure-4. The temperature profile EWI ternary system for Run-1, Run-7 simulation and experimental results

Figure 4 is divided into three areas: one region of activity coefficient for butanol showed declining profile and activity coefficient of ethanol rises so does not affect the temperature profile. Figure 4 shows that the greater the composition of butanol in the feed entering the maximum ethanol influence the activity coefficient when the dimensionless time showed 1.75; after declining in the end a constant profile. The bigger the bait composition of butanol at entry showed a steeper slope of decline in temperature, because the composition of butanol in the feed mixture entering EWI greatly affect the temperature, the greater the composition of butanol in the feed mixture entering the higher temperatures.

For two region temperature profile rose for all the run, the greater the ethanol composition of the feed entry indicates temperature rise slope more gentle. This is because the composition of the ethanol in the incoming feed EWI greatly affect the temperature of the mixture, the greater the ethanol composition of the feed mixture entering the smaller temperature.

For three regions rising temperature profile and most constant, for the temperature rise profile means the separation of a mixture of ethanol EWI still has not evaporated, while for the constant composition profile means it's all the ethanol evaporated. The greater the composition of butanol in the feed mixture entering the temperature the greater and faster the temperature profile constant.

For regions 3.5 dimensionless time temperature profile is not constant, so it takes a dimensionless time greater to obtain a constant temperature profile. If a simple batch distillation continued until the dimensionless time is very large, it will obtain the value of the composition of each mixture negative. Because the ethanol component is volatile components while the components butanol is non-volatile components, when a simple batch distillation process ethanol component in a larger portion is evaporated and the remaining components of butanol. Butanol component has not been approached purely with dimensionless time 3.5 way to obtain a more pure butanol component of the initial composition is to enlarge the composition of butanol in the feed entry.

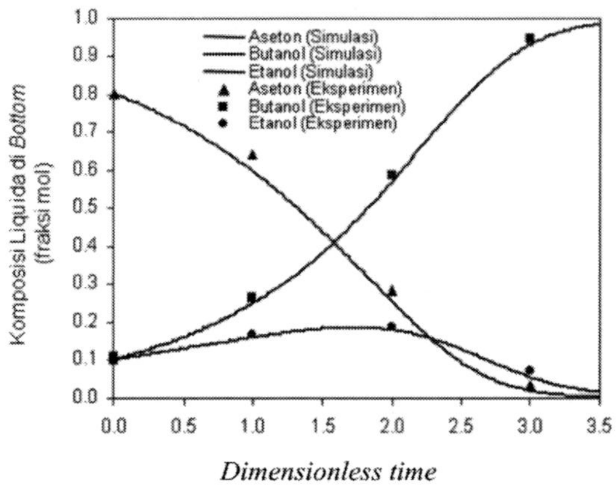

Figure 5. Profile composition of liquid in the bottom ternary systems ABE-1 Run simulation and experimental results.

The composition of the liquid in the bottom profiles and temperature profiles EWI ternary system simulation and experimental basis, to Run -1 indicates that approaching the same profile. For binary systems research has been conducted by Sari, *et al.* (2007), wherein the composition of the liquid in the bottom profile and temperature profile of a binary system in simulations and experiments showed a similar approach, with an average deviation of 0.0063.

CONCLUSIONS

ABE temperature profile as a whole rose to dimensionless time, except at the beginning of the process shows the temperature profile decreases, due to the nature of the characteristics of the mixture.

The composition of the liquid in the bottom profile and temperature profile between the simulation results with experiments show that profiles the same approach, with an average deviation 0.0063

ACKNOWLEDGEMENT

The authors would like to acknowledge the financial support of the Ministry of National Education of the Republic of Indonesia with the National Strategic Competitive Grant, Contract Number: 180/SP2H/PL/DIT.LITABMAS/V/2013.

REFERENCES

Doherty and Perkins. 1978. Model Differential-Algebraic-Equations (DAEs), John Wiley & Sons, Inc., New York.

Handogo, R. and G. Wibawa. 1997. Experiments and Correlations of Vapor-Liquid Equilibria of Acetone-1-Butanol-Ethanol Ternary Mixture, International Conference on Fluid and Thermal Energy Conversion. (pp. 587-592). Yogyakarta, Indonesia.

Henley, E. J., J. D. Seader. 1998. Separation Process Principles (pp. 586-712). John Wiley & Sons, Inc., New York.

Karimi, K. Kheradmandinia S., Taherzadeh MJ. 2006. Conversion of Rise Straw to Sugars by Dilute Acid Hydrolysis. Biomass and Bioenergy, 30, 247-253.

Nibedita Sarkar, Sumanta Kumar Ghosh, Satarupa Bannerjee, Kaustav Aikat. 2006. Bioethanol production from agricultural wastes: An overview, Renewable Energy, 37, 19-27.

Sari N. K., Kuswandi, Nonot S., Renanto Handogo. 2006. Comparation Residu Curve Maps System Ternary ABE with Metanol-Etanol-1-Propanol, Journal: REAKTOR, 13(2), 6-11. Chemical Engineering Department, UNDIP, Semarang, Indonesia.

Sari N. K., Kuswandi, Nonot S., Renanto Handogo. 2007. Separation Binary System Ethanol-Water and Ternary System ABE with Simple Batch Distillation, Jurnal INDUSTRI Jurnal Ilmiah Sains dan Teknologi, 6(5), 28-34. Faculty of Engineering Industrial, ITS Surabaya, East Java, Indonesia.

Saravana Kannan Thangavelu, Abu Saleh Ahmed, Farid Nasir Ani. 2014. Bioethanol production from sago pith waste using microwave hydrothermal hydrolysis accelerated by carbon dioxide. Applied Energy, 128, 277-283.

Prausnitz, J. M. 2001. The Properties of Gases and Liquids, ed. 5, p. A.50-A.51, Mc. Graw-Hill, New York.

Widagdo, S., Warren D. Seider. 1996. Journal Review: Distillation Azeotropic. AIChE Journal, 42(1), 96-130.

PYROLISATOR COAL TO BE COKES (COAL COKES) CASTING METAL INDUSTRY STANDARD

Sukamto*, Didi Samanhudi, Sunardi, Nana Dyah Siswati.
University of Pembangunan Nasional "Veteran" Jawa Timur, Surabaya, East Java Indonesia
E-mail : kamtonep@gmail.com

ABSTRACT

Pyrolisis of coal is partial combustion to reduce total moisture, volatile matter and sulfur contens and increase the calorific value of coal. The results of pyrolysis of coal is coke. At the laboratory level studies, pyrolisis done in batch using different calorie, namely 5800, 6000, 6300 kcal/kg and a time of 15-60 minutes and the temperature 400-800°C. Maximum results obtained total moisture (0.44%), fixed carbon (89%), volatile matter (2.4%), sulfur content (undetected) and ash (7.2%). Then applied to the scale miniplant with continuous processes using multitube pyrolisator which are designed to operate in the temperature range 400-800oC and a flow rate of 240-730 kg/h, obtained coal cokes that meets industry quality standards, namely TM (0.42%), FC (90.40%), VM (2.16%), S (not detected), Ash (6.8%) incalori 6300 kcal/h, a flow rate of 240 kg / h and temperatures between 600-700°C

Keywords: coal, coke, pyrolisis, solvenisasi, total moisture, fixed carbon, volatile matter, sulfur content.

INTRODUCTION

Utilization of coal as an alternative energy source Fuel needed considering Indonesia has coal reserves are enough. Of national coal production on average per year amounted to 131.72 million tons, which is used in the country is only 32.91 million tonnes / year, while the rest, as many as 92.5 million tons exported abroad (Pusdatin, 2014). The types of coal that exist in Indonesia, including anthracite by 0.4%, 14.4% bituminous, sub-bituminous 26.6%, lignite amounted to 58.6%, and the remainder in the form of peat. Utilization of coal used tend to be of high quality coal types of anthracite and bituminous because both have a high calorific value and if used continuously will make coal resources will be depleted. Therefore, researchers are working to increase the economic value of coal into coal cokes.By obtaining coal will reduce or as a substitute for imported that had been totally still imported.

Pyrolysis is the chemical decomposition process of coal by using heating in the absence of oxygen, in which the raw material will undergo breakdown of the chemical structure into the gas phase. In the pyrolysis process will be obtained in the form of solid residues which yield high carbon tar and oil and gas high levels of hydrogen to be used to convert into liquid fuel (Hidayat, 1995). Pyrolysis is a special case of thermolysis. In many industrial applications, the process is performed under pressure and operating temperatures above 430oC (806oF) (Novitasari, W., 2015). In the pyrolysis of coal to produce gas feedstock such as CO, CO_2, NOx, and SOx. Pyrolysis extremes, which only leaves a carbon residue, is called carbonization. Coal carbonization is the process of heating up the temperature and time (ranging from-above 200°C-1000°C in low oxygen conditions to remove the volatile matter content of the coal so that the resulting solids in the form of charcoal or cokes or semi cokes with tar and gas byproduct. Cokes produced is usually marketed as a domestic smokeless fuel which is produced in the form of a lump or powder containing 8-20% volatile matter (daf). Carbonization main function is to increase the calorific value, since the release of the water content, as well as the formation of tar that can function as a coating film that prevents re-absorption of water content.

The study aims to improve the quality of coal into cokes with quality standards industry and households, using coal pyrolisis process continuously in order to obtain optimum or maximum parameter and will generate pyrolisis products such as cokes, light gas, light oil and heavy oil. Parameters used as determinants of quality coal cokes is a total moisture (TM), fixed carbon (FC), volatile matter (VM), sulfur content (S) and Ash.

METHODOLOGY

Materials and equipment

Materials used in this study is coal that has a calorific value of 5800-6300 cal / kg, less glossy black compact, relatively high carbon content, water content, ash, and a little sulfur, obtained from Kalimantan. While the equipment used in this study is choper (crushers), pyrolisator multitube which has been equipped with temperature control, bucket elevators, and a condenser.

Solvenisasi process

Coal calori 5800 cal / kg, 6000 cal / kg and 6300 cal / kg, destroyed up to a size of 2-5 cm, then soaked 1 s / d 2 hours with solvent Amoks Chem.04 at concentrations of 0.4%, then dried to dry, so that the coal has been ready for use as a raw material to be used as a pyrolysis process.

Pyrolysis process

Coal (semi-coke) from the solvenisasi process weighing 200 kg fed into pyrolisator multitube, burned at the temperature varies (400 °C, 500 °C, 600 °C, 700 °C, 800°C), flow rate also varies (240 kg/h, 360 kg/h, 540 kg/h, 730 kg/h).

RESULT

Table-1. Composition of the start of coal at various calorie

Coal calories (kcal/kg)	**Weight(%)**				
	TM	FC	VM	S	Ash
5800	11	41.2	43.18	0.7	4
6000	10.60	41.30	39.00	0.9	6.4
6300	10.80	43.42	38.85	0.9	6.02

Results of analysis of total moisture of cokes in this study on a variety of variables stable, ranging from 0.4 to 0.6%.

Table-2. Fixed carbon(%) various calorie coal and the temperature at a flow rate of 240 kg/h.

Coal calories (kcal/kg)	Temperature (°C)				
	400	500	600	700	800
5800	78.30	86.06	89.00	89.20	89.60
6000	78.00	88.32	90.20	90.46	90.80
6300	78.05	88.00	90.40	90.46	90.82

Table-3. Levels of volatile matter(%) various calorie coal and the temperature at a flow rate of 240 kg/h.

Coal calories (kcal/kg)	Temperature (°C)				
	400	500	600	700	800
5800	17.01	7.8	3.80	2.9	2.2
6000	16.2	5.80	3.42	2.40	2.32
6300	16.02	6.40	3.60	2.39	2.32

Results of analysis of sulfur content of various types of calories, the temperature and the flow rate of cokes, undetected.

Table-4. Ash content (%) of various calorie coal and the temperature at a flow rate of 240 kg/h

Coal calories (kcal/kg)	Temperature (°C)				
	400	500	600	700	800
5800	4.60	6.14	6.70	7.40	7.70
6000	5.80	5.88	5.88	6.68	6.38
6300	5.90	5.60	5.50	6.65	6.36

DISCUSSION

The results obtained from this study compared to the composition of the coking coal industry standards, to determine whether eligible or not.

Table-5. The composition of cokes for Standard Industry

No	Parameter	% Berat
1	Total Moisture (TM), max	1
2	Fixed Carbon (FC), min	85
3	Volatile Matter (VM), max	2.38
4	Sulfur Content (S), max	0.01
5	Ash, max	12

When compared to the industry standard coking coal as shown in Table 5, the results of which are eligible are in calories 6300 kcal / kg, between 600-700°C temperature and flow rate of 240 kg/h. The lowest flow rate allows a long contact time resulting in more complete combustion than at high flow rates (short contact time).

Total Moisture (TM)

As it is known that the water damp or moisture (moisture) is regarded as one of the impurity (impurity or a diluent) were just as ash in the coal that is forming components of coal which can not be burnt (being incombustible) so as to lower the commercial value in the form of parameter values calories (calorific value = CV). Therefore, the effect on the quality of coal must be strictly controlled in order to be sold (salable coal) as a result of the levels that have met the requirements or specifications of coal demanded by the market or consumer. (timorhauniarain, 2013).

Fixed carbon (FC)

Fixed carbon is the fixed carbon contained in the coal after volatile matters separated.Shown in Table 2, the fixed carbon in the cokes produced, the greater as the calorific value of coal used, this is in accordance with Indah, 2009, that the value of the carbon content is increasing along with the increasing quality of the coal.

Volatile Matter (VM)

Volatile Matter is a compound in a volatile coal at a certain temperature under standard conditions.Results of analysis of volatile matter in cokes as shown in Table 3, the greater the calorie coal and the higher the combustion temperature, the volatile matternya getting smaller, it is because the volatile matter is that many substances that are lost when the coal sample is heated at a temperature and a predetermined time (after corrected with moisture content). Factors affecting the results of the determination of volatile matter the temperature, time, velocity, warming, the spread of grain and particle size (Simorangkir, T., 2013). Volatile matter is used as a measure of the quality of coal (Palupi, I., 2012). The higher the coal rank, the lower the levels of volatile matternya (Indra, 2015).

Sulfur Content (S)

Results of analysis of sulfur at various variables is not detected, because sulfur is converted to sulfur oxides during the combustion process that can cause corrosion and scaling of equipment also causes air pollution. Sulfur is used for the evaluation of the use of coal combustion (Palupi, I., 2012).

Ash Content (ash)

Shown in Table 4, that the greater the calorie coal, the ash content produced smaller, or conversely the lower calorie coal, the higher ash content, which is in line with Rismayanti, 2012, that the higher the ash content in the coal type the same, the low caloric value.

Applications on a scale miniplant

For miniplant scale coke production process has been carried out at the optimal operating conditions according to the results of research namely the use of coal 6300 kcal/kg in pyrolisis at temperatures between 600-700°C and a flow rate of 240 kg/h. Process equipment designed by the researchers, is designed on a small scale with supporting equipment (auxilang equipment) as buchet elevators, condensor and blower as gas suction. Equipment and supporting the scheme shown in Figure 1.

DOI: 10.1051/matecconf/20165801015

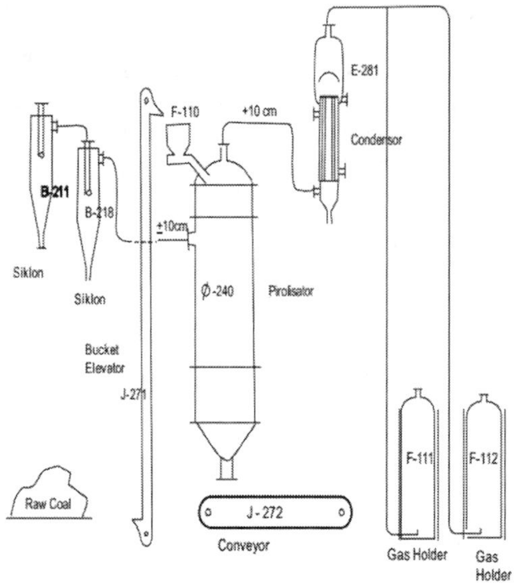

Figure-1. Diagram of the industry process of cokes

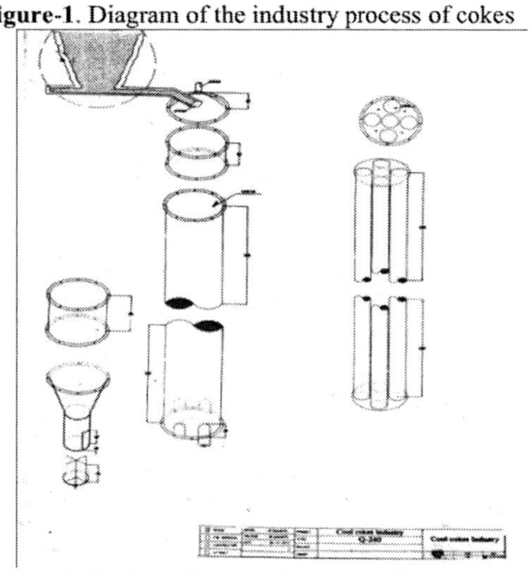

Figure-2. Design of the multi tube pyrolisator

Table-6. Specification of equipment :

No	Name	Size
1.	Disc Head	Stainliess steel 306 3 mm thick. ⌀ = 80 cm, 45 ° tilt angle, and height 20 cm.
2.	Tube evaporation zone	Steinliess steel 306 3 mm thick. ⌀ = 80 cm, height 20cm.
3.	Furnace tube Pyrolisator	Stainliess steel 306 3 mm thick. ⌀ = 80 cm, height 20 cm. - Gas stove 4 units with specifications 50000-150000 calories / kg - Isolator ± 160 m of asbestos and gypsum.
4.	Temperature control	Digital temperature control with a temperature range up to 1100o C.
5.	Pipe (tube) pyrolisator	Stainless steel 306, unit 5, pipes ⌀ 4 ", height 240 cm
6.	Tube cooling zone	- Stainliess steel 306 3 mm thick. ⌀ = 80 cm, height 60 mm. - Conical stainless steel 306 3 mm thick. - Discharge pipe stainless steel 306, equipped with screw discharge angle of 45o ⌀ 8 ".
7.	Motor Drive	Power 1.5 hp, 1800 rpm
8.	Reducer gear 30	Rpm regulator equipped with pully
9.	Bucket Elevator	Height 6.30 cm.
10.	Condensor	
11.	Gas Holder	

Results of analysis of the coke production process, the obtained coke that meets industry standards, as shown in Table 7.

Table-7. Composition of cokes mini-plant scale at a flow rate of 240 kg / h and temperatures between 600-700°C.

Parameter	Coal calorie (kcal/kg)		
	5800	**6000**	**6300**
TM max	0.43	0.40	0.42
FCmax	89.02	90.2	90.40
VM max	2.56	2.37	2.16
S max	undetected	undetected	undetected
Ash, max	7.2	9.80	6.80

CONCLUSIONS

Research that has been carried out, showing that optimum results in the treatment of variables 6300 calories, a flow rate of 240 kg / h and temperatures between 600-700oC with a Total Moisture 0.42%, Fixed Carbon 90.40%, Volatile Matter 2.16%, Sulfur Content undetected and 6.8% Ash. Pyrolisator multitube goes well, by adjusting the operating conditions of temperature and flow rates have been determined. Capacity and quality can be maintained so that the products cokes meet the standard metal casting industry or can be used as a chemical industry to another.This study of the data obtained and the phenomena that occur (at temperatures of 700°C and 800°C), required a more adequate planning, especially at the rate of condensation

REFERENCES

Andre, A., 2012. Analysis of samples of coal from South Sumatra. Chemical Analysis Program of Bogor Agricultural Diploma program.

Anonymous, 2008. Coal-alternative-energy-replacement, Semarangantenan.blogspot.com

Anonymous, 1999. Coal, www.scribd.com

Anonymous, 2012. Literature-materialism Inorganic-In-Coal, www.scribd.com

Anonymous, 2013.http://www.alpensteel.com/article/51-113-energi-lain-lain/2336--reserve-fuel-oil-diminish.html

Bayuseno, A.P.2009. Effect of Physical Properties and Structure of the Local Mineral Coal Combustion properties. Faculty of Engineering, University of Diponegoro, bosstambang.com

Hadijah, NR and Damayanti, R., 2006. Research Abu coal as Pembenah Land, Mineral and Coal Technology Journal, No. 36, Year 14, January 2006

Idhamds, 2008. Analysis of coal. idhamds.wordpress.com
Indah, 2009. Analysis of coal quality. indah4din4t4.wordpress.com

Indra, 2015. The coal quality parameters. www.academia.edu.

Kadir, A., 2010. Energy. Publisher University of Indonesia (UI Press), third edition.

Narozana, 2009. Preparation of liquid smoke by pyrolysis method. asapcairsebagaipengawet.blogspot.com

Novitasari, W., 2015. Pyrolysis. www.academia.edu
Palupi, I., 2012. The analysis of coal samples. irlandapalupi.blogspot.com/2012/03/analisa-sample-batubara.html.

Putrago, 2009. Definition-resource-and-reserve coal, putrago.blog.akprind.ac.id

Sari, NL, 2009. Potential of Coal Indonesia. Journal of Environment, ed. August 2009, uwityangyoyo.wordpress.com/tag/batubara

Simorangkir, TA, 2013. Proximate Analysis, Ultimate Analysis and Miscellaneous Analysis on coal. Mineral Mining Engineering Faculty of Technology Institute of Technology Medanta.

Soetjijo, H.2004. Low Levels of Coal Quality Improvement (Upgrading Low Grade Coal). Research Centre for Geotechnology – LIPI

Suganal, 2009. Coal Briquette Making Process Design nonkarbonisasi small scale of High ash coal. Mineral and Coal Technology Journal, Volume 5, Number 13, January 2009. ISSN 1979-6560

Timorhauniarain, 2013. Characterization of coal quality). timorhauniarain.blogspot.com (2013).

NATURAL MEDIUM FOR GROWING OF ENDOPHYTIC BACTERIA FROM SOLANACEAE IN MALANG-INDONESIA

Arika Purnawati.[1] Wiwik Sri Harijani.[2] and Wiwin Windriyanti.[3]

Faculty of Agricultural UPN "Veteran" East Java, Indonesia
E-Mail: ikahptupn@gmail.com

ABSTRACT

Endophytic bacteria are important microorganisms having potential as biocontrol agents for many pathogens. Until now, the growth of it always uses semi-synthetic or synthetic medium so it was difficult to be used by farmers in the field and it was expensive to have its propagation as biocontrol agents. Based on the problem, this research will study the natural medium as propagation medium of Endophytic bacteria. It had natural ingredients such as soybean, chicken broth, egg, worms, snail, sorghum and they were easy to get by farmers. This study used endophytic bacteria from Solanaceae in Malang- Indonesia. Four isolates of endophytic bacteria were grown in agar and liquid medium with ingredients of corn flour, soybean flour, sorghum flour, snail flour, and worm flour. There is no difference in the incubation period, color, shape, and surface colony. The population in medium with snail flour ingredients at a concentration of 10^7 cfu/ml is the highest and snail flour is the best medium for growing endophytic bacteria.

Keywords: natural medium, endophytic bacteria

INTRODUCTION

Endophytic bacteria have been found in every plant, where they colonize the internal tissues of their host plant and can form a range of different relationships including symbiotic, mutualistic and commensalistic. Endophytic bacteria can also be beneficial to their host by producing natural products that be potential use agriculture, having been isolated from flowers, fruits, leaves, stems, roots and seeds of various plant species (Christy and Sudha, 2014; Kobayashi and Palumbo, 2000).

The growth of it is influenced by the medium and growth factors. Growth factors are required in small amounts by cells because they fulfill specific roles in biosynthesis. The need for a growth factor results from either a blocked or missing metabolic pathway in the cells and can be added to culture media that are used to grow bacteria. The type of the culture medium used depends on the purpose, whether liquid or solid medium. The culture medium may be classified into several categories depending on their composition or use. Chemical (synthetic) medium is one in which the exact chemical composition is known and the semi-synthetic medium is one in which the exact chemical constitution of the medium is not known (Todar, 2012). The growth of bacteria also influenced by the diversity of nutritional types found among bacteria like inorganic salt, carbon, inorganic nitrogen, amino acid, and vitamins (Goyal, 2007) but the medium is usually expensive.

Endophytic bacteria as biocontrol agents of pathogens in Malang-Indonesia need a growing medium with a composition that is easy to use for propagation by farmers in the field and also cheap. Based on the problem, we used mediums from natural materials such as corn flour, soybean flour, sorghum flour, snail flour, and worm flour. All of them contain growth factors such as organic and inorganic elements, vitamins, and other elements (Ahmad, 2002; Syatrawati, 2008).

The object of this experiment is to study the natural medium for growing of endophytic bacteria from Solanaceae in Malang-Indonesia.

MATERIAL AND METHOD

Isolates of endophytic bacteria

Endophytic bacteria used for the experiment was a personal collection Dr. Arika Purnawati. Bacteria culture was purified on NA (Difco) : 8 g / l, pH 7.0, at 24 hours age is used for experiments.

Natural Medium

Corn flour. Corn was shelled then cleaned, dried (1-2 days) at 50^{0}C until having 15-18% moisture content. Provision of flour was done using a sieve size of 50 mesh. **Soybean flour.** Soybean was washed with clean water flow until it is clean. It was then soaked in water (4 hours) while squeezed to clean the epidermis. Soybean then was washed again with water and drained (15 minutes). It was dried (1-2 days), roasted (10-15 minutes) and ground into the flour. **Sorghum flour**. Sorghum seeds were cleaned then dried to 20% moisture content and soaked in water (8 hours) and drained until 16% moisture content. **Snails flour**. Snails fresh meat was removed from the shell then dried through drying in the sun (3 days) or dried using a dryer until having 14% moisture content. After the snail meat was dried, it then was milled using a grinder into flour. **Worm flour.** Worms were washed then boiled in boiling water (3 minutes) then drained. After the leak, the worms were cut into pieces of 1 cm and being washed again. The pieces were dried using an oven at 500C (4 hours) and mashed using a mortar into flour. **Achatina flour**. Achatina was allowed to stand for 2 days and 2 nights, added with salt and stirred (15 minutes) then drained (15 minutes). It was then washed and boiled (20 minutes), drained, and dried. The process was repeated once again: washing and boiling (20 minutes), draining, and drying. The end result was then sliced and ground into flour.

Natural medium test for growing. Before undergoing the experiments, each flour type was weighed as much as 20 g/l. The medium was divided into liquid medium without added agar and solid medium with added agar 20 g/l. Then the medium was sterilized using autoclave at 1210C, 1.5 atm (15 minutes) and then as much as 10 ml was poured into a Petri dish. Endophytic bacteria colonies were inoculated to the medium then incubated at 28^{0}C (48 hours). The observation was performed to investigate incubation period, color, shape, surface colony and the population of bacteria on natural medium using dilution technique at a concentration of 10^7 cfu/ml, at 24, 48, 72, and 96 hours.

DOI: 10.1051/matecconf/20165801016

The experiments used completely randomized design using 2 factors which were: the type of medium (liquid and solid) and the medium ingredients (corn flour, soybean flour, sorghum flour, snails flour, worm flour, and achatina flour) and with 3 replications. The data is then analyzed statistically using LSD 5%.

RESULTS

According to the experiment, the incubation period was 24 hours, whereas the colony bacteria on all natural medium were white in color, having round shape with shiny surface (Fig 1) with the population on each natural medium is shown in (Fig 2)

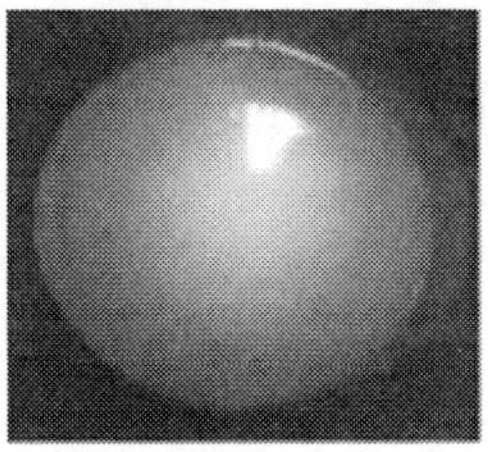

Figure-1. Microscopic colony of endophytic bacteria in natural medium (p.4.5x)

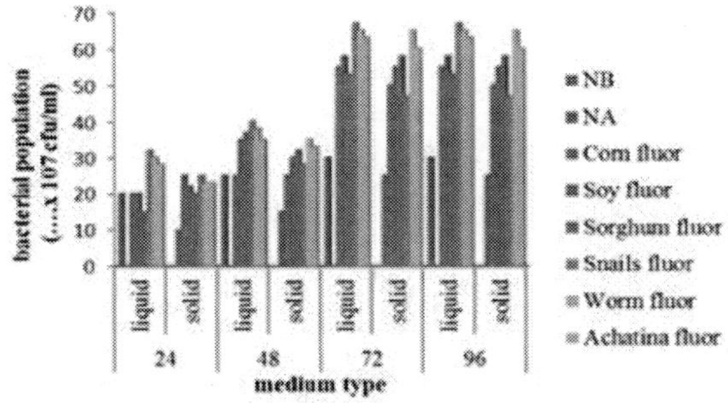

Figure-2. Population of endophytic bacteria in natural medium

According to Fig 2, the population of endophytic bacteria in the solid and liquid mediums at 10^7 cfu/ml is the highest on natural medium with ingredients of snail flour ($7.5x10^6$ cfu/ml) and ($1.1x10^5$ cfu/ml), respectively. The result was followed by the natural mediums with ingredients of soybean flour, corn flour, sorghum flour, worm flour and achatina flour. The population (cfu/ml) in solid natural medium are $6.9x10^6$, $6.7x10^6$, $6.5x10^6$, $7.0x10^6$, $7.4x10^6$, respectively, while in liquid medium are $1.05x10^5$, $1.03x10^5$, $1.02x10^5$, $1.07x10^5$, $1.06x10^5$, respectively. The population in the natural medium is higher than that of in control medium Nutrient Agar (NA) which is $6.0x10^6$ cfu/ml, and in Nutrient Broth (NB) which is $1.0x10^5$ cfu/ml. The difference of bacterial population in the natural medium is significant than control medium, so the bacterial population in the liquid medium is higher than that of solid medium for all natural mediums.

Beside that result, at the 24th hour, for all solid and liquid natural mediums, there was an increase of population and it increased significantly at 48th, but at 72nd there was no increase of population until at 96th.

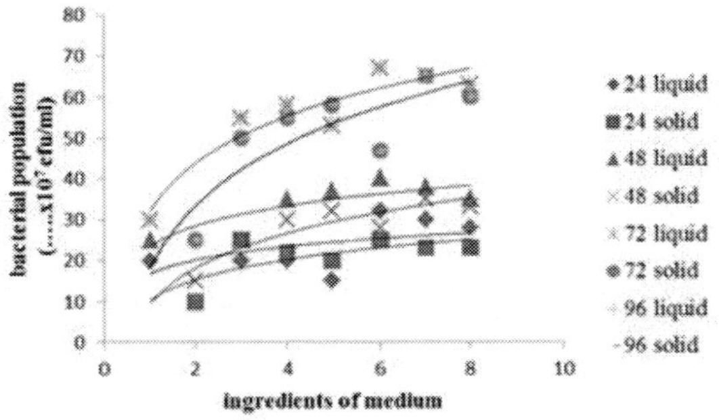

Figure-3. Curve of growing endophytic bacteria

Fig 3 shows that the bacterial growth followed the pattern of the logarithmic curve and the curve is divided into four phases: which are lag phase, log phase, stationary phase and death phase. In Lag phase (24 hours), the bacteria began to divide and perform the metabolic activity. In Log phase or exponential phase (48 hours), the mass and volume of the bacteria were increased. During this period, the growth was balanced and the speed increased. In Stationary phase (72 hours), the population was declining because the number of dead cells increased until a number of the living cells division were equal to the number of dying cells, so the number of living cells is constant, as there was no growing. In Death phase (96 hours), there was an increase in the speed of cell death until the number of living cells decreased. Despite this decline, the number of living cells was not zero; a specified minimum number of microbial cells will remain for a very long time in the medium.

DISCUSSIONS

The incubation period and colony bacteria in all of the natural mediums so in control medium are not different because nutrition in the mediums did not affect the incubation period and colony. The population of endophytic bacteria is different in various media because it was influenced by the nutritional content in the medium. The population of endophytic bacteria in the liquid medium is higher than that of solid medium because in the liquid medium the whole bacteria were mixed with the liquid medium thus affecting the acceleration of cells division and resulted in high population. Todar (2012) stated that the growth of bacteria is affected by major elements (C, H, O, N, S. P, K, Mg, Fe, Ca, Mn) and trace elements (Zn, Co, Cu, Mo). Razzak (2009) stated that bacteria will grow best on agar plates where air readily diffuse into the colony or on a liquid medium that are shaken. Todar (2012) stated that there are other factors such as temperature, pH, and water activity that affected the bacteria growth.

The research conducted by Chikere and Udochukwu (2014) mentioned the use of Nutrient Agar (NA) and Plate Count Agar (PCA) can cause an increase in the counts of endophytic bacteria and in incubation time. The counts obtained in NA were higher than that of PCA followed by the soil extract agar (SEA) which had the lowest count and a longer bacteria lag phase due to the low nutrient content in SEA. The Plate count agar formed more distinct colonies than that of NA and SEA. During the incubation period, colonies still appeared on the NA even on the 7th day. These characteristics of the NA to support a higher number of colonies for a long period of time could be attributed to its nutrient composition. Typical growth curve of the bacteria population can be divided into lag phase, exponential phase (log phase), stationary phase and death phase. In Lag phase: When a microbial population is inoculated into a fresh medium, growth usually does not begin immediately but only after a period of time called the lag phase, which may be brief or extended, depending on the history of culture and growth conditions. In Exponential Phase: This is the phase of the bacterial growth curve in which the bacterial cell numbers doubles during each unit time period. When the cell number from such as experiment is graphed on arithmetic coordinates as a function of elapsed time, one obtains a curve with a constantly increasing slope. The rate of increase in cell number is slow initially but in a later stage the cell numbers increase explosively. The rate of exponential growth varies between bacterial genera (i.e. Genetic characteristics of bacteria) and is also influenced by environmental conditions. In Stationary Phase: exponential growth cannot occur indefinitely because the essential nutrients of the culture medium are used and waste products of organisms were built up in the environment. In stationary phase, there is no net increase or decrease in cell number. The cells functions, such as energy metabolism and some biosynthetic processes, continue on. In Death Phase: the bacterial population reaches the stationary phase and the cells may start dying. Cell death may be due to cell lysis and this is also an exponential process but is much slower than that of exponential growth (microbeonline.com, 2013).

CONCLUSIONS

From the experiment, in all solid and liquid mediums, there is no difference in the incubation period and colony, whereas the population in agar and liquid medium with the ingredient of snail flour at a concentration 10^7 cfu/ml is the highest.

ACKNOWLEDGMENT

Thank you to DP2M Dikti who has supported in funding for this experiment.

REFERENCES

Ahmad, R.Z. 2002. The local media for growth nematophagous fungi as a model. Proceedings of National Seminar on Animal Husbandry and Veterinary Technology. pp : 444-449.

Chikere, C.B. and U. Udochukwu. 2014. Effect of soil bacteria. IOSR J. of Pharmacy and Biological Sciences. 9 : 6-9.

Christy, P.H. and Sudha, S.S. 2014. A study on the potential of endophytic bacteria as biocontrol agent. Scrutiny International Research J. of Biological and Environment. 1(10) : 7-11.

Goyal, D. 2007. Methods of studying microorganisms. Depart of Biotechnology and Environmental Sciences Thapar University. Patiala.

Kobayashi, D. Y., and J. D. Palumbo. 2000. Bacterial endophytes and their effects on plants and uses in agriculture, p. 199-233. *In* C.W. Bacon and J.F. White (ed.). Microbial endophytes. Marcel Dekker, Inc. New York.

Microbeonline.com, 2013. Typical growth curve of bacterial population. Download at August 2014

Razzak, L. 2009. Nutrition and growth of bacteria. www. uobabylon.edu.iq. Download at July 2014.

Syatrawati. 2008. Production of biofungicide compound witn *Gliocladium* sp. active several waste organic media. Agrisistem J. 4 (2) : 121-127

Todar, K. 2012. Nutrition and growth of bacteria. Online Texbook of Bacteriology. www. textbook of bacteriology.net. Download at July 2014.

OPTIMIZATION OF THE SINGLE STAGGERED WIRE AND TUBE HEAT EXCHANGER

I Made Arsana, Susianto, Kusno Budhikarjono, and Ali Altway*
Department of Chemical Engineering, Sepuluh ,Nopember Institute of Technology (ITS), Surabaya, Indonesia
*E-Mail: alimohad@chem-eng.its.ac.id

ABSTRACT

Wire and tube heat exchanger consists of a coiled tube, and wire is welded on the two sides of it in normal direction of the tube. Generally,wire and tube heat exchanger uses inline wire arrangement between the two sides, whereas in this study, it used staggered wire arrangement that reduces the restriction of convection heat transfer. This study performed the optimization of single staggered wire and tube heat exchanger to increase the capacity and reduce the mass of the heat exchanger. Optimization was conducted with the Hooke-Jeeves method, which aims to optimize the geometry of the heat exchanger, especially on the diameter (dw) and the distance between wires (pw). The model developed to present heat transfer correlations on single staggered wire and tube heat exchanger was valid. The maximum optimization factor obtained when the diameter wire was 0.9 mm and the distance between wires (pw) was 11 mm with the f_{ref} value = 1.5837. It means that the optimized design only using mass of 59,10 % and could transfer heat about 98,5 % from the basis design.

Keywords*:* Single staggered wire and tube heat exchanger, heat exchanger capacity, heat exchanger mass, geometry optimization.

INTRODUCTION

Heat exchanger is an equipment used to exchange energy in the form of heat between different temperature fluid circulation which can occur through direct or indirect contact (Pitts and Sissom, 1977). One of the applications of heat exchange principles is in wire and tube heat exchanger (Kumra *et al.*, 2013). This heat exchanger is belonging to the type of extended surface heat exchanger where wire functioning as fin is installed to the tube which channels hot fluid in order to increase heat transfer surface area and then increase heat transfer rate (Srinivasan and Shah, 1997). Mechanically, wire also functions to support tube configuration (Petroski and Clausing, 1999).

Wire and tube heat exchanger consists of coiled tube, and wire welded on the two sides of it in normal direction of the tube (Witzel and Fontaine, 1957). Generally this heat exchanger uses the array of inline wire (simetry) between both its sides (Samana *et al.*, 2012), but this research used the array of staggered wire so that it decreases the resistance of convection heat transfer. The effort was conducted by the researcher in the field of cooler to expand a new design in the certain part which will give the effect of increasing the ability of the heat exchanger and lessen the use of construction materials (Jaluria, Y., 2007). This heat exchanger has been widely used to dissipate heat from hot fluid which flows through a tube as a condenser of a small air refrigeration system (refrigerator) to condense fluid which flows in tube or applied only as fluid cooler that flows in tube without happening to the phase change (Tanda and Tagliafico, 1997). Nevertheless, studies on the optimization of the heat exchanger have not been done a lot.

Bansal and Chin (2003) developed the computer model for condenser on the condition of free convection. This study used the same method as Tagliafico and Tanda (1997) to get the heat transfer coefficient of free convection. For model validation, it was performed by comparing the modelling result with total heat load from condenser. From the result of data verification, it was obtained that the deviation of heat load is ± 10%.

Based on the modelling result above, Bansal and Chin (2003) performed optimization of wire and tube condenser. This optimization proposed optimization factors of heat load ratio permodel condenser mass with heat load per mass from the carly model condenser. By changing diameter wire the distance between wires and diameter tube, it obtained the design of the optimum condenser, which is 3% heat load increase with 6% mass condenser decrease.

Pradeep Kumara (2011) performed optimization of a domestic condenser, that is model NST 200 by using the model developed by Tagliafico and Tanda which aimed to increase heat transfer and decrease the production cost of condenser. The result shows that modified condenser can increase the heat transfer as big as 32,9% and decrease 19% of the production cost from the existed design (present condenser).

Some of the researches above averagely used wire and tube heat exchanger with inline arrangement design, so this research conducted a new study using single staggered configuration between two sides of heat exchanger. With the position of staggered wire, the phenomena of mergering thermal boundary layer will not happen because the position of wires faced each other between both of heat exchanger sides (Incropera, 1990). Next, almost all researchers in making optimization have not entered optimum wire space variable based on thermal boundry layer analysis as the introduction study.

Based on the thoughts above, a study about optimization of wire and tube heat exchanger by using single staggered wire and tube heat exchanger needs to be performed to obtain the optimum design. The optimum design meant is to reach the performance of heat exchanger as maximum as possible by using fabrication material as minimum as possible.

METHODOLOGY

To perform the optimization of single staggered wire and tube heat exchanger, we need a model development first. To get the heat transfer of each element (Q_{ele}), it needs the calculation of the heat transfer coefficient by using semi empirical correlation from Tagliafico and Tanda (1997) for every element. Then, the total heat is the number of the heat transferred from every element. The simulation used the element finite method with the assist of MATLAB program (Kwon Young, 1997), where the element of a heat exchanger is divided into some element units. Each modelling element consists of a tube as long as the pitch wire (pw) and a wire as a fin as long as the pitch tube (pt) with the entrance and exit flows of every element as the following figure-1.

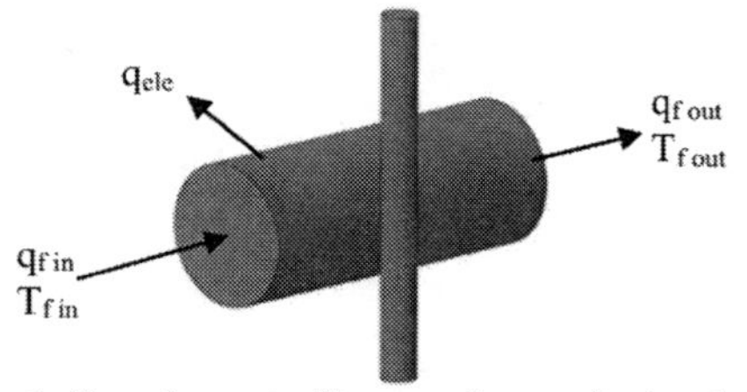

Figure-1. One element of heat exchanger in the element finite method

The heat transferred by every element is (Q_{ele}) and then the number of all Q_{ele} is the total heat transferred by the heat exchanger (Q_{total}). The validation was conducted by comparing the modelling element surface temperature with the experiment results where the element average temperature (T_{ex}) is calculated with the equation as follows:

$$T_{ex} = \frac{(T_{to} + GP.\eta_W (T_{to} - T_\infty) + GP.T_\infty)}{(1 + GP)} \tag{1}$$

Where: T_{ex} = element average temperature
T_{to} = tube outside temperature
GP = geometry parameter
η_W = wire efficiency
T_∞ = room temperature

By developing a new geometry parameter (*GP*) based on the geometry characteristics of the single staggered wire and tube heat exchanger with only one wire, that is:

$$GP = \left(\frac{p_t}{d_{to}}\right)\left(\frac{dw}{pw}\right) \tag{2}$$

For optimization procedure has been used optimization factor as follows (Bansal and Chin, 2003):

$$f = \frac{Q/W}{Q_0/W_0} \tag{3}$$

Where Q is the capacity of optimized heat exchanger and W is the heat exchanger mass, but Q_o and W_o are the capacity and mass of basis design heat exchanger from local design. In this research, the heat exchanger design used as basis is a wire and tube heat exchanger with the specifications as follows:

- Exchanger (H) : 445 mm
- Width of heat exchanger (wire): 431 mm
- Width of heat exchange (tube : 476 mm
- Length of tube : 6416 mm
- Diameter tube outside : 4.8 mm
- Diameter tube inside : 3.2 mm
- Diameter wire : 12 mm

This research has been observed to obtain the maximum condition in order to optimize the optimization factor. The variable designs manipulated are:

- Diameter Wire (dw)
- Distance between Wire (pw)

The optimization theory said that determining the early point at Hooke-Jeeves method very much influenced the optimization result. Therefore, to determine the initial point, this research has used thermal boundary layer analysis (Romero-Méndez, 2000), with this analysis convergence process and the exact of optimum value obtained in maximum. Based on the analysis, it was obtained that the optimum value of pitch wire (pw) was ± 9 mm.

However, wire diameter was 1.2 mm where tube diameter and the distance between tubes were kept constantly, which were 5 mm and 40 mm, respectively. Entrance fluid temperature of heat exchanger was 60 ^{0}C, whereas fluid flow rate (oil) was 0.006 kg·s^{-1} and environment temperature was 30 ^{0}C.

RESULTS AND DISCUSSIONS

In this case, the model used has been validated with the result of experiment performed by using a set of single staggered wire and tube heat exchanger with the specification as follows:

- Height of heat exchanger (H) : 445 mm
- Diameter tube outside (dto) : 4.8 mm
- Diameter tube inside (dti) : 3.2 mm
- Diameter wire (dw) : 1.2 mm
- Distance between wires (pw) : 7, 14, 21mm,
- Number of tube : 12 coils
- Width of heat exchanger (W) : 431 mm
- Distance between tubes (pt) : 40 m

Operation conditions:

- Fluid flow rate : 0.006 kg·s^{-1}
- Entrance fluid temperature : 60 ^{0}C dan 70 ^{0}C
- Envinronment temperature : 30 ^{0}C
- Distance between wires : 7 mm
- Number of tubes : 40 mm
- Material of heat exchanger : Steel

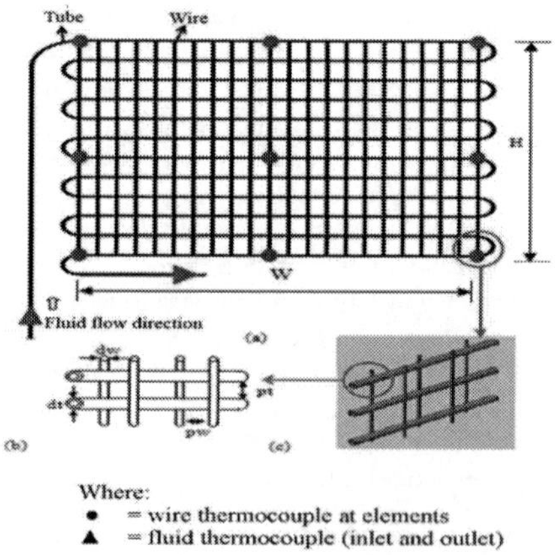

Figure-2. Single staggered wire and tube heat exchanger geometry for validating

DOI: 10.1051/matecconf/20165801017

Validation was performed by comparing with the temperature of experiment result at the certain elements with the calculation temperature at the same elements. For heat exchanger with pw: 7 mm, the model elements are at elements no: 1, 96, 189, 283, 378, 474, 567, 661, and 756. Then for pw = 14 mm, the model elements are at elements no: 1, 49, 96, 144, 192, 241, 288, 336, and 384. But for pw = 21, the elements are at elements no: 1, 34, 66, 99, 132, 166, 198, 231, and 264. The operation condition of calculation must be the same as the experiment condition.

As the early step in performing this validation, we performed the early condition trial of surface temperature distribution of heat exchanger by mapping infrared thermography instrument at the whole heat exchanger surfaces. The result of observation with infrared thermography obtained the early description of temperature distribution as follows.

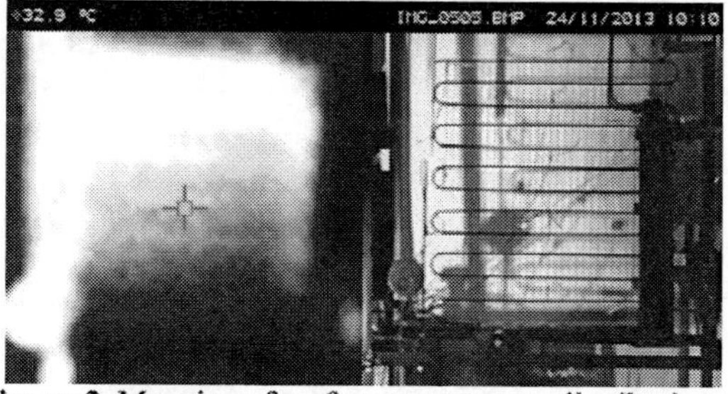

Figure-3. Mapping of surface temperature distribution of heat exchanger using infrared thermography

The Figure above is the result on infrared thermography photo that shows surface hot area of heat exchanger after being flown by hot fluid (hot oil). White area is part of surface with high temperature area because that area is a part of entrance fluid to heat exchanger where the fluid temperature is relatively high. It is then followed by yellowish area which has medium temperature that is a part of wire array in the middle part of overall heat exchanger surface area. But a part of red color is the area which has low temperature because at that part fluid leaves the heat exchanger. From this early description, it can be seen that the temperature decreases together with release of heat performed by the array of wires which functions as fins, and finally cold fluid will leave the heat exchanger.

After being known the trend of temperature characteristic at heat exchanger surface generally, it is then performed the measurement of surface temperature distribution accurately by using thermocouple measurement at each element point of heat exchanger. The following is the temperature of element point (9 measurement points) for heat exchanger with pw (pitch wire) = 7 mm with entrance temperature at 60 ^{0}C.

Table-1. Distribution of element temperature at heat exchanger pw = 7 mm

Element to	Temp. experiment	Temp. calculation
1	49.94649	49.425
96	46.20797	46.78
189	44.43979	44.51
283	42.98776	41.093
378	40.53906	39.608
474	37.79227	37.355
567	36.10708	36.385
661	35.28887	34.8972
756	34.29777	34.255

Based on the observation result data at 9 thermocouple points located at the elements above the difference from modeling, some result data will be obtained. The analysis result shows that the error percentage between the observation and the modeling results are 1.36% and temperature profile of each point based on the observation result and modeling can be presented in the following figure-4.

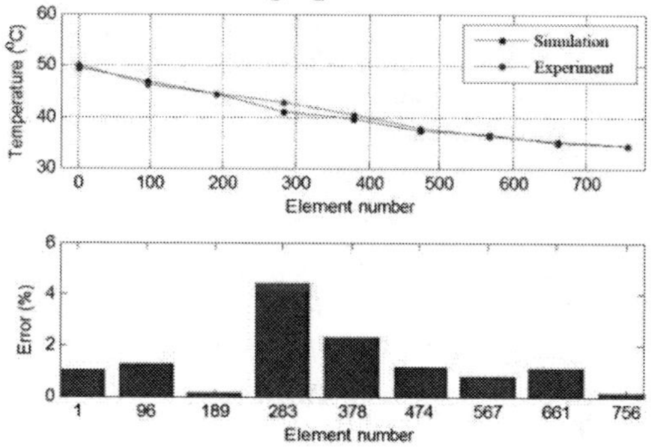

Figure-4. Temperature validation graphic on each element for pw = 7 mm

Heat exchanger with pw = 14 mm obtained the element temperature distribution from 9 points of thermocouple placement at the elements below (table-2). Data was taken at the entrance temperature of heat exchanger of 60 ^{0}C and room temperature of 30 ^{0}C.

Table-2. Distribution of element temperature at heat exchanger pw = 14 mm

Element to	Temp. experiment	Temp. calculation
1	48.75064	49.6487
34	46.00827	47.218
66	42.55736	45.104
99	42.35471	41.84
132	38.34888	40.395
166	36.25591	38.155
198	34.60221	37.17
231	33.52144	35.6335
264	32.14177	34.957

Based on the analysis of the error percentage, we obtained a deviation of 4.96% between data from experiment result and modeling. Temperature profile of each point based on observation result and modeling can be presented in the following figure-5.

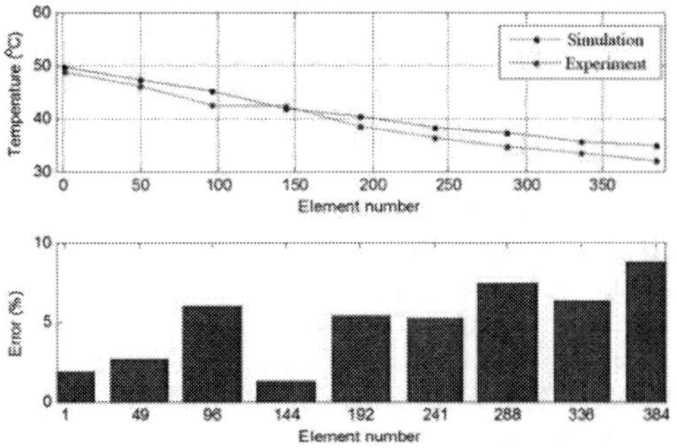

Figure-5. Temperature validation graphic on each element for pw = 14 mm

Based on both data verification performed between the experiments and modeling results, we obtained its error percentage which is relatively small i.e. less than 5%. This case means that the developed model is considered to be good enough and accurate to perform optimization.

The change of diameter wire (dw) and pitch wire (pw) gives a very significant impact to the total heat which can be transferred by wire and tube. Besides, it influences the coefficient of heat transfer at outside part of tube h_o. Increasing or decreasing the number of wires on tubes also influences the area of heat transfer surface. Besides the number of wires, diameter also influences the area of heat transfer surface. By changing geometry parameters, it is expected to obtain maximum total heat which can be transferred to the environment. But only if total heat is maximized, of course theoretically, the increase in the number of many wires will improve heat transfer rate and finally the materials used at fabrication will also increase. Therefore, beside heat transfer rate, it will also be important to consider the mass used in making wires and tube. By performing optimization, it is expected that wire and tube can transfer heat at the maximum with material mass used as little as possible.

In this research, the basis design used as a basis is a heat exchanger with the specifications as follows:

- Height of heat exchanger (H) : 445 mm
- Width of heat exchanger (wire): 431 mm
- Width of heat exchange (tube : 476 mm
- Length of tube (total) : 6416 mm
- Diameter tube outside : 4.8 mm
- Diameter tube inside : 3.2 mm
- Diameter wire : 1.2 mm
- Distance between wires : 7 mm
- Number of tube : 12 coils
- Distance between tubes : 40 m
- Material of heat exchanger : steel

By using the model developed in this research, the obtained result of a basis heat exchanger is tabled as follows:

Table-3. Q_{total}, f_0 for basis heat exchanger

Parameters	Values	Unit
Q_{total}	121.7274	Watt
Massa *wire and tube*	1.0788	kg
$f_0 = Q_{total}$ / massa	112.8359	Watt/kg
$f_{ref} = f / f_0$	1	

Since in this research f_{ref} will be maximized, there is a possibility that Q_{total} will be less than Q_{total} basis. Althought Q_{total} is less; maximum f_{ref} will give a comparison between Q_{total} and the most maximum wire mass. It means that a little mass can transfer high enough heat, in which its value can approach, be equal to or even bigger than Q_{total} basis. The following presented a table (table 4) and graphic (figure 6) of the influence of wire (dw) and distance between wires (pw) to optimization factor (f_{ref}).

Table-4. Influence of pw and dw to f_{ref}

dw\pw (m)	0.005	0.007	0.008	0.009	0.01	0.011	0.012
0.0008	1.4832	1.5531	1.5576	1.56	1.5811	1.5785	1.5528
0.0009	1.4306	1.5322	1.5468	1.5577	1.5828	1.5837	1.5609
0.001	1.3603	1.4968	1.5244	1.543	1.5481	1.579	1.5603
0.0011	1.2974	1.4507	1.4896	1.5174	1.53	1.565	1.5515
0.0012	1.2352	1.3972	1.4459	1.4827	1.5032	1.5427	1.5351
0.0013	1.176	1.3396	1.3961	1.4408	1.4689	1.5131	1.5118
0.0014	1.1208	1.2806	1.3423	1.3936	1.4287	1.4774	1.4826
0.0015	1.0699	1.2221	1.2868	1.343	1.3841	1.4369	1.4485

The yellow highlights on some values in the table show the highest f_{ref} for each diameter wire. From the table above, it is known that the result of maximum f_{ref} is obtained when the diameter wire is the same as 9 mm and the distance between wires is 11 mm with f_{ref} value is 1.5837. In table-4 above, the minimum value pw is limited at 5 mm and dw at 8 mm, this is because of the fabrication limitation in making those tools. If the values of pw and dw are less than the boundary, it will be difficult in fabrication, it is even impossible for the size of dw and pw which are very little. Optimization method of Hooke-Jeevess is the solution of optimization without limitation. Theoretically, maximum f_{ref} is obtained when the values of pw and dw are very small. This case can be seen at the graphic, trending at the red diagonal which increasing more and more because of the decrease of pw and dw. However, because there is a boundary, the maximum f_{ref} can be obtained as it has been explained before. Graphically, the profile of pw and dw's influence to f_{ref} is presented as follows (figure-6):

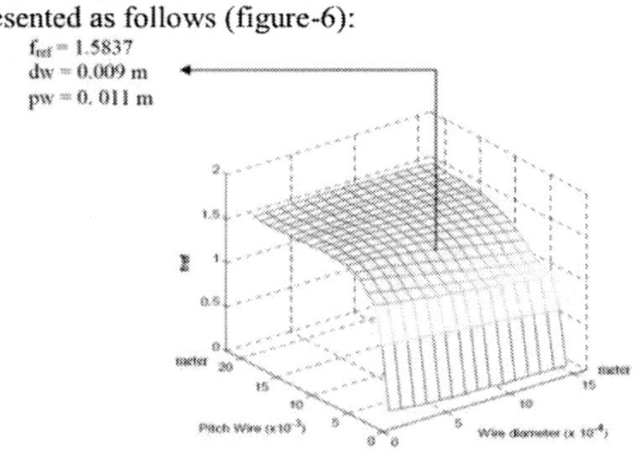

Figure-6. Graphic of dw and pw influence to f_{ref}

It has been explained before that from the graphic above, it seems that the optimum f_{ref} is obtained at dw 0.9 mm and pw = 11 mm with the value of f_{ref} = 1.5837. However, from the graphic above, it cannot be determined how much heat is transferred by heat exchanger. The influence of dw and pw to how much heat transferred will be presented at the following table 5.

Table-5. The influence of pw and dw to Q_{tot}

dw\pw	0.005	0.007	0.008	0.009	0.01	0.011	0.012
0.0008	120.86	120.2507	119.7495	119.3572	119.3819	119.055	118.3943
0.0009	121.513	121.0994	120.6373	120.2519	120.2756	119.9214	119.2145
0.001	121.791	121.8046	121.4058	121.0528	120.5808	120.7312	119.9919
0.0011	122.296	122.3841	122.0586	121.7564	121.3069	121.4778	120.7202
0.0012	122.808	122.8627	122.6061	122.3646	121.9526	122.157	121.3947
0.0013	123.356	123.2683	123.064	122.884	122.5189	122.767	122.0123
0.0014	123.959	123.6277	123.4509	123.3249	123.0101	123.3083	122.5717
0.0015	124.624	123.965	123.786	123.6999	123.4329	123.7839	123.0733

When the condition of f_{ref} is maximum, the heat transferred is only 119.9214 watt, which was only 98.5% from the basis total heat. However, the mass used is less, i.e. 0.6376 kg that means only 59.10% from the basis mass, or on the other hand, it has decreasing material mass used until 40.89%. In table-5, it can be seen that the smaller the distance between wires, the higher the Q_{total} and an increase in diameter will also cause an increase in Q_{total}. On the other hand, if pw $\leq$ dw, the system will be considered as tube with arrangement of fins like a plate, which means that Q_{total} will be smaller.

Figure-7 explains the graphic of the ability of releasing heat from each heat exchanger:

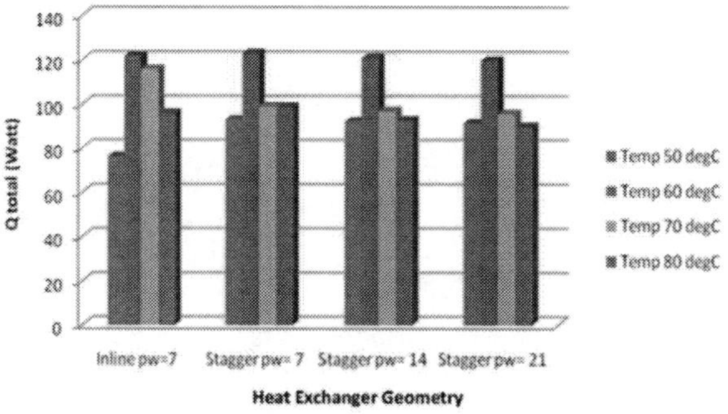

Figure-7. The influence of heat exchanger geometry to heat transfer rate.

From the above graphic, the rate of total heat transfer (watt) released by each heat exchanger can be seen. The heat exchanger with pw = 7 mm inline (basis) releases the highest heat rate because this heat exchanger has the biggest heat transfer surface. Then it is continuously followed by heat exchanger of pw = 7 mm staggered, heat exchanger pw = 14mm staggered, and the last is heat exchanger of pw = 21 mm staggered which are suitable for total surface area of each heat exchanger. The interesting thing from the figure above is the heat exchanger of pw = 7 staggered has less number of wires but it can release heat not much of a different from basis heat exchanger which has much more number of wires. This case means that heat exchanger of pw = 7 staggered has a high optimization factor because it has the ability of releasing high enough heat with low enough material mass.

CONCLUSIONS

Based on the research result data and analysis performed, some important things connecting to the optimization of single staggered wire and tube heat exchanger can be concluded as follows:

A model has been developed to present heat transfer at single staggered wire and tube heat exchanger i.e. one heat exchanger element which consists of one tube as long as pitch wire and a wire which as long as pitch tube was valid. A validation was performed with the approach of element temperature parameter at 9 points thermocouple with the error percentage is less than 5% between the results of experiment and modeling. Based on the optimization study by Hooke-Jeevess, method it could be identified that the maximum optimization factor (f_{ref}) was at the diameter wire (dw) of 0.9 mm and the distance between wires was 11 mm with the value of f_{ref} is 1.5837. It means that the optimized design is only using 59.10% mass and can transfer 98.5% heat from the basis heat exchanger. An increase in the number of wires (N) will increase the heat transfer rate from heat exchanger as long as the convection coefficient is not influenced by pitch fin.

ACKNOWLEDGEMENT

We thank DIKTI and Heat Transfer Laboratory of Chemical Engineering Department of ITS for all helps and supports for this research.

REFERENCES

Bansal, P.K., T.C. Chin. 2002. Design and modelling of hot-wall condensers in domestic refrigerators. Applied Thermal Engineering, Vol. 22, pp. 1601–1617.

Bansal, P.K., and Chin, T.C. 2003. Modelling and optimisation of wire-and-tube condenser. International Journal of Refrigeration, Vol. 26, pp. 601-613.

Incropera, Frank P., DeWitt, David P. 1990. Fundamentals of Heat and Mass transfer. John Wiley & Sons, New York.

Jaluria, Y. 2007. Design and Optimization of Thermal Systems. Second Edition, McGraw-Hill, New York.

Kumara, Pradeep. 2011. Optimization Of Performance And Assessment Of Material Cost Of The Refrigerator Condenser. Master Of Science Thesis, KTH School of Industrial Engineering and Management.

Kumra, A., Rawal, N., and Samui, P. 2013. Prediction of heat transfer rate of a wire-on-tube type heat exchanger: An Artifical Intelegence approach. Procedia Engineering, Vol. 64, pp. 74 – 83.

Kwon Young W., Bang Hyochoong. 1997. The Finite element methode using MATLAB, 3rd edition, CRC Press, New York.

Pitts, D and Sissom, L. 1977. Heat Transfer. McGraw-Hill, inc.

Petroski, S.J., and Clausing, A.M. 1999. An Investigation of the Performance of Confined, Saw-Tooth Shaped Wire-on-Tube Condensers. University of Illinois at Urbana-Champaign, ACRC TR-153.

Romero-Méndez, R., Sen, M., Yang, K.T., and McClain, R. 2000. Effect of on spacing on convection in a plate fin and tube heat exchanger. Int. J. Heat and Mass Transfer, Vol. 43, pp. 39-51.

Samana, T., Kiatsiriroat, T., and Nuntaphan, A. 2012. Air-side Performance Analysis of Wire-on-tube Heat Exchanger Having Oscillating Heat Pipe as Extended Surface under Natural Convection Condition. Heat Transfer Engineering, Vol. 33, issue 12, pp. 1033-1039.

Srinivasan, V. and Shah, R.K. 1997. Fin Efficiency of Extended Surfaces in Two Phase- Flow. Journal of Heat and Fluid Flow, Vol. 18, pp. 419 - 429.

Tagliafico, L., and Tanda, G. 1997. Radiation and natural convection heat transfer from wire-and-tube heat exchangers in refrigeration appliances. International Journal of Refrigeration, Vol. 20, No. 7, pp. 461-469.

Tanda, G, and Tagliafico, L. 1997. Free Convection Heat Transfer From Wire and Tube Heat Exchangers. Journal of Heat Transfer, Vol. 119, pp. 370 - 372.

Witzell, O.W. and Fontaine, W.E. 1957. Design of Wire and Tube Condensers. Journal of Refrigerating Engineering, Vol. 65, pp. 41 – 44.

EFFECTS OF THE OPTIMISED pH AND MOLAR RATIO ON STRUVITE PRECIPITATION IN AQUEOUS SYSTEM

Luluk Edahwati[1,3] , S.Sutiyono[1], Dyah Suci Perwitasari[1,3],
Stefanus Muryanto[2], .J.Jamari[3], A.P.Bayuseno [3]*

[1]Department of Chemical Engineering, University of Pembangunan Nasional "Veteran" Jatim, Surabaya 60294, Indonesia.
[2]Department of Chemical Engineering, UNTAG University, Bendhan Dhuwur Campus, Semarang 50233, Indonesia.
[3]Mechanical Engineering Graduate Program, Diponegoro University, Tembalang Kampus, Semarang 50275, Indonesia.
E-mail: * apbayuseno@gmail.com

ABSTRACT

Struvite ($MgNH_4PO_4.6H_2O$) is one of phosphate minerals, commonly forms into aqueous solutions. It can be precipitated as mineral deposits for optimization of phosphate recovery based on the pH optimum, molar ratio and temperature levels. This paper presents results of a study on the struvite precipitation under the influence of pH variation, at optimized molar ratio and temperature, which were calculated from an experimental design methodology. Based on the methodology, a laboratory prepared struvite, made by mixing solutions to NH_4OH, $MgCl_2$ and H_3PO_4 for a molar ratio of 1: 2: 1 in a 500 mL volume of batch stirred crystallizer at room temperature. The crystallization was done at 200 rpm and the pH variation was adjusted to 8, 9 and 10 with KOH for a time of 70 minutes. The resulting crystals were filtered and dried at room temperature for 48 h and subsequently stored for further analysis. Material characterisasion of the crystals was conducted using XRPD Rietveld method of mineralogical composition. SEM equipped by EDX was employed for investigation of morphology and elemental composition of the crystals obtained. During the experiment, struvite crystals were firstly nucleated and subsequently developed at major value. The increase in pH is assumed to convert some of the struvite phase into struvite (K) and minor sylvite (KCl). It demonstrates that Visual MINTEQ can be employed to estimate the mineral formation out the synthetic solutions.

Keywords : Crystallization, pH optimum, Struvite, Struvite (K) and XRPD Rietveld method

INTRODUCTION

Formation of struvite ($MgNH_4PO_4{\cdot}6H_2O$) scale deposits are one of the key issues in the wastewater treatment plants [Gaterell *et al.*, 2000; Priestley et.al., 1997], because it can be precipitated out the solution and eventually block in the pipes, and thus influencing the efficiency of treatment processes and causing maintenance problems [Whitaker and Jeffery, 1970; Battistoni et.al., 1997; 2000]. Efforts to control of struvite scale formation have been made through the dilution of struvite crystals with water effluents [Suzuki et al, 2005; 2007]; preventive action by chemical additives of iron salts; [Mehta and Batstone, 2013] or addition of chemical inhibitors [Stratful et.al. 2001; Kofina and Koutsoukos, 2005]. In contrast, struvite precipitation has gained interest as a route to phosphorus and ammonium recovery from wastewaters [Doyle and Parsons, 2002].

Basically, the wastewaters contain rich source of nitrogen (N), phosphorus (P) and magnesium (Mg) ions, which provides it a potentially marketable product for the fertiliser industry, while the properties of the final product should be controlled [Tünay et,al., 1997; Shin and Lee, 1997; Schulze-Rettmer, 1991; Snoeying and Jenkins, 1980; Loewenthal et.al., 1994].

Further, phosphorus (P) is one of the main sources of nutrients yielding eutrophication in aquatic systems [Abdelrazig and Sharp, 1988. This can only be prevented by treatment of municipal or agricultural wastewaters for reducing the phosphorus concentrations in the wastewater reaching surface water streams. Although P is considered as a pollutant in a water body, phosphorous is a valuable resource in agricultural fertilizers, food supply, and industrial raw materials [Sarkar, 1991]. However, the phosphorous resources are likely to be limited to the recent enormous utilization. It has been reported previously that [Frost et al., 2004]], phosphorous mineral resources are economically feasible for only 50 years. Thus, P recovery from wastewater provides benefits to prevent water pollution, remove scales of the inner surface of pumps and pipes, facilitate successive treatment steps, and prevent the devasta¬tion of mineral resources [Musvoto et al., 2000].

Here a strategic way for P recovery from the waste water is to develop an effective nucleation and growth of struvite crystals. Struvite nucleation and growth rate are mainly controlled by pH, mixing energy and starting molar ratios (Koralewskaet al., 2009; Nelson et al., 2003; Kofina & Koutsoukos,

2005; Ohlinger et al., 1999; Bouropoulos and Koutsoukos, 2000).

Therefore, the effective technique for optimizing the precipitation process of struvite is to select the optimized parameters which can be adapted to the experimentation work. The purpose of the present study was to examine the effectiveness of struvite crystallisation in a stirred batch crystalliser while examining the influence of pH, temperature and supersaturation on struvite formation. The precipitation products were then characterized by XRPD for the mineral phase composition and SEM/EDX for the morphology and elemental analysis.

METHODOLOGY

Solution preparation

A synthetic wastewater was prepared for the mixing solutions to NH_4OH, $MgCl_2$ and H_3PO_4 for a molar ratio of 1: 2: 1 in a 500 mL volume of batch stirred crystallizer at room temperature. Double-distilled deionized water was used in all experiments. Analytical grades in ammonium hydroxide, magnesium chloride, and phoshate hydroxides were supplied from Merck (Germany). In this work, struvite solution was prepared for mixing NH_4OH, $MgCl_2$ and H_3PO_4 in the mol ratio of 1:2:1 and then stirred at 200 rpm for 70 minutes. Temperature of 40 ^{O}C and an initial pH of 8, 9 and 10 were adjusted. The precipitated deposit obtained was then dried at room temperature for 48 hours and subsequently analyzed by by powder X-ray difraction (XRPD) (Philips 1830/40) scanning electron microscopy (SEM, JEOL JSM 5200).

Chemical equilibrium modeling

The Visual MINTEQ software programs version 3.0 was used in the study to predict solution equilibrium based on input of the activities of the various ions present in the solution. The program is able to calculate the solubility of solids, simulate equilibrium and speciation of inorganic solutes in the prepared solutions. The Visual MINTEQ could predict every precipitated solid phase in oversaturated condition equilibrium. The minerals selected in the model were calculated by MINTEQ program by entering total Mg^{+2}, PO_4, NH_4, K, Ca, Cl and hydrogen (H+) values at variation of pH value (8, 9 and 10). By using these values, the program calculated Mg^{+2}, NH_3, NH_4 , H_3PO_4, $H_2PO_4^{-}$, HPO_4^{-2}, $PO4^{-3}$, and $MgNH_4PO_4$. Moreover model outputs of minerals were compared to the experimental XRDP results. The chemical composition of the synthetic wastewater for the input of the Program is given in Table 1.

Table-1. The approximate composition of the synthetic wastewater

Parameter	Concentration (mol)
Mg	0.005092
NH_4	0.010184
PO_4	0.005092
K	0.21158
Cl	0.010184

X-ray powder diffraction (XRPD) Rieveld analysis

XRPD measurements were conducted on A Philips PW 1710 Diffractometer equipped with a Cu-tube and measurement parameters: 40 kV, 40 mA, 0.02^{O} step, 2 s per step, 5 – 90 ^{O} 2θ. Phase identification was made with (Powder Diffraction File)-PDF-2 Phillips software. Rietveld refinement of the XRPD data was carried out with Program Fullprof-2k, version 3.30 [Rodriguez-Carvajal, 2005], with the crystal structure database from the published data in Mineralogical Society of America (Table 2). Results presented here was obtained by refining fundamental parameters of (i) the 2 θ^{O} scale zero position, (ii) the polynomial fitting for the background with six coefficients, (iii) the phase scale factors, (iv) the cell parameters, (v) the peak asymmetry, (vi) the peak shape functions, (vii) the atomic coordinates, (viii) anisotropic temperature factors, and (ix) the preferred orientation with March-model. Site occupation factors were not refined during Rietveld refinement, because of all struvite peaks are overlapped with peaks of other minerals. The error of the Rietveld quantifications can be estimated to about 2 wt. % (1ρ) per phase. The final Rietveld scale factors were converted to PCs-by-weight using the "ZMV" expressions below:

$$\text{wt.\%(analyte phase)} = \frac{[s(ZMV]\,\textit{analyte phase}}{\Sigma[s(ZMV)]all\ phase}$$

where s, Z, M and V signify the phase scale factors; number of formula units/cell; formula weights and unit cell volumes, respectively.

DOI: 10.1051/matecconf/20165801018

Table-2. Crystal structure models used in Rietveld refinements

Structure models	PDF number	Authors
Struvite [$MgNH_4PO_4$•$6H_2O$]	PDF#712089	Whitaker and Jeffery (1970)
Stuvite-(K) [KMgPO4•$6H_2O$]	PDF#35-0812	Graeser et al. (2008)
Newberyite [$MgHPO_4$•$3H_2O$]	PDF# 76-3368	Abbona et al. (1979)
Sylvite [KCl]	PDF#89-6096	Ott (1926)

RESULTS

Visual MINTEQ results

The parameters presented in Table 1 were entered into Visual MINTEQ to determine the saturation indexes (SI) value for estimating the possibility for the formation of struvite and other minerals out from the solution.

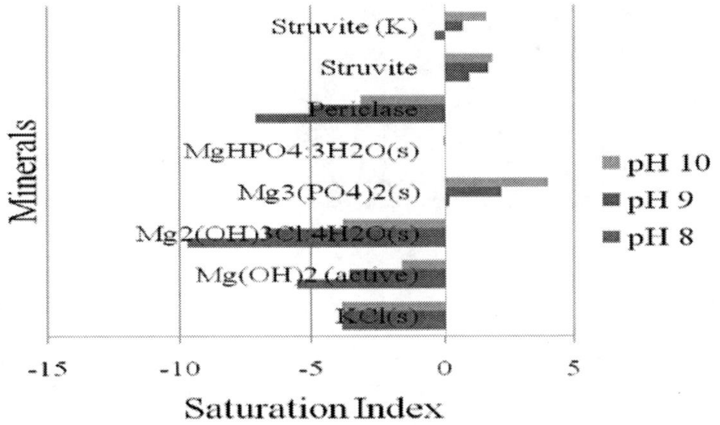

Figure-1. Modeling results of minerals from MINTEQ calculation

Material characterization results

In order to confirm the presence of minerals predicted by Visual MINTEQ, three samples of precipitates obtained at different pH were subjected to X-ray diffraction.

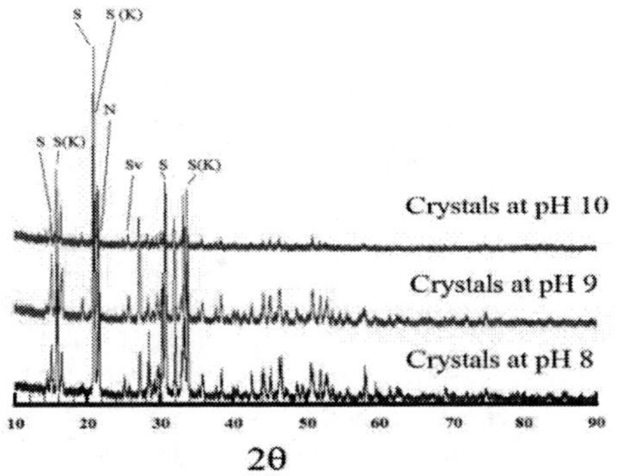

Figure-2. XRPD patterns of the precipitates obtained at temperature of 40 ^{O}C and pH variations.

The morphology of the crystals precipitated spontaneously is shown in the scanning electron micrograph in Fig. 3a. The resulting struvite structures may accommodate cation of Mg^{2+},N^{5+}, O^{2-} and P^{5+}, as shown by the EDX spectrum (Fig. 3b).

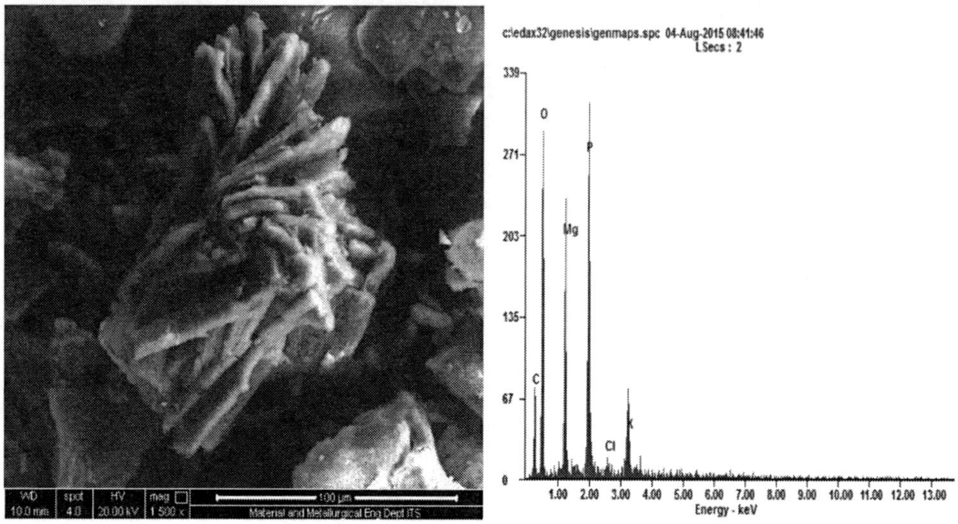

Figure-3. a) Crystal morphology b) EDX spectrum of precipitates obtained at pH 8 and temperature of 40 ^{O}C.

Table-3. XRPD-based mineralogy of the precipitates obtained at pH variation and mixing speed of 200 rpm

pH	8	9	10
Struvite (K)	18.5	32.5	47.3
Struvite	71.8	66.1	51.3
Newberyite	8.9	0	0
KCL	0.8	1.4	1.4

DISCUSSIONS

Figure 1 shows results of the calculated SI as a function of the pH variation. As can be seen, all pH conditions had a positive SI value, which theoretically showed that there was a potential for struvite and other mineral formation. These minerals found in the program were subsequently confirmed in the XRPD data. Thus, calculating the SI on a regular basis can practically forecast possible struvite and other mineral formation.

The struvite formation potential can also be used as the primary indicator in the precipitation analysis. This is a relative value that depends on the SI value of a particular system. The increase in the concentrations of struvite constituent ions (ammonium, magnesium and phosphate) and conditions (pH, temperature and conductivity) at the solution may determine the extent of the formation potential. The Visual MINTEQ programs provided that minerals of struvite-(K), struvite, and $Mg_3(PO4)_2$ should be precipitated. On the other hand, minerals of sylvite were undersaturated.

In Fig 2, a typical XRD spectrum of the solid precipitate obtained as a function pH is shown. Initially, the classical (*albeit computerized*) XRPD search-match method was used to identify the XRPD peaks that all the patterns agree very well with the PDF#71-2089) value of struvite, the PDF#35-0812 value of struvite-(K) [$KMgPO_4{\cdot}6H_2O$], the PDF#70-2345 value of newberyite [$MgHPO_4{\cdot}3H_2O$] and PDF# 76-3368 value of sylvine (KCl). Furthermore, those minerals identified in the search-match procedure was subsequently judged by the full profile Rietveld refinement, as the peaks of phases which have been overlapped in the search match or mistakenly assigned phases clearly stand out in the difference plot of the calculated and the measured diffraction profile [Prince, 1993; Rietveld, 1969; Winburn *et al.*, 2000].

The reflections characteristic for struvite and other minerals are also shown and they were matched with those of reference materials in Table 1. The exclusive formation of struvite at pH 8 is in agreement with reports in the literature concerning the stability diagrams of struvite and newberyite [Abbona et al, 1979; Mehta and Batstone, 2013]. Reports on the spontaneous formation of struvite in supersaturated solutions suggested that the crystal habit of struvite depends on the solution pH and the solution supersaturation [Doyle and Parsons, 2002].

Further, the mineral species predicted by model and found by experimentation are in agreement except for periclase and $Mg_2(OH)_3Cl{\bullet}4H_2O$ and $Mg_3(PO_4)_2$. These minerals may have dissolved to be below the detection of X-ray diffraction. On the other hand, sylvite was found in the sample because it may be precipitated during drying. Quality of struvite crystals formed into the precipitating process as a function of pH at temperature of 40 ^{O}C was evaluated by the XRPD Rietveld method (Table 2). In the resulting precipitate of pH 8, there is an evidence for the major formation of the struvite (71.78 wt.%) and struvite-(K) (18.52 wt.%), but the minor minerals of newberyite (8.9 wt.%) and sylvite (0.8 wt.%).

A decreased amount of struvite was subsequently observed as a result of the increasing pH 9 and 10, compared to those in the previous precipitating solids. While struvite-(K) was still formed out with increasing pH, newberyite disappeared. This is because newberyite is commonly unstable at pH > 6.5 [Abbona et al., 1979]. The significant reduction of the struvite content for the struvite-(K) formation may be attributed the decreased probability of ammonium concentration. Moreover, the major phase arising out of the solution to pH 9 was found to be struvite (66.1 wt .%) and struvite-(K) (32.5 wt.%) , but only minor impurities of sylvite (1.4 wt.%) were unexpectedly produced. Similarly, amounts of the major phases such as struvite (51.3 wt.%) and struvite-(K) (47.3 wt.%) were formed in the precipitating solids with pH 10, while the minor amount of sylvite was still developing.

CONCLUSIONS

The computer model developed using Visual MINTEQ in this investigation accurately predicts equilibrium dissolved mineral constituent concentrations. Accuracy of the tool was examined by comparing the output of calculation using the composition data onto synthetic waste water and phase identification with XRPD Rietveld method. The tool should prove useful for/in engineering and plant operation practitioners interested in preventing Struvite scale damage to infrastructure and in harvesting the nutrients nitrogen and phosphorus from waste water streams. A mixture of Struvite , Struvite (K) and Newberyite is a major mineral controlling MAP ions recovery out of the solution at the temperature of 40 ^{O}C and initial pH 8. The significant amounts of Struvite and struvite-(K) were precipitated out the solution with increasing pH at the same temperature . Over the range of pH examined and the temperature of 40 ^{O}C, the precipitated Struvite and struvite-(K) became major minerals controlling MAP ions recovery . The minor impurity of sylvine was formed into all precipitating solids studied. The Struvite crystals precipitated from the solution to the pH variation showed the same morphology as a function of pH variation.

Acknowledgements

The work was supported by the University of Pembangunan National, Surabaya Indonesia and PhD research projects.

REFERENCES

Abbona F., Boistelle R., Haser R. (1979). Hydrogen bonding in $MgHPO_4{\bullet}3H_2O$ (newberyite). Acta Crystallographica, B35, 2514-2518.

Abdelrazig B.E.I., Sharp J.H. (1988). Phase changes on heating ammonium magnesium phosphate hydrates. Thermochimica Acta 129,197-215.

Abbona F., Lundager Madsen H. E., Boistelle R. (1988). The final phases of calcium and magnesium phosphates precipitated from solutions of high to medium concentration J. Cryst. Growth. 89, 592-602.

Battistoni P., Fava G., Pavan P., Musacco A., Cecchi F. (1997). Phosphate Removal in Anaerobic Liquors by Struvite Crystallization Without Addition of Chemicals: Preliminary Results, *Wat. Res.*, 31, 2925-2929.

Battistoni P., Pavan P., Prisciandaro M., Cecchi F. (2000). Struvite Crystallization: A feasible and reliable way to fix phosphorus in anaerobic supernatants, *Wat. Res.*, 34, 3033-3041.

Doyl J.D. and Parsons S.A. (2002). Struvite formation, control and recovery, Water Research Vol.36, 3925–3940.

Frost R.L., Weier M.L., Erickson K.L. (2004). Thermal decomposition of struvite, J. Therm. Anal. Calorim. 76,1025-1033.

Gaterell M.R., Gay R., Wilson R., Gochin R.J., Lester J.N. (2000). An economic and environmental evaluation of the opportunities for substituting phosphorus recovered from wastewater treatment works in existing UK fertiliser markets, *Env. Technol.*, 21, 1067-1084.

Graeser S., Wostl W., Bojar H.P., Berlepsch P., Armbruster T., Raber T., Ettinger K., Walter F. (2008). Struvite-(K), $KMgPO4{\cdot}6H_2O$, the potassium equivalent of struvite a new mineral. European Journal of Mineralogy 20, 629–633.

Lu H. M. and Hardy J. R. (1991). First-principles study of phase transitions in KNO_3. Physical Review. B 44, 7215.

Loewenthal R.E., Kornmuller U.R.C., Van Heerden E.P. (1994). Struvite Precipitation in Anaerobic Treatment Systems, *In 17th International Symposium on Anaerobic Digestion, Capetown, South Africa,*498-507.

Kofina A.N., and Koutsoukos P.G. (2005). Spontaneous Precipitation of Struvite from Synthetic Wastewater Solutions, Crystal *Growth & Design* Vol.5, 489–496.

Musvoto E. V., Wentzel M. C., Ekama G.A. (2000). Integrated Chemical-Physical Process Modeling-Ii. Simulation Aeration Treatment For Anaerobic Digester Supernatant. *Water Research,* 34, 1868 - 1880.

Mehta C.M. and Batstone D. J. (2013). Nucleation and growth kinetics of struvite crystallization, Water Research Vol. 47, 2890-2900.

Ott H. (1926). Die Strukturen von Mn O, Mn S, Ag F, Ni S, Sn I4, Sr Cl_2, Ba F_2, praezisions messungen einiger alkalihalogenide. Z. Kristallogr. 63, 222-230.

Pernet M., Joubert J.C., Berthet-Colominas C. (1975). Etude par diffraction neutronique de la forme haute pression de FeOOH, Solid State Communications, Volume 17, Issue 12, 15, Pages 1505–1510.

Prince E., (1993). Mathematical Aspects of Rietveld Refinement. The Rietveld Method. Edited by Young, R.A. International Union of Crystallography, Oxford, New York, 43-54.

Priestley A.J., Cooney E., Booker N.A., Fraser I.H. (1997). Nutrients in wastewaters-ecological problem or commercial opportunity. Proceedings of the 17th Federal Convention of the Australian Water and Wastewater Association, Melbourne, **1,** 340-346.

Rietveld H.M. (1969). A profile refinement method for nuclear and magnetic structures. Journal of Applied Crystallography 2, 65–71.

Rodriguez J.-Carvajal. Program Fullprof.2k, version 3.30, Laboratoire Leon Brillouin, France, June 2005.

Snoeying V.L., Jenkins D. (1980). *Water Chemistry*, John Wiley and Sons Inc., New York.

Sarkar A.K. (1991). Hydration/dehydration characteristics of struvite and dittmarite pertaining to magnesium ammonium phosphate cement systems. J. Mater. Sci., 26, 2514–2518.

Schulze-Rettmer R. (1991). The Simultaneous Chemical Precipitation of Ammonium and Phosphate in the Form of Magnesium Ammonium Phosphate, *Water Sci. Technol.,* 23, 659-667

Stumm W., Morgan J.J. (1996). *Aquatic Chemistry,* Wiley-Interscience, New York .

Shin H.S., Lee S.M. (1997). Removal of Nutrients in Wastewater by Using Magnesium Salts, *Env. Technol.,* 19, 283-290.

Stratful L., Scrimshaw M.D., and Lester J.N. (2001). Conditions influencing the precipitation of magnesium ammonium phosphate, Water Research Vol. 35, 4191–4199.

Suzuki K., Tanaka Y., Kuroda K., Hanajima D. and Fukumoto Y. (2005). Recovery of phosphorous from swine wastewater through crystallization, Bioresource Technology Vol. 96, 1544–1550.

Suzuki K., Tanaka Y., Kuroda K., Hanajima D., Fukumoto Y., Yasuda T., Waki M. (2007). Removal and recovery of phosphorous from swine wastewater by demonstration crystallization reactor and struvite accumulation device, Bioresource Technology Vol. 98, 1573–1578.

Tünay O., Kabdasli I., Orhon D., Kolçak S. (1997). Ammonia removal by magnesium ammonium phosphate precipitation in industrial wastewaters, *Water Sci. Technol.,* 36, 225-228.

Whitaker A., Jeffery J.W. (1970) . The crystal structure of struvite, $MgNH_4PO_4{\cdot}6H_2O$. Acta Crystallographica B26,1429-1440.

Winburn R.S., Grier D.G., McCarthy G.J., Peterson R.B. (2000). Rietveld quantitative X-ray diffraction analysis of NIST fly ash standard reference materials. Powder Diffraction 15, 163-172.

PERFORMANCE OF INCLINED BAFFLE COLUMN FOR PECTIN CONTINUOUS EXTRACTION PROCESS FROM COCOA PEEL (*Theobromacacao*)

Soemargono, Sutiyono, Edi Purnomo Sasongko, and Bambang Priyanto
University of Pembangunan Nasional (UPN) "Veteran" Jawa Timur, Surabaya, Indonesia
Email: soemargono@upnjatim.ac.id

ABSTRACT

Pectin from cocoa peel was successfully prepared by extraction method in the bubble column reactor. The process could be carried out in batch and continuous, with co-current or counter current flow. Bubble pipe reactor with inclined baffle is expected to reduce the problem from gas flow patterns, jagged sloping bulkhead serves as a bubble breaker that could replace mechanical stirrer. In this study the reactor column sloping bulkhead was used as extractor to recover pectin in the cocoa peel by citric acid (1 N) solvent. Cocoa peel paste and citric acid were entered to the top of the column at 1 : 1 rate ratio by streamed using dosing pump. Nitrogen gas as a medium stirrer flowed counter current from the bottom of the column. Once extraction was completed, the liquid then precipitated by acid alcohol. Pectin solids were washed with alcohol (96%). The temperature of extraction process is maintained at 70° C. The residence time in the column material was determined by feed flow rate (cocoa paste and citric acid), and the volume of suspension that remains in the reactor (V). Residence time equation obtained is: $t = (7045.86G^{-0.1275})/L$. The yield of yellowish white precipitate reached up to 5.2973% and methoxyl concentration reached 5,21% for 72 minutes. The experiment shows that the continuous extraction process of pectin in baffle columns is better than batch process.

Keywords: pectin, cocoa peel, extraction, citric acid.

INTRODUCTION

Pectin is an important additional component in the food industry, cosmetics, and pharmaceuticals, since its ability to alter the functional properties of food products such as viscosity, emulsions, and gels (Hawley, 1981). Sources of pectin can be obtained from agricultural crops such as papaya, oranges, apples, avocados, pineapples, cocoa, and others. At the plant cell wall pectin binds to calcium ions and serves to reinforce the structure of the cell wall. Therefore, to maximize the extraction process, pectin should be released from calcium ions. The method is to chelate calcium by chelating agent. One of chelating agent that can be used is citric acid. Pectin is a polymer of D-galacturonic acid linked by β-1,4 glycosidic bonds. D-galacturonic acid has the same structure as the structure of D-galactose, the difference lies in the C6 primary alcohol groups which have carboxylic groups (Stephen, A. M. 1995).

Acid in the extraction of pectin is used to hydrolyze proto-pectin to be water-soluble pectin or to release pectin from the bonds with other compounds, such as cellulose. Meyer (1978) states that proto-pectin later become pectin is a macromolecule which has a high molecular weight, formed between pectin molecular chains with one another or with other polymers. Proto-pectin is insoluble in the form of pectinate calcium-magnesium salts. Dissolution process of Proto-pectin occurs by the replacement of calcium and magnesium ions by hydrogen ions or by the severing of ties between pectin and cellulose. The higher the hydrogen ion concentration (pH), the lower the ability to replace calcium and magnesium ions, or cut ties with the higher the cellulose and soluble pectin will increase. In addition to pH, temperature greatly affects the results obtained. Pectin extraction takes place by hydrolysis, meaning there is release of water (H_2O) in the extraction process. It can be said that the extraction takes place, accompanied by a chemical reaction. In such process, in addition to the concentration and temperature, stirring is very influential. Temperature affects the speed of chemical reactions, as well as agitation (Levenspiel, 1972). Eddy on this sloping insulated column reactor occurred by blowing inert gas as a nitrogen gas. The gas will be broken down by the serration of sloping bulkhead mounted. Thus, the concentration of the solvent acid (pH), temperature, and residence time affects result of pectin extraction from cocoa fruit peel. The experiment proved that the continuous process extraction of pectin in columns is better than batch process. They are shown by yield and methoxyl level at continuous process which are larger than batch process obtained at the same condition.

Continuous process for the extraction of pectin has not been done. One of the tools used is a column reactor. Bubble column reactor is very popular due to its ease of operation and simplicity, requiring a relatively narrow ground, and cost inexpensive. The process could be carried out in batch and continuous with co-current or counter current flow. In the continuous process, the flow pattern in the reactor pipe bubbles often impaired especially in the gas stream at high speed. It causes the slug flow patterns and "churn turbulent" (heterogeneous regime) which is undesirable (Shah, et al., 1982). Various type of continuous reactor has been used; one of the simplest is in the form of flow reactor with baffle. Other types, that is column reactor, includes bubble reactor, fluidized bed and moving bed reactor up to tickle-bed reactors is best for homogenous and heterogeneous reaction (Harriott, P, 2003; Zehner, P and Kraum, M, 2000; Rase, H. F., 1977). Shin, S. B., et al (2007) has been done optimization using computer simulation at continuous multi tube reactor for hydrocarbon

 DOI: 10.1051/matecconf/20165801019

reactions. Stangle and Mahalingam (1990) uses a different type of column reactor for heterogeneous reactions which is a Three-Phase Foam-Slurry Reactor. It is intended to expand the contact surface between the liquid and solid reactants. Extractor could also be classified as reactor. The simplest extraction is done using a soxhlet in semi-continuous mode. There are also many types and usage of continuous extraction. As done by Holder, R. W. (1973) who studied fluid-fluid extraction. Also M.C. Goldberg, *et al* (1971) who studied continuous extraction of organic matter from water. Bhornsmithikun, V., *et al* (2010) studied continuous extraction of prebiotics from jackfruit seeds. Kim, K. H., *et al* (2001) studied Continuous Countercurrent Extraction of Hemicellulose from Pretreated Wood Residues, although the process was hydrolysis continued with fermentation to form ethanol. Types of reactor and extractor were studied for their performance to obtain the best condition and output. Bubble pipe reactor with inclined baffle is expected to reduce the problem from gas flow patterns, jagged sloping bulkhead serves as a bubble breaker that could replace mechanical stirrer (Soemargono, 2007). In this study, the rector column sloping bulkhead used as extractor to extract of pectin in the cocoa fruit peel. The performance of the column with oblique sectional results obtained. The results are expected to be used to design the reactor / extractor mini scale plant when needed.

The experiment proved that the continuous process extraction of pectin in columns is better than batch process.

METHODOLOGY

Pectin extraction in a pilot scale is performed continuously. The cocoa leather raw materials were obtained from the Association of Farmers Group "Guyub Santoso" as a partner in Blitar. The cocoa peel was chopped before pulverized using a grinding or blender with water in the weight ratio of 1:10. Citric acid was used for solvent and baking soda ($NaHCO_3$) as regulator of acidity (pH). Alcohol 96% is used for precipitation of pectin. The chemicals were obtained from chemical suppliers in Surabaya.

The tool used was assembled together like the sketch in figure 2. The main tool was an inclined bulkhead reactor fitted with heating jacket, a feed regulator valve, air intake and result outtake, thermometer, and rotameter. Heating jacket used circulated water diverted from water heater tank. Research reactor was made from a glass that is resistant to temperature > 80 ° C, 5 cm in diameter, heating jacket diameter of 10 cm and 150 cm tall columns that bulkhead mounted tilted, alternating and jagged. It is intended to enable good mixing. Barrier cut elliptical with short axis equals to the diameter of the column, and a long axis 2x diameter. Those ellipses were cut into two along the short axis, and then this side sliced serrated. To keep the barrier position steady and not easily driven by a liquid or gas flow, a piece of stiffener was mounted in the middle arch. Elliptical tip was connected with a second reinforcing rod at an angle of 30° to the wall (Figure 1). At the top of the reactor there was the entry point of the feed, a thermometer, and gas outlet, while the bottom of the reactor was connected to gas / air intake and result outlet. Columns were surrounded by larger-diameter pipe that serves as a heater with a diameter of 10 cm using water media. Hot water was supplied to the heating jacket and came back out on the water heater tank. Heater tank was equipped with nickelin heater and temperature controller. Nitrogen gas was used as a mixing medium for perfect mixing to occur during the reactor column.

Extraction process

The process continued in a inclined bulkhead column with the observation of the results influenced by: The speed of material flow, the ratio of raw materials and solvent, solvent pH, process temperature, and processing time.

Method

1. Preparation of raw materials.

Dried cocoa peel was washed by water then chopped and mashed by the blender then screened with wire screen of 10 mesh passes. Powdered cocoa peel (200 grams) was added with water until two liters of volume, or solids contained were 10%. This experiment used concentrate of citric acid of 0.1 N.

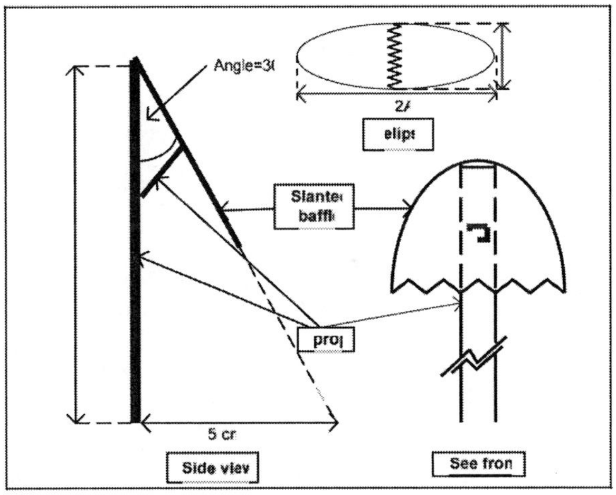

Figure-1. Insulation slant

2. Extraction.

Slurry mixture of 1 liter cocoa peel and 1 liter citric acid with certain normality was loaded into the reactor until over half of the height of the columns, followed by air flow at a certain speed. At the same time, fluid was flowed from the feed tank with a ratio of 1: 1 between the suspension of the cocoa fruit peel and a certain normality citric acid and water heater was turned on. Feed rate of each stream was set with the help of a "dosing pump". Temperature was controlled with a water heater set with a controller. Fluid speed was regulated with feed valve set, while the air flow rate was set with a valve and its value was read on the gas rotameter. Fluid velocity can also be seen from the fluid coming out of the reactor. Because of the barrier were jagged and counter current flowed between the liquid and the air, stirring occured and mixing between the reactants can take place properly. Process was carried out until a certain process condition was achieved, i.e. the desired temperature has been reached and a steady pH at the output.

The residence time of the liquid in the reactor, t minutes, determined by liquid flow rate, L cm3 / min, and the volume of liquid remaining in the reactor, V cm3, in the form of the equation (Soemargono, 2007):

$$t = \frac{V}{L} \quad \text{...} \quad (1).$$

V depends on the flow rate of the gas, G cm3 / min. The volume of suspension remained in the reactor was calculated from the difference between the initial liquid volumes in the reactor before the gas flowed, V0 cm3, with the liquid that comes out after the gas is bubbled. Correlation of V and V0 to G can be expressed in the form of equation:

$$\frac{V}{V_0} = G^a \quad \text{...} \quad (2).$$

Combined equations (1) and (2) to:

$$t = \frac{V_0 G^a}{L} \quad \text{.......................................} \quad (3).$$

In the reactor bubble, different sizes of bubbles provide different reactor performance (Levenspiel, 2007). Referring to this study, the slant barrier was able to break the bubbles into small size and evenly throughout the column. It shows the perfect turbulence in the slurry, so that the contact between cocoa pulps with citric acid chelating would be better.

The use of a paddle impeller types required a power ratio between laminar to turbulent regions. The force required to achieve turbulent pattern required a power mixer that is big enough too, especially if the slurry has a high density and a high viscosity. When using nitrogen gas stream, the pressure is only 1.2 cm H2O. In terms of energy extraction, using bubbles is much more efficient with reactors. After that the process was continued for 20-30 minutes to accommodate the results, then stopped. After the extraction process was complete, the results continued with the process of fluid coagulation, sedimentation, and drying of pectin.

3. Thickening.

The liquid was heated at a temperature of 95 - 97 °C, stirred until its volume was reduced by half. The result obtained is called the filtrate concentrated and carried out the precipitation process.

4. Precipitation.

Precipitation pectin done with the addition of acid alcohol, 96% of alcohol acidified per 100 mL by adding 1 mL of concentrated citric acid. The addition of alcohol to acid in the liquid filtrate as much as 1.5 times that of existing materials. Then the solution is allowed to stand for 10-14 hours (overnight) to obtain precipitated pectin. After that, precipitated pectin is separated from the liquid using a suction strainer. The results obtained so-called acid pectin.

5. Laundering Pectin.

Into acid pectin, it was added with 96% alcohol while stirred. Then the filtering with a suction strainer was done. This was done several times until the pectin is white.

6. Drying.

The washed pectin was dried at 60°C for 6-10 hours. The results obtained so-called dry pectin and weighed to obtain the yield and analyzed for its methoxil levels.

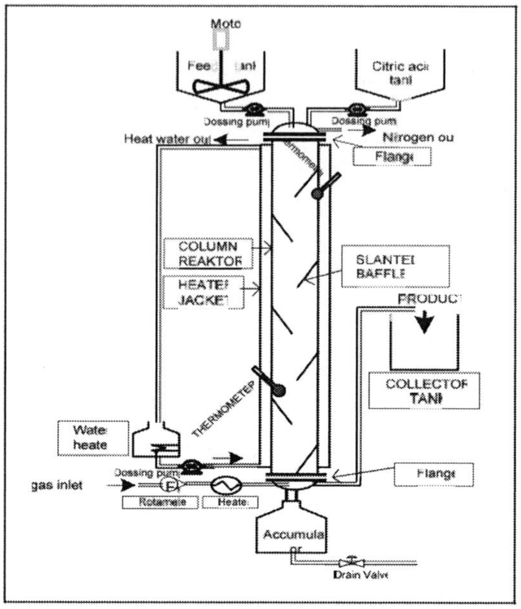

Figure-2. Baffle column reactor diagram for experiment

The residence time of the material in the column was calculated by the equation (1). Results pectin solids were weighed and analyzed for its yield and was calculated for methoxil levels.

Yield

Yield is calculated by the equation:

$$\text{Yield} = \frac{weight\ of\ solids\ result}{initial\ weight}\ x\ 100\% \quad \text{..............} \quad (4).$$

In the continuous process, the initial weight of solids is calculated by:

The initial solid = time x velocity cocoa pulp.x 10% .... (5).

Concentration of methoxyl

Fifteen milligrams of pectin solid was added with 25 mL of 0.25 N NaOH solutions, shaken and let stand for 30 minutes. The indicators of penophthalin and 25 mL of 0.25 N HCl solution was added then titrated with 0.1 N NaOH until it changed color (to pink) and known the volume of NaOH required. Methoxyl levels were calculated by the equation:

$$\%\text{methoxyl} = \frac{mL\ alkali\ x\ N\ alkali\ X\ 31}{sample\ weight}\ x\ 100\% \quad \text{...........} \quad (6).$$

RESULTS AND DISCUSSIONS

Solving eq. 1 to 3, it was conducted an experiment to find the correlation of V with G. Table 1 corresponding to Figure 3 shows the correlation of the volume of suspension that residence in the reactor column with nitrogen gas flow rate. Plot of ln V versus ln G would be forming a straight line with the equation.

$$\text{Ln } V = -0.1275 \ln G + 8.8602 \quad \text{.................... (7).}$$

modification of eq.7, obtained the equation

$$V = 7045.86G^{-0,1275} \quad \text{.................................. (8).}$$

where, the correlation coefficient is 0.9694.

Tabel 1. The correlation of reactor volume and gas flow rate

G mL min^{-1}	2,880	3,010	3,180	3,420	3,750
V_{avg}, mL	2,560	2,530	2,518	2,495	2,470

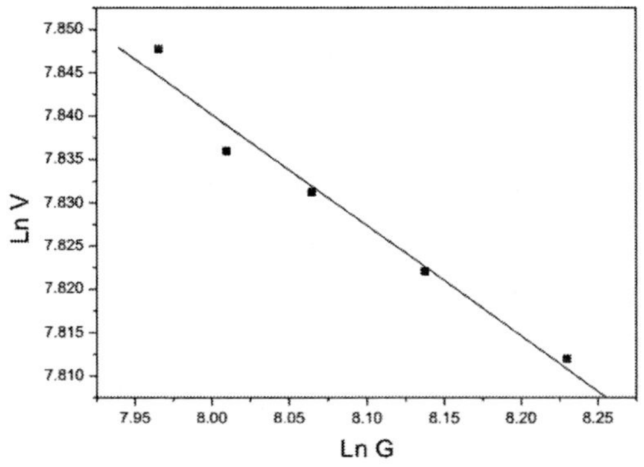

Figure-3. The correlation of volume and gas flow rate

If the eq. (8) substituted into the eq. 3 obtained residence time equation:

$$t = (7045.86G^{-0.1275}) / L \quad (9).$$

Eq. 9 is used to determine the residence time in the baffle reactor.

The continuous process was conducted with a steady gas flow of 2880 mL min^{-1} so that the suspension volume in the column is about 2650 mL. The residence time of the suspension in the column only depends on the feed flow rate. Figure 4 shows the effect of temperature on the yield of pectin prepared by batch and continuous process.

From the observation, stirring generated by blowing nitrogen gas in a column equipped with a jagged slant bulkhead is perfect compared to the way batch with stirring. In the process of batch in a stirred tank equipped with barriers (baffle) it can be seen that perfect eddy occurs only near the pedal stirrer. This is due to the high viscosity of the suspension of cocoa peel.

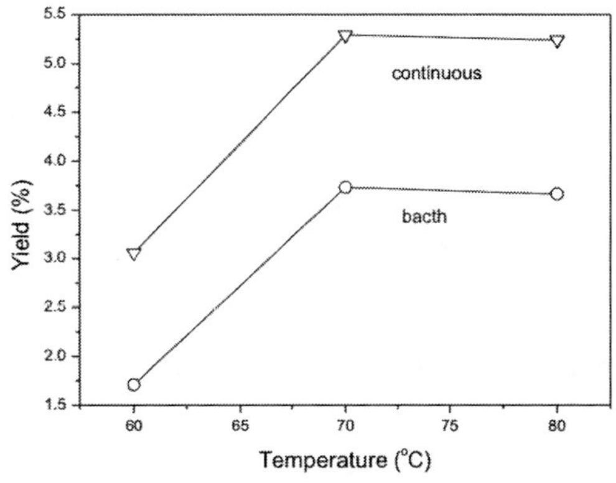

Figure-4. Effect of temperature on the yield

In the continuous process, the yield of pectin produced was slightly higher than those of batch process with the same conditions. Figure 5 shows the effect of temperature on methoxyl product, although at 60°C the difference is quite distant.

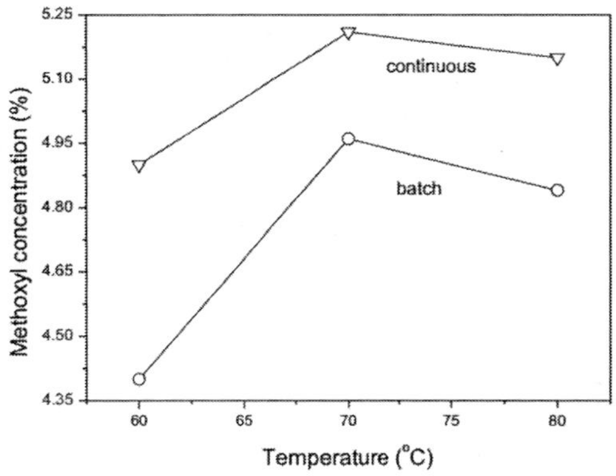

Figure-5. Effect of Temperature on the methoxyl concentration

In the continuous process, the liquid obtained looks clear without solids. It is more advantageous than the batch process. The first advantage is that the process for obtaining pectin can be continued without filtering. The second advantage, there is no loss of liquid entrained in the fabric filter. The result of field trials and fluid loss in the filtering process results ending up in cloth filter and process the rest of the cocoa solids peels is quite high, which is between 20% to 27.5%.

CONCLUSION

The experiment showed that the continuous process extraction of pectin in baffle columns is better than batch process.

ACKNOWLEDGEMENT

On this occasion, the researchers would like to thank the Ministry of Research, Technology and Higher Education who have given the research fund to DP2M.

REFERENCE

Bhornsmithikun, V., Chetpattananondh, P., Yamsaengsung, R., and Prasertsit, K. 2010. Continuous Extraction Of Prebiotics From Jackfruit Seeds. Songklanakarin J. Sci. Technol. 32 (6), 635-642.

Harriott, P. 2003. *Chemical Reactor Design*. Marcel Dekker, Inc, New York.

Kim, K. H., Melvin P. Tucker, M. P., Fred A. Keller, F. A., Aden, A., and Nguyen, Q. A. 2001. Continuous Countercurrent Extraction Of Hemicellulose From Pretreated Wood Residues. Applied Biochemistry and Biotechnology Vol. 91–93, pp. 253-267

Levenspiel, O., 1972, *Chemical Reaction Engeneering*, 2nd edition, John Wiley & Sons, New York.

M. C. Golberg, L., DeLong, and Kahn, L. 1971. Continuous extraction of organic materials from water. Journal of Environmental Science and Technology, vol. 5, issue 2, pp. 161-162

Rase, H.F. 1977. *Chemical Reactor Design for Process Plants, Volume One: Principles and Techniques.*John Wiley & Sons, New York.

Shin, S. B., Han, S. P., Lee, W. J., Im, Y. H., Chae, J. H., Lee, D. I., and Lee, W. H. 2007. Optimize Terephthaldehyde Reactor Operations. Process Systems Enterprise Ltd., London, United Kingdom. Hydrocarbon Processing April 2007

Soemargono dan Billah, M. 2007. Pembuatan Kalsium Karbonat dari Bittern dan Gas Karbon Dioksida secara Kontinyu. Journal of REAKTOR Teknik Kimia FT-UNDIP, Vol.11,No.1 Juni 2007:14-21

Shah, Y.T., Kelkar,B.G., Godbole, S.P.,and Decker, W.D.,1982, Journal Review: Design Parameters Estimations for Bubble Column Reactors", *A.I.Ch.E Journal*, *28*, 353-379.

Stangle, G.C., and Mahalingam, R., 1990, "Mass Transfer with Chemical Reaction in a Three-Phase Foam-Slurry Reactor", A. I. Ch. E. Journal,36, 117-125.

Stephen, A. M. 1995. *Food Polysaccharides and Their Applications.* Marcel Dekker, Inc. New York.

Zehner, P and Kraum, M. 2000. Bubble Columns. http://www.researchgate.net/publication/228033687

MATEC Web of Conferences **58**, 01020 (2016) DOI: 10.1051/matecconf/20165801020
BISSTECH 2015

PHYSICOCHEMICAL AND ORGANOLEPTIC PROPERTIES OF DRIED SYNBIOTICS YOGHURT FROM LESSER YAM TUBERS (*Dioscoreaesculenta* L.)

Sri Winarti[1*]and Erwan Adi Saputro[2]

[1]Department of Food Technology,
[2]Department of Chemical Enginering,
[1,2]University of Pembangunan Nasional "Veteran" JawaTimur, Surabaya, Indonesia
E-mail: swin_tpupn@yahoo.com

ABSTRACT

There are many variations of Yam (*Dioscorea* spp.) which spread out in the world including Indonesia. Those variations could be categorized based on their shape, size, colour, and flavour of their tuber. The genus of *Dioscorea* spp. has more than 600 species, such as *Dioscorea hispida*, *Dioscorea esculenta* (lesser yam), *Discorea bulbifera*, *Dioscorea alata* (purple yam), *Dioscorea opposita* (white yam), *Dioscorea villosa* (yellow yam), *Dioscorea altassima*, and *Dioscorea elephantipes*. At the present, the utilization of yam tubers is seen to be limited. The purpose of the research was to evaluate the properties/characteristics of dried synbiotics yoghurt from lesser yam tubers. The treatment on this study was performed by the number of substitution of lesser yam tubers and the fermentation time. The results show that the dried synbiotics yoghurt that most preferred by consumers was G_2F_3 (lesser yam 40% and fermentation time 22 hours) with an average value 3.65. It is followed by the total LAB (lactic acid bacteria) 8,15 log cfu/g, pH 4.27, total lactic acid 1.02%, yield of dried synbiotic yoghurt 32.30%, and total soluble protein 4.53%.

Keywords: synbiotic dried yoghurt, lesser yam, *Dioscorea esculenta*, lactic acid bacteria

INTRODUCTION

One of the food products being developed recently is a functional food. Functional food is defined as the food that provides beneficial effects for health apart from the fulfillment of basic nutritional needs. One example of a functional food is the one that is containing probiotic bacteria and prebiotic compound.

Probiotics are foods supplements which contain live microbial that have beneficial effects on the health of its host. The function of probiotics includes the following; improving gastrointestinal health, suppresses the growth of pathogenic bacteria, synthesizes vitamins and protein, helps in the absorption of nutrients, aids to overcome lactose intolerant, as well as stimulate immune function (Pompei, et.al, 2008).

Prebiotics is food components that cannot be digested in the upper gastrointestinal tract. They selectively stimulate the growth and activities of beneficial bacteria in the digestive tract, such as Bifidobacterium and lactobacilli, so as to improve the health of the host (Gibson, 2004; Pompei et al., 2008; Gaggia et al., 2010). One prebiotic that is widely used in food formulations is inulin. Lesser yam tuber contains the highestinulin among other *Dioscorea*spp. That is 14.77% (db) (Winarti et al., 2011).

Food products containing probiotics and prebiotics are called synbiotic. The synbiotic product needs to be developed in order to increase the level of consumption of food containing probiotic bacteria and prebiotic compound. Various synbiotic food products such as ice cream, yoghurt, yakult, a fermented beverage and kefir can be developed.

Yoghurt is a fermented milk product made by lactic acid bacteria fermentation. Yoghurt can be made from any milk, including soy milk. In the process of fermentation, the milk sugar (lactose) hydrolyzes to produce lactic acid, which contribute to the flavor and odor of yoghurt. Milk proteins produce a gel-like texture and a unique smell on yoghurt.

This research will improve the utilization of lesser yam to produce synbiotics yoghurt. Yoghurt synbiotic from lesser yam tubers containing inulin as prebiotic compound and combined with probiotic bacteria (bifidobacteria and lactobacilli) will produce new products that are beneficial to health.

In this study, the properties of dried synbiotic yoghurt using fermented milk as a carrier of probiotic bacteria and lesser yam extract as a source of prebiotic were evaluated. Probiotic bacteria were used in this study *Bifidobacterium breve* BRL-131 and *Lactobacillus casei* FNCC-90 that has been shown the prebiotic activity 1.071 and 1.21respectively (Winarti, et .al. 2013).

The purpose of this study was to evaluate the levels of lesser yam tubers and fermentation time on the quality and consumer acceptance of dried synbiotic yoghurt from lesser yam tuber.

METHODOLOGY

Materials

Lesser yam tuber (*Dioscoreaesculenta* L.) was obtained from traditional markets in Surabaya East Java.

Supporting materials, i.e. Na-pyrophosphate, eggs, whipping cream, skim milk, sugar, cocoa powder, essence, cooking oil, ovalet, salt and add coloring.

Probiotic bacteria consisted of *Bifidobacterium breve* BRL-131 and *Lactobacillus casei* FNCC-90 obtained from the Food Nutrition Culture Collection, Food and Nutrition Study Centre, Universitas Gadjah Mada. That media used to the growth of bifidobacteria and lactobacilli were MRS broth and MRS agar.

Equipments

Spectrophotometer Spektronic 21D, autoclave, cabinet dryer, centrifuge, water bath shaker, mixer, viscometer, pH meter, incubators, and Quebec colony counter. Equipment for the processing of lesser yam tubers includes; flour grinding machine (disk mill), dryers (cabinet dryer), the packaging tool (sealer), oven, blender, mixer, stove, baking molds and plastic tools.

Procedures for making dried synbiotics yoghurt

Procedures for making synbiotic yoghurt lesser yam tubers are as follows: firstly, mixing all of the ingredients which include lesser yam tubers filtrate, sucrose, and skim milk. Secondly, heating (pasteurization) at a temperature of 70-80°C for 15 seconds. Then, mixing and stirring all of the ingredients. Last step was cooling off at a temperature of ± 40-45°C. The mixture was inoculated with starter culture using *Bifidobacteria breve* BRL-131 and *Lactobacillus casei* FNCC-90 and incubated at 37°C for 18, 20 and 22 hours.

Calculation of total lactic acid bacteria

One g of the sample was homogenized with 1:10 PBS buffer (w/v) 10%; then glycerol was added, and the samples were frozen -80°C, before the calculation of the total cells. Total lactobacilli calculated by pour plate method using MRS agar medium (MRS broth plus 1.2% technical agar and 0.8% of $CaCO_3$) (Chun et al., 2007). Total bifidobacteria calculated using BSM-Agar (Bifidus Selective Medium-Agar), total *E.coli* calculated using media TBX (Tryptone Bile X-Glucuronide) with a pour plate method.

Total acid (Fardiaz, 1992)

A level of total acid was calculated as total lactic acid. The analysis is as follows: weighing 3g of the sample and add 30 ml of distilled water. The solution was titrated with 0.1 N NaOH using indicators PP 1% until it changes color into pink. Lactic acid was calculated by the formula:

$$\frac{\text{mlNaOH} \times \text{N NaOH} \times 90}{\text{g sample} \times 1000} \text{ X } 100\%$$

Total soluble protein

Total soluble protein analysis by the Lowry method. In this method combines the biuret reagent with another reagent (the Folin-Ciocalteau phenol reagent) which reacts with *tyrosine and tryptophan* residues in proteins. This gives a bluish color which can be read some where between 500-750 nm depending on the sensitivity required. There is a small peak around 500 nm that can be used to determine high protein concentrations and a large peak around 750 nm that can be used to determine low protein concentrations.

A consumer test dried synbiotic yoghurt

The quality of food can be determined in three ways: chemical, physical and sensory/organoleptics. The acceptance of the food products by consumers was primarily determined by the organoleptic quality. Organoleptic properties is the nature of the materials began using human senses are senses of sight, smell and taste. The organoleptic properties of dried synbiotic yoghurt wich determined was the consumer acceptance to the overall quality included of color, flavor, aroma and softness.

RESULTS AND DISCUSSIONS

Total lactic acid bacteria

The results show that the fermentation time and the levels of lesser yam tubers effect on the total lactic acid bacteria in dried symbiotics yoghurt were produced (Figure 1). Increasing in the fermentation time and levels of lesser yam tubers can lead to increase in the total lactic acid bacteria in the yoghurt. The increase of total lactic acid bacteria was caused by nutrition is still available in fermentation 18 and 20 hours, so the bacteria's growth is maximized. Lesser yam tubers contain inulin as prebiotic compounds that may stimulate the growth of bacteria. This result supported by Roberfroid (2005), that dietary inulin can be used to support the health and performance (well-being) and reduce the risk of diseases, and inulincan increase bacteria growth such as *Lactobacillus* spp. and *Bifidobacterium*spp.

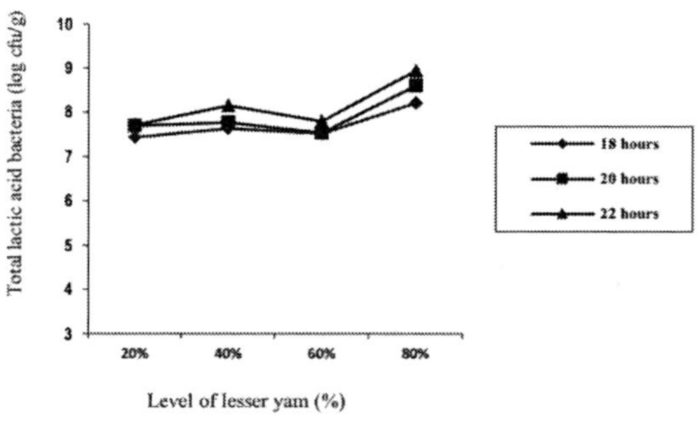

Figure-1. Effect of lesser yam tubers and fermentation time to total lactic acid bacteriaon driedsynbioticyoghurt

Inulin can be increases the growth of *Bifidobacterium adolesentis, Bifidobacterium infantis, Bifidobacterium breve, Bifidobacterium longum, Lactobacillusplantarum, Lactobacillus rhamnosus, Lactobacillus reuteri, Lactobacillus delbruechii* and inhibit the growth of *E. coli* and *Clostridia* (Pompei et al. 2008).

PH value of synbiotic yoghurt

The results show that increasing of fermentation time and levels of lesser yam extract can cause a reduction in the pH of synbiotic yoghurt (Figure 2). The decrease in the pH yoghurt was caused by lactic acid which was produced by the bacteria. This result supported by Roberfroid, (2006) who states that the pH value is affected by the acid produced, due to the increase of lactic acid will lead to the increasing number of H ions are released. Increased lead-free H ions decrease the Ph.

Bifidobacterium spp. and *Lactobacillus* spp. are a group of lactic acid bacteria that produce lactic acid and other organic acids during fermentation. The acid produced by the bacteria group will lower the pH of the growth medium (Kaur et al., 2011).

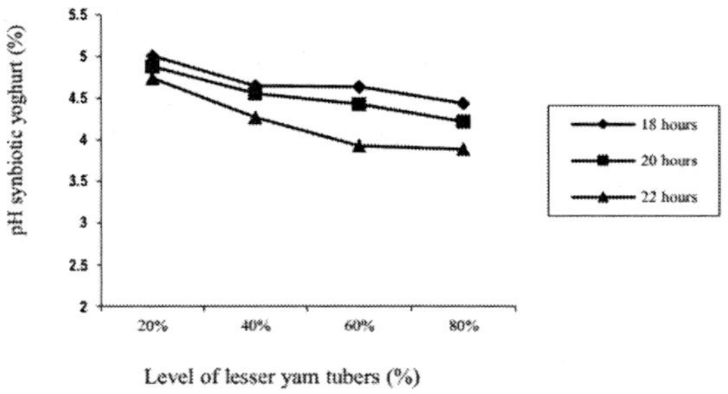

Figure-2. Effect of fermentation time and level of lesser yam extract to pH synbiotic yoghurt

Total acid of synbiotic yoghurt

The results show that increasing fermentation time and levels of lesser yam extract increased the total acid of synbiotic yoghurt (Figure 3.). Increasing total acid was caused by increasing the total lactic acid bacteria. This result supported by Matsuzaki in Farnworth (2003), who states that homo fermentative lactic acid bacteria that produce lactic acidis the main product in the fermentation process.

Inulin is fructan, metabolized by lactic acid bacteria through the glycolytic pathway to produce pyruvate, and then pyruvate converted into Acetyl-co-A, lactate and succinate. Acetyl-co-A can be converted to acetate, butyrate, whereas succinate converted to propionate and formate converted to H_2, CH_4 and/or H_2S gas (Macfarlane and Macfarlane, 2003).

Yield of dried synbiotic yoghurt

Effects of the fermentation time and level of yam tubers to yield dried synbiotic yoghurt can be seen in Figure 4.

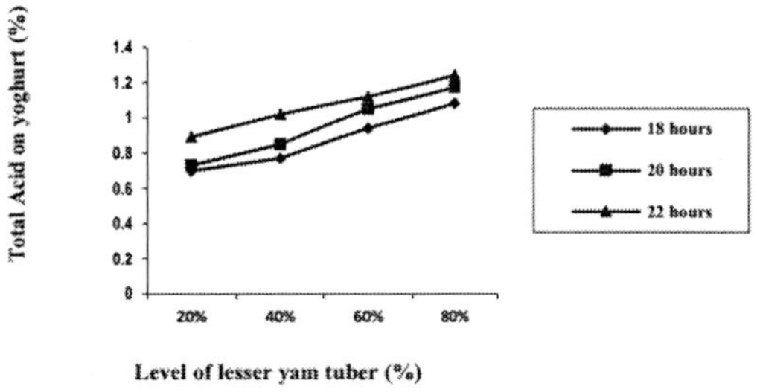

Figure-3. Effect of fermentation time and level of lesser yam extract to total lactic acid in synbioticyoghurt

Increasing the level of lesser yam tubers can increase the yield of dried synbiotic yoghurt whereas increasing fermentation time reduced the yield of dried synbiotic yoghurt. That is occurred because the increasing lesser yam tubers will increase the total soluble solids, so that after drying process, the total soluble solids were added to the total weight of dried synbiotic yoghurt. Increasing fermentation time causes more complex compounds hydrolyzed/broken down by bacteria into simple compounds with smaller molecular weight. The small molecules have the ability to trap water lower than large molecules so that more water can be evaporated in drying process and reducing total dry weight in dried synbiotic yoghurt.

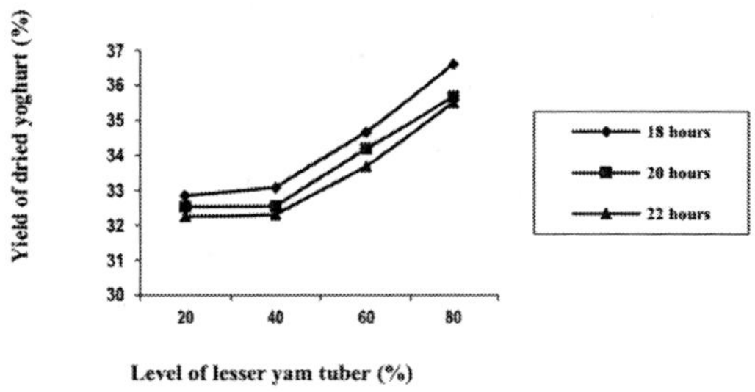

Figure-4. Effect of fermentation time and level of lesser yam extract to yield of dried synbiotic yoghurt

Inulin is fermented by anaerobic bacteria colonies of *Bifidobacterium* spp. because some of *Bifidobacterium* spp. produce extracellular inulinase enzymes that hydrolyzed the β-(2,1)-fructosil-fructose bond to produce fructose that can be used as an energy source for the growth of *Bifidobacterium*spp. (Robertfroid, 2005).

Total soluble protein of dried synbiotic yoghurt

The results show that the increasing fermentation time increased the levels of total soluble protein in dried synbiotic yoghurt while increasing lesser yam tubers reduced the levels of total soluble protein. The average value of soluble protein content on dried synbiotic yoghurt can be seen in Figure 5.

 DOI: 10.1051/matecconf/20165801020

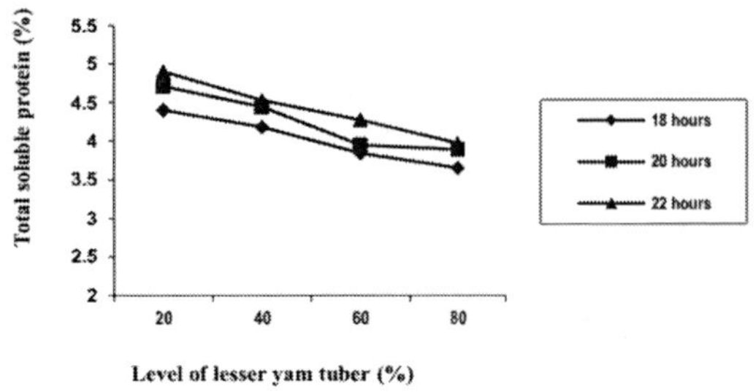

Figure-5. Effect of fermentation time and level of lesser yam extract to total soluble protein of dried synbiotic yoghurt.

Increasing the fermentation time can increase the levels of soluble protein; this is caused by increasing growth of lactic acid bacteria (LAB), which have the ability to hydrolyze proteins to produce small peptide compounds. These compounds are easier to dissolve in water. Widodo (2003), who states that in order to grow properly, lactic acid bacteria, must be able to ferment lactose, hydrolyze proteins and produce amino acids from proteolysis process. Protein degradation is done by lactic acid bacteria with proteolytic enzyme complex including the proteinase and peptidase enzyme. The first stage of protein degradation by the proteinase enzyme occurs on the surface of bacterial cells thus producing oligopeptides and amino acids which would then be further degraded by peptidase enzymes to produce amino acids.

Increasing the lesser yam tubers reduced the levels of soluble protein in the dried synbiotic yoghurt. It was becaused the largest component in the lesser yam tubers are carbohydrates. In accordance with the results of the first phase, it is known that starch content of lesser yam tuber is 82.82%.

Consumers preference of dried synbiotic yoghurt

The quality of food can be determined in three ways: chemical, physical and sensory. The acceptance of food products by consumers is determined by the quality of sensory factor. Sensory properties are the natural of the starting material using human senses such as the sense of sight, smell and taste. The consumer's acceptance of the sensory properties of dried symbiotic yoghurt is due to the overall quality of color, flavor, smell and softness.

The results of research show that fermentation time and levels of lesser yam tubers have an effect on consumer preference to dried synbiotic yoghurt. Effect of fermentation time and levels of lesser yam extract to synbiotic yoghurt can be seen in Figure 6. The dried synbiotic yoghurt most preferred by consumers was for the treatment of lesser yam tubers 40% and fermentation time 22 hours (G_2F_3).

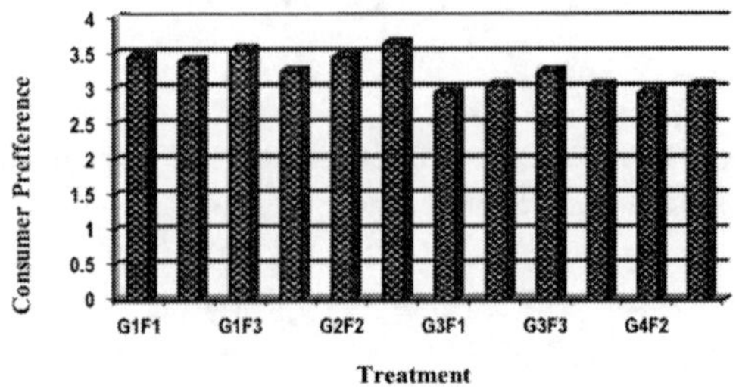

Figure-6. Effect of fermentation time and levels of lesser yam extract to consumer preference of dried symbiotic yoghurt

CONCLUSION

The research can be concluded that the levels of lesser yam tubers and fermentation time have significant effect on the quality of dried synbiotics yoghurt. The most consumers preferred of dried synbiotics yoghurt was G_2F_3 (the addition of lesser yam 40% and fermentation time 22 hours), with an average value of 3.65; and has the following characteristics: total LAB (lactic acid bacteria) of 8.15 log cfu/g , pH 4.27, a total lactic acid of 1.02%, yield of dried synbiotic yoghurt 32.30%,and 4.53% of total soluble protein.

REFFERENCES

Arrizon, J., Morel, S., Gschaedler, A. and Monsan, P.(2010). Comparison of the water-soluble carbohydrate composition and fructan structures of *Agave teguilana* plants of different ages. *Food Chemistry*122:123-130.

Berghofer, E., Cramer, A., Schmidt, V., and Veighl, M. (1993). Pilot-scale production of inulin from chicory roots and its use in food stuffs. In. Inulin and Inulin-containing crops. Elsevier Science, Amsterdam.

Gibson, G.R., Beatty, E.R., Wang X., and Cummings J.H. (1995). Selective stimulation of *Bifidobacteria* in human colon by oligofructosa and inulin. *Gastroenterology*; 108:975-982.

Gibson, G.R., Cartney, A.L.M. and Rastall, R.A. 2005. Prebiotics and resistance to gastrointestinal infection. The British Journal of Nutrition, (93), Suppl 1, pp 531-534.

Huebner, J., Wehling, R.L. and Hutkins, R.W. (2007). Functional activity of commercial prebiotics. *International Dairy Journal*. 17:770-775.

Kaur, A., Rose, D.J., Rumpagaporn, P., Peterson, J.A. and Hamaker, B.R. 2011. In vitro batch fecal fermentation comparation of gas and short-chain fatty acid production using "slowly fermentable" dietary fibers. Journal of Food Science, (04):111-116.

Macfarlane, S. and Macfarlane, G.T. 2003. Regulation of short-chain fatty acid production. Proceedings of the Nutrition Society, 62:67-72.

Oliviera, R.P.D.S., Perego, P., Oliviera, M.N.D. and Converti, A. (2011). Effect of inulin as a prebiotic to improve growth and count of a probiotic cocktail in fermented skim milk. *Food Science and Technology* 44: 520-523.

Pompei, A.,Cordisco, L., Raimondi, S., Amaretti, A. and Pagnoni, U.M. (2008). In vitro comparation of the prebiotic effect of two inulin-type fruktans. *Anaerob*14:280-286.

Roberfroid, M.B. (2005). Introducing inulin-type fructans. *British Journal of Nutrition* 93**:** Suppl.1,S13-S25.

Sardesai, VM. 2003. *Introduction to Clinical Nutrition.* 2nd Ed. USA: Marcel Dekker, Incon: Herb.

Toneli, J.T.C.L., Park, K.J., Ramalho, J.R.P., Murr, F.E.X. and Fabbro, I.M.D. (2008). Rheological characterization of chicory root (*Cichoriumintybus* L.) inulin solution. *Brazilian Journal of Chemical Engineering*25**:**03: 461-471.

Wang, X., G.W. Yuan, Z. LiMing, X. PeiGen, Y. LiPing, L.Yi, L.KeFeng and X.W.Guang. 2008.Study on the morphology, crystalline structure and thermal properties of yam starch acetates with different degrees of substitution. Sci China Ser B-Chem, Vol.51 no.9:859-865.

Widodo, 2003. Biotechnology Milk Industry, Kanisius Press, Yogyakarta.

Winarti, S.; Harmayani, E and Nurismanto, R. 2011. Characteristics and Profile of Inulin from Various Yam Tubers (*Dioscoreaspp.*). AGRITECH, Vol.31, No.4: 378-383

Winarti, S., E.Harmayani and R.Nurismanto. 2011. Ekstraction of inulin from various yam tubers (*Dioscorea spp.*), article was presented in AFC (Asian Food Conference), Bangkok 15-19 June 2011.

Winarti, S., Harmayani, E., Marsono, Y., and Pranoto, Y. 2013. Effect Foaming on Drying Inulin from Lesser Yam Tubers (*Dioscoreaesculenta*) to Phisichochemical Characteristics Prebiotics Activity. AGRITECH, 33(4):311-319.

Winarti, S., Harmayani, E., Marsono, Y., and Pranoto, Y. 2013. Effect of inulin isolated from lesser yam (*Dioscoreaesculenta*) on the growth of probiotics bacteria and SCFA formation during fermentation. International Research Journal of Microbiology (IRJM), Vol 4(2): 53-63.

Winarti, S., Harmayani, E., Marsono, Y., Pranoto, Y., Nishi, K., and Sugahara, T. 2014.Immunostimulatory and Prebiotic Activities of Inulin Extracted from Lesser Yam Tuber (*Dioscoreaesculenta*). Bali International Seminar on Science and Technology (BISSTECH 2). UPNV Jatim and STIKOM Bali. Denpasar,2-4 September 2014.

Winarti, S., Harmayani, E., Marsono, Y., and Pranoto, Y. 2014.Quantification of Colonic Microbiota *Sprague Dawley* Rats with Diet Containing Lesser Yam Inulin by Florescent in Situ Hybridization (FISH) Method. International Converence Food for a Quality Life. SEAFAST CENTER and PATPI. Jakarta, 15-16 Oktober 2014.

POTASSIUM SILICATE FOLIAR FERTILIZER GRADE FROM GEOTHERMAL SLUDGE AND PYROPHYLLITE

Srie Muljani, Bambang Wahyudi, Ketut Sumada, Suprihatin
University of Pembangunan Nasional "Veteran" Jawa Timur
Surabaya, Indonesia
E-Mail : eboy_tk@yahoo.com

ABSTRACT

Potassium silicate fertilizer grade were successfully produced by direct fusion of silica (SiO_2) and potasium (KOH and K_2CO_3) in furnaces at temperatures up to melting point of mixture. The geothermal sludge (98% SiO_2) and the pyrophyllite (95% SiO_2) were used as silica sources. The purposes of the study was to synthesise potassium silicate fertilizer grade having solids concentrations in the range of 31-37% K_2O, and silica in the range of 48-54% SiO_2. The weight ratio of silicon dioxide/potasium solid being 1:1 to 5:1. Silica from geothermal sludge is amorphous, whereas pyrophylite is crystalline phase. The results showed that the amount of raw materials needed to get the appropriate molar ratio of potassium silicate fertilizer grade are different, as well as the fusion temperature of the furnace. Potassium silicate prepared from potassium hydroxide and geothermal sludge produced a low molar ratio (2.5: 1 to 3: 1). The potassium required quite small (4:1 in weight ratio), and on a fusion temperature of about 900 °C. Meanwhile, the potassium silicate prepared from pyrophyllite produced a high molar ratio (1.4 - 9.4) and on a fusion temperature of about 1350 °C, so that potassium needed large enough to meet the required molar ratio for the fertilizer grade. The product potassium silicate solid is amorphous with a little trace of crystalline.

Keywords: geothermal sludge, pyrophyllite, fusion reaction, potassium silicate solid, fertilizer

INTRODUCTION

Potassium or sodium water glasses are generally produced on an industrial scale by melting together quartz sand and sodium/potassium carbonate in suitable furnaces at temperatures in the range of 1400° to 1500° C with the splitting-off of carbon dioxide. This high-temperature melt process is, however, very costly both in equipment and as regards the amounts of energy required and leads moreover to not inconsiderable emissions, such as dust, nitrogen oxides, and sulfur oxides (US 5238668 A). The alkali extraction and acid precipitation method, low energy method, has also been successfully used to produce silicate solution (Muljani et al, 2014), but this method is less precise when solid silicate product is desired. The hydrothermal reaction of quartz sand with aqueous potassium hydroxide obtained potassium silicate solutions which have SiO_2: K_2O molar ratios of less than 2.75:1. However, in this case, hydrothermal reaction must go through a two-stage process: stage quartz reaction with KOH solution at a temperature of 300 C and quartz reaction stage at a temperature up to melting point (over 1100 °C).

SiO_2 structural differences between amorphous and crystalline silica can lead to differences in dissolution behavior. But the physical properties of amorphous silica remains elusive in comparison to crystalline silica based on the more comprehensive quartz dissolution studies. There are also known hydrothermal processes for the production of aqueous potassium silicate solutions that are described in a number of patent applications. Gunnarsson, et al (2010) reported that the solubility of amorphous silica at 1 bar and a temperature of about 200 °C is higher than the solubility at higher temperatures. Some mixtures of silica and alkali metal silicates have somewhat lower melting temperatures, than the alkali metal carbonates used (US 2823098 A).

In this study, the use of amorphous silica and alkaline hydroxide for potassium silicate production is expected to reduce the energy while reducing air pollution, especially carbon dioxide. The mixture of alkaline hydroxide and silica may lower the melting point of the mixture. It is caused also by the solubility of amorphous silica that is lower than crystalline silica. So in the use of alkaline hydroxide the temperature should be maintained below the melt temperature of the product. In a previous study, amorphous silica purification from geothermal sludge in furnace at temperatures up to 1000 °C, it is known that the impurities contained in the geothermal sludge were much reduced with the increase in temperature furnace (Muljani et al, 2011).

Potassium silicate provides an excellent source of soluble silicon for plants and also provides supplemental potassium, a plant macronutrient (Jayawardana et al, 2014). Potassium silicate is used to strengthen plant resistance to pathogens and to enhance agricultural productivity. Here a potassium silicate is proposed that can be adapted by adjusting the physico-chemical properties as requested by the formulation of fertilizers. The potassium silicate is available both in liquid and in powder form with high solubility to easily produce liquid fertilizers. The powder form can be used to make fertilizers in powder form. Yao et al, (2014) synthesized a potassium calcium silicate mixture, by fusing a mixture of K_2CO_3, $CaCO_3$, and SiO_2. Produced from steel-making slag, fertilizer potassium silicate compounds have received considerable attention as slow-release potassium fertilizers beneficial for crops.

Soluble potassium is completely water soluble and can be used as foliar fertilizers. The foliar fertilizer application of soluble Si has been reported to reduce powdery mildew of cucumber, mask melon, zucchini squash, grapes, and angular leaf spot of beans. The foliar spraying of Si may offer practical and viable mean of reducing plant diseases with low cost (Nolla et al, 2008). It can be synthesized by reacting potassium with silicon dioxide following this reaction

$$SiO_2 + 2\,KOH \rightarrow K_2SiO_3 + H_2O \quad (1)$$

$$SiO_2 + K_2CO_3 \rightarrow K_2SiO_3 + CO_2 \quad (2)$$

The rate of the reaction depends on the ratio of alkali metal carbonate to silicon dioxide and silica structural. Crystalline quartz will only dissolve very slowly in hot water alkaline solutions, while amorphous silicon dioxide, will be readily dissolved at room temperatures (Robert et all, 1977). Although much research on the production of potassium silicate has been done but there is no comprehensive information related to energy efficiency and economy.

This study was to investigate the effect of the raw materials and the need for potassium which can reduced the energy and raw materials. The aim of this study was to provide a process for production of potassium silicate foliar fertilizer that have potassium concentrations in the range of 31-37 % K_2O, and silica in the range of 48-54 % SiO_2 by the fusion reaction of silicon dioxide with potassium. The crystalline silicon dioxide (pyrophyllite) and amorphous silicon dioxide (geothermal sludge) are used in the weight ratio of silicon dioxide/potassium. The molar ratio used for a specific application depends on the desired properties and economic impact.

METHODOLOGY

Potassium silicate production

Silicon dioxide in this experiment was obtained from geothermal sludge and pyrophylite. Geothermal sludge is the solid waste from the geothermal plant located in Dieng, Central Java, Indonesia, while pyrophyllite was mined from Malang East Java, Indonesia. X-ray fluoroscence analysis showed successive silica content of about 98 % SiO_2 in the geothermal sludge and about 85 % SiO_2 in the pyrophyllite. To increase levels of silica in pyrophyllite it is necessary to use hydrochloric acid leaching. 0.5 N HCl was used to remove impurities in pyrophyllite so that its purity reaches 95% SiO_2.

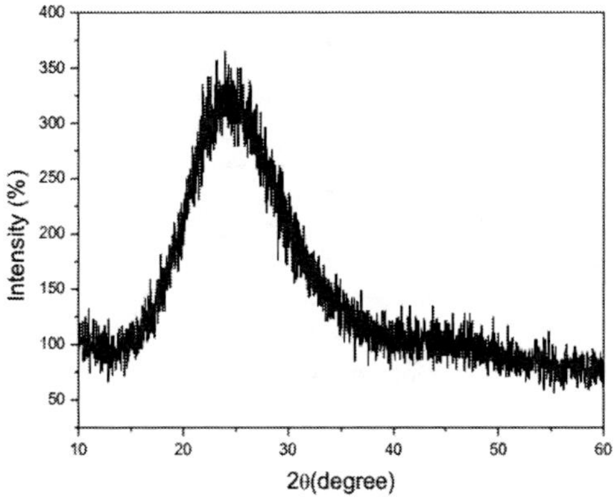

Figure 1. X-ray diffraction (XRD) patterns of silica from geothermal sludge.

X-ray diffraction (XRD) pattern have shown that the silica from geothermal sludge is amorphous (Fig. 1) and silica from pyrophyllite is crystalline (Fig. 2). The Silica obtained from the pyrophyllite sample agree very well with JCPDS data of 46-1045. However, the graphs obtained from this essay demonstrated that the pyrophyllite had a low reactivity due to the large amount of crystallinity peaks found in the charts.

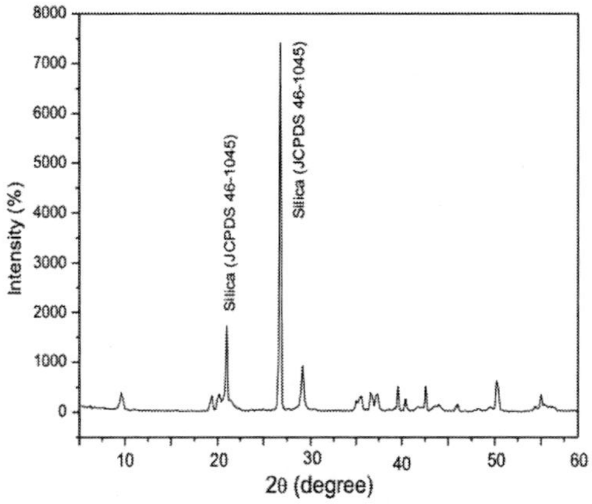

Figure 2. X-ray diffraction (XRD) patterns of silica from pyrophyllite.

Pyrophylite and geothermal sludge was crushed and sieved through 100-mesh sieves. KOH and K_2CO_3 is a source of potassium for the formation of potassium silicate.

Silicon dioxide and potassium mixed by variation of the weight ratio of silica versus potassium in the range of 1 : 1 to 5 : 1. The potassium silicate obtained in different grades should have a chemical formula of $K_2 SiO_3$. Fusion in the furnace is done such that the resulting product is only melted but still solid form. Fusion reactions over a range of temperatures below the melting point of the mixture until it reaches the melting point. The powder of silica (geothermal sludge or pyrophyllite) and potassium were mixed and then sprinkled by water until the mixture was slightly clump so not much lost powder when hot air that blows touching the mixture. The mixture was reacted in the furnace at temperature in the range of 600 °C to 1000 °C.

Characterization

X-ray diffraction (XRD) patterns of the potassium silicate were obtained using an X-ray diffractometer (X'pert, Philips). The composition of silicon dioxide and potassium silicate was analyzed by X-ray fluoroscence spectrometry (XRF) and energy dipersive X-ray fluoroscence spectrometry (EDXRF, Minipal 4, PANalytical). Scanning electron microscopy (SEM) was used to observe the morphology of the product samples.

RESULT AND DISCUSSION

Fig. 3 shows the correlation of raw material ratio (SiO_2/K_2O) to the concentrations of silica and potassium prepared by K_2CO_3. The concentration of SiO_2 and K_2O on potassium silicate from geothermal sludge and pyrophylite followed the change of raw material ratio of SiO_2/K_2O.

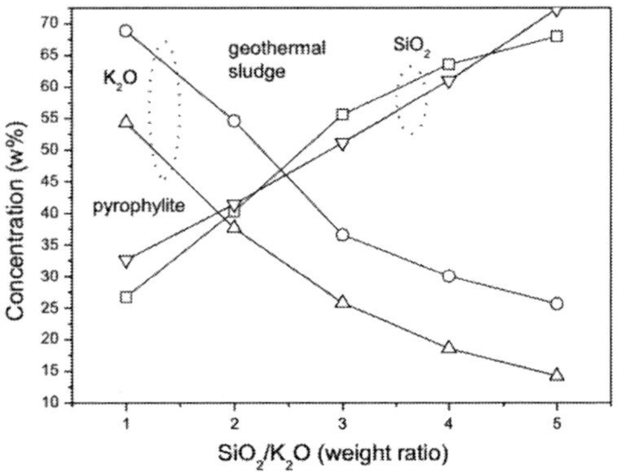

Figure 3. The correlation of raw materials ratio (SiO_2/K_2O) to the concentrations of silica and potassium in potassium silicate product prepared by K_2CO_3.

There was no significant difference for SiO_2 concentration in the potassium silicate produced from these two types of raw materials, but there were significant differences in the concentration of potassium in potassium silicate product. This shows that the potassium silicate produced from geothermal sludge and pyrophylite have different grade ratio in the same raw material ratio (SiO_2/K_2O).

Fig. 4 shows the effect of raw materials ratio on concentration of K_2O in potassium silicate prepared by KOH and K_2CO_3. KOH provides the amount of K_2O in potassium silicate which is slightly larger than K_2CO_3. The concentration of K_2O in potassium silicate produced from pyrophylite was lower than that of geothermal sludge. In raw materials ratio 3:1 the concentration of K_2O is 44.7% prepared by geothermal sludge and KOH, and 24.8% prepared by pyrophyllite and KOH. This indicated that the consumption of pottassium prepared by pyrophyllite larger than those prepared by geothermal sludge for the same expected molar ratio.

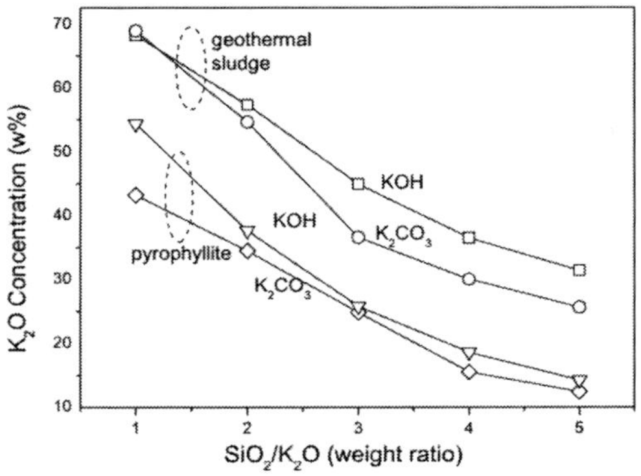

Figure 4. The effect of raw materials on the concentrations of potassium silicate solids.

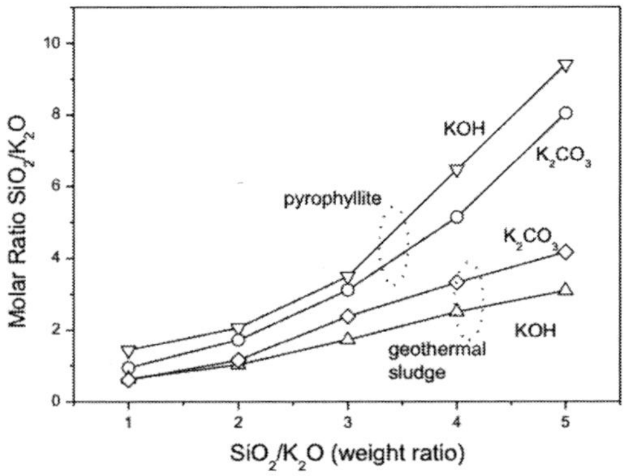

Figure 5. The effect of raw materials ratio on the molar ratio SiO_2:K_2O

Fig. 5 shows the effect of raw materials ratio on the molar ratio SiO_2:K_2O. Potassium silicate prepared by geothermal

DOI: 10.1051/matecconf/20165801021

sludge and KOH has a molar ratio in the range of 0.6 - 3.1, which is less than the molar ratio of potassium silicate from pyrophyllite which is in the range of 1.4 - 9.4. While K_2CO_3 produced potassium silicate with a molar ratio that is slightly larger than the KOH in the range of 0.6 - 4.1. Amorphous silica from geothermal sludge facilitate the process of mixing and reaction.

Potassium silicate solid was prepared by geothermal sludge in accordance with the grade of fertilizer produced on the raw material weight ratio of 4 : 1 with a molar ratio to 2.5 : 1 for the use of KOH, and the raw material weight ratio of 3 : 1 with a molar ratio of 2:33 : 1 for the use of K_2CO_3.

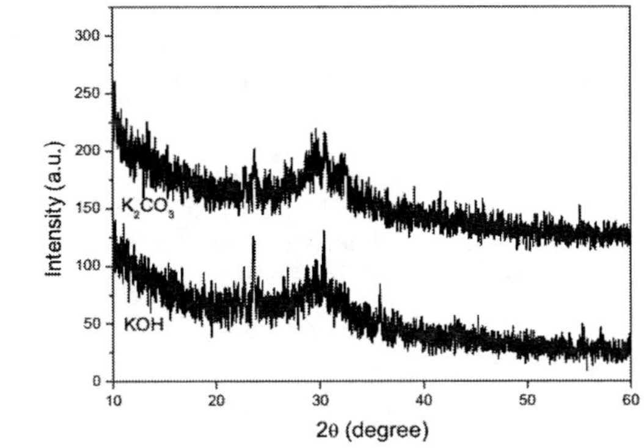

Figure 6. X-ray diffraction (XRD) patterns of potassium silicacate solids on the 1:1 weight ratio

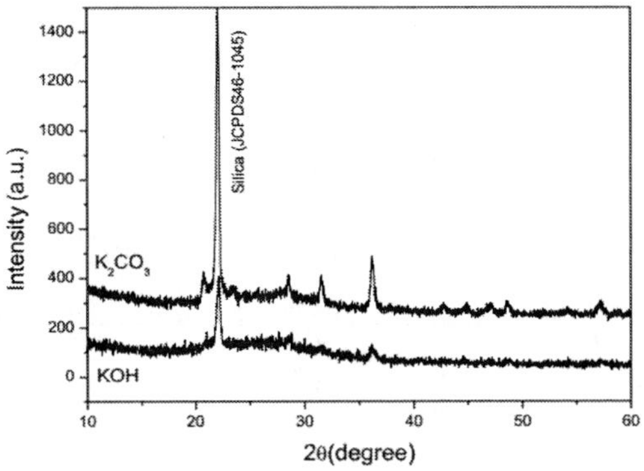

Figure 7. X-ray diffraction patterns of potassium silicacate solids prepared by the 4:1 weight ratio

Diffraction pattern of solid potassium silicate produced from geothermal sludge using KOH and K_2CO_3 in the weight ratio of 1 : 1 is shown in Fig. 6. Both samples showed amorphous nature but the little silica peaked at 22 ° (JCPDS 46-1045). The contrast when the ratio increased to 4 : 1 as diffraction pattern shown in Fig. 7. The Silica intensity is very high (1300) at 22° prepared by K_2CO_3 than prepared by KOH is 200 only. The potassium silicate from KOH samples appears largely amorphous with little traces of crystalline phase.

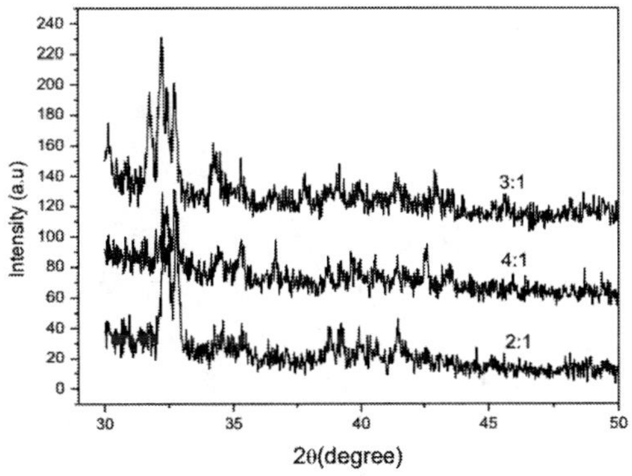

Figure 8. X-ray diffraction patterns of potassium silicacate solids prepared by pyrophyllite

Diffraction pattern of solid potassium silicate prepared by pyrophyllite using K_2CO_3 in the weight ratio of 2:1, 3:1, and 4:1 are shown in Fig. 8. Transformation of crystalline structure (Fig.2) into an amorphous (Fig 8.) occur during the fusion reaction between pyrophyllite (crystalline) with potassium underway. Amorphous structure on potassium silicate product can increase its solubility.

The yield of potassium silicate solid produced from geothermal sludge (70-80%) is lower than those produced from pyrophyllite (91-97%) prepared by KOH. In the other hand, the yield of potassium silicate from geothermal sludge prepared by K_2CO_3 (60-70%) is lower than prepared by KOH (70-80%).

The melting point of potassium carbonate is 891 °C and potassium hydroxide is 406 °C, while the melting point of silica around 1600 to 1730 °C. The difference between the melting point of silica and potassium lead to the melting point of a mixture of both to be different depending on the composition ratio of SiO_2/K_2O. Fig. 9 shows the effect of raw material ratio (SiO_2/K_2O) on melting temperature. Because potassium hydroxide has the lowest melting point, the melting point of the mixture is in the range of 800 ° C to 1100 °C. Based on the melting point of a mixture of geothermal silica and potassium in the molar ratio 3: 1 to 4: 1 the melting temperature up to 900 °C using KOH and up to 1100 °C using K_2CO_3. As for the mixture of pyrophyllite and potassium in the molar ratio 2: 1 reaches the melting temperature 1200 °C using KOH and 1350 °C using K_2CO_3.

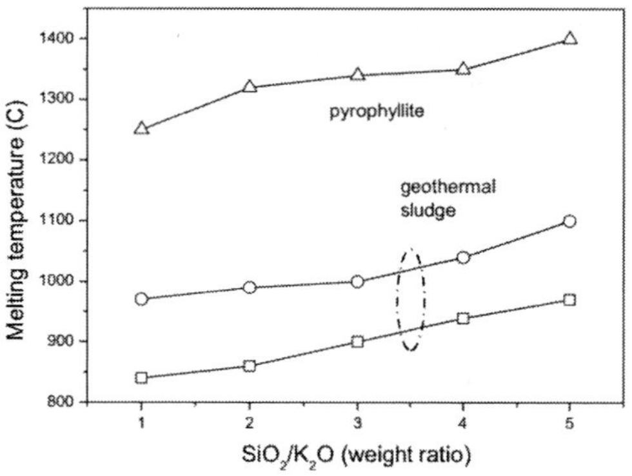

Figure 9. Effect of raw materials ratio SiO_2/K_2O on melting temperature

The SEM images shown in Fig. 10 confirm that the particles of potassium silicate prepared by KOH were smaller and more uniform than those prepared by K_2CO_3. The size distribution of the particles, as deduced from the SEM images, shows that particle size prepared by KOH is in the range of 62-145 nm and that prepared by K_2CO_3 is in the range of 40-220 nm.

(a) (b)

Figure 10. SEM images of potassium silicate solid from geothermal sludge prepared by (a) KOH and (b) K_2CO_3.

CONCLUSION

Characteristics of potassium silicate which is characterized by a molar ratio is influenced by the type of raw material and potassium salts used.
The amount of raw materials needed to get the appropriate molar ratio of potassium silicate fertilizer grade are different, as well as the temperature of the furnace.

Potassium silicate prepared by potassium hydroxide and amorphous silica from geothermal sludge showed in accordance with fertilizer at a low molar ratio (2.5:1), potassium requirement is quite small and a temperature of about 900 °C. Meanwhile, the potassium silicate prepared from pyrophyllite produced a high molar ratio (1.4 - 9.4) and on a fusion temperature of about 1350 °C, so that potassium needed large enough to meet the required molar ratio for the fertilizer grade.

The product of potassium silicate solid is amorphous with a little trace of crystalline.

ACKNOWLEDGMENT

The Authors are grateful to the Ministry of Research and Technology, Indonesia, for its financial support through Hibah Penelitian Unggulan. We also thank Mr. Ermawan of PT Geodipa Energi for providing the geothermal sludge and Mr Haryanto of PT Gunung Bale for providing pyrophyllite. Thank to Pradipta Utama and Dwi Agus Setiawan for assisting with the experiments.

REFERENCES

Gunnarsson, I.,Arnorsson, S., 2000, Amorphous silica solubility and the thermodynamic properties, Geochimica et Cosmochimica Acta, 64(13):2295-2307

Jayawardana, H,A,R,K., Weerahewa H.L.D., and. Saparamadu, M.D.J.S., 2014, Effect of Root or Foliar Application of Soluble Silicon on Plant Growth, Tropical Agricultural Research, Vol. 26 (1): 74 – 81

Muljani, S., Setyawan, H., Sumada, K., 2011, Production of pure amorphous silica from geothermal sludge of Dieng, Indonesia, BISTECH-1 proceed ISSN 978-979-986231-0, D21:1-5.

Muljani, S., Setyawan, H., Wibawa, G., Altway, A., 2014, A facile method for the production of high-surface-area mesoporous silica gels from geothermal sludge, Advanced Powder Tech. 25(5): 1593-1599.

Nolla, A., Coelho, L., 2008, Potassium Silicate as Foliar Spray and Rice Blast Control, Journal of Plant Nutrition, 31: 231–237

Robert O. Fournier, R, O., Jack, J, R., 1977, The solubility of amorphous silica in water at high temperatures and high pressures, American Mineralogis, 62: 1052-1056.

Process for production of potassium silicate solutions by the addition of tempered quartz to hydrothermally reacted quartz and KOH, US 5238668 A

Production of silicates, US 2823098 A

Yao Y., Hamada E., Sato K., Takashi Akiyama T., Yoneyama T., 2014, Identification of the Major Constituents of Fused Potassium Silicate Fertilizer, ISIJ International, 54: 4, 990-993

REMOVAL OF Mg^{2+}, K^{+}, SO_4^{-2} IONS FROM SEAWATER BY PRECIPITATION METHOD

Caecilia Pujiastuti, Ketut Sumada, Yustina Ngatilah, Prasetyo Hadi
University of Pembangunan Nasional "Veteran" Jawa Timur, Surabaya, Indonesia
Email : caeciliapujiastuti@gmail.com

ABSTRACT

Removal of Mg^{2+}, K^{+} and SO_4^{-2} ions in seawater has been successfully done by precipitation in a mixing tank method. This study aims to remove the content of magnesium ions (Mg^{+}), potassium (K^{+}) and sulfate (SO_4^{-2}) in sea water with the addition of chemicals disodium phosphate (1.2 % volume), calcium chloride (2 % volume) and sodium hydroxide (2% volume). Stirring is performed at 100 rpm and the pH solution is adjusted to 9. Disodium phosphate serves to bind magnesium ions and potassium, $CaCl_2$ serves to bind the sulfate ion, while sodium hydroxide is used to adjust the pH of the solution mixture and also reacted with magnesium ions. In total, the removal efficiencies of Mg^{2+}, K^{+} and SO_4^{-2} ions in seawater were 97%, 96%, and 92%, respectively. The precipitated solids contains component of PO_4^{-} (14.5%), Mg^{2+}(13.8%), $SO4^{2-}$ (28.2%), Ca^{2+} (24.1%) and K^{+} (1.9%) ions.

Keywords: sea water, disodium phosphate, calcium chloride, precipitation, removal ions

INTRODUCTION

The average waters of the world's oceans contain 3.5% dissolved salts, 96.5% water and average salinity is about 35 g/kg. When salts dissolve in water they usually react with water and dissociate (break apart) into ions, positively and negatively charged atoms or groups of atoms. These ions are either negatively charged anions, or positively charged cations. Six major ions make up > 99% of the salts dissolved in seawater: four cations sodium (Na^{+}), magnesium (Mg^{2+}), calcium (Ca^{2+}), and potassium (K^{+}); and two anions: chloride (Cl^{-}) and sulphate (SO_4^{2-}). These six ions and the five next most common ones make up the major constituents of seawater and comprise 99.99% of dissolved materials. Sodium chloride (NaCl), makes up 86% of dissolved ions in seawater. Many other elements (trace elements) are dissolved in seawater in concentrations less than one part per million (ppm).

The molar concentration of magnesium in sea water is about five times higher than calcium. In fact, both magnesium and calcium are in a delicate equilibrium where slight charge in alkalinity and carbon dioxide tension may cause precipitation (Irving, 1926). Jijun et al, (2012) developed the precipitation of calcium and magnesium from seawater using CO_2. Mg ion can rarely incorporate into the solid precipitates, compared, for example, to calcium salts. The rate of Mg phosphates nucleation is several orders of magnitude lower than that of Ca phosphates, at the same values of pH and total phosphate concentration (Golubeva, 2001). The removal rate of Ca is higher than the removal rate of Mg in pH value between 8.0 and 9.0. In the Mg-free seawater system, the removal rate is high enough in pH value between 8.0 and 9.0. Meanwhile, in the Ca-free seawater system, there is no precipitation between pH 8.0 and 8.3 (Jijun et al, 2012).

Precipitation of phosphate is high enough in pH range from 8 to 10. However, it decreased as pH is increased from 10 to 12, probably because of competition between phosphate and hydroxyl OH^{-} ions (Liu and Warmadhewanti, 2009). Fernando and Lerida (2005) developed a method of binding magnesium and potassium ions in bittern by disodium phosphate to produce a phosphate-based fertilizer following the reaction:

$$MgCl_2 + KCl + Na_2HPO_4 + NaOH \rightarrow MgKPO_4\downarrow + 3\,NaCl + H_2O \quad (1)$$

Magnesium chloride ($MgCl_2$) was more effective than magnesium oxide (MgO) in inducing precipitation reactions between magnesium and phosphate. The magnesium- potassium- phosphate ($MgKPO_4$) precipitation is an efficient method for the simultaneous removal of P and K to yield multi-nutrient products (Xu et al, 2011). In addition, the removal of SO_4 ions from bittern using $CaCl_2$ is using the following precipitation reaction: (6,776,972. AS)

$$MgSO_4 + CaCl_2 \rightarrow CaSO_4\downarrow + MgCl_2 \quad (2)$$

$CaCl_2$ solid is not ionized, but in aqueous solution only $Ca(aq)^{2+}$ and $Cl(aq)^{-}$ exist. The salvation and ionization of the cations are chemical processes responsible for the heat evolved. Magnesium hydroxide will precipitate from the alkalinity part and calcium sulfate will precipitate from the calcium part. Basically, any solutions of a soluble calcium salt and a soluble sulfate salt can be used to prepare calcium sulfate (Yun-Wei et all, 2011) Most sulfates (SO_4^{-2}) are soluble. $CaSO_4$ is slightly soluble so the yield is may be a little lower than the theoretical.

In the previous study, to improve the quality of the salt is done by purification with the addition of binder impurities. In the process of recrystallization the salt that has been formed then dissolved by water to form the solution of salt. The removal of impurities in salt solution performed by the addition of certain chemicals such as NaOH, $BaCO_3$, and Ca $(OH)_2$. This process becomes less efficient due to the salt production process requires two stages of crystallization (Sumada et al, 2012)

Some separation process was carried out for the recovery of salt from sea water which is a chemical treatment (Fernandez, 1993), nanofiltration (Telzhenky et all, 2011), and ion exchange.

This study developed a purification of salt through the removal of ions in sea water by precipitation method. The precipitation is also influenced by the pH of the mixture due to t quite diverse he components in sea water. It is important to adjust pH in the precipitation process.

METHODOLOGY

The sea water used in this study is obtained from Segoro Tambak, Sedati, Sidoarjo, East Java Indonesia. Seawater sampling was conducted during different times to obtain the average number of components in sea water. Table 1 shows the concentration of some components of sea water using AAS and Spectrophotometer. In the present study, ions removal from seawater process is done in a stirred tank at 100 rpm, and at room temperature. The concentration of Na_2HPO_4 is 15% w, $CaCl_2$ is 15% w and residence time is 10, 20, 30 minutes. The first stage of the preliminary process is carried out using each Na_2HPO_4-free seawater system and $CaCl_2$-free seawater system while the second stage is done using these two chemicals.

The first stage: 1) sea water 1000 ml was entered in a stirred tank and then added with Na_2HPO_4 solution in the range of 8-16 ml, stirred for 10 minutes and then added with NaOH 2N solution until the mixture solution reaches pH about 9, then stirred until a specified time. 2) In the same way as point (1) but using the $CaCl_2$ solution in the range of 10-30 ml. For the second stage: 1000 ml sea water was entered in stirred tank then added with Na_2HPO_4 solution in the range of 8-16 ml, stirred for 10 minutes then added $CaCl_2$ solution in the range of 10-30 ml; stirred for 10 minutes then added with NaOH solution until the mixture reaches a pH 9 then stirred until a specified time. Figure 1 showed the flow diagram of experimental procedures.

Table-1. The Concentration of some components of sea water (Segoro Tambak, East Java).

No	Parameters	Concentration (mg/L)			Avarage
		I	II	III	
1	NaCl (%)	1.97	1.99	2.03	1.997
2	$MgCl_2$	842.75	872.16	869.42	861.44
3	$MgSO_4$	2526.23	1978.56	2164.4	2223.06
4	KCl	308.5	288.06	301.56	299.37
	Total avarages				3657.68

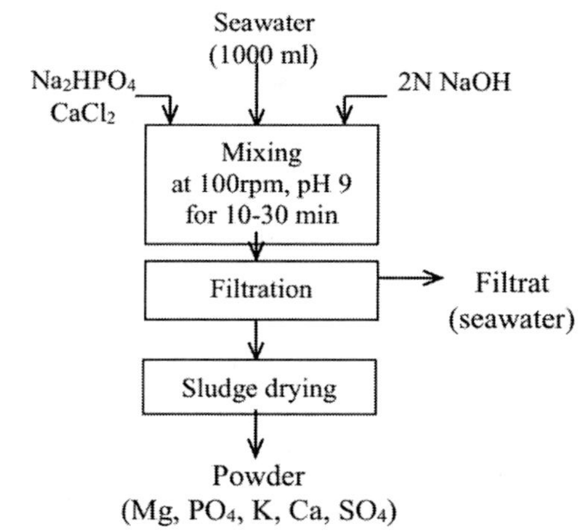

Figure-1. Schematic of experimental procedure for ions removal from seawater

The analysis of components of Na, K, Ca and Mg ions used Absorption Atomic Spectrophotometry (AAS), while for components Cl, SO_4 and PO_4 ions, a spectrophotometer. EDAX and X-ray fluorescence (XRF) were used for component analysis in the precipitated solid product.

RESULT AND DISCUSSION

Figure 2 shows the effect of the volume of Na_2HPO_4 solution on the removal of magnesium ions. The results showed that the larger volume of the solution of disodium phosphate the bigger of magnesium ions are removed until it reaches a constant. The effect of reaction time on the percentage removal rates of the constituent ions of Mg^{2+} dosed at Na_2HPO_4 12 ml with only 20min when pH value was about 9. This suggests that Mg^{2+} is formed homogeneously under certain conditions and that nucleation is rapid. Due to the abundance of foreign solid phases in a seawater treatment, the principal precipitation mechanism is likely to be heterogeneous, however, precipitation is still expected to be rapid.

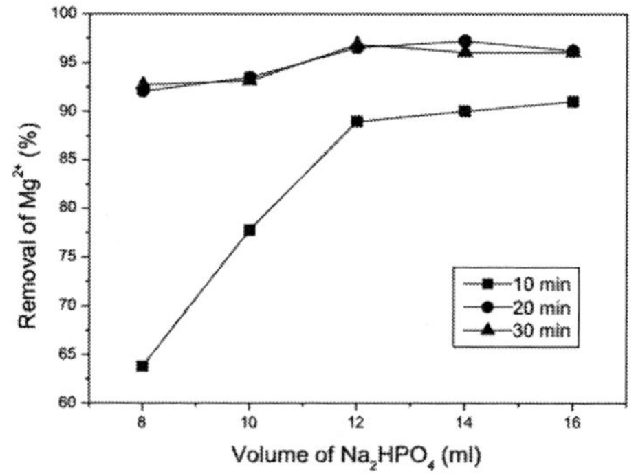

Figure-2. Effect of disodium phosphate (Na_2HPO_4) solution on the removal of magnesium ions.

The reaction time of 30 min is no longer effective because the same ions removal results are obtained with a reaction time at 20min. It indicates the maximum condition in which the Mg^{2+} ions could be separated in addition solution of 12 ml and a stirring of 20min while the removal percentage of Mg^{2+} ions reached 97.15%.

Figure 3 shows the effect of the Na_2HPO_4 solution on the removal of K^+ ions. The same trend as the removal of Mg^{2+} ions occurs in the removal of K^+ ions. K^+ ions concentration in sea water almost unchanged on the addition of 12 ml to 16 ml Na_2HPO_4 and in residence time from 20 minutes to 30 minutes. The removal rate of K^+ ions reached 96.26 % in 20min and pH 9.

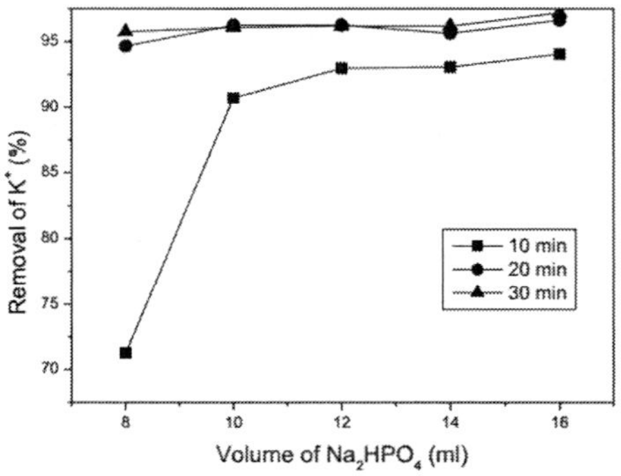

Figure-3. Effect of disodium phosphate (Na_2HPO_4) solution on the removal of potassium ions.

Figure-4 shows the effect of $CaCl_2$ solution on the removal of SO_4^{-2} ions. The results showed that the larger the volume of the solution of calcium chloride, the greater the removal of SO_4^{-2} ions, until it is relatively unchanged on the addition of 15 ml to 30 ml calcium chloride and in a residence time from 20 minutes to 30 minutes. The percentage of SO_4^{-2} ions removal reached 92.29 %.

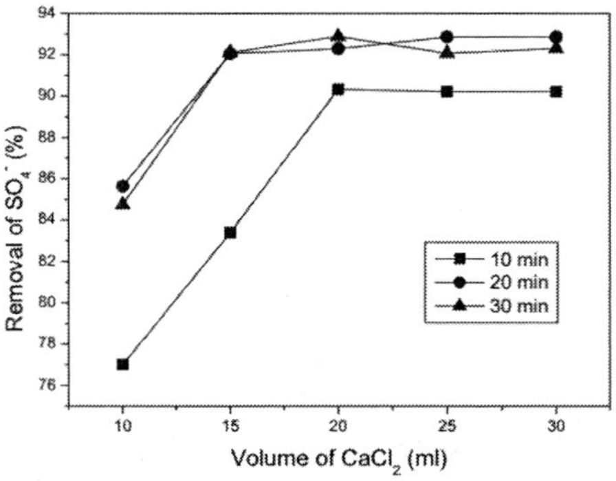

Figure-4. Effect of calcium chloride ($CaCl_2$) solution on the removal of sulfate ions.

Figure 5 shows the XRD pattern for precipitated solid produced from seawater and $CaCl_2$. Whereas Figure 6 shows the XRD pattern for precipitated solid produced from seawater and Na_2HPO_4. This indicated that the precipitated solid product prepared by Na_2HPO_4 is amorphous with a little trace crystalline. The precipitate solid prepared by $CaCl_2$ shows more peaks, it proves that there are several types of ions contained in the precipitate.

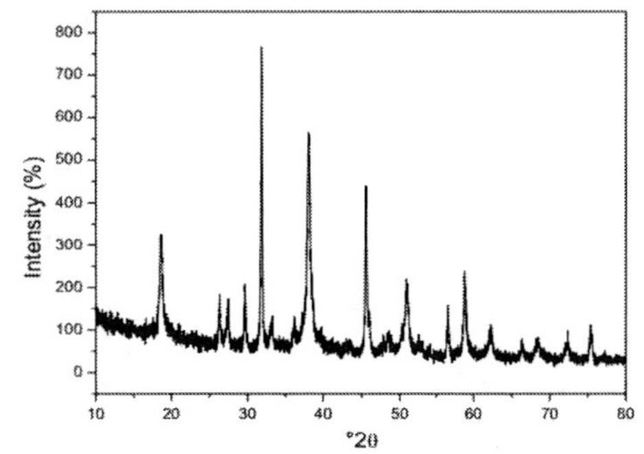

Figure-5. X-ray diffraction (XRD) patterns of precipitated solid produced from seawater and $CaCl_2$.

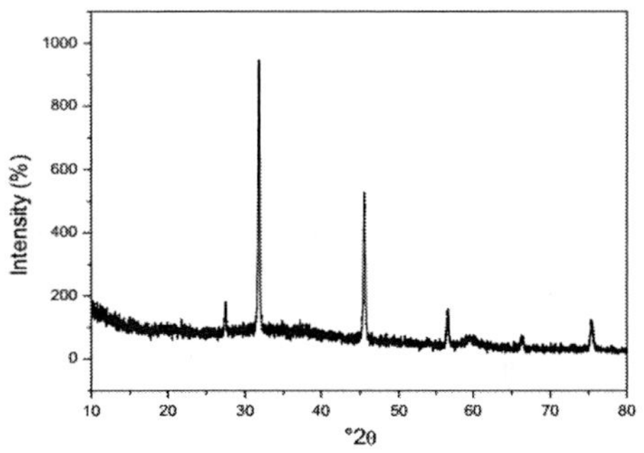

Figure-6. X-ray diffraction (XRD) patterns of precipitated solid produced from seawater and Na_2HPO_4.

Figure 7 shows ions distribution in the precipitated solid by EDAX analysis. The results show that the precipitated solids consist of PO_4^- (14.5%), Mg^{2+}(13.8%), $SO4^{2-}$ (28.2%), Ca (24.1%) and K (1.9%) ions. The removal rate of Mg^{2+} ions reached 95% and the removal rate of SO2- ions reached 93% when the pH value is about 9.0. The precipitated solid in the $CaCl_2$-free seawater system mainly consisted of PO_4^- (41.4%), Mg^{2+}(37.3%), $SO4^{2-}$and K (7.2%) ions. Meanwhile, the precipitated solid in the Na_2HPO_4-free seawater system mainly consists of Mg^{2+}(9.3%), $SO4^{2-}$ (38.8%), and Ca (37.9%) ions.

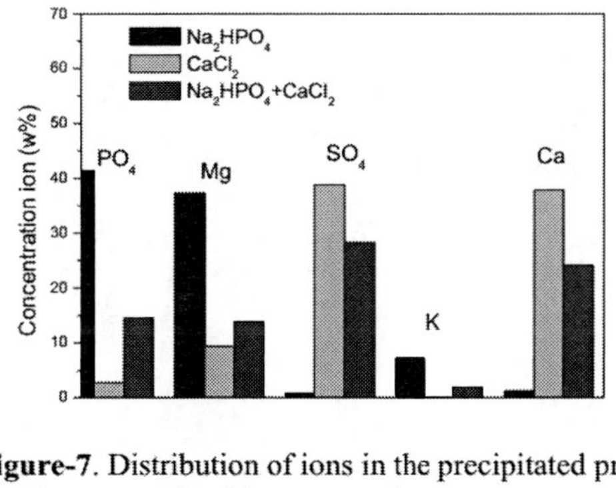

Figure-7. Distribution of ions in the precipitated product as reaction proceeds of ions removal.

CONCLUSION

The results showed that it is possible to remove Mg^{2+}, K^+ and SO_4^{-2} ions in seawater by precipitation method using Na_2HPO_4 and $CaCl_2$. The removal efficiencies of Mg^{2+}, K^+ and SO_4^{-2} ions in seawater were 97%, 96%, and 92%, respectively.

The solid precipitated can be used as some fertilizer components (Mg, P, K).

REFERENCES

Fernandez, L, J.A., 1993, Production of potassium sulphate double salt and magnesium chloride rich solution from seawater bitterns, Chem.Eng, 52, 89-92

Fernando, J, A., and Lerida S., 2005, Multinutrient Phosphate-Base Fertilizer From Seawater Bitterns, Interciencia, 27(9), 496-499

Golubeva, S.V. Pokrovskyb, O.S., Savenko V.S., 2001, Homogeneous precipitation of magnesium phosphates from seawater solutions, J.Cryst. Growth, 223, 550–556

Irving, L., 1926, The precipitation of calcium and magnesium from sea water, J. Mar. Bio., 14(2).

Jijun, Z., Junsheng, Y., Xia,L., 2012, Study on the removals of calcium and magnesium from seawater using CO_2, Chem. Ind. and Eng. Prog., 31(03): 681-686.

Liu, J., and Warmadewanthi, 2009, Selective Precipitation of Phosphate from Semiconductor Wastewater. J. Environ. Eng., 135(10), 1063–1070.

Sumada, K., Caecilia P., dan Widodo,L,U., , Study of impurities removal by recrystallization method, Proceed. Seminar Nasional Teknik Kimia Soebardjo Brotohardjono IX ,UPN "Veteran" Jawa Timur, Surabaya (2012

Telzhensky, M., Birnhack, L., Lehmann, O., Wndler, E., Lahav, O., 2011, Selective separation of seawater Mg2+ ions for use in downstream water treatment processes, Chem. Eng., 175, 136-143

Xu, K., Wang, C., Liu, H., Qian, Y., 2011, Simultaneous removal of phosphorus and potassium from synthetic urine through the precipitation of magnesium potassium phosphate hexahydrate, Chemosphere, 84 (2), 207–212

Yun-Wei, W., Yi-Yeoun, K., Hugo K. C., Fiona C. M., 2011, A New Precipitation Pathway for Calcium Sulfate Dehydrate, Electronic Supplementary Material (ESI) Journal: The Royal Society of Chemistry

A process for recovery of common salt and marine chemicals from brine in an integrated manner, 6,776,972. AS

REMOVAL OF ORGANIC LOAD IN COMMUNAL WASTEWATER BY USING THE SIX STAGE ANAEROBIC BAFFLE REACTOR (ABR)

Minarni Nur Trilita, Novirina Hendrasarie and Iwan Wahjudijanto
University of Pembangunan National "Veteran" Jawa Timur, Surabaya, East Java, Indonesia
E-Mail: trilita69@yahoo.com

ABSTRACT
The reduction of water quality in the urban drainage is a crucial problem to overcome because it can affect the health of community. This fact encouraged the researcher to conduct the research in efforts to increase the water quality in the drainage. One of the solutions to increase the water quality in the drainage is that the domestic wastewater must be treated at first before it is flown through the drainage. Furthermore, the wastewater treatment was conducted by employing the communal wastewater processor. The research was aimed at knowing the capability of Anaerobic Baffle Reactor with the six-stage design in communal wastewater processor in efforts to decrease the organic load. This research was conducted in a laboratory scale. Meanwhile, the sort of waste used was taken from the domestic wastewater of settlement by varying its discharge and waste concentration flowing into the waste processor. Finally, the research result showed that the reduction of organic load of COD was reaching up to 92%, N was 85% and Phosphate was 50%.

Keywords: suspended microbiology, anaerobic process, domestic wastewater

INTRODUCTION

In developing countries, especially in the urban areas, the reduction of water quality in the drainage is often found. This reduction of water quality in the drainage might cause some nuisance to the community's health. Therefore, the increase of water quality in the drainage is required. One of the solutions is by making use of communal wastewater processor to make the wastewater flowing in the drainage already fulfill the water quality permitted in efforts to increase the self purification of environment.

The process commonly used in Indonesia to treat the domestic waste water is septic tank which combines anaerobic and infiltration. Septic tank is a simple process and quite cheap, and it is only appropriate for the less crowded population. If implemented in the urban areas with crowded population, it might cause some contamination of the water under the ground. However, septic tank usually treats the wastewater from the lavatory only. Meanwhile, the other wastewater such as the laundry wash, kitchen and bath room, they are usually directly thrown out without being treated. This might cause some contamination in that area. Furthermore, the Centralized Sewage Treatment System requires much money. (Benefield and Randall, 1980; Grady and Lim, 1980; Metcalf and Eddy. 2003). Therefore, the process communal wastewater treatment is an important and realistic option in efforts to treat the domestic wastewater around the river bank, so that the disposal will not contaminate or pollute the self purification of the river.

Anaerobic Baffle Reactor (ABR) is anaerobic suspended treatment system in a baffle reactor. Meanwhile, the suspended growth is more advantageous than the attached growth because it does not need supporting media and is not easily plugged up. ABR is an Upflow anaerobic Sludge Blanket (UASB) installed in series, but it does not need any particle in its operation, so that the shorter start-up period is required. Furthermore, a series of vertical partition is put in the ABR to make the wastewater flow under and over from inlet to outlet, so that there will be a contact between wastewater and active biomass. The concentration profile of organic compounds varies along the ABR. This might cause the population growth of microorganism different between that of put in one compartment and another. This depended on the condition of specific environment produced by the compound as a result of decomposition (Malakahmad et.al., 2011, Wang et.al., 2004, Shanmugam and Akunna, 2010; Liu et.al., 2007; Kuşcu and Teresa, 2007; Jamshidi and Khalesidoost, 2014; Young and Young, 1988; Grobicki, Aand Stuckey, D.C., 1992).

The bacteria in the bioreactor will float or settle in accordance with the characteristics of flow and gas produced. However, it can move horizontally to the top of reactor slowly so that it can increase its cell retention time. Furthermore, while the contacted wastewater with active biomass was flowing, the effluent was free from biological solids. This configuration could show the high rate of COD elimination (Sinan Uyanik, 2003; Wang, Jianlong, et.al., 2004).

Furthermore, communal wastewater system design is used in this research. To make the performance of ABR more efficient, pre-treatment is added. The pre-treatment used in this research includes equalizing tank and primary clarifier. The function of equalizing tank is to make the stream's flow of the wastewater to be treated become stable its fluctuation. Meanwhile, primary clarifier is used to reduce the solid matter to be removed in the wastewater gravitationally so that the ABR reactor does not work harder.

Some advantages of ABR are its capability of separating the asidogenesis process and metanogenesis longitudinally that enables its reactor have two-phase system. This means that this phase does not need any control and does not spend much money. Its design is also quite simple, it does not need mechanical mixer, it spends less money for its construction cost. Besides, its biomass does not need a specific characteristics of sedimentation, it contains low sludge, and its hydraulic system of ABR is capable of reducing the bacteria from losing.

Considering with the low cost and it is easy to be implemented in a urban community especially in developing countries, biological communal wastewater processor of anaerobic baffle reactor was chosen in this research. Meanwhile, the anaerobic baffle reactor with the six-stage design was expected to be capable of reducing the total of nitrate and phosphate load in the domestic wastewater. Therefore, this research was aimed at knowing the capability of communal wastewater processor in efforts to reduce/decrease the organic load in the domestic wastewater.

METHODOLOGY

The research was conducted by designing Anaerobic Baffle Reactor (ABR) with 6 stages applied in the communal wastewater. The tool design of the experiment was based on the calculation. Meanwhile, the reactor of ABR was made of fiber plastic (Figure 1).

Figure-1 Schematic Anaerobic Baffle Reactor

The more specific about it was explained in Table 1.

Table-1. Specifications of Reactor

Description	ABR
The width of Settling Tank (m)	0.33
Depth (m)	0.43
Up flow velocity (m/hour)	1.8
The number of chamber	5
Depth of outlet (m)	0.75
L : D	0.5 : 0.75

Based on the research scheme, it could be explained in Figure 2.

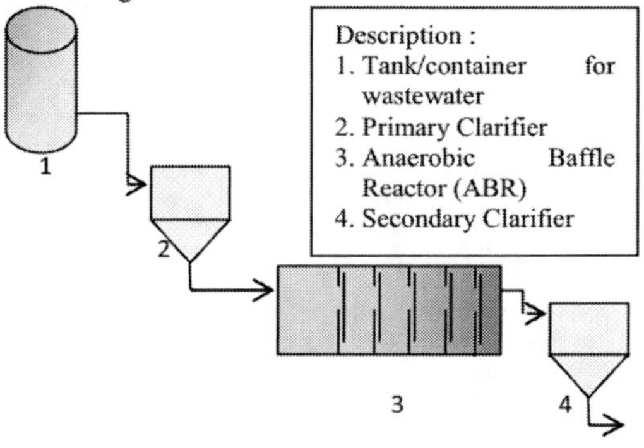

Figure-2. Schematic Experimental Set Up

The wastewater used in this research was communal wastewater with the former characteristic of wastewater of COD = 150 – 800 mg/L, BOD5 = 75 – 400 mg/L, TSS = 134-786 mg/L, pH = ± 7 N Total = 250 – 700 mg N/L, PO4 = 20 – 60 mg/L. The bacteria to be breed was taken from its own communal wastewater. This was done by varying the inflow discharge as the following: 120 mL/minute, 140 mL/minute, 160 mL/minute, 180 mL/minute, and 200 mL/mnt. Based on this experiment, it could be obtained that HRT was about 16.47 – 27.45 hours.

RESULTS AND DISCUSSIONS

The capability of eliminating the organic substance of the communal wastewater employing ABR could be seen from the following discussion

Seeding Stage and Acclimatization

There were 4 (four) stages in the metabolism of bacteria growth: hydrolysis, asidogenes, asethogenesis, and methanogenesis. These four stages of anaerobic processes must be conducted in efforts to make the reactor of ABR be ready to use. Furthermore, these four stages were conducted during seeding process, so that it would be obtained the appropriate bacteria. At first, when the seeding process happened, there was pH decrease (hydrolysis, asidogenesis and asethogenesis) and it will become stable if the range of pH is 6.6 – 7.6 (methanogenesis stage). However, during the seeding process, some nutrient was required for microorganism growth or bacteria with its comparison as C : N : P = 100 : 5 : 1, added with some glucose as a source of C, Urea (fertilizer) as a source of N, and Kh2PO4 as a source of P. The initial result of analysis was known that COD = 560.64 mg/L, total N = 493.3 mg/l and Phosphate = 24.51 mg/l. Based on this result (nutrient comparison), there should have been some more glucose addition and phosphate, so that the score for COD would become COD = 6451.2064 mg/l.

Furthermore, aclimatization process was conducted in efforts to adapt the bacteria available in ABR with the wastewater to be treated. Aclimatization was conducted with the wastewater concentration of 50% up to 100%. Aclimatization was stopped when the outlet of ABR got its score < 50% from the initial wastewater. To reduce the COD content about 50% - 75%, it was required 1 day and 2 days for the concentration of 100% with the percentage of COD decrease as much as 81%. This indicated that the reactor was ready to use and operate.

The removal of Organic Substance of Communal Wastewater

The analysis result of organic substance of communal wastewater that had been treated in ABR showed that the best elimination percentage of input COD concentration was 184.52 mg/l, with the discharge of 140 ml/mnt, HRT = 18.86 hours in ABR. This score had already fulfilled in accordance with the established regulation/rule. The regulation employed was the Governor's Decree East Java, No 72 in the year 2013 about the standard of wastewater containing the maximum effluent of COD of 50 mg/l, TSS = 50 mg/l, and pH = 6-9.

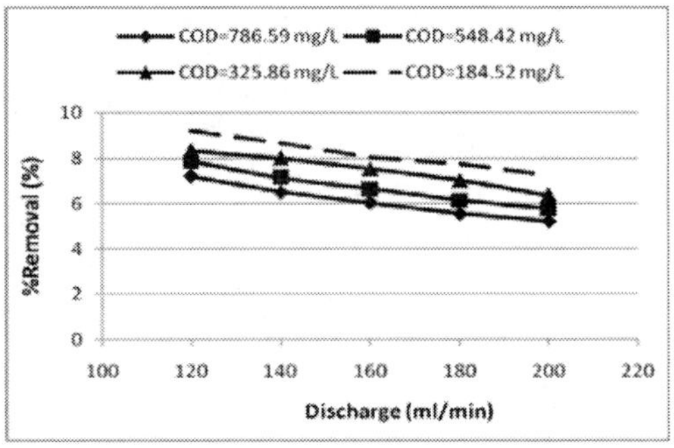

Figure-3a. Percentage of removal COD in the Primary Clarifier

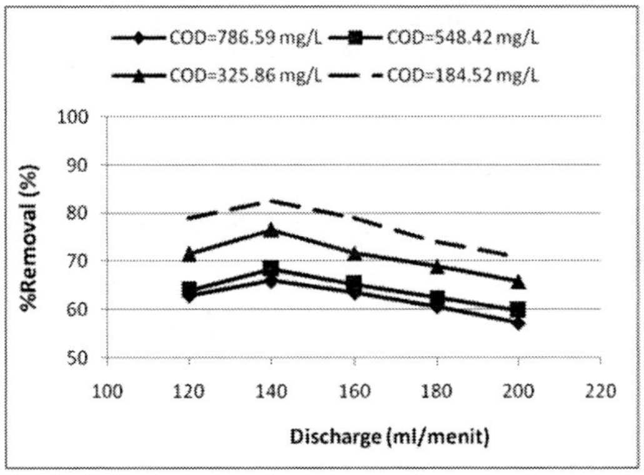

Figure-3b. Percentage of removal COD in the ABR

Based on the experimental result, the best discharge where ABR was set up, the primary clarifier was put before ABR and the secondary clarifier was put after ABR. This composition resulted in the highest elimination percentage of COD at the discharge of 140 ml/minute with its total of HRT = 23.53 hours. Figure 3 showed the elimination percentage of COD of each reactor.

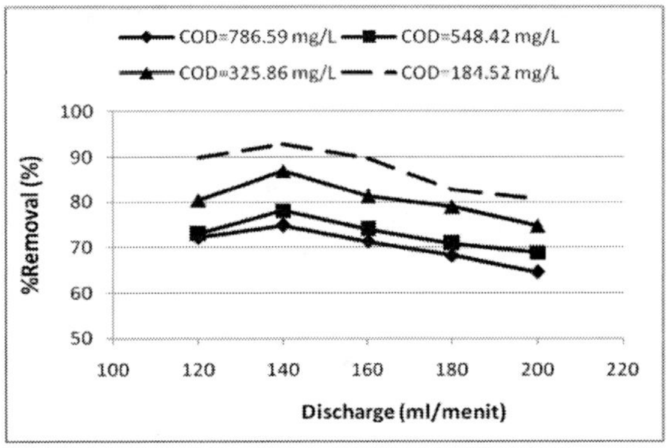

Figure-3c. Percentage of removal COD in the Secondary Clarifier

To reduce the content of Total Suspended Solid (TSS), the best score was obtained from secondary clarifier with its concentration variation of COD = 134.52 mg/l with its discharge = 140 ml/minute (HRT 18.86 hours), elimination percentage = 95.67%. Meanwhile, the concentration score of initial TSS of 1010 mg/l, the score of final TSS was 44 mg/l. This score had already fulfilled the standard established, that was less than 50 mg/l (Figure 4).

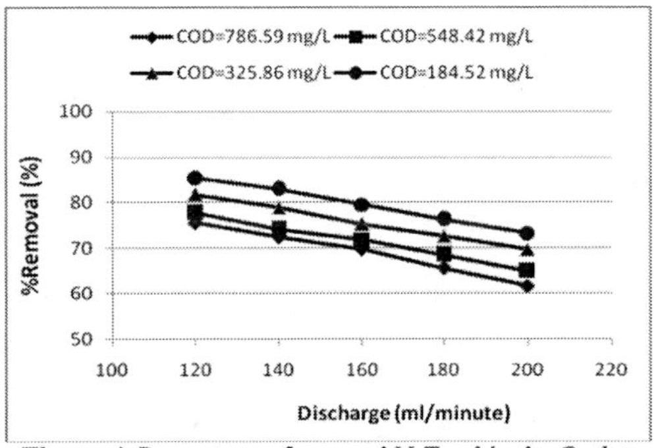

Figure-4. Percentage of removal N-Total in the Outlet ABR

Meanwhile, to reduce the total content of Nitrogen, the best score was obtained at the discharge of 120 ml/hour as much as 85%. The reduction percentage of N-Total was almost the same as the reduction percentage of COD and TSS which reached up to 70-80%. Furthermore, the best reduction percentage of Phosphate at the discharge of 120 ml/hour as much as 50.54%. However, the reduction percentage of Phosphate was only reaching up to 50%. This score was smaller than the reduction percentage of COD and TSS which was reaching up to 70-80% (Figure 5).

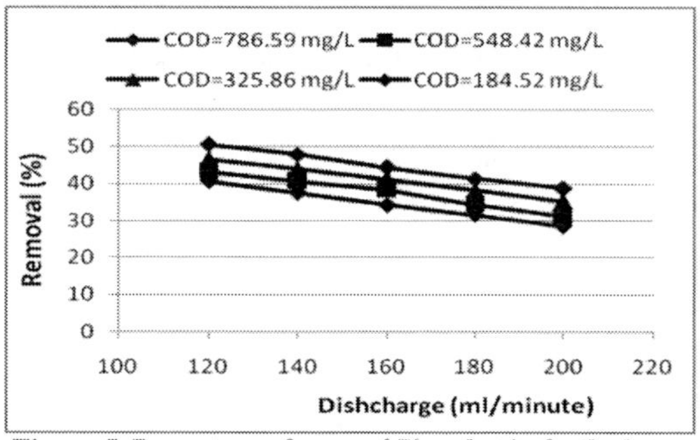

Figure-5. Percentage of removal Phosphat in the Outlet ABR

CONCLUSION

Based on the research, it was found that the Anaerobic Baffle Reactor was capable of reducing the organic parameter of communal wastewater. This might not make the water quality flowing in the drainage be polluted. The reduction capability of organic load of domestic wastewater for COD was about 90%, Total of Nitrogen was 80% and Phosphate was 50%. This showed that the pre-treatment design (equalizing tank & primary clarifier) and Anaerobic Baffle Reactor (ABR) could be used in communal treatment system.

ACKNOWLEDGMENTS

This work was supported by research grants (PUPT) from DIKTI (Department of Higher Education Indonesia) in the year 2015 – 2016.

REFERENCES

Malakahmad A, et.al, 2011, Study on Performance of a Modified Anaerobic Baffled Reactor to Treat High Strengh Wastewater, Journal of Applied Sciences, Vol. 11, Issue 8. 1449-1452.

Benefield and Randall. 1980. Biological Process Design for Wastewater Treatment.

Grady, L..C.P., and Lim, C.H., 1980, Biological Wastewater Treatment, Pollution Engineering and Technology, New York.

Grobicki, A., and Stuckey, D.C., 1992, Hydrodynamic Characteristics of The Anaerobic Baffled Reactor, Water Research, Vol. 26. Issue 3, 371-378.

Jamshidi, S., Khalesidoost, M., 2014, Biogas Production Yield of Anaerobic Baffled Reactor in Low pH and HRT, Journal of Middle East Applied Science and Technology (JMEAST), Vol. 15, No. 4, pp. 765-770.

Kuşcu, Ö.S., Teresa, D., 2007, Effects of Nitrobenzene Concentrations and Hydraulic Retention Time on The Treatment of Nitrobenzene in sequential Anaerobic Baffled Reactor and Continuously Stirred Tank Reactor System, Water Science & Technology, Vol. 55, No. 10, pp. 227 – 236, IWA Publishing.

Liu, Xiao-lei, Ren, Nan-qi, Wan, Chun-li, 2007, Hydrodynamic Characteristics of a Four-compartment Periodic Anaerobic Baffed Reactor, Journal of Environmental Science (Elsevier), Vol. 19, Issue 10, p1159-1165, 7p.

Metcalf and Eddy. 2003. Waste Water Engineering Fourth Edition. NewYork: Mc .Graw-Hill Company.

Shanmugam, A.S, Akunna, J.C. 2010. Modelling Head Losses in Granular Bed Anaerobic Baffled Reactors at High Flows During Start-up, Water Research, Vol. 44 Issue 18, p5474-5480, 7p.

Uyanik S., 2003, A Novel Anaerobic Reactor: Split Fed Anaerobic Baffled Reactor (SFABR), Turkish J. Eng. Env. Sci., Vol. 27, 339-345, EBSCO Publ.

Young, H.W, and Young, James C., 1988, Hydraulics Characteristics of Upflow Anaerobic Filters, Journal Environmental Engineering, Vol. 114, No. 3, 621-638.

Wang, Jianlong, Huang, Yongheng, Zhao, Xuan, 2004, Performance and characteristic of an Anaerobic, Journal Bioresources Technology, Vol. 93, Issue 2, p205, 4p.

THE BENEFIT OF CACAO PEEL's LIGNIN AS AN ADHESIVE USING MULTI FUNCTION EXTRACTOR

Mu'tasim Billah[1], Titi Susilowati[2], Susilowati[3], and Diah Hari Suryaningrum[4]
University of Pembangunan Nasional"Veteran" Jawa Timur, Surabaya, Indonesia
E-Mail: tasimbillah60@yahoo.co.id, tasimbillah60@gmail.com

ABSTRACT

Lignin of Cacao Peel can be used as lignin adhesive by cooking process with sodium hydroxide liquid in the multifunction extractor. The benefit of this study is to minimize the waste of cacao peel that is plentiful. Cacao peel was cleaned then dried and pounded. Powder of cacao peel (200 gr) was cooked with water at 70 °C for an hour and then leaked. Sodium hydroxide liquid 1000 mL is used in various concentration (%) ; i.e : 8,10,12 and 18 for different times (60, 90, 120 and 180 minutes). After that, the mixture of liquid and solid cooking was separated to get lignin and its deposit was disposed. The black liquor was added with water until it reached a pH of 8-9. The sedimentation process was then performed by adding Chloride acid 10 % until pH reached 2-3, during 8-10 hours. The Lignin formed is separated and dried. Lignin powder (5 gr) was mixed with 2.5 gr resorcinol and 10 mL of formaldehyde solution, agitated by adding NaOH 10% and 50 %, respectively, each of that 20 mL, until lignin gum is formed. The result showed that 600 mL of black liquor provided 29.8 gr lignin powder and lignin adhesive. It's adhesive capacity is 3.54 kg/sq.Cm and the color is near to cacao peel's color.

Keywords: Cacao peel, black liquor, lignin adhesive.

INTRODUCTION

One ton of cacao seed produces approximately 10 tons of fresh cacao peel, resulting in generating piles of cacao peel's waste. One of the methods to solve the problem is processing cacao peel to become lignin products which have a sale value that is beneficial for humans. Lignin is the primary component in chemical's wood beside cellulose and hemicellulose. The existence of lignin is plentiful in the world and regarded as the second polymer organic component. Lignin has been studied as an adhesive for more than 100 years, but there are only a few industrial applications. The reason for the current interest is the high availability and low price of lignin, (Hemmila, *et.al.*, 2013).

According to Research and Development Center in Department of Forestry, as much as 20-60% of production cost in wood industries was used to buy adhesives.The industries that depend on this gum i.e.: plywood industry, particle board industry, vinyl, fiberboard industry, and connected board. Most of the adhesives were obtained from petroleum that is non-renewable. Lignin as an adhesive has been the subject of several reviews (Rowell, 2005) (Pizzi, 2006) (Hu. *et al.*, 2011) (Imam. *et al.*, 2012) (Mankar. *et al.*, 2012).

The simplest way to use lignin as an adhesive is to use it as a partial replacement for phenol in PF resins. A lot of work has been done on this for different types of lignin and in various percentages and P/F ratios, and it can be concluded that for applications that require fast curing, the use of unmodified lignin is not cost-effective (Danielson and Simonson 1998, Sarkas 2000, Zhang 2013).

Lignin is a material that is obtained from wastes resulted from delignification process of cacao peel that reacted with formaldehyde resorcinol to be lignin resorcinol formaldehyde adhesive. The objective of this study is to minimize dependence on adhesive demand, to minimize environmental pollution, and also to minimize production cost of adhesive products. This study hopes that the results of this research can be developed in industrial scale in Indonesia which will be able to open up working opportunity for Indonesian society. Due to the important role of lignin in fiberboard manufacture, several studies have investigated the use of lignin as a natural adhesive and the possibility of replacing fibers with lignin. Angles et.al. (2001) tried to use lignin as a natural adhesive in fiberboard manufacture by adding different types of technical lignins. They found that replacing up to 20 % of the fibers with kraft lignin improved the water resistance and mechanical properties of the boards without significantly affecting their density. Subsequently, Velasquez *et.al.* (2003) studied the use of untreated or unpurified kraft lignin resulting from the kraft pulping process to the steam exploded Miscanthus sinensis. Thus, in this study we try to put lignin from cacao peel with cooking liquid of sodium hydroxide effectively in the process and use chemical agents efficiently and then lignin powder can be used as a material of lignin adhesive production. The resulted adhesive is close to the capacity of commercial adhesive and can be diluted using water.

METHODOLOGY

The study used cacao peel from International Coffee and Cacao Research Center (ICCRI) in Jember Regency. The variables of this study are: sodium hydroxide concentration (i.e. 8%, 10%, 12% and 18 %); with the cooking time of: (60, 90, 120, 180) minutes. The weight of cacao peel is 200 g, and oven temperature is 105 °C during 2-3 hours. We used the extractor, to cook and extract lignin from cacao peel, and dried the lignin in the oven. The cacao peel to be used was cleaned, and cut into small sizes about 0.5-1 cm; it was pounded until a fiber formed, and then dried in the sun. A total of 200 g dried fiber was cooked with water for an hour as the softening process. A sodium hydroxide solution was added to the soft fiber with a variable concentration of (1:5), then cooked appropriately with the variable timing at a temperature of 100 oC to obtain the black liquor; and then it was separated from its pulp by a screening process. The black liquor obtained was diluted with water until pH was 8-9, and then it was added with HCL 10 % until pH reached 2-3. The sedimentation process took 8-10 hours, and then it was separated from its liquid. The next step was drying the lignin in an oven at a temperature of 105oC for 2-3 hours. The dried lignin was pounded until it became a powder ready to use as the material of lignin adhesive. A quantity of 5 g lignin powder was mixed with 2.5 g of resorcinol and 10 mL of formaldehyde solution; then it was agitated continuously and added with NaOH 10 % and 60 % (20 mL respectively) until it formed a lignin adhesive.

Testing of the Adhesive capacity of Lignin adhesive

This study determined the adhesive capacity of lignin adhesive by the following measurement: The wood was cut into two parts of 8 Cm x 3 Cm and glued together with the lignin adhesive; with a position of 3 Cm x3 Cm. It was then allowed to dry out then was pulled hard until it broke off. The result of the adhesive capacity of the lignin adhesive was 3.54 kg/sq.Cm.

Lignin Quantity Determination

A total of 200 g of cacao peel was boiled with water in order to soften the cacao peel and then it was screened. The softened cacao peel was then cooked with 1 L of NaOH 10 % for 2 hours. It resulted in 600 mL of lignin, and then it was diluted with 3 L of water and it was heated at 60 °C. After that, it was cooled down and the measured pH was 8-9. Then it was deposited by adding a 300 mL of sulfate acid. The deposit of lignin was dried in the oven at a temperature of 105 °C for 24 hours and resulted in 29.8 g of lignin.

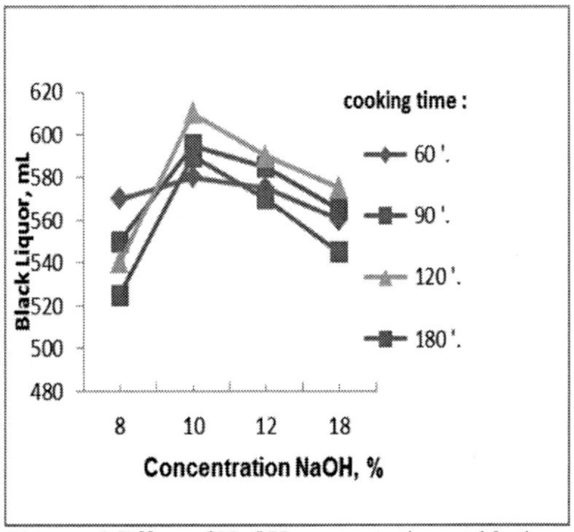

Figure- 1 Effect of NaOH concentration on black

RESULTS AND DISCUSSIONS

The results obtained during the study have been analyzed for its density of black liquor and lignin powder quantity.

Table-1. Data of Black Liquor towards Cooking Time and Concentration of NaOH cooking liquid

Cooking time, min.	Concentration NaOH	Black Liquor	Lignin Powder
60	8	570	28.3
	10	580	28.8
	12	575	28.5
	18	560	27.8
90	8	550	27
	10	595	29.5
	12	585	29
	18	565	28
120	8	540	27.8
	10	610	29.8
	12	590	29.3
	18	575	28.5
180	8	525	26
	10	590	29.3
	12	570	28.3
	18	545	27

Based on table 1, the higher the concentration of NaOH at variations of cooking time, the fewer the black liquor obtained, even though at low concentrations it was increasing. The reason was that the increasing of NaOH concentration would scatter the lignin in cacao peel. However, in the condition of cooking liquids of NaOH with higher pH, the lignin will be ionized and salt was formed.

At a concentration of NaOH 10 % and cooking time of 120 minutes, as much as 610 mL black liquor can be obtained, and then we can take its solid lignin.

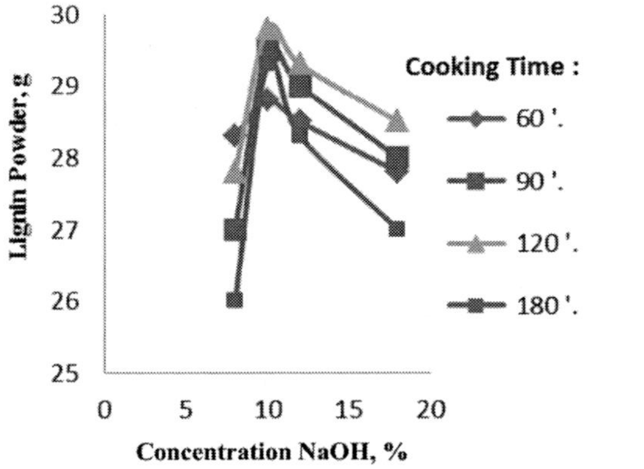

Figure-2. Effect of NaOH concentration on yield Of lignin

The longer the cooking time at variable concentrations of cooking liquid, the more lignin powder

to be obtained. The best result can be found at a cooking time of 120 minutes with a concentration of NaOH 10 %, which was 29.8 g. After that, it will decrease because the longer the cooking time, the lignin separated from cacao peel will be dissolved and mixed with its cacao peel.
Compared to the research from the previous researchers, this study makes it relatively easier to separate lignin from cacao peel by an earlier softening method of cacao peel's material and adjusting pH of liquid before going to the sedimentation process of lignin with HCL solution 10 %. After that, we analyzed the adhesive capacity of the lignin adhesives, and gave a result of 3.54 kg/ sq.Cm, where the adhesive's color is dark brown, near to cacao peel's color. Compared to the commercial adhesive that possess an adhesive capacity of 3.65 kg/sq.Cm, the lignin adhesive produced by this research is nearly equal to the capacity of commercial adhesive. Although the color is still not perfect, we are making progress on it.

CONCLUSIONS

Black liquor can be obtained from cacao peel and is deposited to lignin powder that is ready for use as a material for adhesives production. The result of lignin powder is 29.8 g at concentration NaOH 10 % and cooking time 120 min. Adhesives capacity of lignin is 3.54 kg/sq.Cm, and its color nearly equal to the color of cacao peel.

REFERENCES

Angles, M.N., Ferrando, F., Farriol, F, X., Salvado, J.,2001. "Suitability of steam exploded residual softwood for the production of binderless panels. Effect of the pre-treatment severity and lignin addition, Biomass & Bioenergy, 21(3), 211-224.

Danielson, B., Simonson, R., 1989. Kraft lignin in phenol formaldehyde resin. Part I. Partial replacement of phenol by kraft lignin in phenol formaldehyde adhesive for plywood. Journal of Adhesion Science and Technology 12:923-936

Hemalia, V., Trisschler, J., Sandberg, D., 2013. Lignin – An Adhesive raw material of the future or waste of research energy. Brischke, C. & Mayer, L.(Eds,) Proc.9th Meeting of the Northern European Network for wood Science and Engineering (WSE), (pp.98-103) Hannover, Germany, September 11-12,2013.

Helio Faustino, Nuno Gil, Cecilia Baptista and Ana Paula Duarte, 2010. Antioxidant Activity of Lignin Phenolic Compounds Extracted from Kraft and Sulphite Black Liquors. Molecules 2010,15, 9308-9322;doi:10.3390/molecules 15129308.

Hu, L., Pan, H., Zhou, Y., Zhang, M., 2011. Methods to improve lignin's reactivity as a phenol substitute an as a replacement for other phenolic compounds: a brief review. BioResources 6(3).

Imam, S.H., Gordon, S.H., Mao, L., Chen, L., 2001. Environmentally friendly wood adhesive from a renewable plant polymer: Characteristics and optimization. Polymer Degradation and Stability 73:529-533.

Mawardi, 2009. "Quality of Particle Board from Palm Coconut Wood Based on Polystyrene, Adhesive , Journal of Machine Engineering, 11 (2) 91-96.

Mankar S.S, Chaudhari, A.R. & Soni, I.2012. Lignin in phenol-formaldehyde adhesives. International Journal of Knowledge Engineering 3:116 – 118.

Nihat Sami CETIN, Nilgul OZMEN, 2003. Studies on Lignin-Based Adhesives for Particleboard Panels.Turk J Agric For 27 (2003) 183-189 @ TUBITAK.

Pizzi, A. 2006. Recent development in eco-efficient bio-based adhesives for wood bonding: opportunities and issues. Journal of Adhesives Science and Technology 20:829-846.

Sarkar, S., Adhikari, B., 2000. Lignin-modified phenolic resin: synthesis optimization, adhesive strength, and thermal stability. Journal of Adhesion Science and Technology, 14(9):1179-1193.

Rowel, R.M.,2012. Hand Book of wood chemistry and wood composites. Bio-based adhesives.CRC Press.

Velasquez, J.A., Ferrando, F., Farriol, X., and Salvado, J.,2003. " Binderless Fiberboard from Steam exploded Miscanthus sinensis,Wood Science and Technology 37 (3-4), 269-302.

Yao Chen, CharlesR. Frihart, Zhiyong Cai, Linda F. Lorenz, Nocole M, Stark, 2013. Lignin-based Phenol-Formaldehyde Resins from Purified CO_2 Precipitated Kraft Lignin (PCO_2 KL). International Conference on Wood Adhesives, page 601. U.S. Departement of Agriculter, Forest Service, Forest Product Laboratory, One Giffort PinchotDrive, Madison, WI53726-2398, United State.

Zhang, W., Ma, Y., Wang, C., Li, S., Zhang, M., Chu, F., 2013. Preparation and properties of lignin-formaldehyde resins based on different biorefinery residues of agricultural biomass. Industrial Crops and Products 43:326-333.

THE EFFECT OF FLUORESCENT PSEUDOMONADS APPLICATION ON THE RESISTANCE OF CHILI PLANTS AGAINST THE ATTACK OF *RALSTONIA SOLANACEARUM* AND *FUSARIUM OXYSPORUM* IN THE FIELD

Yenny Wuryandari, Sri Wiyatiningsih, and Agus Sulistyono
University of Pembangunan Nasional "Veteran" Jawa Timur
Surabaya, East Java, Indonesia
E-Mail: yennyupn@yahoo.com

ABSTRACT

Chili (*Capsicum annuum L.*) is one of the vegetable commodities with high economic value. Until now, the necessity rate of chili is still high but the production is still low. One of the obstacles, which highly influence chili production, is wilt disease due to *Ralstonia solanacearum* and *Fusarium oxysporum*. This study aims at identifying the best fluorescent pseudomonads isolate in suppressing the development of the primary wilt disease due to *Ralstonia solanacearum* and *F. oxysprorum* in the field. This study employs Completely Randomized Design that comprises 3 treatments. The three treatments are the type of fluorescent pseudomonads which is Pf-122 isolate, Pf-160 isolate, and Pf-B isolate. The observation parameter on suppressing the disease development is perceived from incubation period of the initial appearance of symptoms, disease index and the change of stem tissue color. The result of the study shows that the longest incubation period is treatment with fluorescent pseudomonads of Pf-122 isolate and Pf-160 isolate. The chili is treated with fluorescent pseudomonads of Pf-122 isolate and Pf-160 isolate, the initial appearance of symptoms at the 8th day after inoculation. In the treatment with fluorescent pseudomonads of Pf-122 isolate and Pf-160 isolate compared to control, is able to delay the appearance of symptom for 3 days. The disease development starts from 10th to 30th day, the treatment with Pf-122 isolate shows the slowest development compared to other treatment and control. In the last observation, the disease index with treatment of Pf-122 isolate is the lowest which is 37%, then it is followed by Pf-B and Pf-160 which are 39% and 40% respectively, whereas in the control it reaches 67%. For the color change on stem tissue, there is no real difference of the three treatments of fluorescent pseudomonads application but it differs and smaller from the control.

Keywords: fluorescent pseudomonads, *Ralstonia, Fusarium*

INTRODUCTION

Chili productivity in Indonesia is currently low. The factors causing low production are the use of less qualified seeds, non-efficient cultivation technique and the chili cultivar cultivation which is not resistant to pests and diseases (Soelaiman & Ernawati, 2013). The primary barrier which highly influences the chili production is the main disease, wilt disease *Ralstonia solanacearum* and *Fusarium oxysporum*. Disadvantages from wilt disease of *R. solanacearum* are quite great (Elphistone, 2005). In Indonesia, this disease causes a very big disadvantage and the disease intensity can reach 75% (Fortman & Martin, 1998).

The disadvantage due to this disease is increasingly high for in the field it commonly attacks chili joint together and synergic. Numerous attempts to overcome the main disease of being wilt disease on chili have been sought to, but an effective and efficient controlling method has not been discovered. It is because both pathogens can withstand in the soil and both of them synergize to attack chili. Therefore, an attempt to overcome the pathogen attack and to improve the chili resistance is necessary.

Biological control can improve the result by suppressing pathogen inoculum, protecting the plant against infection, and improving the plant resistance from pathogen. Microorganism will block pathogen at the root. Useful biological control with microorganism can induce root growth will be increasingly performed (Podile & Kishore, 2007).

Among Plant Growth Promoting Rhizobacteria (PGPR), fluorescent pseudomonads gain more attention. Certain strain from fluorescent pseudomonads shows its ability in suppressing the development of several plant diseases due to pathogens and inducing plant growth (Raaijmakers *et al.,* 1995). Primary mechanism of suppressing disease by biological agents are including antagonism and or inducing plant resistance (Hyakumachi, 2013). Bacteria are able to induce plant systemic resistance by producing phytohormone, solubilizing inorganiz phosphate, improving fastening of Fe and siderophore (Podile & Kishore, 2007). As suggested by Pinton *et al.* (2001), rhizobacteria is also able to induce plant growth by producing growth regulator such as auxin derivatives. Most of *Pseudomonas*, *Bacillus, Azotobacter* rhizobacteria produce growth inducing compound such as Indol Acetic Acid (IAA), gibberellins, and other substances such as cytokinin (Brimecombe *et al.*, 2001). Antibiotics produced by fluorescent pseudomonads contributing in biological control are phenazines, 2.4-diacetylphloroglucinol, pyoluteorin, pyrrolnitrin, lipopeptides, and hydrogen cyanide (Haas and Keel, 2003). Antibiosis is activity of response generally from several biological agents such as *Pseudomonas* spp, *Bacillus* spp, *Streptomyces* spp, and *Trichoderma* spp. (Alabouvette *et al.*, 2006).

Wuryandari *et al.*, (2007) showed that fluorescent Pseudomonad of Pf-122 isolate can inhibit bacterial wilt disease caused by *Ralstonia solanacearum*. In the last observation (30 days) the disease index is only 49.9% at greenhouse scale. To determine the ability of fluorescent Pseudomonad to improve the plant resistance toward 2 pathogens working in synergy in the field (*R.*

solanacearum and *F. oxysporum*), research needs to be done by using fluorescent pseudomonad to induce chili plant resistance toward the primary wilt disease due to *R. solanacearum* and *F. oxysporum* in the field.

METHODOLOGY

The study is conducted in land commonly used for *Solanaceae* plant cultivation and frequently exhibited the symptom of wilt disease, which is in Lebo village, Sukodono, Sidoarjo, East Java, Indonesia. The study employs Completely Randomized Design comprising 3 treatments. The three treatments are the types of fluorescent pseudomonads isolate which is Pf-122 isolate, Pf-160 isolate, and Pf-B isolate. Each treatment has three repetitions. Control treatment as the comparator is chili without treatment with fluorescent pseudomonads

The root of chili seeds aged 30 days are immersed into fluorescent pseudomonads suspension for 30 minutes, then it is planted. Negative comparator consists of chili seed immersed into sterile water for 30 minutes and is planted in the soil. The concentration of fluorescents pseudomonads used is 10^{10} cfu/ml. Each treatment is repeated three times. To identify plant resistance toward wilt disease due to *R. solanacearum* and *F. oxysporum*, observation parameter includes incubation period, disease index, and the color change of stem tissue.

Incubation period is observed from pathogen inoculation until the appearance of wilt disease symptom on chili. The disease index is observed in the development of wilt disease symptom 5 days / plant until 30th day after inoculation. For the color change of stem tissue is observed at the 30th day, by splitting length wise the stem and measuring the length of color change in vascular tissue.

The weight of attack is estimated according to the following scale:

0 = no symptom, 1 = 1 to 10% wilt leaves, 2 = 11 to 30% wilt leaves, 3 = 31 to 60% wilt leaves, 4 = 61 to 99 % wilt leaves, and 5 = 100% wilt leaves

The magnitude of disease index is estimated by using a formula (Arwiyanto, 1995):

$$I = \frac{\sum_{i=0}^{k} k.nk}{ZxN} \times 100\%$$

The formula description is as follows; I = disease index, nk = the number of plant having disease symptom with scale k (0, 1, 2, 3, 4, 5), N = total number of inoculated plant, and Z = highest category of attack.

RESULTS AND DISCUSSION

a. Incubation period

If it is seen from the results of observation on incubation period of wilt disease, the three fluorescent pseudomonads show different result. Incubation period of wilt disease is fastest in the treatment of Pf-B isolate then in Pf-122 isolate and Pf-160 isolate which are 7 days and 8 days. Control in incubation period is 5 days. If compared to control then the treatment with Pf-122 isolate and Pf-160 isolate it is able to delay the emergence of the symptom for 3 days, whereas with Pf-B isolate it is only 2 days (Figure 1).

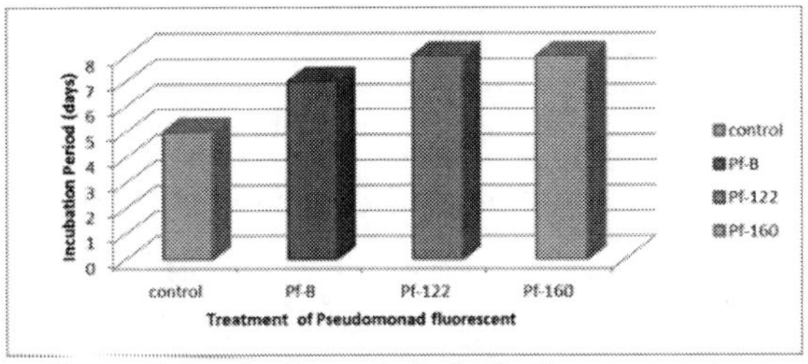

Figure-1. Incubation period of wilt disease symptoms on chilli

Short delay of incubation period of wilt disease by fluorescent pseudomonads rhizobacteria, it is possible due to lack of fluorescent pseudomonads population, which may inhibit pathogen attack. The dose of pseudomonad fluorescents rhizobacteria used to plant submergence is probably still less high, so that it is not quite able maximally in inhibiting pathogen to infect chili. The result of a study conducted Wuryandari *et al.* 2012, that chili treated with fluorescent pseudomonads of Pf-160 isolate, Pf-B isolate and Pf-122 isolate and is inoculated with *F. oxysporum*, able to delay the disease symptom longer which is 13 days, 12 days, and 6 days. The difference of incubation period delay is possible because beside pathogen which has to be inhibited is not only one, but also because different application scale. Previous research application is in greenhouse, whereas this study is conducted in the field. Field condition with more complex microorganism so that it can influence the ability of pseudomonad fluorescents in expressing the ability in producing inhibiting compound to resist pathogen and the ability in producing growth hormone for chili. Biological control of pathogen by rhizobacteria can be based on bacteria metabolite such as siderophores, antibiotics, and hydrogen cyanide. Rhizobacteria can also induce the plant growth by producing growth regulator such as auxin derivatives (Pinton *et al*, 2001).

b. Disease Index

The development of disease stating from inoculation until 30th day of observation after inoculation can be seen in Figure 2.

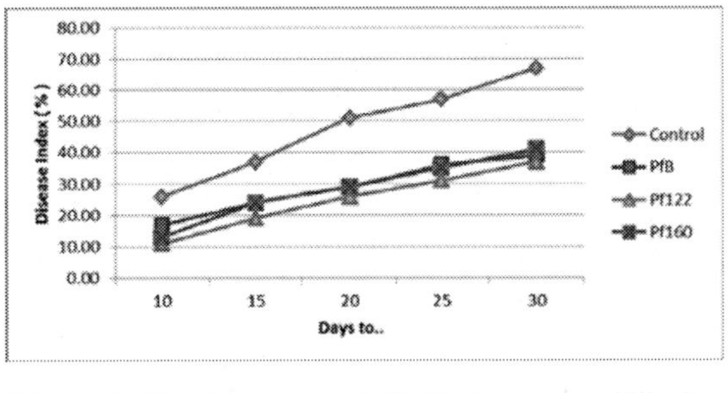

Figure-2. The development of wilt disease on chili after treated with fluorescent pseudomonads rhizobacteria

The result of disease index from the appearance of wilt disease symptom to the 30th day of observation, it is seen that the development of wilt disease on plant with fluorescent pseudomonads from the three isolates is slower compared to control. Based on the three fluorescent pseudomonads rhizobacteria, Pf-122 isolate shows the slowest or the lowest disease development compared to other isolate fluorescent pseudomonads.

In the last observation, the treatment with Pf-122 isolate shows the lowest disease index which is 37% then it is followed by Pf-8 isolate and Pf-160 isolate which are 39% and 41%, whereas in the control it reaches 67%. If it is compared to control, then the submergence by Pf-122 isolate is able to suppress the disease development by 30%, whereas it is with Pf-B and Pf-160 isolates only able to suppress the disease development by 28% and 18% (Table 1)

Table-1. The index of wilt disease on chili with treatment using fluorescent pseudomonads rhizobacteria at 30th day

Number	Treatment	Disease Index on day 30
1	Fluorescent pseudomonads of Pf-122 isolate	37 a
2	Fluorescent pseudomonads of Pf-160 isolate	41 a
3	Fluorescent pseudomonads of Pf-B isolate	39 a
4	Control	67 b

The result shows that the application of fluorescent pseudomonads can suppress the wilt disease by *Fusarium* and *Ralstonia solanacearum*. Fluorescent pseudomonads of Pf-122 isolate is the best in suppressing the wilt disease. It is possible because fluorescent pseudomonads is able to produce a compound which can inhibit pathogens of *Fusarium* and *Ralstonia solanacearum* and can induce chili growth so that it can be resistant to the pathogen attack. Primary mechanism in suppressing disease by biological agents includes antagonism and or inducing plant resistance (Hyakumachi, 2013).

The application of fluorescent pseudomonads in chili rooting, it probably causes fluorescent pseudomonads colonize root area so that it inhibits the pathogen population causing wilt disease. Mechanism of biological control of pathogens by rhizobacteria may be in form of bacteria metabolite production. Bacteria of *Pseudomonas* genus also produces antimicrobial compound such as phenazine, pyrrolnitin, pyluteorin, phloroglucinols, cyclic lipopeptides, and hydrogen cyanide (Haas & Defago, 2005). Application of pseudomonad fluorescens also allows the plant growth to be induced so that it can be more resistant to pathogen attack of *Fusarium* and *Ralstonia solanacearum*. As suggested by Pinton *et al.*, (2001), rhizobacteria is also able to induce plant growth by producing growth regulator such as auxin derivatives. Most of *Pseudomonas, Bacillus, Azotobacter* rhizobacteria produce growth inducing compound such as Indol Acetic Acid (IAA), gibberellins, and other substances such as cytokinin (Brimecombe *et al.*, 2001). The sufficient existence of fluorescent pseudomonads in rooting area is allegedly able to dissolve phosphate plays a role in repairing the plant. In accordance to Beattie (2007), one way to repair phosphorus deficiency on plant is by soil inoculation with phosphate solubilizing microorganism. Several saprophytic bacteria playing role as biological agents of wilt disease control is fluorescent pseudomonads bacteria on tomato, potato, and eggplant, and on chili (Saddler, 2005). Bacteria are able to induce plant systemic resistance by producing phytohormone, solubilizing inorganiz phosphate, improving fastening of Fe and siderophore (Podile & Kishore, 2007; Haas & Defago, 2005).

c. Color change on stem tissue

Based on the observation result on symptom of color change in stem vascular tissue, it is seen that the average symptom of color change in stem with the treatment using fluorescent pseudomonads, is smaller than control. Application using Pf-160 isolate shows the smallest change of color then followed by Pf-122 and Pf-B isolates (Figure 3). It is probably that fluorescent pseudomonads rhizobacteria applied is able to colonize in chili root so that it can inhibit microorganism in soil including pathogen causing wilt disease of *R. solanacearum* and *Fusarium* sp. The ability of pseudomonad fluorescents rhizobacteria of Pf-160 isolate in inhibiting disease is seemingly better compared to other fluorescent pseudomonads rhizobacteria. It is seen from the slow or small vascular tissue in the stem. Although fluorescent pseudomonads of Pf-160 isolate is better in inhibiting the damage on vascular tissue, it is apparently not different from other isolate fluorescent pseudomonads in arising symptom of wilt disease on plant. According to Baharuddin *et al* (2005), the result of his study explains that *Pseudomonas* spp. fluorescent group in plant rooting is able to colonize well so that it can suppress the attack by *R. solanacearum*. The application of fluorescent pseudomonads expected to be able to inhibit pathogens causing wilt disease which is *R. solanacearum* and *Fusarium* sp. living in rooting area before the pathogens entering the plant root tissue.

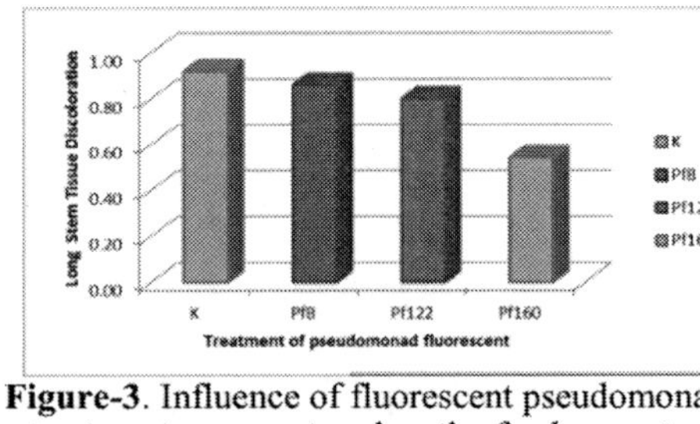

Figure-3. Influence of fluorescent pseudomonads treatment on symptom length of color on stem

CONCLUSION

Application with pseudomonad fluorescent of Pf-122 isolate compared to control is able to postpone the emergence of the symptom for 3 days. However, if it is seen from the disease index, compared to other isolate and control, Pf-122 isolate shows the highest inhibiting ability.

ACKNOWLEDGEMENT

The author would like to thank Director of Research and Public Service, Directorate-Generals of Higher Education, Ministry of Education and Culture of

Republic of Indonesia for funding this research through National Strategic Research Grant No. 180 / SP2H / PL / Dit.Littabmas /V/2013.

REFERENCES

Alabouvette C, Olivain C, and Steinberg C. Biological control of Plant Disease : the European situation. European Journal of Plant Pathology. 114 : 329 – 341.

Arwiyanto T and Hartana I. 1999. Biological Control of Tobacco Bacterial Wilt Disease: 2. Experiments in Greenhouse. J. Plant Protection Indonesia.5 (1) 50-99.

Beattie G A. 2007. Plant-Associated Bacteria: Survey, molecular phylogeny-genomics and Recent Advances. In : Gnanamanickam, S.S. (ed.). Plant-Associated Bacteria. Springer The Netherlands

Brimecombe M J, De Leij F A, and Lynch J. 2001. The Effect of Root Exudates on Rhizosphere Microbiae Populations. In: Pinton, R., Varanini, Z., Nanipieri, P. The Rhizophere. Biochemistry and Organic Substances at the Soil-Plant Interface. Marcel Dekker, Inc. New York – Basel.H. 95-140.

Elphinstone J G. 2005. The Current Bacterial Wilt Situation: A Global Overview. p. 9-28. In : Allen C, Prior P, Hayward A C. Bacterial Wilt Disease and the *Ralstonia solanacearum* Species Complex. The American Phytopathological Society Press.

Haas D and Defago G. 2005. Biological Control of Soil Borne Pathogens by fluorescent pseudomonads. Nat Rev Microbial. 3.307-319.

Fortman B A and Martin S B. 1998. Disease Management Strategies for Control of Bacterial Wilt of Tobacco in the Southeastern USA. p. 394-402. In: Prior P H, Allen C and Elphinstone J (eds.). *Bacterial Wilt Disease Molecular and Ecological Aspects*. Springer-Verlag Belin Heidelberg, INRA Paris.

Haas D And Keel C. 2003. Regulation of Antibiotic Production in Root-Colonizing Pseudomonas spp. And Relevance for Biological Control of Plant Disease. Annual Review of Phytopathology. 41, ProQuest p: 117 – 153.

Hyakumachi M. 2013. Research on Biological Control of Plant Diseases : Present state and Perspectives. J Gen Plant Pathol 79: 435 – 440.

Pinton R, Varanini Z, and Nanipieri P. 2001. The Rhizophere as a site of Biochemical Interactions Among Soil Components, Plants, and Microorganisms. In: Pinton, R., Varanini, Z., Paulo Nanipieri, P.. The Rhizophere. Biochemistry and Organic Substances at the Soil-Plant Interface. Marcel Dekker, Inc. New York–Basel. P. 1 – 18

Podile A R and Kishore G K. 2007. Plant Growth-Promoting Rhizobacteria.p. 195-230. In : Nanamanickam, S.S. (ed.). Plan-Associated Bacteria. Springer

Raaijmakers J M, Leeman M, Van Oorschot M M P, Van der Sluis I, Schipper B and Bakker P A H M. 1995. Dose-response relationships in biological control of fusarium wilt of radish by *Pseudomonas* spp. *Phytopathology* 85: 1075-1081.

Saddler G S. 2005. Management of Bacterial Wilt Disease. p.121-132. In: In : Allen, C., Prior, P., Hayward, A.C. Bacterial Wilt Disease and the *Ralstonia solanacearum* Species Complex. The American Phytopathological Society Press.

Soelaiman V and Ernawati A. 2013. Growth and Development of Chilli (*Capsicum annuum* L.) in *in vitro* at some concentrations of BAP and IAA. Bul. Agrohorti 1(1): 62 - 66.

Wuryandari Y, Purnawati A, Arwiyanto T, and Hadisutrisno B. 2007. Effect of pseudomonads fluorescens to development of wilt disease due to *Ralstonia solanacearum* in Greenhouse. Journal Saintek Kopertis Region VII. East Java. 4 (2): 48-52.

Wuryandari Y, Wiyatiningsih S, and Sulistyono. 2012. Control of *Fusarium oxysporum* wilt disease using Pseudomonas fluorescens Rhizobacteria. Proceedings of the National Seminar PERHORTI. p.507-512.

THE EXTRACTION PROCESS OF *TRIMETHYL XANTHINA* IN VITRO CULTURE OF CALLUS *CAMELLIA SINENSIS* WITH ETHYL ACETATE SOLVENT

Sutini [1], Susilowati [2], Djoko Agus Purwanto [3], M. Rasjad Indra[4]
[1]Agrotechnology Department of Agriculture Faculty UPN "Veteran" Jawa Timur, Indonesia
[2]Chemical Engineering Department of Industrial Technology Faculty UPN "Veteran" Jawa Timur, Indonesia
[3]Department of Pharmaceutical , Faculty of Pharmacy, Airlangga University, Jawa Timur, Indonesia
[4]Department of the school physiology Faculty of Medical Brawijaya University, Malang, Indonesia
E-mail: tien_basuki@yahoo.com

ABSTRACT

Trimethyl xanthina is one of the compounds contained bioactive culture in vitro *Cammelia sinensis* callus which is widely used in the field of food, beverage, agriculture and health industries. The presence of *trimethyl xanthina* on food, beverages and health is needed in a certain amount depending on the use which is achieved by the user. To get a certain amount of *trimethyl xanthina* from callus culture of *Cammelia sinensis*, the extraction process is performed on the water solvent, as well as non-solvent water / organic solvent such as ethyl acetate. The purpose of this study was to obtain profile of *trimethyl xanthina* in the extraction of *Cammelia sinensis* callus. The experimental methods used consisted of dissolution, filtration, extraction with water solvent and ethyl acetate, then followed by identification of *trimethyl xanthina* using HPLC. The results shows the profile form of *trimethyl xanthina* of *Cammelia sinensis* callus have similarities with the standard form of *trimethyl xanthina*.

Keywords: *cultured in vitro, Cammelia sinensis* callus, *trimethyl xanthina.*

INTRODUCTION

Trimethyl xanthina is one of the bioactive compounds found in vitro culture of callus *Cammelia sinensis* (Shervington et.al., 1998). Trimethyl xanthine, according to a study of Wei et.al. (2008), could be raised in production by culture in vitro by addition of precursors of purine. Research conducted by Li et.al.., (2008) shows that gene *trimethyl xanthina* could expressed on new *Cammelia sinensis* leaves. According to Shane et.al. (2013) on new *Cammelia sinensis* leaves, *trimethyl xanthina* is synthesized in the chloroplasts of cell during photosynthesis than subsequently transported to vascular tissue for plant defense of against pathogens and predators (Ministry for primary Industries Manato Ahu matua New Zealand Government. 2012). *Trimethyl xanthina* is widely used in many fields of food and beverage industry, agriculture, and health.

For the health, *trimethyl xanthina* plays a role in physiological effect on increasing the freshness of the human body (Gramza-Michałowska 2014). According to Rebecca et.al. (2013), *trimethyl xanthina* is classified as food and medicine with a particular dose. Toxic dose is 100 mg for an average adult per day. But according to Krueger and Howard (2011), dose of *trimethyl xanthina* most likely to be effective without causing undesirable side effects is between 100-600 mg.

In the field of food-beverage industry, bioactive *Trimethyl xanthina* increase levels of dopamine, which will be activated the body's metabolism (Susita 2014). In the beverage industry, *trimethyl xanthina* of *Cammelia sinensis* leaves presence is expected and loved by the British people (Kato 1989).

In the agricultural industry, *trimethyl xanthina* is used to eradicate the beetle (Hewavitharanage *et.al.*., 1999). Almost all varieties of tea plants containing *trimethyl xanthina*, but the content of this *trimethyl xanthina* varies depending on the age of the plant, harvesting techniques, the particle size of the harvested leaves, (Hyong et.al.., 2007) and the equipment used in the manufacture of liquid extraction (Astill et.al.., 2001). According Komes et.al.., (2009) the content of *trimethyl xanthina* has been associated with the origin of plants and plant growth conditions.

The use of liquid in liquid-liquid extraction method includes the separation of *trimethyl xanthina* compound that soluble in chloroform solvent, is partially soluble in the solvent ethyl acetate, but in this study in addition to extraction using chloroform also using ethyl acetate solvent. The use of another solvent is ethanol at the optimum temperature (temperature of 60° C) extraction time of 240 minutes, resulting in the highest *trimethyl xanthina* biomass. While the extraction temperature of 40 ° C, 15 min extraction time decreased biomass (Setyo Pratomo 2014).

The purpose of this study was to obtain profile of *trimethyl xanthina* found in in vitro callus cultures of *Camellia sinensis* with the method of callus extraction using ethyl acetate solvent and using HPLC method for the analysis.

METHODOLOGY

Callus *Cammelia sinensis* was made from leaf explants of tea plucked from PT Perkebunan Nusantara (PTPN) XII Lawang, Malang, East Java. Distilled water, chloroform, ethyl acetate, and *trimethyl xanthina* raw material supplies from specialized sales agents from Sigma.

The experimental method was performed as follows: (1) in vitro callus induction by growing tea leaf explants on media with growth regulating substances and the maintenance (2) harvesting callus followed by weighing callus and callus morphological observation. (3) Isolation and *trimethyl xanthina* bioactive extraction (4) qualitative identification of *trimethyl xanthina* of tea callus by HPLC.

Tools for analysis included High-performance liquid chromatography (HPLC) Agilent 1100 with the specification: detector spectrophotometer UV-ST diode array, with the column RP 18 Waters μ Bondapak 10 μm, 3.3 x 300mm. (2) filter "Nylon membrane f ilter "0.2 μm (3) analytical balance (Shimadzu) with 0.001 mg sensitivity (3) pumpkin separating funnel, 5 ml volumetric flask (4) rotavapour.

Callus induction *in vitro*

The results of in vitro callus induction was obtained callus growth starts from explant changes from inflated shape of tea leaf tip, morphological changes such as the formation of callus from the cut edge of the wound leaves until the entire surface filled with callus.

Harvesting callus

Formed callus was harvested then weighed and the callus morphology is observed in micro and macroscopicly (Figure 1).

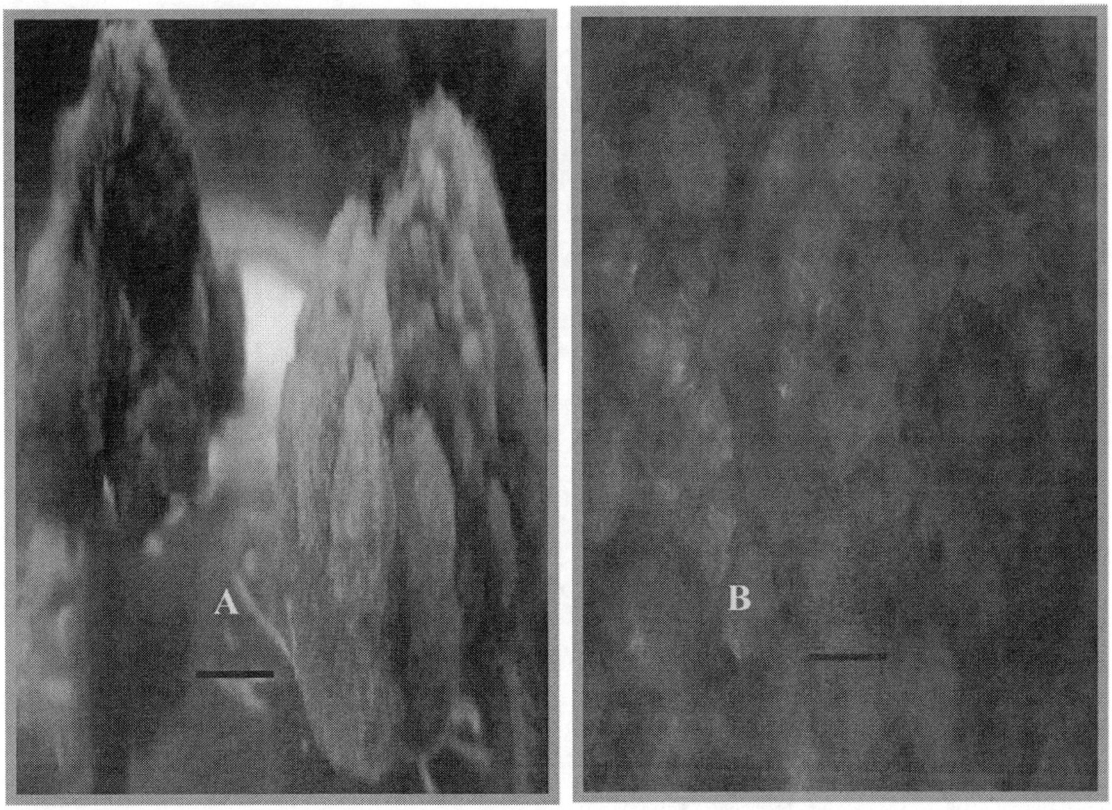

Figure-1. The shape of macroscopis callus (A). Microscopic callus with microscope triokuler with 400x magnification (B). (Bars = 1cm)

Isolation and extraction

Isolation and extraction of *trimethyl xanthina* was done by carrying out the extraction of tea callus by smoothing callus then carefully weighed with ± 415 mg (predetermined water content by gravimetric method) and adding hot distilled water with temperature of 70-80 ° C 25.0 m L, settling for ± 30 minutes (Shirai et.al.., 1994). After that filtered, and put into a 50.0 mL volumetric flask. The dregs of tea callus was rinsed with 10 mL of hot water 2 times, settling ± 30 minutes and then filtered. The second and the third extract results were subsequenly collected in the same flask, then add distilled water up to 50.0 m L. Put 25.0 mL

Tea extract solution was then gently shaken with 25 mL of chloroform in a separating funnel, it will form two layers, the bottom and the top. Bottom layer is the chloroform phase while the upper layer is the water phase. Mixing with chloroform was repeated twice. From obtained water phase, take 25.0 mL then extract with 25 mL of ethyl acetate 3 times to form two layers. Bottom layer is the water phase and the upper layer is the ethyl acetate phase. Ethyl acetate phase is accommodated and rotavapored to obtain a dry extract.

Identification by HPLC

The obtained dry extract was dissolved in methanol and then injected into the HPLC (Karlina 2006), then it will obtain the form of *trimethyl xanthina* profile with retention time (RT) which is approximately equal to the standard (Figure 2).

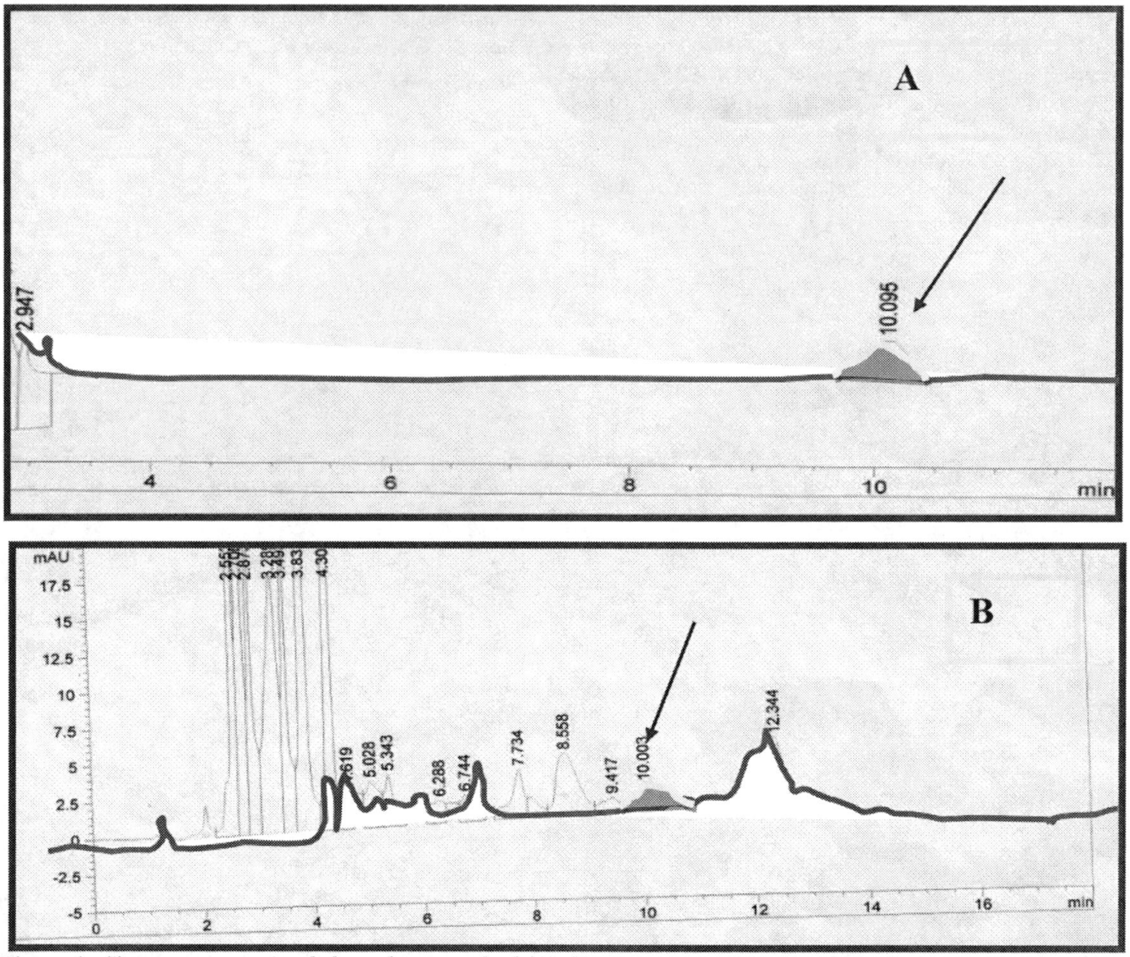

Figure-2. Chromatogram *trimethyl xanthina standard* (A) Chromatogram tri *trimethyl xanthina* methyl xanthina callus (B) with R is approximately equal.

RESULTS AND DISCUSSIONS

Greenish white callus induction, along with greenish white growth on the wound surface is response to damage. According to Sutini (2010) when the tea leaf is cut, there will be thickening wound called callus.

According to research by Borzabad et.al. (2010) on new leaves of the Artemisia vulgaris L plant could induce callus and regenerated with MS medium with growth regulators containing optimum concentrations of 1.0 mg^{-1} BAP and 3.0 mg^{-1} NAA.

Morphology of callus is microscopically obtained irregular shapes since the cells have not differentiated. Final extraction uses ethyl acetate to get *trimethyl xanthina* in a certain amount as research need.

Results of research conducted by Amra et.al. (2006) says that the solvent extraction using water, acetone, ethanol, methanol, acetonitrile at a temperature of 60 to 100 C for 240 minutes produces 36 grams of *trimethyl xanthina* / caffeine / kg dry component.

Identification using HPLC obtained chromatogram with the standard retention time is the same with sample retention time. This study using in *vitro culture* techniques which is relevant to Maria (2013) that *in vitro culture* can be as an alternative for the production of secondary metabolites *Camellia sinensis.*

CONCLUSIONS

Trimethyl xanthine compound is secondary metabolite compounds that are bioactive which can be produced through in vitro culture.

ACKNOWLEDGEMENTS

The author would like thank Directorate General of Higher Education and National Department of Education Indonesian, which has funded the research Competence Grants 2013-2015.

REFERENCES

Amra P. U., Mojca S K., Knez M. (2006). Extraction of active ingredients from green tea (Camellia sinensis): Extraction efficiency of major catechins and caffeine. J. Food Chemistry 96, 597–605.

Astill C., Mark R. B., Dacombe C.(2001). Factors Affecting the Caffeine and Polyphenol Contents of Black and Green Tea Infusions. J. Agric. Food Chem,49, 5340-5347

Borzabad, Karami R, Sudarshana. 2010. In vitro Plant Regeneration from Leaf Explants of Artemisia vulgaris L. - A Medicinal Herb. J. Modern Applied Science (4): 130

Gramza-Michałowska A (2014). Caffeine in tea Camellia sinensis — Content,absorption, benefits and risks of consumption. Abstract. Journal of nutrition, health & aging (18): 143-149

Hewavitharanage P., Karunaratne S., Kumar S., 1999. Efect of ca}eine on shot!hole borer beetle "Xyleborus fornicatus of tea "Camellia sinensis. J. Phytochemistry (40): 24-30

Hyong S.K., a, Hyung K. C., Sung J.L. (2007). Effect of mass transfer on the removal of caffeine from green tea by supercritical carbon dioxide. J. of Supercritical Fluids (42): 205–211. www.elsevier.com/locate/foodchem

Kato (1989). *Camellia sinensis* L. (Tea): In Vitro Regeneration. Abstract. J. Biotechnology in Agriculture and Forestry (7): 82-98

Karlina 2006. Determination of Levels Epigallocatechin gallate (EGCG) in tea leaves with HPLC method, Thesis Faculty of Pharmacy Universitas Airlangga Surabaya.

Komes D., Horžlc D., BelšcAK, A. 2009. Determination of Caffeine Content in Tea and Maté Tea by using Different Methods. Czech J. Food Sci. 27:213-216

Krueger G., and Howard M.L. 2011. Effects of Psychoactive Chemicalson Commercial Driver Health and Performance: Stimulants, Hypnotics,Nutritional, and Other Supplements A Synthesis of Safety Practice. www.national-academies.org. Diakses 22-3-2015.

Li Y, ogita S, Charman A K, (2008). Expression of caffeine biosynthesis gene in tea (Camellia sinensis). J. Springer Verlag. 267-270

Wei W D, Yeyun Li Y, Ogita S, Ashihara H. (2008). Fine control of caffeine biosynthesis in tissue cultures of *Camellia sinensis*. Abstract. Phytochemistry (1): 195–198

Maria John M, Praveen N, Muthu T, Abul Kalam A M. (2013). Enhancement of the Productivity of Tea (*Camellia sinensis*) Secondary Metabolites in Cell Suspension Cultures Using Pathway Inducers. J. Crop Sci. Biotech.(2) : 143-149

Ministry for primary Industries Manato Ahu matua. New Zealand Government. 2012. www.mpi.govt.nz

Rebecca LJ., Seshiah C.,Tissopi T. 2014. The annals of "valahia" university of Targoviste Extraction of caffeine from used tea leaves. J. Department of Industrial Biotechnology, Bharath University.

Setyo pratomo P. 2014. Extraction of phenolic compounds from green tea Using ethanol. ARPN Journal of Engineering and Applied Sciences. (9): 1516-1521

Shirai, T., Sato. A., and Hara Y. (1994). Epigallocatechin gallate: the Major Causative Agent of Green Tea-Induced Asthma, Chest. 106 (18): 01-05.

Shervington A, Leroy A. Shervington F A, Mohamed A. E. (1998). Caffeine and theobromine formation by tissue cultures of *Camellia sinensis*. Abstract. J. *Phytochemistry*, (47): 1535-1536

Susita, S. (2014). 10 Fakta Menarik Tentang Kopi.http://id.she.yahoo.com/10- fakta-menarik-tentang-kopi- 194840452.html. Diakses 22-3-2014.

Shane V. B., Chris F. M., Robbertse H., Apostolides Z. (2013). Immunohistochemical localization of caffeine in young Camellia sinensis (L.) O. Kuntze (tea) leaves. J. Planta (237): 849–858.DOI 10.1007/s00425- 012-1804.

Sutini. 2010. Production of epigallocatechin gallate Through Callus Camellia Sinensis L, With Induction elicitor, Cu2+, Salicylic Acid and Precursor Phenylalanine. Dissertation. Graduate Program of the Faculty of Agriculture Universitas Brawijaya.

THE INFLUENCE OF VIBRATION ON $CaCO_3$ SCALE FORMATION IN PIPING SYSTEM

W. Mangestiyono[a], S. Muryanto[b], J.Jamari[c], A.P. Bayuseno[c]
[a]Vocational Program in Mechanical Engineering, Diponegoro University,
[b]Department of Chemical Engineering, UNTAG University in Semarang, Semarang Indonesia
[c]Graduate Program in Mechanical Engineering, Diponegoro University, Semarang Indonesia
E-mail : rhombo_hedral@yahoo.co.id

ABSTRACT

Carbonate scale is a common problem found in a piping system of industrial process. The presence of mechanical equipment such as turbine, compressor, blower, mixer and extruder produce a mechanical vibration on the piping system which is placed near these equipments. The influence of vibration on the $CaCO_3$ scale formation in the piping system was experimentally investigated in the present study. The aim of the research was to understand the effect of vibration on the kinetics, deposition rates and the crystals formation in the synthetic solution. The solution was prepared using $CaCl_2$ and Na_2CO_3 for concentration of calcium of 3.500 ppm, while the induction time, deposition rate, crystal growth were investigated at temperature of 25^0C. In generating the vibration force, the mechanical equipment consisting of electrical motor, crankshaft, connecting rod and a vibration table were employed, including four coupons inside the pipe for investigating the scale formed. Frequency of the vibration was set at 0.00, 1.00 Hz and 2.00 Hz, respectively. A dosing pump with two inlets and two outlets was used to circulate the solutions at flowrate of 30 ml/min from each vessel to the coupons. After running for three hours, the induction time was recorded at 17; 10 and 8 minute with vibration frequency of 0.00; 1.00 and 2.00 Hz, respectively. The scale formed was then characterized using SEM/EDX for crystal morphology and elemental analysis. The results show that the deposition rates were 0.9457 and 3.3441 gram/h for the frequency of 1.00 and 2.00 Hz. The carbonate crystals found in coupon and filter were vaterite.

Keywords : Carbonate scale, mechanical vibration, crystals, vaterite, aragonite, calcite

INTRODUCTION

The $CaCO_3$ crystals have been widely known as a scale formed into the piping system of industrial installations (Abdel and Sawada, 2003; Tzotzi et al, 2007) which could lead to industrial disturbances namely reduction of mass transport and heat transfer including brake-down of the industrial process (Setta and Neville, 2011; Jamialahmadi and Steinhagen, 2007). Generally, the growth of scale depends on the chemical factors such as solution concentration and the type of impurities (Curcio et al, 2010; Euvrard et al, 2004) as well as physical factors; flowrate and temperature (Ang et al, 2006). The two previous factors have been investigated widely by a large number of researchers, however, the influence of vibration on $CaCO_3$ scale formation has not been well documented in the literature.

The mechanical vibration is generally found in many mechanical industrial equipment during operation of such turbine, generator, blower, mixer, extruder, compressor (Ruiz et al, 2015), in particular, it will affect on the scale formation in the piping system. The vibration of a pipe is commonly caused by the fluid flowing inside the pipe and some external force (Liang and Bangchun, 2014). However, the occurrence of vibration on the scale formation may significantly affect the scale formation in the piping system and thus need to be controlled.

The vibration will drive an exitation force to the solution in the pipes and providing more energy to the system (Liang and Bangchun, 2011). Excitation force in the system increases the kinetic energy. If the amount of the kinetic energy is in excess of the activation energy of the ions, the collision of the ion would be more effective (Shubashini et al, 2014). On the other hand, the vibration is also potential to remove newly-formed scale inside pipes since the attachment of such scale to the pipe walls is not strong enough. Thus vibration may have two influences on scale formation, i.e. increasing or decreasing mass of the scale. Efforts to predict the calcium carbonate formation under the influence of the mechanical vibration needs a complete materials characterization of scale.

The aim at the present research was to examine the effect of vibration on the kinetics, deposition rate for the $CaCO_3$ crystal formation. In the present study, the $CaCO_3$ crystals were grown in the prepared solution under the influence of mechanical vibration. The crystals obtained were then characterized by SEM analysis with EDX for elemental analysis and morphology.

METHODOLOGY

$CaCO_3$ scale formation was made by mixing $CaCl_2$ and Na_2CO_3 solution according to the following reaction :

DOI: 10.1051/matecconf/20165801027

$CaCl_2 + Na_2CO_3 \longrightarrow CaCO_3 + 2NaCl$ (1)

The Phreeqc program was subsequently used to predict the potential for scaling. In operating the program, the composition input of the solutions was based on the calculated concentrated namely Ca: 1.750; Cl: 3.1625; Na : 2.0125 and CO_3: 2.625 g/kg water. The program was run and the potential for scaling was obtained. Here the saturation index (SI) of 1.49 was aragonite, while 1.63 of calcite and the specific conductance of 13,448 μS/Cm. The prediction showed that the dominant crystal of the scale was calcite.

Deposition rates

The method of calculating deposition rates has firstly to be formulated because of additional vibration effect on the system. It has been reported that, deposition rates include particle scale, precipitation scale and removal scale (Sheikholeslami, 2000) and given in the equation as follow :

$$W_{net} = A.W_{precipitation} + B.W_{particulate} + W_{removal} \quad (2)$$

Where $W_{precipitation}$ is deposition rate in the coupon, $W_{particulate}$ is deposition rate due to the existence of particle in the solution, $W_{removal}$ is deposition rate of removal crystal, A and B are dimensionless constant.

Particle scaling can be defined as deposition process of particles carried by a flowing fluid as well as by matters generated in the solution. The term "particle" is general and may refer to particulate matter, bacteria, corrosion product (Young, 2004). In this present work, by straining the solution with strain paper 0.22 μm, all particles in the solution are separated and deposition rate of particle scaling can be neglected. More over, deposition rate of removal ($W_{removal}$) was captured by straining paper namely deposition rate of strainer (W_{strain}). Thus, equation 2 can be modified into the new equation to acomodate the influence of vibration on deposition rate, as shown in equation 3.

$$W_{vib} = (W_{prec}+W_{strain})_{vib} - (W_{prec}+W_{strain})_{without\ vib} \quad (3)$$

Where W_{vib} is deposition rate as the influence of vibration and called as deposition rate of vibration. $(W_{prec} + W_{strain})_{vib}$ is all deposition rate measured when the system running with vibration frequency 1.00 and 2.00 Hz. $(W_{prec}+W_{strain})_{withhout\ vib}$ is all deposition rate measured when the system running without vibration or in this term it has frequency 0.00 Hz. W_{precip} is the scale mass precipitated and attached to the coupon. W_{strain} is deposition rates, which was measured in the filter paper and has the amount as the mass of the bulk crystallization diminished by the mass of precipitation.

Mechanical vibration

The vibration used in this research was the force vibration. It means that all parameters of vibration such as frequency, amplitudo, acceleration, displacement were induced by mechanical equipment. The mechanical equipment consisted of several components such as: crankshaft, connecting rods and table vibration. Force given to the system was calculated with the Newton'second law F = m . a, where *m* is the mass of solution to the pipe, *a* is the acceleration in root mean square (RMS).

The vibration was produced by an electrical motor in an intermittent manner i.e. the vibration given in 10 minute every 20 minute long. Furthermore, the energy given by vibration to the piping system was called as Excitation Energy in kJ/hr, i.e. 350 and 700 kJ/hr for the low level test and the high level test, respectively. The vibration parameter of the experiment is given in Table 1.

Table-1. Vibration parameter of the experiment

Vibration parameter	Control Test	Low Level	High Level
Frequency (Hz)	0	1	2
Acceleration(m/s^2,RMS)	0	11	22
ExcitationEnergy(kJ/hr)	0	350	700

Variables

Independent variabel of this experiment was energy excitation driven by frequency of the vibration, while dependent variables included induction time, deposition rate and polymorph of the crystals.

The effect of flowrates, concentration and temperature were netralized to strenghten the influence of vibration on the scale formation. The temperature was set at 25^{0}C. The concentration of Ca of 3,500 ppm, and the flowrate of 30 ml/min and atmospheric pressure were selected for all testing. This vibration-induced scale formation was conducted using the experimental rig (**Figure 1**).

Experimental rig and materials characterization

The experimental rig was schematically shown in Figure 1. It consists of two vessels (1) and (2) which contained $CaCl_2$ and Na_2CO_3 solutions, respectively. The concentration of Ca in solution was set at 3,500 ppm, while temperature of 25^{0}C was selected. To get the homogeneous solution, a stirrer was attached in each vessel and run at 30 rpm. All solution in each vessel was pumped by dosing pump CHEM FEED Ca-92683 (3) similarly, with 30 mL/min of flowrate to the test pipe section (4). In this test pipe, four pairs of coupons was inserted in, where scale was needed to form. The pipe was attached to a table (5) and vibrated by electrical motor HF-12207 (6).

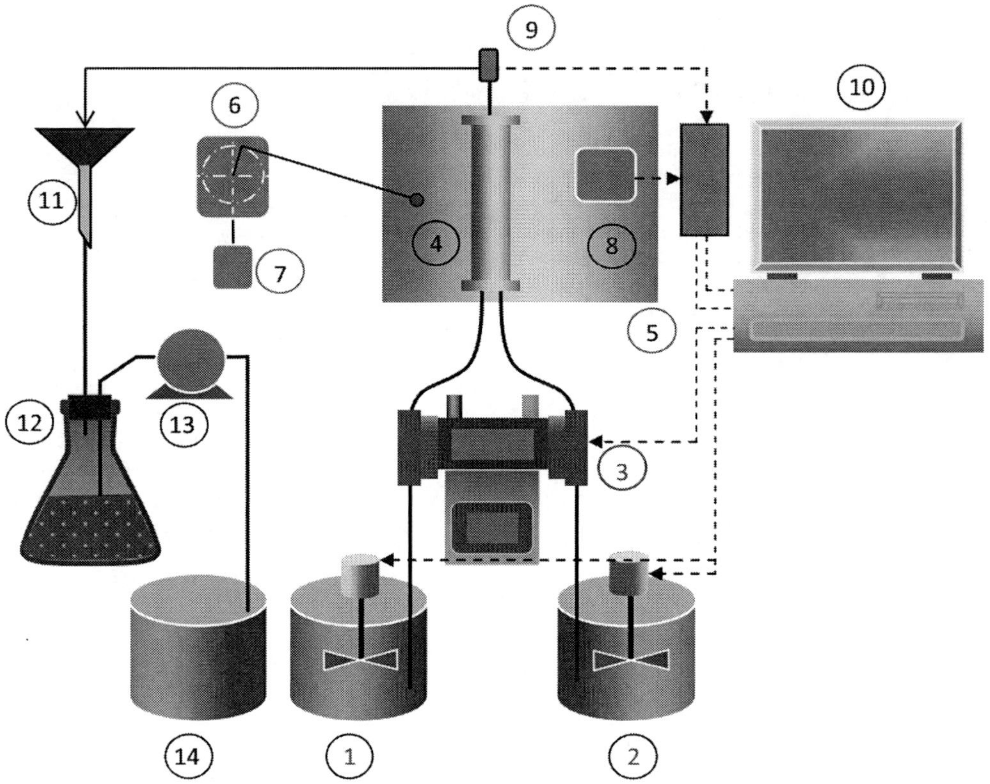

Figure-1. Experimental rig used in the research work

The vibration frequency was set by frequency regulator (7) to find the frequency needed i.e. 0.00; 1.00 and 2.00 Hz. Vibration meter Lutron VT-8204 (8) was used to observe the vibration parameter such as displacement peak to peak, velocity and acceleration in RMS. The conductivity of the two solution after leaving the coupon was measured by TDS meter produced by Komatsu (9), acquisition data of conductivity and vibration was held by data logger and computer program (10) to fulfill the accuracy. A filter paper of 0.22 µm (11) was used to capture the mass of scale which will be removed from the pipe wall or the mass brought by flowing force. A vacuum tube (12) and pump (13) was also employed.

Further, scanning electron microscopy (SEM) on a JEOL DSM apparatus equipped with an EDX was employed for characterizing the microstructure and morphology. For the observation, the powder crystals were mounted on a circular metallic sample holder and subsequently sputtered with gold. To suck the water in the fllter, the fllter paper was replaced every ten minute to keep the formation of the crystal remain of its phase. The scaling process was done in three hours duration of every running.

RESULTS AND DISCUSSIONS

After running in three hours duration all the data needed was formed, namely induction time and deposition rate. Moreover, the scale was characterized by SEM/EDX to obtain morphology and chemical elements of the crystal.

Induction time

During the scaling formation, the conductivity of solution after leaving the coupon was measured with TDS meter and recorded by the computer program every one minute. Measurable change of conductivity showed that a number ions in solution have reacted and precipitated as scale. The time of first measurable change of conductivity is called as an induction time. Figure 2 shows a variation on conductivity of the solution to the time. The induction time found this work was 17; 10 and 8 minute for frequency of vibration 0.00; 1.00 and 2.00 Hz respectively. The data proved that the vibration controlled the $CaCO_3$ scale formation by decreasing nucleation time and increasing mass of scale production.

No references were found to explain the mechanism of vibration in affecting the nucleation time. However, it could be caused by the increasing of ion collision, in which for frequency 2.00 Hz of vibration will produce 14,000 times per hour of mechanical collision. Furthermore, this result does not contrary to other previous research in which vaterite will be formed after 8 minutes (Chen, 2014).

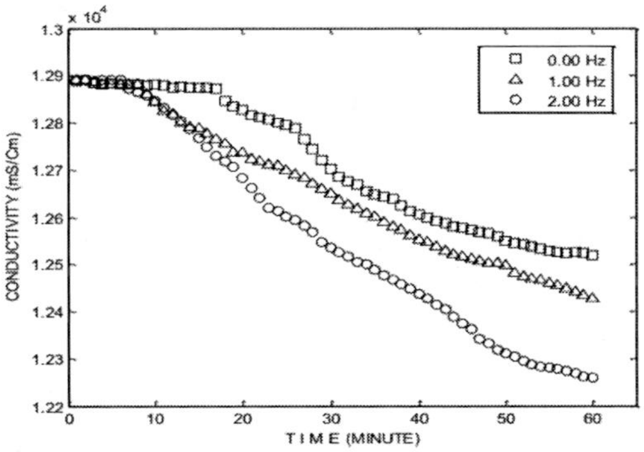

Figure-2. Conductivity of the solution over the time

Mass of scale

As it has been discussed before that the mass of scale formation in this system was classified into deposition rates of precipitation (W_{precip}), deposition rate in the strainer (W_{strain}) and deposition rate of vibration (W_{vib}).

Table-2. Deposition rate of $CaCO_3$ (gram/h)

Frequency	$W_{precipitation}$	$W_{strainer}$	$W_{vibration}$
0.00 Hz	0.0507	5.2157	0.00
1.00 Hz	0.0722	6.0892	0.9457
2.00 Hz	0.1207	8.4391	3.3441

The deposition rate of precipitation was measured in the coupon, deposition rate of the strainer was measured in the strainer for every running. The deposition rate of vibration was calculated as the difference between all deposition rates with vibration to all deposition rates without vibration. All of the deposition rates are presented in Table 2. It shows that the deposition rates of vibration in 2.00 Hz of frequency was calculated as : (0.1207 + 8.4391) – (0.0507 + 5.2157) = 3.3441.

This calculation as well as the logical consequency that scaling without vibration has been functioned as a control test to investigate the influence of vibration it self. Moreover the correlation between deposition rate of precipitation ($W_{precipitation}$); deposition rate of the strainer (W_{strain}) and deposition rate of the vibration ($W_{vibration}$) was plotted in Figure 3 to show the influence of vibration on the deposition rate.

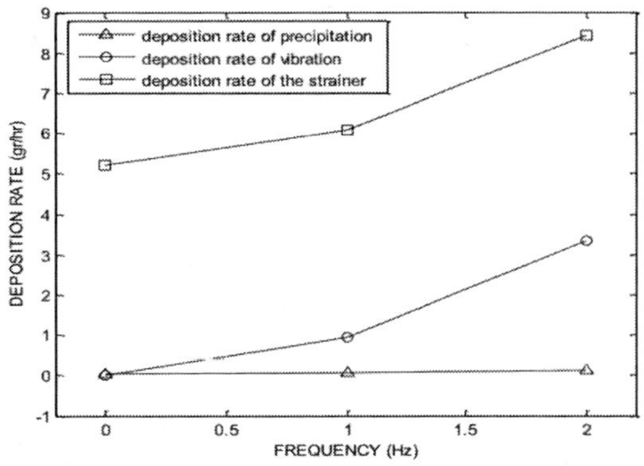

Figure-3. Deposition rate of scale formation

This graphic justifies that the vibration increases the deposition rate of $CaCO_3$ scale even precipitation rate, vibration rate and deposition rate in the fllter.

Crystal morphology

Morphology of $CaCO_3$ crystal was investigated by SEM/EDX for the specimen with 2.00 Hz of frequency obtained after running in three hours (Figure 4).

Figure 4. Crystals polymorph of precipitation scaling

Crystal of the precipitation scaling which is shown in Figure 4 was vaterite with the absence of aragonite and calcite. This fenomenon was unusual and showed the contradiction to the prediction by the phreeqc program in which it has predicted that the polymorph of the crystal were aragonite and calcite. The difference between the prediction of the phreeqc program to the crystal resulted by the research could be explained as follow : In the phreeqc program, calcite was predicted to dominate the crystals polymorph. Furthermore, crystallization of $CaCO_3$ may be very changeable, because the crystalline polymorph of calcium carbonate i.e. calcite, aragonite and vaterite are highly sensitive to the local condition (Su, 2009). The term "local condition" in this investigation was the vibration. Vibration produced excitation energy to the system and affect the crystaline polymorph.

More explanation to answer this fenomenon as it has published in previous research, among vaterite, aragonite and calcite, only aragonite is stable at room temperature (Heberling, 2014). This is the reason why calcite was not appear since aragonite was not formed. So, the absence of calcite on the precipitation could be accepted.

Figure-5. Crystals polymorph on the filter

Crystal found in the filter paper was vaterite with sperical shaped morphology. This fenomenon could be understand clearly because of replacing time of the filter was done no more than 10 minutes, so vaterite has not enough time to transfer into aragonite or calcite. The low temperature and pressure may also bring vaterite at its stability condition (Heberling, 2014). That's why all crystals in the filter was vaterite.

Chemical composition

Composition of the crystal was characterized by SEM/EDX JEOL JSM-6510JA providing the composition in % weight i.e. 17.26; 57.16 and 25.33 for C, O and Ca respectively. The peak shown in Figure 6 proved that $CaCO_3$ crystal was pure. It indicates that the downstream process of this research such as washing, filtrating, drying and powder storaging was done very well (Ralf Beck, 2009).

The differences between the theoretical and the actual value in % weight of crystal composition are 5.26 ; 9.16 and 15.33 for C, O and Ca respectively. These differences could be driven by the presence of vibration to the system.

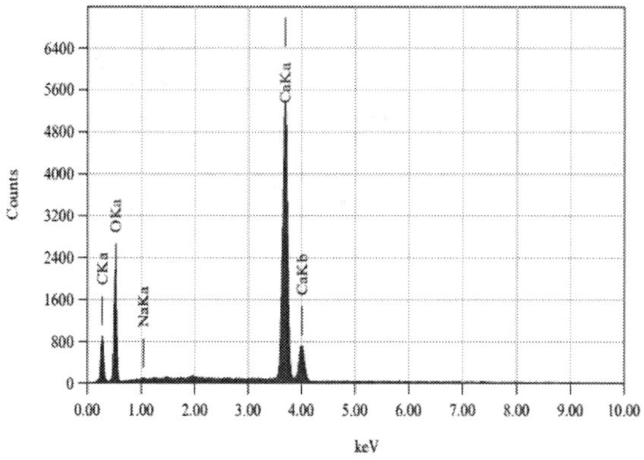

Figure-6. Composition of aragonite crystal obtained after running in three hours and frequency 2.00 Hz

CONCLUSIONS

Vibration has positive correlation between the kinetic and deposition rate.This can be proved in the graph of conductivity of the solution to the time and deposition rates of scale formation. The results confirm that, when the vibration increases, the kinetic and deposition rate also increase. Crystals polymorph was not really the same to the prediction by phreeqc program in which all the crystals of precipitation scaling was vaterite.

Acknowledgment

This research was supported by the Laboratory of Metrology and Instrumentation at the Vocational Program in Mechanical Engineering, Diponegoro University. The laboratory staffs who has helped for intensive tutorials and practice in calibration and checking standard of error of all instruments was gratefully acknowledged.

REFERENCES

Abdel-Aal,N and Sawada,K. 2003. Inhibition of adhesion and precipitation of $CaCO_3$ by aminopolyphosphonate, Journal of Crystal growth, 256, 188-200.

Ang H A., Muryanto S. and Hoang T. 2006, Gypsum scale formation control in pipe flow system: a systemic study on the effect of process parameters and additives, Curtin University Of Technology, Perth, Australia.

Beck R., Hakkinen A. and Didrik.M-S. 2009, The effect of crystalization condition, crystal morphology and size on pressure filtration of Lglutamic acid and an aromatic amine, Separation and Purification technology, 66, 549-558.

Chen T., Neville N., Sorbi K. and Zhong.Z. 2009. Insitu monitoring the inhibiting effect of polyphospinocarboxyliq acid on CaCO3 scale formation by syncroton XR-Difraction, Chemical Engineering Science, 64, 912-918.

Curcio E., Ji X. and Di Profio G. 2010, Membrane distillation operated at high sea water concentration factor : role the membrane on $CaCO_3$ scaling in presence of humic acid, Journal of Membrane Science, 346, 263-269.

Euvrard M., Membrey F., Filiatre C. and Foissy A. 2004, Crystallization of Calcium Carbonate, Journal of Crystall Growth, 265, 322-330.

Heberling F., Bosbach D. and Eckhart J D. 2014, Reactivity of Calcite-water-interface from molecular scale process to geochemical engineering, Applied Geochemistry, 45, 158-190.

Jamialahmadi M. and Muller-Steinhagen H. 2007, Heat exchanger fouling and cleaning in the dehydrate process for the production of the phosporic acid, Chemical EngineeringResearch And Design, 85, 245-255.

Liang F. and Wen B., 2014, Force vibrations with internal resonance of a pipeconveying fluid under external periodic excitation, Journal of Hydrodynamics, DOI : 10.1016/S101-5068 (14) 60086-X 2014.

Miwa S., Mori M., and Hibiki T. 2015. Two phase flow induced vibration in piping system, Progress in Nuclear Energy, 78, 270-284.

Muryanto S., Bayuseno A P., Ma'mun H., Usamah M. and Jotho. 2014, Calcium carbonate scale formation in pipes : effect of flowrate, temperature and malic acid as additives on the mass and morphology of the scale, International Conference and Workshop on Chemical Engineering UNPAR 2013, Procedia Chemistry, 9, 69-76.

Ruiz C., Jaramillo V H., Mba D., Ottewil J R. and Cao.Y. 2015, Combination process and vibration data for improve condition monitoring of industrial system working under variable conditions, Mechanical System and Signal Processing, 66-67, 699-714.

Scaussberger P., Mustafa G M., Leslie G. and Friedel.A. 2009, Scaling prediction based on thermodynamic equilibrium Calculations-scope and limitation, Desalination, 244, 31-47.

Setta F A. and Neville A. 2011, Eficiency assesment of inhibitors on CaCO3 precipitation kinetic in the bulk and deposition on a stainless steel surface, Desalination 281, 340-347.

Sheikoleslami. 2000, Calcium sulfate fouling-Precipitation or particulate : A proposed Composite model, Heat Transfer Engineering, 21(3), 24-33.

Sinir B G. and Demir D D. 2015, The Analysis of non lineair vibration of pipe conveying an ideal fluid, European Journal of Mechanic Fluids, 52, 38-44.

Su Y., Yang H., Shi W, Guo H. and Zhao Y. 2009, Cristallization and morphological control of calcium carbonate by functionalized tri block polymers, Colloid and Surfaces A : Psychochemical engineering Aspects, DOI:10.1016/J.Colsurfa.2009.12.002.

Shubashini O., Ning H., Ferguson I. and Lu N. 2014, Effect of surface treatment on thermal stability of the hemp PLA composities : Correlation of activation energy with thermal degradation, Composities Part B,67, 227-232.

Suai Jun L., Gong-min L. and Wei-Tao K. 2014, Vibration analysis of pipe conveying fluid by transfer matrix method, Nuclear Engineering and Design, 266, 78-88.

Tzotzi, Pahiadaki, Yiantsios, Karabelas, and Andritsos. 2007. A study of $CaCO_3$ scale formation and inhibition in RO and NF membrane processes, Journal of Membran Science, 296, 171-184.

Young I., Li S H., Kim W. and Suh S. 2004, Physical Water Treatment for the Mitigation of Mineral Fouling In Cooling-Tower Water Applications, 2003, ECI Conference on Heat Exchanger Fouling and Cleaning: Fundamentals and Applications Santa Fe, New Mexico, USA.

EFFECT OF HUMIC ACID ON SOIL CHEMICAL AND PHYSICAL CHARACTERISTICS OF EMBANKMENT

Munawar Ali[1] and Wanti Mindari[2]

[1]Department of Environmental, Faculty of Civil and Environmental Engineering, University of Pembangunan Nasional "Veteran" East Java, Indonesia
[2]Department of Agrotechnology, Faculty of Agriculture, University of Pembangunan Nasional "Veteran" East Java, Indonesia
E-Mail: munawar1960@gmail.com[1]

ABSTRACT

The effectiveness of the treatment of pathogens disease in fish using chemicals is very limited because of the stress of toxic ions. A treatment of humic acid of 50-90 mg /L on a farmland has been able to reduce illness and death compared to a mixture of formaldehyde and blue-green methylene. Humic acid is suspected to be able to increase yield, through improved conditions and resistance to diseases, health, and cultural vitality, either by itself or combined with cation species toxic. Humic acid can balance the soil cation so that the soil pH reached 7-8, through a chelate of Al, Fe, Ca or exchanged with NH_4, Na and K. Humic acids were extracted from compost plants with a weak base of 0.1 N NaOH and precipitated at pH 2. The concentration of AH 0- 400 ml was applied to three soil types with an area of 0.12 m^2 and 15 cm thickness. The results showed that the application of 100-200 ml HA/0.12 m^2 gave optimum yield in improving the physical-chemical characteristics of the soil embankment. Soil pH reached a value of 7-8, cation exchange to 60% saturated, and soil bulk density was reached in the range of 1.1- 0.97 g / cm^3.This condition is suitable for fish or shrimp in the embankment.

Keywords: cation, embankment, humic acid, pH, soil

INTRODUCTION

The low levels of fertility and quality of soil embankment are due to fluctuations in salt content, organic sediment, plankton, and soil structure (Machmudin, 2009). The pH of water is not affected since high biological activity rarely reaches > 8.5, but the pH of water of fish or shrimp embankment may reach 9 or more (Boyd, 2002). The seven primary ions: sodium, potassium, potassium, magnesium, chloride, sulfate and bicarbonate have a great contribution towards salinity within (Boyd, 1990). Calcium ions (Ca), potassium (K), and magnesium (Mg) is the most important ions in propping up the level of living shrimps (Davis et al. 2004). The ideal range of saturation of Ca, Mg and K are 60-80%, 10-25%, and 3-5%, respectively, of the CEC soil. Magnesium is recommended to be given when saturation is below 15%, or the ratio of Mg/K < 2:1. Osman *et al* (2010) found that the concentration of chemical compounds increases in the Outlet and declines in the Inlet. Heavy metals and inorganic anion varies significantly ($P < 0.05$) depending on the type of fish and the location of the relative abundance of cultivation. Sequence elements: sediment > > fish water. The order of heavy metals and inorganic anions in water: Fe-Mn-Zn > Pb >Cu > Cd, whereas in sediments are: Fe >Mn-Zn > Cu >Pb > Cd.

Alkalinity is a buffer important to control embankment soil pH and changes of dissolved cations. Ion solubility or salinity in the embankments water for shrimp cultivation is in a range of 15-30 g/l. The addition of organic fertilizers has increased the production of shrimp in 2011 only by 5% while in 2012 there was an increase of 22 %. An alternative solution is to use a buffer to keep the pH value from changing (Tan, 1988). Alkalinity acts as a buffer for the pH fluctuations as well as stabilizing the tapping solution of pH and determines the natural fertility of the waters (Boyd, 2002). Humic acid (HA) is a major component of humic substance, produced from the biodegradation of dead organic matter, containing carboxyl and phenolic so that it behaves functionally as dibasic acid or sometimes as a tribasic acid. Functional groups which most contribute to surface charge and reactivity. The presence of carboxylic groups and phenolic gives the ability to form a complex with HA ions such as Mg^{2+}, Ca^{2+}, Fe^{2+}, and Fe^{3+}. The ability of humic acid to adsorb cations follows the lipotrophic sequence, i.e., $Al^{3+} = (H^+) > Fe^{3+} > Fe^{2+} > Ca^{2+} > Mg^{2+} > K^+ = NH_4^+ > Na^+$ (Tan, 1998). Sorption of NH_4^+ is similar to Na^+ (Nursyamsi et al., 2009). Sorption and maximum buffering capacity of the NH_4^+ and Na+ are relatively different. Cation adsorption by HA occurs through the exchange of cations in solution or that adsorbed by clay-humic. Adsorption of cations or metals by HA can be through (a) direct adsorption (Ca^{2+} that release PO_4^{3-}), (b) complexation of Cu^{2+} or outer-sphere interactions for hydrated Mg^{2+} , (c) serving as a cation bridge through direct or indirect chelation, and (d) interaction with Ca^{2+}-HA aggregates or with amine groups (Sharma and Kappler, 2011). Clay or humic materials have a strong affinity to weak acids containing phenolic hydroxyl, a carboxyl group, or amino sulfonyl. Alkaline cations (Na^+, K^+, Ca^{2+}, Mg^{2+}) are primarily retained by simple cation exchange with COOH groups (RCOONa, RCOOK) (Zhang et al., 2013).

Humic acid (HA) is a derivative product of decomposed organic material that is soluble in alkali but insoluble in acid (Mikkelsen, 2005; Pena-Méndez et al., 2005). A typical structure molecule of HA may consist of six carbon aromatic rings of the basis of di- or tri-hydroxyl phenols linked by -O-, -NH-, -N-, - S-, and contain -OH group and quinone (O- C_6H_4-O-) (Tan, 1998). Humic acid (HA) is a cyclic organic compound having high molecular weight, long-chain, and active carboxyl group (- COOH) and phenolic (-OH), which are amphoteric, binding of cations/anions at certain pH dependent charge (Stevenson, 1994; Bohn et al., 2001; Pena-Méndez et al., 2005; Khaled and Fawy, 2011). Carboxylate of some carboxyl group is released below pH 6 leaving a negative charge on the functional group: R-COOH = R-COO- + H^+ (Pettit, 2011). Dissociation of H+ from amide (= NH) can also increase the negative charge. Humic acid levels in compost, manure, straw, and others are relatively low (<1%), while that in coal deposits "Leonardite" is relatively very high (~ 15%) (Humintech, 2012). Mindari *et al.* (2013) obtained HA extracts from compost, coal and peat with 0.5 N NaOH and its deposition to pH 2, respectively of 2.6%, 4.6%, and 7.6%. Goff (1982), Lebo *et al.* (1997), Anaya-Onala (2009), and Chen *et al.* (2009) obtained HA at least 60-80% and has a high solubility. Nur Hasinah et al. (2008) reported that reduction of the duration of the extraction time from 24 hours to 12 hours gave AH a similar character. Characteristics of humic acids can be analyzed from the ratio of the value of E4 / E6 which is the value of absorbance at 465 nm (E4) and the absorbance of 665 nm (E6). Kononova (1966) and Chen *et al.* (1977) believed that the ratio of E4 / E6 was associated with the degree of condensation of aromatic carbon lattice. Weak ratio values indicate weak condensation of high levels of humic aromatic components while strong ratio indicates the proportion of higher aliphatic structure. The ratio value of E4 / E6 is primarily governed by the size of the molecule or molecular weight or particle, which is correlated with the concentration of free radicals with O, C, CO_2H and total acidity (Chen et al., 1977), but it does not depend on the concentration of humic acid or fulvic. The structure of humic acids has a range of E4 / E6 ratios about 3.3-5.0 (Pansu and Gautheyrou, 2006).This study was aimed to examine the effect of humic acid on the soil cation concentration, pH, EC, and soil bulk density (SBD) at the embankment.

METHODOLOGY

The experiment was conducted from April 2015 to July 2015 in the glasshouse and laboratory, consisting of two activities.

Extraction and characterization of humic acid as buffering agent

Humic acid were extracted from compost with 0.5 1 NaOH solution (1:5) through the 24-hour agitation and precipitation with 6 N H_2SO_4 up to pH 2 in accordance with the method of Stevenson (1994). The extracted humic acid was purified by adding a mixture of HCl and HF (2:1 by volume) which were then forwarded with water up to 3 times by centrifugation or settling in the separator tube. The humic acid precipitate was heated at 40°C to obtain concentrated HA. The humic acid having a pH 2 adjusted to a pH of 6 for the application.

Organic-C content of the humic acid was determined using the method of Walkey and Black. The percentage of humic acid was calculated by the gravimetric method at a temperature of 100°C. The value of E4 and E6 were analyzed by diluting humic acid with 0.05 N $NaHCO_3$ (1 mL of humic acid: 10 mL), and then each of which was analyzed at a wavelength of 465 and 665 nm using a Spectro Pharo 100. The ratio of E4 / E6 was obtained by dividing the value of E4 with E6 (Kononova, 1966; Chen , *et al*, 1977). The E4 / E6 ratio value of less than 5 shows the character of humic acid while that of more than > 5 is fulvic acid (Tan, 2003). CEC value of the humic acid was analyzed by saturation 1N NH4OAc at pH 7. The C organic content and CEC were analyzed according to the previous method in organic materials that have been oven-dried at 70°C as proposed by Pansu and Gautheyrou (2006). CEC was analyzed by saturation 1N Ammonium Acetate. The pH and EC values were analyzed in pasta humic acid 1:1.

Experiment: Effect of humic acid on soil pH, cation, and bulk density

The experiment was arranged according to a factorial completely randomized design (CRD) where the first factor 1 was five dose of humic acid : 0,100, 200, 300, 400 ml/1200cm^2 (100 ml/1200cm2= 889 mg/kg). The second factor was three types of embankment. Each treatment was repeated three times. The soil used for this experiment was topsoils (0-20 cm depth) collected from Gununganyar Village of Surabaya. Soil sampling sites located as far as 2-3 km from UPN Surabaya. They were taken from three different levels of fertility, i.e.: slow to moderate permeability (2.2-25.5 mL / h), clay texture (60- 62%), pH 7.9-8.55, EC 6.8-7.8 mS / cm, and 1.88–2.96% organic C. Samples were taken when the embankment is still saturated with water, and then placed in a plastic chamber of 30x40cmx15 cm^3 capacity.

The buffer made of humic acid extracts was blended with NPK fertilizer. The buffer-based humic acid liquid was given according to the dosage then evenly mixed and incubated for 1 month at a room temperature. Water with a salinity of less than 1 mS/cm was added to the soil to approximate field capacity. After incubation, soil samples were taken for analysis of cations of soil and soil pH. The chemical characteristics of the embankment: pH, EC, heavy metals Pb, Cu, Cd, bases, and heavy soil content farms. Character humic acids were analyzed by NMR, pH, EC and Redox with soil paste method, and cations using ammonium acetate extract. After 1 month of incubation, the soil samples were analyzed for changes in pH, EC, and redox. At 2 months of incubation, the soil samples were analyzed for their soil bulk density change.

After a month of incubation, 50g of soil subsample was collected from each pot, air-dried and sieved to pass through 0.5 mm sieve for chemical analysis. The soil chemical characteristics analysis includes pH, EC, and soil bases.

Data analysis

Observations data were summarized in the chart treatment of the results. A regression treatment was performed on the results to assess the optimal dose to the buffer embankment fertility levels. LSD 5% was used to assess the best materials for fish embankment land buffer. The statistical analysis was performed using Excel. The regression and correlation treatment of the results were used to assess the buffer dose optimization.

RESULT AND DISCUSSION

Characteristics of Humic Acid

The humic acid used in the experiment is presented in table 1 and figure 1. Humic acid was extracted by NaOH and precipitated with H_2SO_4 until pH reached 2. This is similar to the characteristics of peat containing higher organic- C, humic acid, CEC, and smaller E4 / E6 ratio than others (Mindari, 2013).

DOI: 10.1051/matecconf/20165801028

Table-1. Characteristics of humic acid materials

Sources of Humic Acid	pH	Org-C (%)	H-NMR ppm	Humic Acid (%)	Ratio E4/E6	CEC (me/100g)
Compost Of leaf	6	25.15	20.000	5.50	3.31	85.72

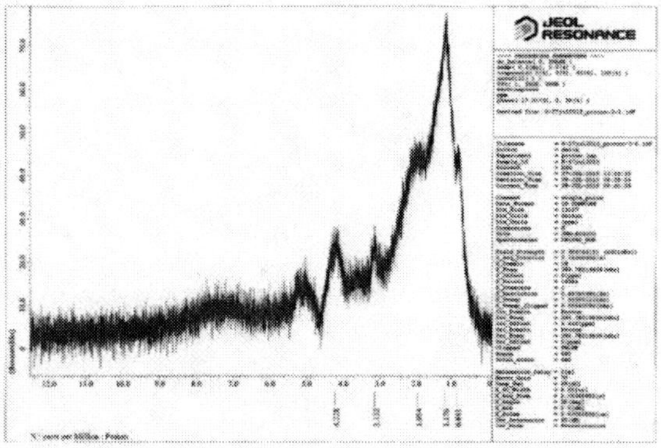

Figure-1. H-NMR humic acid contains concentration of H

Compost saturated with NaOH caused the **site** to be oxidized to molecules saturated with Na, which was ready to be exchanged with all ions in the soil. A highly oxidized organic matter, in which parts of its chemical structure are oxidized, will create site adsorption to bind micro nutrients, microflora, and the molecules. According to Tan (2003), Miklesen (2005), positive ions bind to oxidized site adsorption to provide space for the entry of negatively charged molecules which causes them to absorb micronutrients.

The addition of NPK to compost -HA is suspected that AH-NH_4^+ and -K^+ exchanged to soil -Na^+ and soil-Ca^{+2} caused reduced soil salinity. This reduction of Na was because of K^+ replaced them on the surface of adsorption colloid so the proportion of K was increased. Because the three ions have a similar valence, their exchangeability is determined by the affinity of the cations (Tan, 2003). Fosfat Ion ($H_2PO_4^-$) adsorbed in the positive charge of HA will eventually be released into the soil solution if needed by plants or biota. In line with the increasing charge, the addition of cations also increased CEC. The ability of K^+ to exchange H^+ potential of missel made the solution become baser. These humic pH were around 6, which was the same as the original pH buffer of 6.3-6.5. The addition of K^+ -NPK fertilizer formed K^+ -humic, that were easier or more preferably adsorbed to the colloid surface than NH_4^+ - NPK fertilizer (Nursyamsi et al., 2009). Because of the ability of colloid to absorb NH_4^+ or K^+ is similar, the addition of one of the ions will exchange the others in the same amount.

The higher the dose of humic acid, the higher the CEC value. This was because of the addition of cations will increase them in the mineral surface and between minerals. Colloids do not only adsorbed ions but also absorbed water, so that increased water reserves. HA absorb more than absorbents used to date (Pena-Méndez et al, 2005).

Effect of humic acid on soil pH, cation, and bulk density

The value of humic acid pH was adjusted to 6 for application to soil embankment. The application of humic acid+NPK significantly affected soil pH (Figure 1), soil cation exchange (Ca, Mg, K, Na) (figure 2 and Table 2) and soil EC on three types of the fish embankment.

The soil reaction greatly affects the availability of nutrients to plants. Under neutral soil pH, nutrients are available in considerable amounts. However, if the soil pH is more than 8.0, nitrogen, iron, manganese, boron, copper, and zinc will be less available to plants (Tan, 1998).

Saline intrusion on soil causes (a) fixation or absorption of other nutrients in the soil by the compounds and silica carbonate or oxide Fe, Ca, and Mg, and (b) disturbance in the balance of Ca^{2+}, Mg^{2+}, Na+, and K^+ in the soil, among others, further strengthen the aggregate stability (Mikkelsen, 2005;

The downward trend in soil pH of all three types of soil is almost the same, but the first soil changed faster compared to the second and third ones. The reduced pH value of the soil was due to the replacement of the soil solution of salt ions with H + humic acid which caused the pH of the solution is lowered (Khaled and Wafy, 2011).

The applications of AH 0-400 ml dose /0.12m^2 lowered the soil pH to 0.8 units on soil 1, 0.6 units on soil 2, and 0.7 unit on soil 3. The addition of humic acid will exchange H + cations to cause lower soil pH. The soil cation exchange has a capability to greatly influence the content of clay and organic matter. The higher the content of clay and organic matter, the smaller the pH change, as in the case on soil 2. The EC value of the soil greatly influenced the accumulation of salt concentration in the soil solution. EC values are correlated with soil pH, wherein if the soil pH dropped due to the decreased levels of salt concentration as shown in Figure 2. The high value of farm land EC may be due to the high content of Na, K, Ca and Mg soil. The addition of humic acid will release H + into the soil solution H + humic where its position was replaced by cation salt, then decrease the salt concentration in the solution so that the value of EC was reduced. Therefore, if the suitability for pond soil pH is between 7 and 8, the pH of HA administration of 100 ml is sufficient for embankment repairs.

The dose of application of HA was up to 400 ml for 5kg **soil** significantly decreased soil pH, cations exchange. The results of analysis of cations embankments with the application of a humic acid with a dose of 400 ml /12m^2 or an equivalent of 3-6g/kg soil have a pH value between 7 and 8 which is suitable for embankments. Cation exchange was a little bit changed and can reach the portion corresponding to the cation saturation. There was an interaction between the dose and the type of soil in affecting the soil cation exchange. The best treatment combination was 400 ml to soil type 3, 100-200 ml HA to soil type 2, and 200 ml to soil type 3 to reach a saturation ca 60% ideal for soil.

Along with the release of H by humic, the cations present in the solution and adsorbed colloid surface soil will be exchanged. The value of the exchanged cations and anions causes particles to loose or ease the incorporation of solid particles added to the soil pore space. The changes in soil porosity affect the flow of water and soil nutrients. Overall, the trend of changes in the soil cation decreases with increasing humic acid applications. The soil 1 embankment with the provision of another soil humic acid was suspected to have ugly structures that need some improvement. Humic acid can increase the aggregate stability (Pena-Méndez *et al*, 2005) by improving the saline soil physic.

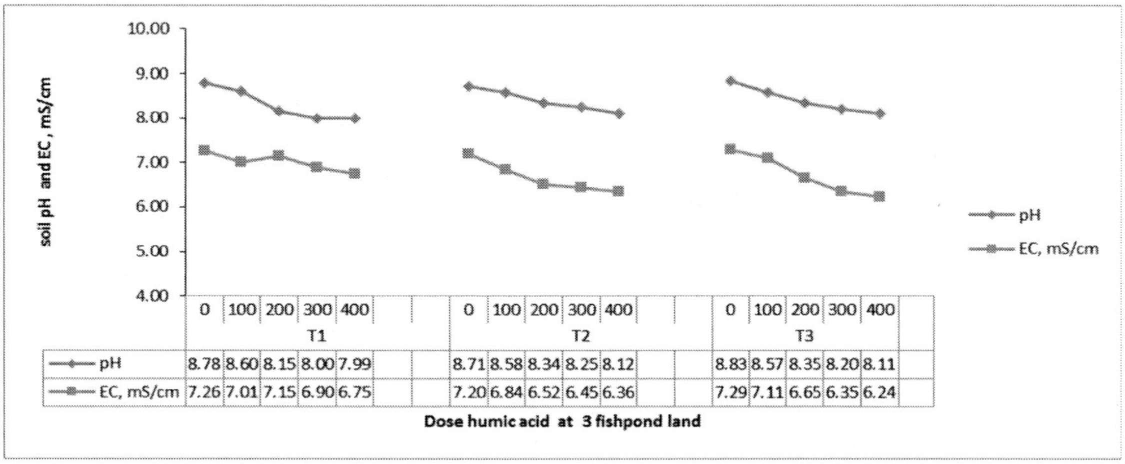

Figure-2. The value of soil pH and EC at 1 month after humic acid +NPK application

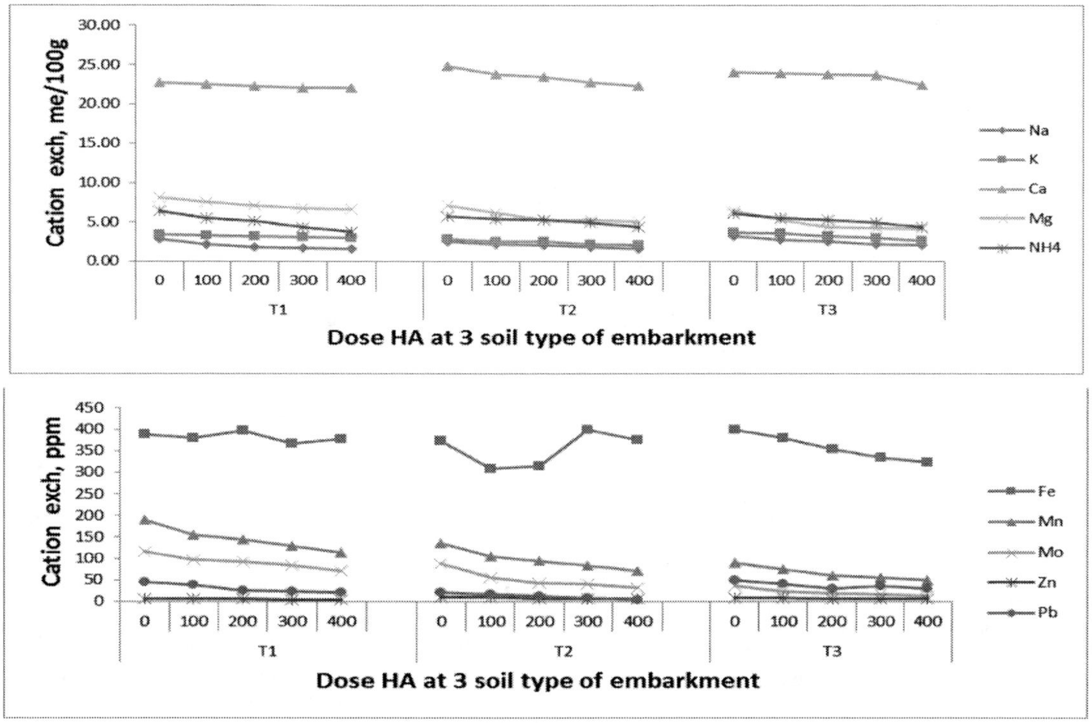

Figure-3. The value of soil cation after 4 weeks of HA application at 3 types of embankment (Above: macro cations, Bottom: micro cations)

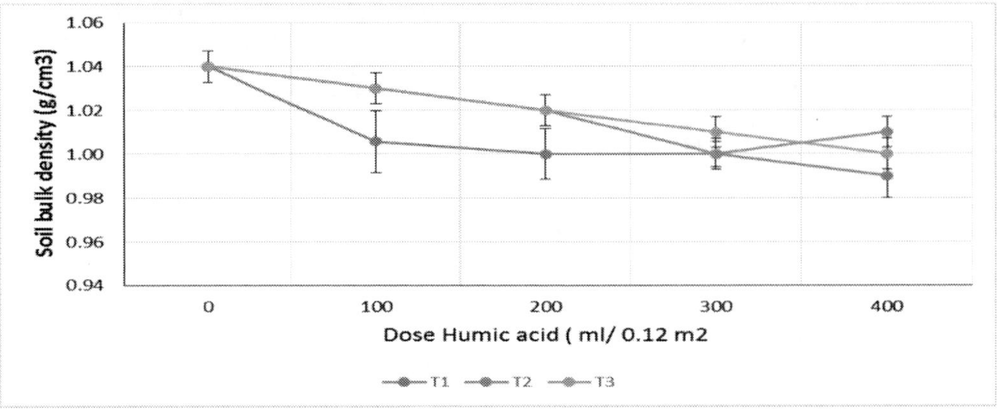

Figure-4. Soil bulk density of embankment after application of HA +NPK

Table-2. Percent saturation cations on 3 type of soil embankment after application AH 0-400 ml / 100g

Type of embarkment	Soil cation (me/100g)	Dose of humic acid (Ml/0.12 m2)				
		0	100	200	300	400
1	Na	6.62	5.32	4.67	4.59	4.45
	K	7.96	8.05	8.16	8.11	8.08
	Ca	52.17	54.80	56.30	58.21	59.29
	Mg	18.56	18.35	17.93	17.66	17.99
	NH4	14.70	13.48	12.94	11.43	10.18
2	Na	5.94	5.45	5.38	4.95	4.67
	K	6.53	6.32	6.45	5.90	5.73
	Ca	57.62	59.29	60.83	61.51	63.07
	Mg	16.49	15.55	13.74	14.25	14.29
	NH4	13.42	13.39	13.59	13.39	12.25
3	K	7.27	6.80	6.41	5.83	5.86
	Na	8.44	8.67	8.32	7.96	7.28
	Ca	55.50	58.28	60.76	62.06	62.88
	Mg	14.76	12.91	11.15	11.15	11.65
	NH4	14.03	13.33	13.37	13.00	12.33

CONCLUSION

The humic acids extracted from compost has a CEC of 60-156 me/100g, organic-organic C content of 20-30%, pH value of 6.0, in black color, and slow soluble in water. HA applications up to 400 ml for 5kg soil significantly lowered the soil pH, cation exchange, and bulk density. The pH value of about 8 and about 60% Ca saturation was achieved in the administration of 400 ml HA for soil type 1, 100-200 ml for soil type 2, and 200 ml for soil type 3.

Acknowledgements
We would like to thank the Directorate General of Higher Education, Ministry of Education and Culture of Indonesia for financial support for this research. Thanks are also due to laboratory and staff of the Faculty of Agriculture UPN "Veteran" East Java and Soil Science Department, Brawijaya University for their strong support and encouragement to undertake this analysis.

REFERENCES

Anaya-Olvera, A. 2009. Method of Obtaining A Concentrate Humic Extract from Organic and Inorganic Sources. Patent No: Us 7,510, 590, B2.

Bohn, H., McNeal, B.L. and O'Connor, G.A. 2001. *Soil Chemistry*, Third Edition. John Wiley & Sons.Inc. 307p.

Çelik, H., Katkat, A.V., Aşik, B.B., Turan, M.A. 2010. Effect of humus on growth and nutrient uptake of maize under saline and calcareous conditions. *Agriculture* 97(4): 15-22.

Chen,Y, Senesi, N. and Schnitzer, M. 1977. Information provided on humic substances by E4:E6 ratios. *Soil Science Society of America Journal* 41: 352–358

Chen, C.H.H., Ahmed, O.H., Majid, N. M.Ab. and Jalloh, M.B.. 2009. Reduction of isolation period of coal humic acids. *American Journal of Applied Sciences* 6 (7): 1327-1330.

Goudarzi M. and Pakniyat, H. 2008. Comparison between salt tolerance of various cultivars of wheat and maize. *Journal of Applied Sciences* 8 (12): 2300-2305

Khaled, H. and H.A. Fawy. 2011. Effect Of Different Levels Of Humic Acids On The Nutrient Content, Plant Growth, and Soil Properties Under Conditions Of Salinity . Soil & Water Res., 6(1): 21–29

Kononova, M.M. 1966. *Soil Organic Matter, Its Nature, Its Role in Soil Formation and in Soil Fertility*, 2nd English ed. Pergamon Oxford, 544 p

Lebo Jr. S.E., Kevin. W.R., Stephen, W. and Dickman, L. 1997. Production of oxidized humic acids by an oxygen-alkali. *Process* 5:688-999.

Mikkelsen, R.L. 2005. Humic Materials for Agriculture. Better Crops 89 (3): 6-10.

Mindari, W., Kusuma, Z. and Syekhfani, 2013. Isolation and Characterization of Humic Acid of Various Waste Materials on Saline Soil and Their Effects to Paddy. *Proceeding of International Conference on Green –Agro Industry,* Ina Garuda Hotel, Yogyakarta 12-14 November, : 255-233

Nursyamsi, D., Idris, K. Sabiham S., Rachim D. A., dan Sofyan, A. 2009. Jerapan dan pengaruh Na , NH4 , dan Fe3+ terhadap ketersediaan k pada tanah-tanah yang didominasi mineral liat smektit. *Jurnal Tanah Tropika* 14 (1): 33-40.

Osman, MA , MAM Mohamed, MHH Ali dan ADG Al-Afify 2010. Assessment of Agriculture Drainage Water Quality to be Used for Fish Farm Irrigation. Nature and Science;8(8)

Paksoy, M., Türkmen, Ö. and Dursun, A. 2010. Effects of potassium and humic acid on emergence, growth and nutrient contents of okra (*Abelmo schusesculentus*) seedling under saline soil conditions. *African Journal of Biotechnology* 9 (33): 5343-5346.

Pansu, M. and Gautheyrou, J. 2006. *Handbook of Soil Analysis Mineralogical, Organic and Inorganic Methods*. Springer-Verlag Berlin Heidelberg. 993p

Pena-Méndez, E.M., Havel, J. and Patocka, J. 2005. Humic substances, compounds of still unknown structure: applications in agriculture, industry, environment, and biomedicine. *Journal of Applied Biomedicine* 3:13-24.

Pettit, R.E. 2011. Organic Matter, Humus, Humate, Humic Acid, Fulvic Acid, and Humin: http://www.calciumproducts.com/articles/dr._pettit_humate.pdf

Sharma, P. and Kappler, A. 2011. Desorption of arsenic from clay and humic acid-coated clay by dissolved phosphate and silicate. *Journal of Contaminant Hydrology* 126 (2011) 216–225

Sparks, D. L. 2003. *Environmental Soil Chemistry*. Second Edition. University of Delaware. Academic Press. 345 P.

Stevenson, F.J. 1994. *Humus Chemistry: Genesis, Composition, Reaction*.2nd Ed. John Wiley & Sons, Inc. New York.. p. 36.

Tan, K. H. 2003. *Humic Matter in Soil and the Environment. Principles and Controversies*. University of Georgia. Athens, Georgia, U.S.A.

Tan, K.H. 1998. *Principles of Soil Chemistry*. 3rd Ed. 521 P.

Turan, M.A., Asik, B.B., Katkat,A.V. and Celik, H. 2011.The effects of soil-applied humic substances to the dry weight and mineral nutrient uptake of maize plants under soil-salinity conditions. *Notulae Botanicae Horti Agrobotanici* Cluj-*Napoca* 39(1):171-177

Zhang, W.Z. Chen, X.Q., Zhou, J.M. Liu, D.H., Wang, H.Y. and Du,C.W. 2013. Influence of humic acid on interaction of ammonium and potassium ions on clay minerals. *Pedosphere* 23(4): 493–502.

MATEC Web of Conferences **58**, 01029 (2016) DOI: 10.1051/matecconf/20165801029
BISSTECH 2015

CALCIUM CARBONATE SCALE FORMATION IN COPPER PIPES ON LAMINAR FLOW

Raharjo, S*, Bayuseno, Ab, Jamarib, Muryanto, Sc
*Department of Mechanical Engineering, Muhammadiyah University in Semarang, Indonesia
bDepartement of Mechanical Engineering, Diponegoro University, Semarang Indonesia
cDepartment of Chemical Engineering, UNTAG University in Semarang, Semarang Indonesia
*Corresponding author. Tel. +62-8164242537, +62-8445768
E-mail address: *samraharjo2@gmail.com; apbayuseno@gmail.com

ABSTRACT
Calcium carbonate is commonly precipitated as a scale in the transportation pipes of water. The presence of this mineral deposit becomes problematic, because it can block the pipes and lead to a decline in piping performance. Calcium carbonate precipitation from the synthetic solution was experimentally investigated in the present study. The aim of research was to predict the occurrence of precipitates and characterize the scale precipitated from the solutions. The synthetic solutions were prepared using $CaCl_2$ and Na_2CO_3, which was mixed with distilled water (H_2O). The concentrations of Ca^{2+} at 2000, 3000, 4000 and 5000 ppm. in the solution were adjusted and the solution flow in the Cu pipes at the different flow rate of 30, 40 and 50 ml/min. It was found that in all the experiments, the conductivity decreased abruptly after a certain induction period. Higher temperature produced more mass of the scale indicating that the increasing temperature promote scale formation. SEM analysis showed that the scale was rhombohedral, while EDS revealed that the elemental composition of the scale consisted of Ca, C and O. The crystalinity of the scale was found to be mostly calcit as shown by the XRD

Keywords: calcium carbonate, induction time, prototipe scale formation, scale morfology

INTRODUCTION

Scaling is one of the main serious engineering problems in process industries as the scales my hinder the flow of solution in industrial pipes or tubes. Accurate estimation of scale formation is thus very important. Calcium sulphate, carbonate potentionaly deposited add clogged pipping system (Azimi and Papangelakis, 2010).

The calcium carbonate scale formation was influenced by several aspects such as flow rate, concentration of solution, pH value, temperature, pressures, and ionic strength (Bahadori, 2010). In laminar turbulence,the higher flow rates,the more calcium carbonate scale mass indicating the fluid flow enchances the scale formation (Muryanto et al., 2012).

Previous studies done by Gourdon, have shown that $CaCO_3$ scale formation can be found in fully laminar and fully turbulent. In fully laminar the formation was in a slow growth while in fully turbulent the formation was easily found in the angels (Amor et al., 2004). The flow of the fouling is described as having pronounced rippling and surface waves of partially laminar and partially turbulent nature.As the turbulent also will enhance the mass transfer,the rate of crystal deposition can also be promoted (Gourdon, 2011).

Other researchers indicated that scale growth does not always remain linear with time,as higher velocity can sometimes reduce scale deposition (Basim et al., 2012). Crystallization occurs due to the lateral growth of the scale deposit on the membrane surface, resulting in flux decline and surface blockage (Zhang, 2002). Thus,the scaling formation is actually the crystallization phenomenon.Bulk crystallization arises when crystal particles are formed in the bulk phase through homogeneus crystallization and my deposit on membrane surfaces as sediments/particles to form a cake layer that leads to flux decline (Zhen et al., 2010). In addition, supersaturated scale forming conditions leads to scale growth and agglomeration (Rabizadeh et al., 2014). This is due to the random collision of ions with particles and secondary crystallization occurs on the surface of these foreign bodies present in the bulk phase (Alice et al., 2011).

In electrolyte solutions, the transition temperature of calcium carbonate solubility may vary because of variation in the water activity (Fathi et al., 2006). The solubility of common scales in water is shown in Figure 1, solubility of $CaCO_3$, $BaSO_4$ and $CaSO_4$ as a function of temperature.

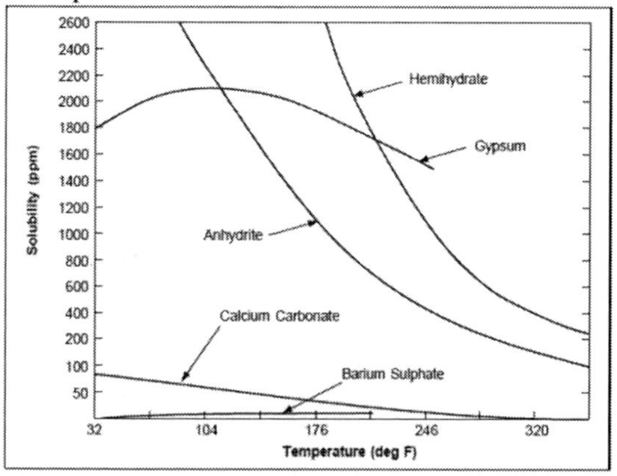

Figure-1. Solubility of Common Scalesas a function of temperature (Amer, 2009).

The aim of the present research was to examine the nucleation and growth of calcium carbonate crystals, and identify calcium carbonate formation and the scaling potential in pipes. In the present study, calcium sulfate crystals were grown by the prepared solution containing calcium carbonate composition and then characterized by XRD for the mineral determination and SEM analysis with EDX for elemental analysis and morphology.

METHODOLOGY

Materials and Apparatus

All solutions used in this study were prepared by dissolving reagent grade chemicals directly without further purification. The solution was prepared under saturated calcium sulfate by mixing directly into glass containers two equimolar solutions of calcium chloride dihydrate ($CaCl_2{\cdot}2H_2O$, minimum assay 99%, Sigma) and anhydrous sodium carbonate (Na_2CO_3, analytical grade, Sigma). Both reactants were dissolved in aqueous solutions containing 10 L of water and the solution concentration of Ca^{2+} was adjusted to be 2000, 3000, 4000 and 5000 ppm respectively. The selected range of temperatures was 30 to 40 °C. Each solution was then poured into a tube made of stainless and stirred at low speed. Two of the tube is connected with two brands of Iwaki magnetic pump long and trial test was conducted for 8 hours for each treatment with different flow rates. Experiments using a scaling simulator are shown in Figure 2.

Further, induction time of precipitation was measured by a conductivity method. This method was employed in the present study to measure the induction time for nucleation of calcium sulfate precipitation. The conductometer with a Consort C831 analyzer was used for each experiment at specified condition of the repeated three times. Therefore, the induction times and precipitation results presented in this study are the average of three values. In order to minimize costs of EDX analysis, the three samples were mixed and analyzed for crystal shape, crystal structure, and density. Each stock solution was then filtered through 0.22 µm paper filters to eliminate the presence of impurities, and stored at room temperature before subsequently analyzed by SEM/EDX.

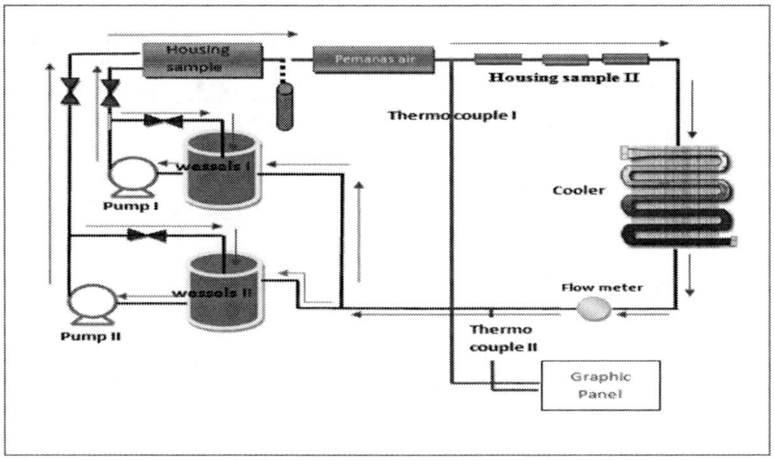

Figure-2: Experimental setup used in the $CaCO_3$ precipitation

Materials Characterization

The characterization of the precipitating scales was conducted through scanning electron microscopy coupled with energy dispersive X-ray analysis (SEM-EDX). SEM apparatus equipped with an EDX was used for characterizing the microstructure and morphology. For the observation, the powder crystals were mounted on a circular metallic precipitate holder and finally sputtered with gold.

RESULTS

Induction Time of crystallization

Induction time can be defined as the time required for between the generation of supersaturation and appearance of nucleic detected through concentration change. This can be seen since the onset of nucleation showing a reduction in the concentrations of crystallizing ion and molecules take place due to their association into the solid phase, thus the conductivity which is proportional to the concentration will undergo a reduction. This reduction in conductivity could be measured as the induction time. However, the measured induction time presents the sum of the nucleation time for appearance of critical size nuclei. Additionally, growth time of nucleic from critical size to the size possessing enough mass at solid phases makes a reduction of the solute concentration in the solution at the detectable extent through conductometry.

In the present study, the reaction was continued for about 25 minutes after the induction time until the nuclei were formed into the coupon and observed in computer monitors. Figure 3 shows the changes in conductivity during the course of nucleation experiment at room temperature and flow rates of 30, 40, and 50 ml/min.

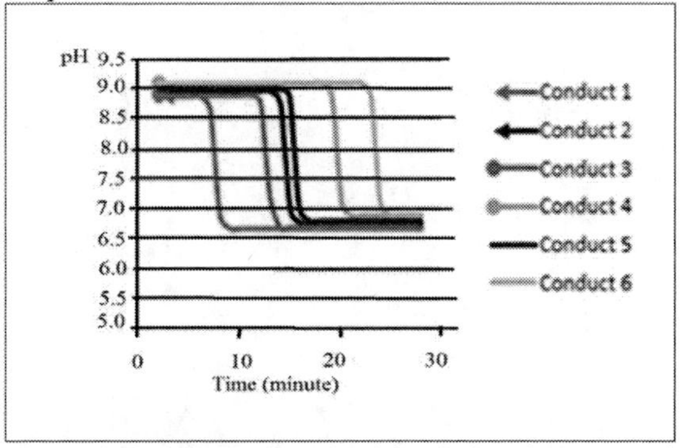

Figure-3 Variations of conductivity during the course of precipitation

Experimental results showed that the induction time was essentially under the influence of carbonate concentration. As Figure 3 shows, the induction time decreased as the pH solution decreased from 9 to 6.5. This effect was assumed to be due to an increase in supersaturation through an increase in hydrogen concentration. However, pH of the solution was higher than 9, a longer induction time was observed. The experimental results of the present study showed that the induction time for precipitation increased from 20 minutes to 25 minutes. At higher supersaturations, a lower value of induction time is expected; however, the temperature will be more influential than supersaturation due to its order and logarithmic function of effectiveness (Mullin, 2004).

Flow rates influence on the mass of precipitates

The mass of precipitated calcium sulfate particles per kg of water at different flow rates and concentration is presented in Figure-4. As the Ca+2 concentration of 2000 ppm is fixed and the flow rates for the solution increases in 30 to 50 ml/minutes. An increase in Ca^{+2} concentration will increase the supersaturation in the solution due to the removal of free water from the availability of the solute species. This will increase the precipitation yield of calcium cabonate. However, the trend of reduction in precipitation is similar to occur to the different flow rates

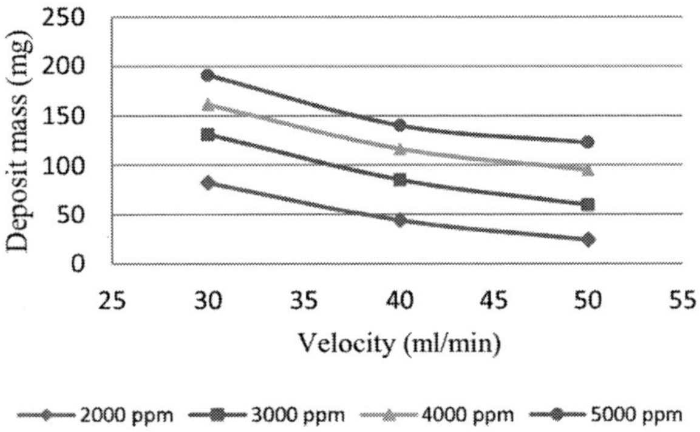

Figure-4. velocity of deposit formed graph

Characteristics and morphology of precipitates

SEM examination of the CaCO3 scale exhibited a wide images of calcite and the scales reveal crystalline, there are scales in the structures rhombhohidral is shown in figure 5. Morphology pyramid formed a) and Compositions chemistry b).

(a)

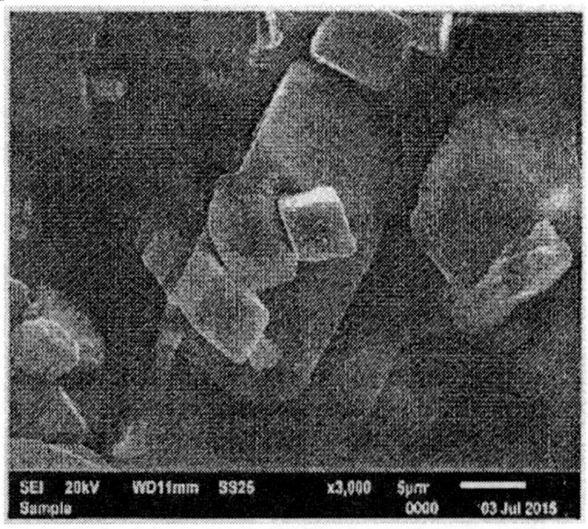

(b)

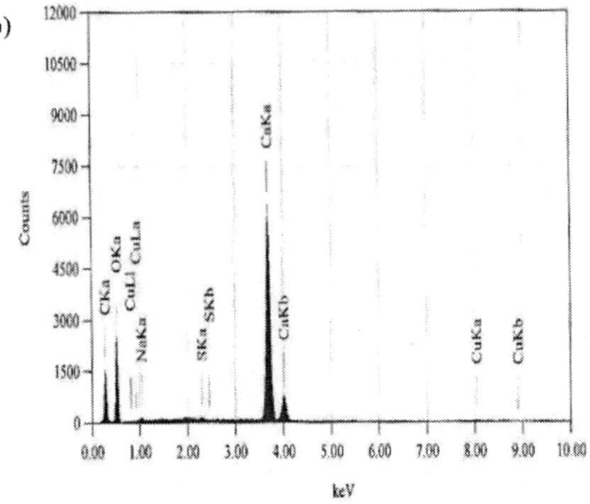

Figure-5. Morphology feature is pyramid formed a), and Compositions chemistry b).

According to theoretical calculations on the percentage of weight of CaCO3 Ca content should be 40%, while the results of the analysis of micro Ca content = 39.8 % so as to have difference of 0.2 % . For levels of carbon (C) shall be 12 wt % while result of analysis of micro C 10.5 % of micro analysis results that have a difference of 1.5% . For oxygen levels should 64 %, while the results of the analysis of micro direct 60.06 % so as to have difference of 3.94%.From the analysis of SEM-EDX test is concluded, its crust really crystal CaCO3.

DISCUSSIONS

The EDX analysis of calcium carbonate as found in this study of precipitates in Cu pipes is probably rhombohidral. It should be considered here that precipitates samples to be observed by SEM provides strong intensities of Ca, O and C elements which is more frequently found in SEM studies of calcium carbonate. However, the conditions used for EDX analysis in the present study and the detected amount of impurities produced a difference in mineral speciation. It is also important the observations of morphologies by SEM in the study of a calcium carbonate with apparent rhombohedral habit in the crystal. It can be assumed from SEM morphology and EDX analysis (Ca, C) to be. In contrast, the EDX analysis alone cannot distinguish whether the analyzed crystals observed by SEM are calcit, i.e. calcium carbonate valerite, aragonite and only by additional XRD techniques could be identified these minerals. In this case, the precipitation mechanisms of calcium carbonate mineral phases (calcit, valerite and aragonite) remain large unexplored. Therefore the role and implications of valerite as a stable precursor phase to calcit precipitation observed using XRD method, will provide interesting results to natural and industrial processes.

CONCLUSIONS

It was found that in all the experiments, the conductivity decreased abruptly after a certain induction period. Higher temperature produced more mass of the scale indicating that the increasing temperature promote scale formation. SEM analysis showed that the scale was rhombohedral, while EDS revealed that the elemental composition of the scale consisted of Ca, C and O. The crystalinity of the scale was found to be mostly calcit as shown by the XRD.

REFERENCES

Amer, B. M. Abu A. M. Y. 2009 Solubility of Common Oil Scales of Injection Water and Hight Barium Concentration and Hight Salinity Formation Water. UniversitasTeknologimalaysia.Jurnal Malaysia Teknology, 50: 67 – 77

Amor, M. B., Zgolli, D., Tlili, M. M., Manzola, A. S. 2004. Influence of water hardness, substrate nature and temperature on heterogeneous calcium carbonate nucleation. Desalination, 166:79-84.

Alice A., Jor H. L., Stephen G., Amy E, Childress, Pierre Le-Clech, Greg Lesli. 2011. Scale Formation and Control in High pressure Membrane Water Treatment systems, Journal of Membrane Science, 383 : 1 -16

Azimi G and Papangelakis V. G. 2010. Thermodynamic modeling and experimental measurement of calcium sulfate in complex aqueous solutions. Fluid Phase Equilibria 290: 88–94.

Basim O. Hasan, Graham J. Nathan, Peter J. Ashman, Richard A. C. 2012. The Effects of temperature and Hydrodynamics on The Crystallization Fouling Under Cross Flow conditions, Journal. Elsevier aplied Thermal Engineering. Volume, 36 : 210-218

Fathi, A., Mohamed, T., Claude, G., Maurin, G., Mohamed, B. A. 2006. Effect of a magnetic water treatment on homogeneous and heterogeneous precipitation of calcium carbonate. Water Research, 40(10):1941-1950.

Gourdon. 2011. The Effects of Flow Velocity on Crystallization Fouling in Falling FilmBlack Liquor Evaporators. Proceeding of International Conference on Heat Exchanger Fouling and Cleaning, 4 : 23-30

Mullin J.W. 2004. Crystallization. Butterworth Heinemann : Boston, MA. 280:185-198.

Muryanto S., Bayuseno A.P., Sediono W., Mangestiyono W and Sutrisno. 2012. Development of a versatile laboratory project for scale formation and control, Education for Chemical Engineers 7:78-84.

Rabizadeh, T., Caroline., Reacock, Liane G. B. 2014. Carboxilic acid : Effective for Calcium Sulfate Precipitation. Mineralogical Magazine, 78(6):1465 - 1472.

Zhang Y. 2002. Nucleation and Growth Kinetics in Synthesizing nanometer calcite. Journal of Crystal Growth, 245: 309-320

Zhen Wu, Jane H D, Lorraine F. 2010. Effect of water chemistry on Calcium Carbonate deposition on metal and polymer surfaces, Journal of Colloid and interfacescience 343 : 17-187

A STUDY ON HOME BASED ENTERPRISES IN *KAMPOENG* PANDEAN AS SUPPORTING SUSTAINABLE ARCHITECTURE

Muchlisiniyati Safeyah, Eva Elviana
University of Pembangunan National "Veteran" Jawa Timur
Surabaya, East Java, Indonesia
E-Mail: yatisafe@ymail.com

ABSTRACT

Home Based Enterprises (HBEs) provide an enormous impact on the lives of the citizens and the environment. The impacts include: increase income and welfare of the family, provide job opportunities, improve the quality of homes and the environment, and ensure life sustainability. The existence of the business leads changes to the house. Those changes that made to the house are often ignore the comfort of home space and the environment as living space. This study aims to look at the development of HBEs performed by community in *Kampoeng* Pandean. The measurement items used are architectural sustainability factors, ie economical sustainability, social sustainability, and enviromental sustainability. The study is located in *Kampoeng* Pandean Sidoarjo. The method used is a combination of qualitative and quantitative method. The results show that HBEs in *Kampoeng* Pandean have not fully supported the sustainable architecture. Environmental sustainability has not been met, due to the density of the environment, the high percentage of building area to land area, and the construction of business space does not consider the comfort factor.

Keywords: Home Based Enterprises, house changes, sustainable architecture, *Kampoeng* Pandean.

INTRODUCTION

House has a multi-dimensional function. It is a place for foster in home, individual development, social activities facilities, for earning money, and others. Laquian (1993) stated that in the lives of slum, squatter com-munities, and even if the house is not only for domestic life course, a house is a place for production, marketing, entertainment centers, financial institutions, and also a place for solitude. Houses with household activities that is also as a place to work as a home-based business are called Home Based Enterprises (HBEs). HBEs are growing rapidly in developed countries, developing countries and the poor countries. HBEs are able to provide a source of income, increases revenues and ensure the sustainability of living. In Indonesia, HBEs have existed for a long time, both in rural and urban areas. HBEs are growing in formal housing as well as in the slums. HBEs has a very big contribution to improving people's lives. HBEs in the city of Lagos, according to Lawanson (2012) were able to be the main source of income. HBEs were able to guarantee the sustainability of community living and able to support sustainable development. This is due to HBEs ability to provide income, increase income, provide opportunities and employment, improve family welfare, improve housing and the neighborhood, contribute to economic development and social progress (Lawanson & Olanrewaju, 2012; Safeyah, 2015).

The presence of business in residential area will be taken by the occupants in various ways. Such as accommodating work into the house, and changing the house in order to accommodate the jobs. Turner (1976) revealed that there were two works done by the occupants to their house, that are housing adjustment and housing adaption. In the lower middle class society, they often made the changes without considering the comfort factor, either in the comfort of home and the environment (Safeyah, 2015).

At the moment, environmental support is getting more limited. Therefore, the application of sustainable architecture becomes very important. Sustainable architecture is the architecture that meets the needs of the present without compromising the ability of future generations to meet their own needs, and those needs are best derived from related communities. Sustainable architecture includes environmental sustainability, social sustainability, and economical sustainability. Environmental sustain-ability is the development that takes natural resources into account in order to last longer, and establish integration between the ecosystems. Environmental sustainability is related to the age of the vital potential of natural resources and human ecological environment, such as climate, biodiversity, and industrial. The expectation of exploitation minimization of natural resources of the earth is that the earth will not lose its potential to support human life. Social sustainability is a development that is able to maintain the character or culture, and improve the social quality of the local community, while the economical sustainability is a productive development in quantity and quality, as well as to provide employment opportunities and other benefits for all of society. The principles of sustainable building design includes site, completion of the building that maximizes ventilation, cooling system and natural lighting, the efficiency of use of natural resources and energy, health and comfort of the interior, sewage treatment, transportation efficiency and habitat restoration (Iyengar, 2015; Bay & Ong, 2006; Al-Hagla, 2008; Cairns Regional Council of Australia, 2011).

There are seven parts to achieve sustainable neighborhood: government, transportation and connectivity, services, environment, economy, housing and the built environment, sociology and culture. Open space has the potential to deliver social benefits, economic benefits, either directly or indirectly. Definition of open space is a place that can hold certain activities of community activities individually or in groups. Examined from the events that occurre, open space is divided into active open space and passive open space. Active open space contains the elements of the activities in it, while the passive open space contains no human activity. Open space can serve as a place of relaxations, recreations, park or play area, social relationship, green open space, sports facilities, historical landscapes, wildlife habitats, track circulation, physical and mental health improvements, welfare and other improvements. The percentage of openness on the land, yard and open space, is an important factor in the home and the environment comforts. (Al-Hagla, 2008; Hamouche, 2008; Tardin, 2010; Khan, 2014; Chiesura, 2004).

Kampoeng is a typical form of settlements in rural or urban areas of Indonesia. *Kampoeng* Pandean is part of the village of Ngingas located in the district of Waru, Sidoarjo, East Java province. *Kampoeng* Pandean is known as *Kampoeng Logam* (metal village). Most community in *Kampoeng* Pandean work as metalworkers, which is a hereditary occupation. Metal craft work is done inside and around the house. Considering the great potential of HBEs, ways of house development due to business, and the importance of the application of sustainability architecture, thus the studi is aimed to explore how the factors of sustainable architecture in *Kampoeng* Pandean is conducted. Issues to be resolved are to investigate the characteristics of HBEs in *Kampoeng* Pandean is and the potential and obstacles exist in the implementation of sustainable architecture in *Kampoeng* Pandean. The study results are expected to be used as a development of HBEs in accordance with sustainable architecture, particularly in *Kampoeng* Pandean.

METHODOLOGY

The approach used in this study is descriptive qualitative. It is an approach that emphasizes the understanding of the problems of social life under the conditions of reality, or natural settings that are holistic, complex and detailed and a focus on the interpretation of meaning (Groat and Wang, 2002). The method used is a combination of qualitative and quantitative methode, quantitative research method is used to facilitate qualitative research (Brannen, 1992). Data collection is done by means of interviews, questionnaires and observation system. Structured and open-ended interviews are done in order to obtain a more detailed explanation. Open questionnaire is done to get a clear picture of conditions in the field. Observation is carried out by shooting, sketches, and observations of activities (Leedy, 2002). Data analysis is conducted on the quantitative and qualitative data. Qualitative data is processed and analyzed according to its contents by using interpretation. Quantitative data is described qualitatively based on percentage level.

Data exploration in this study is obtained from various sources. Primary data is obtained from metalworkers in *kampoeng* Pandean, as well as from the wives and children of the respondents. Primary data that is not listed in the questionnaire is obtained from the village officer, such as chairman of youth organization, elders of the *Kampoeng*, and others. Secondary data is traced through literature, books, journals, internet and research document. The number however metalworkers in Pandean are around 56 people. The number of respondents that is used as the object of study are 25 people.

The collected demograhic data is as follows: 1) Profile of the respondents, which is the general data of the respondents, include; age, education, number of occupants, number of families in a single house, ethnicity, religion. 2) a business profile, that is characteristic data of metal handicraft business, among others; business status, length of effort, motivation of business, type of business, the characteristics of employees, revenues, business constraints. 3) Home and businesses profile, such as; building area, the number of repair, home improvement motivation, home space completeness, the look of the house. 4) The opinion of respondents on the environmental conditions, namely; the standpoint of occupants on the potentials and constraints they have during their settlement as well as the hope of the inhabitants for the environment.

RESULTS AND DISCUSSION

Kampoeng Pandean is a unique form of settlement. Almost 100% of its inhabitants are Pandean indigenous people who are Javanese. The socioe conomic status of Pandean inhabitants are averagely categorized as medium. As the characteristic of another *Kampoeng*, the kinship among inhabitants of *Kampoeng* Pandean is very close and very strong. *Kampoeng* Pandean can be referred to as a productive *Kampoeng* (*Kampoeng* Based Enterprises), which is a *Kampoeng* dominated by metal craft business activities carried out within or around the house. Most households have a job that is homo-geneous, ie as a metal craftsman. Existing metal craft business types in the *kampoeng* Pandean include agricultural tools, building tools, electrical components, telecommunication components, factory machinery parts, automobile parts, appropriate machinery, household appliances, and others fit the customer's order.

Metal craft business is a main job, it is based on the opinions of 24 respondents (96%). HBEs in *Kampoeng* Pandean have been running for long enough, with the details of over 45 years of three respondents (12%), 35-45 years 20%, 12% 25-35 years, 15-25 years 20%, 5-15 years and 5 respondents (20 %), and less than 5 years only by 4 respondents (16%). The motivation rank in opening a metal craft business is as follows, the first motivation is the desire to earn money, stated by 12 respondents (48%), second is to continue their parents' business, stated by 8 respondents (32%), and the rest is due to the business opportunity they see. Metal crafts in *Kampoeng* Pandean is able to provide enough average income. Respondents who earn 2-4 million Rupiahs per month are 11 respondents (44%), those who earn 1-2 million Rupiahs are seven respondents (28%), 4-10 million Rupiahs or more are 6 respondents (24%), one respondent stated that his income is uncertain, it is highly dependent on the orders obtained. Monthly business turnover is quite high due to the quite expensive price of iron as base material. Monthly turnover of 2-5 million per month by 11 respondents (44%), 5-10 million five respondents (20%), 10-20 million third of respondents (12%), 20-30 million four respondents (16%), and 30-60 million two respondents (8%). These metal craft is capable to provide employment. This can be seen from as many as 20 respondents have employees. Craftsmen who have employees 4-6 employees dominate by 10 respondents (40%), the Craftsmen who have employees 1-3 people are 7 respondents (28%), the Craftsman who have employees 7-9 people is 1 respondent (4%), and Craftsmen who have more than 9 employees are 2 respondents (8%). This is in accordance with the opinion of Lawanson (2012), Lawanson & Olanrewaju (2012) and Safeyah (2015), that the house is able to be the main source of income, able to provide an income earning, provide opportunities and employment.

Out of ± 350 residential houses, there are ± 56 HBEs (16%), houses those are used for household activity as well as for metal craft business. In conducting metal crafts business activities, there are household spaces that are used as a terrace or a room in the house, and there is also a special space built in the courtyard of the respondents. Respondents who conduct their business activities on the terrace are 4 respondents (16%), who conduct inside the house are 3 respondents (12%), and who conduct in the yard (front, side or rear) are 15 respondents (60%), see Figure 1. A total of three respondents (12%) run their business in other places, such as rental house or yard, his own land separated by the road or together with the their parents in law's business place. The presence of business addressed by the occupants with housing adjust-ment and housing adaptation (Turner, 1976). When there is a possibility to developed the house, they performed house adjustment, and when there is no possibility to changes the house to accomodate the bussines activities, then the housing adaption is carried out.

(a) (b) (c)

Figure-1. Business activites are conducted (a) at terrace, (b) inside the house, (c) at courtyard.

The people of Pandean are very religious in their daily activities. Starting from children, teenagers, up to fathers and mothers have related activities to religion. Almost every week people of Pandean conduct this social religious activity. Activities for children and teens conducted through associations "Jamiyah". "Yasinta" and "Manaqib" is the association for the fathers. There are several religious associations for the mothers of Pandean, such as, Muslimat, Fatayat, Qodmil Qur'an, Jamiyah, Burdah and Sewelasan. Other than Manaqiban that is held in mosques, all religious activities are done in turns from house to house. However, if host owner has an obstruc-tion, the activity can be done in the mosque. Spaces that are used are the living room, terrace, yard or street of the *kampoeng*. Social activities in the field of religion is a factor that supports social sustainability (Chiesura, 2004; Iyengar, 2015).

Kampoeng Pandean has an area of ± 1.3 ha which is home to about 370 heads of household. *Kampoeng* has very high density of 1,500 people/ha. Existing open spaces are in the form of narrow streets, fields, and some of the home yards with a limited area. The percentage of open space to the existing area is ± 15%. The quantity propor-tion of existing open spaces do not meet the minimum standards of green open space requirements, ie minimum 20%. This is not in accordance with the opinion of Khan (2014) and Al-Hagla (2008) that for comfort, openness percentage factor on the field is important. Similarly in quality, it has not met the minimum eligibility standards. *Kampoeng* Pandean main road has a width of ± 3 m, a width of supporting road between ± 1-2 m. Existing road conditions are less supported by adequate infrastructure, such as sewers. *Kampoeng* roads have multiple functions, namely as the circulation of vehicles and pedestrians, children's playground, a place to trade, and a wide range of social activities; such as recitals, celebration of Indepen-dence Day, celebration of Islam holidays, and so on, see Figure 2.

(a) (b) (c)

Figure-2. Various kinds of use of the road in *Kampoeng*: (a) as a place of recitals, (b) as a place of celebration of Islam holidays, (c) as a place of celebration of Independence Day.

Open space in the form of a field in *Kampoeng* Pandean serves as a place to exercise, play game, do refreshing, and a place to socialize. Every afternoon, *Kampoeng* Pandean society, especially men, ranging from children to fathers take advantage of this field in turns. On the anniversary of the independence day, the courts functions as a place to play game/competition. The existing field is bordered by houses on the three sides, with less greening conditions. The social function of the field is very prominent compared to the environmental function.

The presence of metal handicraft business in residences, results in a change in the residences. Residen-tial spaces have additional space to conduct business activities. Five respondents (20%) lose their yard spaces as it is used for business activities. This shows that the house yard has a very high economic function. Completeness of residential space owned by the respondent does not change much after the business starts. 15 respondents (60%) used to have complete space for habitable activities before the business started, such as yard (front, side, back), terrace, living room, bedroom, kitchen and bathroom. Five respondents (20%) have a complete residential space with an additional living room or dining room, while 1 respondent (4%) has additional study room, and 4 respondents (16%) have a prayer room, see Table 1. Ways that made respondents to change the home less attention to the comfort factor.

Table-1. Type of room before and after the business starts

No.	Type Space	Before the Business Starts (%)	After the Business Starts (%)
1.	Yard, terrace, living room, bedroom, kitchen, bathroom	60	36
2.	Yard, terrace, living room, bedroom, kitchen, dining room, bathroom	20	20
3.	Yard, terrace, living room, bedroom, kitchen, dining room, study room, bathroom	4	4
4.	Yard, terrace, living room, bedroom, kitchen, dining room, bathroom	16	20
5.	Terrace, living room, bedroom, business room	0	20

Initial building areas with the width under 61 m^2 are owned by the 8 respondents (32%), whereas the initial building areas with the width above 120 m^2 are only owned by 3 respondents. The average areas of the initial building owned by the respondents are ranged from 61-120 m^2. The details of the initial area ownership are as follows: initial building areas with the width of 61-80 m^2 are owned by 6 respondents (24%), the width of 81-100 m^2 are owned by 4 respondents, and 101-120 m^2 are also owned by 4 respondents. With the business activities and increase the number of occupants, the building area is currently experiencing an increase. The number of respondents who have a building area below 61 m^2 are only 3 respondents (12%), while those with a building area over 120 m^2 are 8 respondents (32%). Respondents who have building areas between 81-100 m^2 are 4 respondents (16%), those who have building areas between 101-120 m^2 are 5 respondents, and those who have building areas between 61-80 m^2 are five respondents, see Table 2. Total floors of the initial house increases after the start of their business activities. There were 22 respondents (88%) who had one storey initial houses before the start of the business, and there are 16 respondents (64%) at the moment who have one storey initial houses after the start of the business. There were 3 respondents (12%) who had 2-storey houses, and after the start of the business, there are now 8 respondents (32%) who have 2-storey houses, and only 1 respondent (4%) who has a 3-storey house. It appears that the presence of business is addressed by the occupants with changes in their houses (Turner, 1976). Land density is higher than before the business starts. Environmental sustainability is not met.

Table-2. Building area before and after the start of the business

No.	Building Area (m^2)	Before the Business (%)	After the Business (%)
1.	< 61	32	12
2.	61 – 80	24	20
3.	81 – 100	16	16
4.	101 – 120	16	20
5.	121 – 140	4	12
6.	> 140	8	20

Ways that are performed by the respondent in conducting additional business space or living space, mostly done without consideration to the factors of space and environmental comfort. The percentage of open space becomes smaller due to a yard that is used for business space is a closed by roof. This causes less light in the residential spaces. Likewise, the ventilation in the house is not optimal since there is no air circulation. Vegetation in *Kampoeng* Pandean is still limited to a few houses that have a yard, and there is no vegetation at all on the *Kampoeng* streets. It appears that the handling of the climate associated with solar radiation, wind direction, precipitation and temperature have not been handled properly. It appears that the handling of the climate associated with solar radiation, wind direction, precipitation and temperature have not been handled properly, not in accordance with the opinion of Iyengar (2015) and Hamouche (2008).

Noise occurs due to production process. Metal craft businesses that use machine for shaping or cutting metals, as well as the use of tools such as hammers bat, are causing noises. Basically, occupants are disturbed by these conditions, but this does not become a problem, because they are used to the condition. 11 respondents/occupants (44%) are disturbed by business activities, 12 respondents (48%) state that there is no interruption and do not answer, and there are 2 respondents (8%) state that they are disturbed by the inadequate environmental conditions. 14 respondents/occupants are not annoyed by business activities because they are already accustomed to it, and 11 respondents /occupants (44%) do not feel disturbed and do not answer. To increase the comfort of living, the noise that occurs should be minimized (Cairns Regional Council of Australia, 2011)

CONCLUSIONS

A study on Home Based Enterprises in *Kampoeng* Pandean as Supporting Sustainable Architecture has been conducted. The result showed that HBEs in *Kampoeng* Pandean generally helped to strengthen economic activities. Metal craft HBEs was capable of being the main source of income, providing adequateearnings, and offering employment. The open space has very high economic and social benefits.The economic benefits were derived from metal craft business, while the social ones were from the high frequency of social activities carried out in *Kampoeng* Pandean. Hence, the criteria for sustainable architecture in Kampung Pandean was obtained from economic and social sustainability. Meanwhile, environmental sustainability has not yet been met.

ACKNOWLEDGMENTS
Acknowledgements are addressed to the Directorate of Research and Community Service, the Directorate General of Higher Education, Ministry of Education and Culture, that have funded the research in 2015 Competitive Grant scheme, No. SPP/0045/Lit/LPPM/II/2015.

REFERENCES

Laquian, A.A. 1993. Basic Housing, Policys for Urban Sites, Services and Shelter in Developing Countries. IDRC.

Al-Hagla, K. 2008. Towards A Sustainable Neighborhood: The Role of Open Spaces, International Journal of Architectural Research Archnet-IJAR. Volume (2), Issue (2): 162-177.

Bay, J.H. and Ong, B.L. 2006. Tropical Sustainable Architecture Social and Environmental Dimensions, Firts Edition. Elsevier Ltd. Oxford.

Brannen, J. 1992. Mixing Methods: Qualitative and Quantitative Research. Aldershot. Brookfield, USA.

Cairns Regional Council of Australia. 2011. Sustainable Tropical Building Design, Guidelines for Commercial Buildings.

Chiesura, A. 2004. The Role of Urban Parks for The Sustainable City. Elsevier B.V., Landscape and Urban Planning. 68:129-138.

Groat, L.N. & Wang, D. 2002. Architectural Research Methods. John Wiley & Sons, Inc., USA.

Hamouche, M.B. 2008. Climate, Cities, and Sustaiability in the Arabian Regio: Compactness as a New Paradigm in Urban Design and Planning. International Journal of Architectural Research Archnet-IJAR. Volume (2), Issue (2): 196-208.

Iyengar, K. 2015. Sustainable Architectural Design. Routledge, New York.

Khan, M. 2014. Study of Open Spaces in the Context of Dhaka City for Sustainable Use: A Syntactic Approach. IACSIT International Journal of Engineering and Technology. Vol. 6, No. 3: 238-243.

Lawanson, T. 2012. Poverty, Home Based Enterprises And Urban Livelihoods In The Lagos Metropolis. Journal of Sustainable Development in Africa. Volume 14, No.4: 158-171.

Lawanson, T. and Olanrewaju, D. 2012. The Home As Workplace: Investigating Home Based Enterprises in Low Income Settlements of the Lagos Metropolis, Ethiopian Journal of Environmental Studies and Management EJESM Vol. 5 No. 4: 397-407.

Leedy, P.D. 2002. Practical Research: Planning and Design. 6th Edition. Merrill, an Imprint of Prentice Hall. New Jersey, USA.

Safeyah, M. 2015. Local Wisdom in the Construction of Place Identity for A Community-Based Enterprise of Pandean Sidoarjo. Proceedings of The 2 nd ECO Architecture Conference: 35-42.

Tardin, R. and Costa, L.M.S.A. 2010. System of Open Spaces and Territorial Planning: Toward A Sustainable Development. 46th ISOCARP Congress: 1-10.

Turner, J.F.C., 1976, Housing by People, Towards Autonomy in Building Enviroments. Pantheos Books. New York, USA.

ACCESSIBILITY TO PUBLIC TRANSPORT SERVICES (CASE STUDY OF TABANAN REGION, BALI-INDONESIA)

Putu Alit Suthanaya
Civil Engineering Department, Udayana University
E-Mail: suthanaya@rocketmail.com

ABSTRACT
Motorization is becoming an issue of concerns for urban governments and citizens in Indonesian cities, including in Bali Province. The cities in Bali Province have not been served well by public transport services. Bali consists of 9 regions and this study focused on a typical low income region in Bali i.e. Tabanan Region as a case study. One way to measure public transport service quality is by measuring accessibility to public transport routes. The objective of this study is to investigate the accessibility of residents to public transport services in Tabanan Region. Accessibility to public transport services in Tabanan Region was analyzed based on travel distance and travel time to reach the closest public transport routes. Based on Z-score diagram, it was found that six zones were in quadrant A which means that these zones were in high accessibility category (Bongan, Gubug, DauhPeken, DelodPeken, DajanPeken and Denbatas. On the other hand, five zones were identified in quadrant D. These zones (Sudimara, Subamia, Buahan, Tunjuk and Wanasari) have a low accessibility values to reach public transport routes. Therefore, improvement of infrastructures and public transport services need to prioritize these five zones with low accessibility values.

Keywords: accessibility, public transport, travel time, distance, Z-score

INTRODUCTION

Rapid growth of motor vehicles has been experienced by many cities. Motorization is increasing rapidly in Indonesia. The increase of population and economic has been followed by the increase of the number of motor vehicles. Motor vehicle increases mobility and improved accessibility particularly in rural areas. However, the growth of motorization has caused negative impacts such as increases traffic jam, air pollution, traffic noise and accidents.

Many researches have been focused on the impacts of motorization and urbanization (for example, see Newman and Kenworthy, 1999; Kitamura, et al., 2003; Joewono, et al., 2013). In sparsely populated area, car is valued as the most optimal mode for commuting (Sandow and Westlin, 2010). Unlike the developed countries, most developing countries are not served by proper mass public transport system. Low public transport services have caused a high increase in the use of private car. Sprawling development without being served by a good public transport system has caused the increasing dependency on the use of motor vehicles among urban residents in developing countries. Residents in a city have become highly dependent on the use of private motor vehicles. This rapid growth of motor vehicles has made the whole transport system become overloaded, thus obstructing the mobility. This high motor vehicle's ownership has caused a high increase in energy consumption and air pollution. From the view point of global warming effects, this motorization will cause serious problems. Therefore, recent transport policies in some developing Asian countries have started to focus on raising public awareness about sustainable transport dan promoting public transport use (Pardo, 2006). A serious action to reduce our dependency on motor vehicles is required, otherwise the urban environment will deteriorate in the future.

The increase in the use of private motor vehicles is also related to the land use planning in a city. Urban sprawl and decline in the city centers have been experienced by many Asian cities including Indonesia. In many cities, such as Bangkok, Manila and other major cities, urban redevelopment plans were introduced by promoting better public transport planning. The target of sustainable city is to create a sustainable and low-carbon society through the urban redevelopment (Kikuchi, et al., 2011). According to Yu, et al. (2011), there is a tendency of increasing energy consumption in the Asian region. They stated that since 1990, the energy consumption has increased by two thirds, which is mainly caused by middle-income countries such as China and India. They predicted that the increase of energy consumption in Asia is estimated to account for 46% in the total world energy increase in the year 2030.

Several studies have investigated the behavior of people towards the use of public transport. Sekar, et.al. (2015) investigated the most preferred mode from three types of public transport, i.e. metrorail, monorail and bus rapid transit in 19 cities. By applying Markov Chain Process and Analytical Hierarchy Process they found that metrorail is the best mode. Lay and Chen (2011) stated that public transit service quality influences the behavioral intentions of public transit passengers. In a similar study, Redman, et.al. (2013) found that public transport service reliability and frequency are two important variables that can attract private motor vehicle users. Other variables such as waiting time, cleanliness and comfort are most valued by users (Dell'Olio, et al., 2011). Study in Phnom Penh found that bus fare and comfort to be the most important aspects for passengers (Choocharukul and Ung, 2011). Japan International Cooperation Agency (JICA, 2011) conducted a research study on practical approach for urban transport planning. They undertook the collection and typological analysis of basic data on current urban situation and traffic patterns of urban agglomerations or cities in the world. Van and Fujii (2011) conducted a study on the attitudes toward car and public transport in six Asian countries (Japan, Thailand, China, Vietnam,

Indonesia, and the Philippines). Based on the principle component analysis, they found that there are three factors of attitudes toward car and public transport, i.e. Symbolic affective, Instrumental, and Social orderliness. By comparing the attitudes toward car and public transport among countries they found that there were great divergences across countries in the all three dimensions of attitudes. In another study, Koizumi, et.al. (2013) have analyzed the relationship between urban socio-economic indicators and traffic patterns particularly in the cities of developing countries. They concluded that the metro operation timing is closely related to the achieved level of socio-economic development of cities. The larger the population the more likely they are to develop a metro system, even though GDP per capita is not enough.

Similar to other cities in developing countries, cities in Indonesia have faced the increasing traffic congestion and air pollution. Environmental degradation has continued to occur which is influenced by poor public transport service and increasing dependence on using private vehicle especially motor cycle. Susilo, et.al. (2007) investigated the influences of motorization on the urban residents' travel patterns in the Jakarta Metropolitan Area (JMA). It has been found that motorization is increasing rapidly in Indonesia and especially in the Jakarta metropolitan area (JMA). They found that between 1985 and 2000, there were significant increase in the car ownership, number of trips and travel distance. The lack of good public transport services has caused travelers to shift to private vehicles. The increase of dependency on the use of motor vehicles is much higher in developing countries than in developed countries. Therefore, traffic congestion has increased rapidly and the public transport system is only used by captive users. Muthukannan and Thirumurthy (2008) stated that low accessibility is one of the main reason for people in Chennai, India for not using public transport. Other variables influencing the use of public transport are location of bus stop, fare and waiting time.

Motorization is becoming an issue of concerns for urban governments and citizens in Indonesian cities including in Bali Province. Bali is well known as one of the famous tourist destination in the world. Unfortunately, the cities in Bali Province have not been served well by public transport services. This poor traffic condition has become an important issue that may reduce the attractiveness of Bali. Motor vehicle ownership has increased at the rate of 10 percent per year. On the other hand, the quality of public transport services becomes worse from year to year. Bali consists of 9 region and this study focused on a typical low income region in Bali i.e. Tabanan Region as a case study. One way to measure public transport service quality is by measuring accessibility to reach the city center by public transport. Ryan (2009) stated that walkability influences transit ridership. Exel and Rietveld (2010) identified the importance of public transport travel time. Majority of the previous studies were conducted in large cities with medium to high income condition. None of those studies have classified zonal accessibility level. This study used accessibility concept to analyze accessibility to public transport services for a tipical small and low income region in developing countries. The objective of this study is to investigate the accessibility of residents to public transport services in Tabanan region. The results of the analyses will provide a basis for the government to plan the strategy for improving public transport services.

METHODOLOGY

Bali is one of the famous tourist destination in Indonesia. The Province of Bali has an area of 5,634.40 km2 and population of about 4 million. Tabanan is one of the regency in Bali Province which consists of 11 districts. The area of Tabanan Regency is 839.33 km2 or 14.9% of the area of Bali Province. The population are 441.900 persons with density of 526.49 persons/km2 (Tabanan Bureau of Statistics, 2013). The study area was divided into 11 zones, as shown in Figure 1.

The primary data used in this study are distance and travel time to reach the closest public transport routes. The secondary data were obtained from Tabanan Bureau of Statistic (2013). Microsoft Access was used to compile relevant cross-tabulations and summaries. Tabulations were then converted to Microsoft Excel spreadsheets to allow presentation and modification of format.

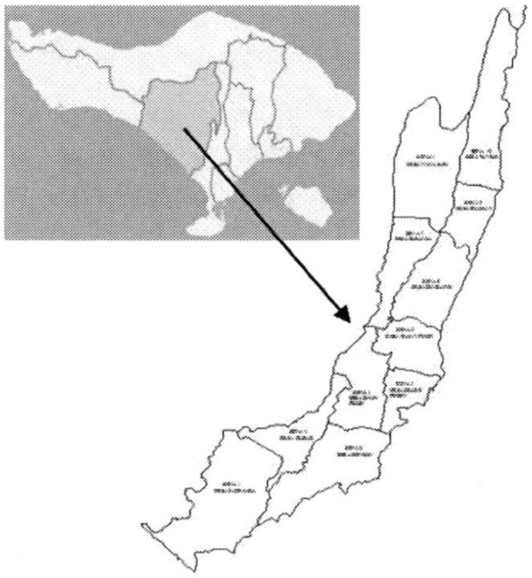

Figure-1. Study area zoning system in TabananRegion, Bali-Indonesia

Accessibility to public transport services were measured based on travel distance and travel time. Based on the travel distance and travel time to reach the closest public transport routes, a Z-score is calculated. There is an infinitely large number of normal curve – one for each pair of values for μ and σ. The standard normal distribution is a transformation of the normal distribution. In the standard normal curve, $\mu = 0$ and $\sigma = 1$. This standard normal curve can be displayed in terms of Z-scores as indicated in Figure 1. The standard normal curve can be used to simplify the calculation of probabilities for normally distributed populations. Because not all normally distributed random variables have $\mu = 0$ and $\sigma = 1$, it is needed to transform the variable so that $\mu = 0$ and $\sigma = 1$. It can be done by using the Z-score, which is calculated as follows (Lee, 1993):

$$Z = \underline{X - \mu} \qquad (1)$$

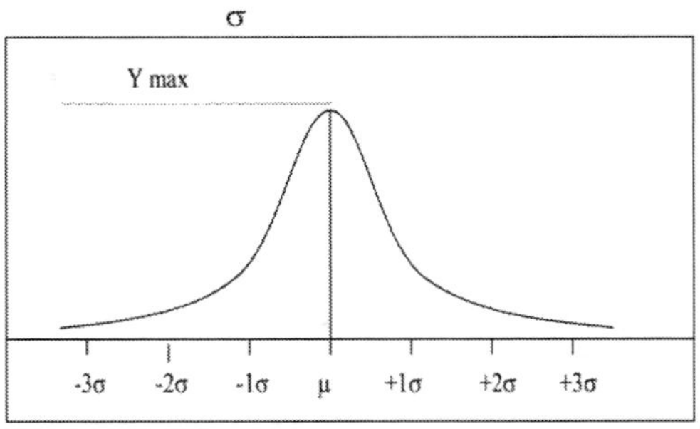

Figure-2. Normal probability distribution with $\mu = 0$ and $\sigma = 0$

The zones were grouped into low, medium and high accessibility zones. The zones were then classified further based on percentile values by using formula as follows:

$$Pi = \text{value}\frac{i(n+1)}{100} \quad (2)$$

Where:

Pi	=	percentile -i
i	=	1,2,3,.....,99
n	=	number of zones

To facilitate a visual inspection, a scatter-plot for the Z-score was developed. This Z-score values were plotted in a graph based on the distance and travel time.The Z-score values were then plotted in scatter plot diagram as shown in Figure 3 where the horizontal axis indicates travel time to reach public transport services while the vertical axis indicates distances. Using scatter plot of Z-score, a visual inspection can be done to identify outliers and clusters of zones. From the four quadrants in Figure 3, the upper right quadrant (quadrant D) indicates the cluster of zones with high accessibility values based both on distances and travel time. Zones in the lower left quadrant (A) are zones with low accessibility values for both distances and travel time. The upper left quadrant (B) indicates a cluster of zones with a low travel time but have relatively short discances to reach public transport routes. The lower right quadrant (C) is for zones that have a high accessibility values based on travel time but low accessibility values based on distances.

Table-1. Public transport routes in Tabanan

No	Public transport route	No of fleets (unit)	Length of route (km)
1	Kediri-Tabanan-Pesiapan	56	5
2	Tuakilang-Tabanan-Pesiapan	21	5
3	Kediri-Tabanan-Tuakilang	61	7
4	Tabanan-Kukuh-Marga	3	9
5	Kediri-Tanah lot	4	11
6	Tabanan-Buahan-Marga	2	12
7	Pesiapan-Tabanan-Bongan	13	6
8	Kabakaba-Kediri-Tabanan	3	9
9	Pesiapan-Tabanan-Yehganga	5	6
10	Tabanan-Kediri-Taman Ayun	-	-

Source: Tabanan Department of Transport, 2014

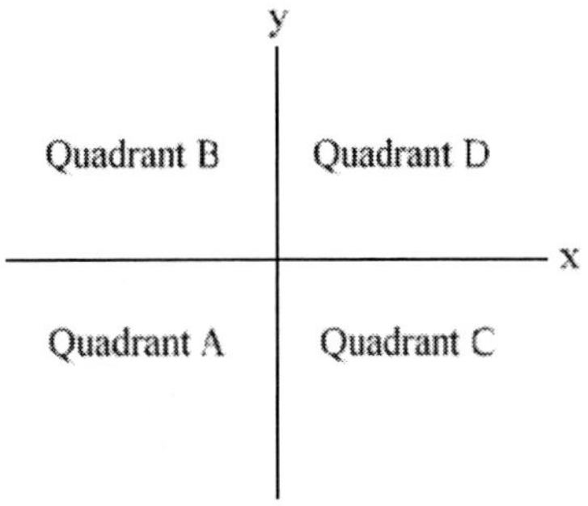

Figure-3. Z-score diagram of accessibility to public transport services

RESULTS

Public transport routes

Table 1 shows public transport routes in Tabanan Region. There are 10 routes available to serve movement of people. The number of fleets varies from 3 units to 56 units per route. The length of the routes varies from 5 to 12 km.

Travel time to reach the closest public transport routes

Travel time analyses include walking travel time to the closest public transport route and passenger's waiting time from each zone. Based on the survey, it was found that the average walking speed was 71.76 m/minute. Table 2 shows walking distance and walking travel time required to reach the closest public transport route. The walking distances vary from 0.17 to 4.6 km with the average walking distance was 1.84 km. Walking travel times were varies between 2.42 to 64.71 minutes. The shortest walking travel time was experienced by zone 5 (Delod Peken) i.e. 2.42 minutes while the longest was experienced by zone 10 (Tunjuk) i.e. 64.71 minutes. With the assumption of ideal walking distance of 500 m, Table 3 presents the values of walking accessibility. It can be seen that 45.46% of zones were in high accessibility category, 9.09% were in medium accessibility and 45.46% were in low accessibility category.

Table-2. Walking distance and travel time to reach the closest public transport routes

Zone No	Zone/ district	Walking distance (km)	Walking travel time (minutes)
1	Sudimara	3.93	56.14
2	Bongan	0.36	5.14
3	Gubug	0.23	3.28
4	DauhPeken	0.53	7.57
5	DelodPeken	0.17	2.42
6	DajanPeken	0.48	6.85
7	Subamia	2.65	37.85
8	Denbates	0.29	4.14
9	Buahan	4.3	61.42
10	Tunjuk	4.6	65.71
11	Wanesari	2.7	38.57

 DOI: 10.1051/matecconf/20165802002

Table-3. Walking accessibility to reach closest public transport routes

Walking comfort level (meter)	Zone number	Percentage (%)
0 -500 (high)	5	45.46
500 - 1000 (medium)	1	9.09
>1000 (low)	5	45.46

Passenger's waiting time

Passenger's waiting time is total time required by passenger to wait public transport arrival. Table 4 shows that the passenger's waiting time varies from 0.96 to 4.7 minutes.

Table-4. Passenger's waiting time

Zone no	Zone/district	Passenger's waiting time (minutes)
1	Sudimara	4.709
2	Bongan	4.250
3	Gubug	4.709
4	DauhPeken	0.966
5	DelodPeken	0.961
6	DajanPeken	0.964
7	Subamia	1.383
8	Denbates	1.429
9	Buahan	1.063
10	Tunjuk	1.429
11	Wanesari	1.429

Total Passenger's Travel Time to Reach the Closest Public Transport Routes

Total passenger's travel time to reach the closest public transport route consists of walking time and waiting time. Table 5 presents total passenger's walking travel time to reach the closest public transport routes from each zone in Tabanan Regency. The longest travel time was experienced by zone 10 (Tunjuk) i.e. 67.14 minutes with a distance to be traveled of 4.6 km. On the other hand, the shortest travel time was experienced by zone 5 (DelodPeken) i.e. 3.39 minutes with distance of 0.17 km. The average walking travel time to reach the closest public transport routes was 28.40 minutes.

Table-5. Total walking travel time to reach the closest public transport routes

Zone no	Zone/district	Distance (km)	Total walking travel time (minutes)
1	Sudimara	3.9	60.85
2	Bongan	0.4	9.39
3	Gubug	0.23	7.99
4	DauhPeken	0.53	8.53
5	DelodPeken	0.17	3.39
6	DajanPeken	0.48	7.82
7	Subamia	2.6	39.24
8	Denbates	0.29	5.57
9	Buahan	4.3	62.49
10	Tunjuk	4.6	67.14
11	Wanesari	2.7	40.00

Accessibility classification based on traveltime to the closest public transport routes

Table 6 and Figure 4 show accessibility classification based on travel time to reach the closest public transport routes in Tabanan Regency. It can be seen that three zones (Zone 5, 6, 8) were in high accessibility category, three zones were in medium category (2, 3, 4), three zones in low category (1, 7, 11), and two zones in very low category (9, 10).

Table-6. Accessibility classification based on total travel time to reach the closest public transport routes

Zone no	Zone/district	Walking travel time (minutes)	Accessibility classification
5	DelodPeken	3.39	High
8	Denbates	5.57	High
6	DajanPeken	7.82	High
3	Gubug	7.99	Medium
4	DauhPeken	8.53	Medium
2	Bongan	9.39	Medium
7	Subamia	39.24	Low
11	Wanesari	40.00	Low
1	Sudimara	60.58	Low
9	Buahan	62.49	Very low
10	Tunjuk	67.14	Very low

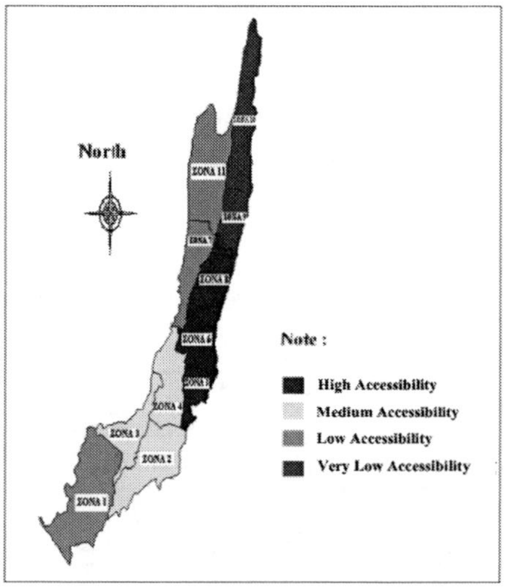

Figure-4. Map of accessibility classification based on walking travel time to reach the closest public transport routes

Accessibility Based on Travel Distance

DOI: 10.1051/matecconf/20165802002

Table 7 and Figure 5 show accessibility classification based on travel distance to reach the closest public transport routes in Tabanan Regency. Three zones were in high accessibility category (zone 3, 5, 8), three zones were in medium category (zone 2, 4, 6), three zones in low category (zone 1, 7, 11) and two zones in very low category (zone 9, 10).

Table-7. Accessibility classification based on travel distance to the closest public transport routes

Zone no	Zone/district	Walking travel time (minutes)	Accessibility classification
5	DelodPeken	0.17	High
3	Gubug	0.23	High
8	Denbates	0.29	High
2	Bongan	0.4	Medium
6	DajanPeken	0.48	Medium
4	DauhPeken	0.53	Medium
7	Subamia	2.6	Low
11	Wanesari	2.7	Low
1	Sudimara	3.9	Low
9	Buahan	4.3	Very Low
10	Tunjuk	4.6	Very Low

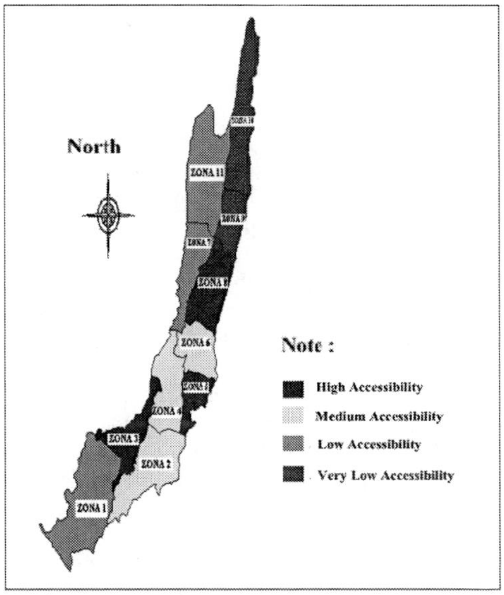

Figure-5. Map of accessibility classification based on travel distance to the closest public transport routes

Z-Score Diagram

Z-Score diagram is used to show the position of each zone in terms of its accessibility. Horisontal axis (Zx) represents Z-score for accessibility based on walking travel time, while vertical axis (Zy) indicates Z-score for accessibility based on travel distance to the closest public transport routes. Z-score positif indicates that the accessibility of the zone above the average standard deviation for all zones while Z-score negative indicates the accessibility below the average standard deviation. Table 8 shows the Z-score for all zones and Figure 6 shows position of each zone in the Z-Score diagram.

Table-8. Z-score for accessibility to public transport services

Zone No	Zone/ district	(Zx) Z-score Time	(Zy) Z-score Distance	Position in Z-score diagram
1	Sudimara	1.252	1.150	D
2	Bongan	-0.733	-0.794	A
3	Gubug	-0.788	-0.888	A
4	DauhPeken	-0.767	-0.722	A
5	DelodPeken	-0.965	-0.922	A
6	DajanPeken	-0.794	-0.750	A
7	Subamia	0.418	0.427	D
8	Denbates	-0.881	-0.855	A
9	Buahan	1.316	1.372	D
10	Tunjuk	1.495	1.538	D
11	Wanesari	0.447	0.483	D

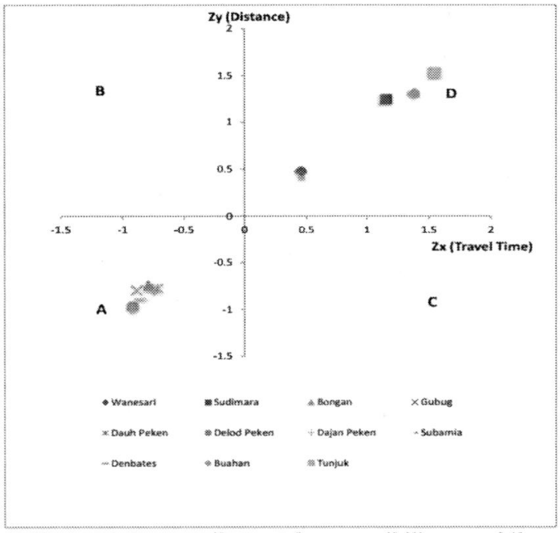

Figure-6. Z-score diagram for accessibility to public transport services

DISCUSSIONS

Quadrant A indicates zones with negative Z-score values for both travel time and distance (the value below the average standard deviation for all zones). Quadrant D indicates zones with Z-score values for both travel time and distance were positive (the value above the average standard deviation for all zones). Table 9 shows that there were 6 zones in quadrant A including Bongan, Gubug, Dauh Peken, Delod Peken, Dajan Peken and Denbates. Table 10 shows that there were 5 zones in quadrant D, i.e. Sudimara, Subamia, Buahan, Tunjuk, and Wanesari. Public transport users tend to be higher in those zones with high accessibility values. This agrees with the previous studies which stated that accessibility is an important factor to influence people to use public transport. Based on these results, those zones that have negative Z-score values need attention in re-structuring public transport network in the future in order to increase their accessibility values.

Table-9. Zones in quadrant A

Zone No.	Zone/ district	Zx (Z-score time)	Zy (Z-score distance)
1	Bongan	-0,733	-0,794
2	Gubug	-0,788	-0,888
3	Dauh Peken	-0,767	-0,722
4	Delod Peken	-0,965	-0,922
5	Dajan Peken	-0,794	-0,75
6	Denbates	-0,881	-0,855

Table-10. Zones in quadrant D

Zone No.	Zone/ district	Zx (Z-score time)	Zy (Z-score distance)
1	Sudimara	1,252	1,15
2	Subamia	0,418	0,427
3	Buahan	1,316	1,372
4	Tunjuk	1,495	1,538
5	Wanesari	0,447	0,483

CONCLUSIONS

Accessibility to public transport services in Tabanan Region was analyzed based on travel distance and travel time to reach the closest public transport routes. Based on Z-score, it was found that six zones were in quadrant A which means that these zones were in low accessibility category (Bongan, Gubug, Dauh Peken, Delod Peken, Dajan Peken and Denbatas). On the other hand five zones were identified in quadrant D. These zones (Sudimara, Subamia, Buahan, Tunjuk and Wanasari) have a high accessibility values to reach public transport routes. Therefore, improvement of infrastructures and public transport services need to be given priority in those six zones with a low level of accessibility values. The actions required include: developing new public transport routes network to serve people living in low accessibility zones; improving the quality of public transport services; increasing the number of public transport fleets in order to reduce headway; provide subsidy to the public transport user in order to reduce fare and attract people to use public transport.

REFERENCES

Choocharukul, K. and Ung, M. H. 2011. Propensity to Switch to a New Bus Service in Phnom Penh using Stated Preference Analysis. Asian Transport Studies, 1(3), pp. 277-287.

Dell'Olio, L., Ibeas, A. and Cecin, P. 2011. The Quality of Service Desired by Public Transport Users. Transport Policy, 18(1), pp. 217-227.

Exel, N.J.A.V. and Rietveld, P. 2010. Perceptions of Public Transport Travel Time and Their Effect on Choice-Sets among Car Drivers. Journal of Transport and Land Use 2 (3/4), pp. 75-86.

Japan International Cooperation Agency (JICA). 2011. The Research on Practical Approach for Urban Transport Planning. Japan.

Joewono T. B., Lauw, B. Z., and Hendy, H. 2013. Motorcycle in the West Java Province, Indonesia: Its Growth and Characteristics, Civil Engineering Dimension, 15(1), 61-70.

Kikuchi, K., Arimura, M., Tamura, T., and Saito, K. 2011. Empirical Study on Optimum Allocation toward Reducing the Length of Journey to Work Trip: The Case of Sapporo City. Proceedings of the Eastern Asia Society for Transportation Studies, 8.

Koizumi, Y., Nishimiya, N., and Kaneko, M. 2013. Urban Transportation Characteristics and Urban Mass Transit Introduction in the Cities of Developing Countries. Journal of the Eastern Asian Society for Transport Studies, 10, pp. 81-99.

Lay, W. T. and Chen, C. F. 2011. Behavioral Intentions of Public Transit Passengers-The Roles of Service Quality, Perceived Values, Satisfaction and Involvement. Transport Policy, 8(2), pp. 318-325.

Lee, C. F. 1993. Statistics for Business and Financial Economics, D.C. Heath and Company, Lexington.

Muthukannan, M. and Thirumurthy, A. M. 2008. Modeling for Optimization of Urban Transit System Utility: A Case Study. ARPN Journal of Engineering and Applied Sciences, 3(2), pp. 71-74.

Newman, P. and Kenworthy, J. 1999. Sustainability and Cities: Overcoming Automobile Dependence.Washington: Island Press.

Pardo, C. F. 2006. Raising public awareness about sustainable urban transport. In Deutsche Gesellschaftfür Technische Zusammenarbeit (GTZ) (Eds.), Sustainable transport: A sourcebook for policy-makers in developing countries (Module 1e).

Redman, L., Friman, M., Garling, T. and Hartig, T. 2013. Quality Attributes of Public Transport that Attract Car Users: A Research Review. Transport Policy, 25, pp. 119-127.

Ryan, S. 2009. Pedestrian Environments and Transit Ridership. Journal of Public Transportation, 12(1), pp. 39-57.

Sandow, E. and Westin, K. 2010. Preferences for Commuting in Sparsely Populated Areas. Journal of Transport and Land Use, 2(3), pp. 87-107.

Sekar, K. R., Padmakrishna, B., Neeli, R. T. and Srinivasan, T. C. 2015. Determining Transportation Industrial Metrics Using Random Process Theory and Analytical Hierarchy Process. ARPN Journal of

Engineering and Applied Sciences, 10(12), pp. 5240-5249.

Tabanan Bureau of Statistics. 2013. Population, Socio-Economic and Transport Data. Tabanan. Bali.
Tabanan Department of Transport. 2014. Public Transport Routes in Tabanan Region. Tabanan.Bali.

Yu, B., Zhang, J., and Fujiwara, A. 2011. Comparative Analysis on Household Energy Consumption Behavior in Asia Megacities by Considering the Effect of Car Ownership and Self-selection. Proceedings of the Eastern Asia Society for Transportation Studies.

MATEC Web of Conferences **58**, 02003 (2016) DOI: 10.1051/matecconf/20165802003
BISSTECH 2015

ANALYSIS OF BALANCE SCORECARDS MODEL PERFORMANCE AND PERSPECTIVE STRATEGY SYNERGIZED BY SEM

Minto Waluyo[1], Syamsul Huda[1], Noer Soetjipto[2], Sumiati[1], Handoyo[1]
[1]University of Pembangunan Nasional "Veteran" Jawa Timur, Surabaya, East Java, Indonesia,
[2]University Muhammadiyah Gresik.
Email: mintowaluyo_ti@yahoo.co.id, syamsulhuda-prof10@yahoo.co.id,
noersoetjipto@yahoo.co.id, busumiati_tic@yahoo.com, handoyo_ti@yahoo.com

ABSTRACT

The performance assessment analysis after the economic crisis by using Balanced Scorecard (BSC) method becomes a powerful and effective tool and can provide an integrated view of the performance of an organization. This strategy led to the Indonesian economy being stretched positively after the economic crisis. Taking effective decisions is not spared from combining four BSC perspectives and strategies that focus on a system with different behavior or steps. This paper combines two methods of BSC with structural equation modeling (SEM)) because they have the same concept, which is a causal relationship, where the research model concept SEM variables use BSC variable. The purpose of this paper is to investigate the influence of variables that synergized between balanced scorecard with SEM as a means of strategic planning in the future. This study used primary data with a large enough sample to meet the maximum likelihood estimation by assessment scale of seven semantic points. This research model is a combination of one and two step models. The next step is to test the measurement model, structural equation modeling, and modification models. The test results indicated that the model has multi colinearities. Therefore, the model is converted into one step model. The test results after being modified into a model of the goodness of fit indices showed a good score. All BSC variables have direct significant influence, including the perspective of strategic goals and sustainable competitive advantage. The implication of the simulation model of goodness of fit-modification results are DF = 227, Chi-square =276.550, P =0.058, CMIN/DF = 1.150, GFI = 0.831, AGFI = 0.791, CFI = 0.972, TLI = 0.965 and RMSEA = 0.039.

Keywords: balanced scorecard variable, perspective of strategic objectives, SEM, center of small enterprises

INTRODUCTION

The monetary crisis in 1997-1998 has made a number of foreign observers predict that Indonesia would be scattered as such in Europe, which after years was repeatedly hit by the economic crisis. However, Indonesia is stronger that it did not take long to be able to crawl up and come out of the crisis. The lower level of the economy at that time, such as the traditional markets, minimarkets, and even supermarkets were still running well. SMEs were still running well because they do not use the capital facility of bank loans, but this can not be allowed because it would cause SMEs become unable to capture a greater opportunity. The Government of Indonesia has provided a policy where major employers have to live with the aid of liquid funds from Bank of Indonesia to bridge SMEs and the national economy so that they become more exist. The direct impact of the monetary crisis hit large businesses whose capital came from bank loans.

One example of the performance of SMEs discussed in this paper is the field of slippers and furniture industries, in which the further research is on the evaluation of balanced scorecards for small and medium enterprises (SMEs), and the continuation of a previous research on Sustainable Competitive Advantage (X_5) entitled "Analysis on Performance Model Using Balanced Score Card, Strategic Objective and Sustainable Competitive Advantage Synergized with SEM."

This strategy has made the economy react positively. This effective decision included combining four perspective steps of BSC that focused on a system with different behaviors. (Kaplan & Norton, 1996b) created the BSC method with the concept of causality so that researchers having BSC mindset can be synergized with SEM, using large enough data to support it (Baggio and Sainaghi, 2011). The previous research has suggested that customers and financial performance are causally related to each other (Nielsen, et al, 2012). BSC is applied to the marketing of a comprehensive performance management to demonstrate the relationship between the marketing performance measurement system and the performance of the company as a condition of marketing alignment and knowledge-based market (Taylor and Baines, 2012).

The advantage of BSC is due to the balance between financial and non-financial variables, move his problems focus on financial variables, customer and business process / internal focus on the future of learning and growth variables (Banker et al, 2004). The year of 2004 was a crucial moment, which the election in Indonesia that ran peacefully became the energy to drive the wheels of the economy to move faster. The achievement of the Indonesian leaders within a decade was able to maintain the high level of economic growth in the range of 4-6%. It is only China and India that could reach that level. However, in terms of stability of economic growth, Indonesia is number one in worldwide. Indonesian leaders in the decade of 2004-2014 were the pioneers of stability post the economic crisis, and the Indonesian leaders in the decade of 2014-2019 are expected to be able to maintain and improve the programs to become more positive, realistic, accountable, and credible. Although Indonesia was struggling at the beginning of its administration, but it has high flight hours tips executed confidently to overcome the problem, besides having the spirit and motivation plus prayers by the people of Indonesia.

THEORETICAL AND BACKGROUND CENTER OF SMALL ENTERPRISES

The center of SME is a small area that has certain characters where production process exists and it is a more specific area for economic activity commodity that has been formed naturally and supported by facilities in order to develop the products or services. It consists of micro, small, and middle-level entrepreneurs. There is a physical union of functions in the center land, geography, infrastructure, institutions, and human resources that have potentials in developing the economy activity under the market influence of a product with higher selling point and competitiveness. This theory of matter has been discussed in the paper entitled The evaluation of balanced scorecards for small and medium enterprises, with eight and nine hypotheses for this study.

Balanced Scorecard (BSC).

According to (Johnson and Kaplan, 1987), in the need for better performance measurement system, BSC is a flexible performance measurement method that can be used for measuring the performance of the enterprise or institution which consists of four perspectives: financial, customer, internal business processes, learning and growth (Kaplan & Norton, 1996a).

This study became different because the learning and growth variables are made into separate variables, in addition to researchers who led Strategic Objective variable as a variable of the ultimate goal supported by a strong theory. BSC can define long-term strategic objectives of each perspective (outcomes) and mechanisms to achieve these objectives (performance driver). This approach combines long-term strategic objectives due to the impact of short-term events. The four processes translate the vision, mission and strategy of the company in determining the size of the performance. The organization's vision is outlined in goals and objectives (Kaplan and Norton, 1996a).

The objective was then explained in the strategic objectives along with the measurement of its achievement. The manager is able to make better decisions by using four perspectives of BSC (Lipe dan Salterio, 2000). The empirical research of BSC combines different steps: financial and operational, qualitative and quantitative (Ittner at. al, 1998; Dess dan Shaw, 2001). Sangjae at, al. (2013) synergizes BSC with PLS by using data from 101 companies to examine the causal relationship between the four categories of BSC. Zahirul Hoque (2014) reported a paper which synergizes BSC with partial least squares (PLS) using limited primary data to examine the causal relationship between the four categories of BSC. Tobia (1999) posited that PLS is useful in screening out negligible factors affecting the dependent variable.

The research framework can be seen in figure 1 and the model is in figure 2. Furthermore, it also provides the theoretical background of BSC, which leads to the development of the hypothesis. Sim and Koh (2001) verified the causal relationship between several indicators of individual performance among the four categories of BSC through relationships and regression analysis. A hypothesis was constructed in relating to four perspectives, against the perspective of strategic goals and sustainable competitive advantage, ultimately the main themes discussed, conclusions, managerial implications taken as a step.

Wu (2012) used a multiple criteria analysis tool to determine causal relationships between main performance indicators for each BSC perspective to create strategy maps visualized in a logical link to improve performance. BSC organizational strategy is represented by the connected performance via a causal relationship because BSC is able to look to the future / long term.

BSC model of measurement results will undertake business presence information. The model presented can be used as the basic model of measurement models in the following year to evaluate the performance of the personnel concerned and short term input events to achieve long-term strategies in accordance with the vision. Gomes et al. (2013) and Mehrdad et al. (2013), through changes in the organization, viewed the purpose and size of the BSC from four perspectives. Bhattacharya, et al (2014) stated a causal relationship involving organizational commitment and traditional financial measures. Ratnasingam (2014) made a report on a small business, the lack of financial capital and technical or managerial skills, as a strategic management system. Elaboration of the four perspectives are as follows: Financial perspective, providing financial goals need to be achieved by the organization in realizing its vision, customer perspective provides a description target market segments and customers along with demands served by the organization in an effort to target financial strategy, internal and Business Process Perspective. Giving a picture of the process that must be built to serve the customer and the Learning and Growth perspective is a booster to build the competence of personnel, information system infrastructure and atmosphere of the work environment needed. Referral approaches of (Grigoroudis at. al, 2012; Huang, 2009; Mehrdad at. al, 2013; Mendes at. al, 2012; and Ratnasingam, 2014).

THE VARIABLE OF FORMULATION STRATEGY OF FOUR PERSPECTIVES OF BSC

BSC method provides a comprehensive framework for outlining a vision into strategic goals. Comprehensive strategic objectives can be formulated since BSC method (Kaplan and Norton, 1996a) uses four perspectives: financial, customer, internal business processes, learning, and growth. The summary descriptions of the four perspectives are as follows: Financial perspective provides financial goals need to be achieved by the organization in realizing its vision. Customer perspective provides a description of targeted market segments and customers as well as their demands will be served by the organization in an effort to achieve the financial strategic objective. Internal business process perspective provides an overview process shall be built to serve customers, and Learning and Growth perspective is a booster to build the competence of personnel, infrastructure of information systems necessary and working environment atmosphere. Reference of approach are (Grigoroudis at. al, 2012; Huang, 2009), Mehrdad at. al, 2013; Mendes at. al, 2012; Ratnasingam, 2014).

DOI: 10.1051/matecconf/20165802003

FINANCIAL PERSPECTIVE

In BSC, financial perspective consistently becomes one concern because the financial measurement was one economic consequence happened due to the decision and stable economic activity. The evaluation/measurement of financial performance showed if planning, implementation, and application as well as strategy provided some basic improvement both in gross operating income and return on investment. (As can be seen in Tabel 1) with an approach of reference (Behn at. al, 1999; Bhattacharya at. al, 2014; Grogorodius at. al, 2012; Huang, 2009; Tseng, 2010; Nielsen at. al, 2012)

CUSTOMER'S PERSPECTIVE

The benchmarks of customer's standard can be classified into two groups: core measurement group and customer's value proposition. The first group consists of market share, customer acquisition, customer retention, customer's satisfaction, and customer's profitability. Meanwhile, the proposition group is classified into 3 (three) sub-groups: production attributes (price, quality, and function), customer's relationship, image, and reputation.

Endogen variable of customer's perspective is formed by two formation variables. The first is Increasing customer's Trust. The references are (Huang, 2009; Wu, 2012; Nielsen at. al, 2012)

PERSPECTIVE OF INTERNAL BUSINESS PROCESS

Efficiency measurement and company's effectiveness in producing goods and services perspective is required to measure all activities conducted by the manager and his employees in a company in efforts to fulfill the customer's satisfaction, the owner of the company, and businessmen. Every company has its own certain series of processes in efforts to create a certain score/point for customers (Kaplan and Norton, 1996b).

Internal business process is established from two formation variables:

The first is Increasing the quality of customer's service process and The second formation variable is State of the Art Technology (Tabel-1). Reference approach: (Baggio et al, 2011; Bhattacharya et al, 2014; Huang, 2009; Wu et al, 2009; Kaplan & Norton, 1992; Malina et al, 2001; Martinsons et al, 1999; Mehrdad et al, 2013; Tseng, 2010; Naranjo et al, 2009; Nikzadet et al, 2011; Ratnasingam, 2014; and Sabah et al, 2012).

LEARNING AND GROWTH PERSPECTIVE

Measuring the company's ability to develop and utilize human resources is required so that the company's strategic objectives could be achieved for the present and the future. Kaplan and Norton (1996b) revealed that it is important for the company to pay attention to its employees, to monitor the welfare of employees and to increase the knowledge of their employees, because with increased knowledge of employees, it would support the significant improvement of the performance of employees in achieving the company's goals (Tabel-1).

The four perspectives have a close relationship and affect each other. It also could not be separated in implementation. This is when the researcher use SEM tool so that the world of publication have different colors. Reference of approach (Bhattacharya at. al, 2014; Chen at. al, 2011; Grigoroudis at. al, 2012; Huang, 2009; Nikzad at. al, 2011).

METHODOLOGY

Measurement Indicators

Constructs/variables are measured through indicators, and most of the previous researchers used secondary data. This research used primary data, where measurement indicators written in the form of statements contained in the questionnaire were given to respondents using a scoring system from strongly agree to strongly disagree with seven (7) points (semantic scale). The tool used to manage is Amos 22 (Arbuckle, 2013) and the same measurement is done by (Lee at. al, 2009). The latent variables and indicators of measurement used in the research are shown in Table - 1.

The research variables consist of five steps, including learning and growth, internal business processes, customer performance, financial performance, strategic goals, and excellence balanced scorecard perspective. Steps are based on the adaptation of previous researchers (Butler et al, 1997; Epstein et al, 1998; Harvey et al, 1999; Lipe et. al, 2000; Martinsons et al, 1999; Mayo, 2000; Petty et al 2000; Yawson et al, 2006) and so forth.

Data Collection

The questionnaires were distributed to 110 entrepreneurs and craftsmen who are competent at the association of shoes and sandals in Wedoro Sidoarjo as well as furniture craftsmen in Pasuruan Indonesia. However, after a re-checking process, there were only 100 questionnaires that contained complete data. Therefore, the number of samples used in this research is 100 samples. Therefore, it is necessary to the choose metrics of in-out by assuming variable formations. Internal Business Processes is to improve the quality of customer service process, State of the art technology, enhance customer trust, and speed of service. These four constructs cause multi-colinearity. Indicators of the four constructs are functioned as indicators of the internal business process. It is also done on the variable of the customer. Performing invalid variable disposal is done on empowerment and alignment indicators, Financial and Comprehensive Perspective. In this process occurs a new discovery of the combined concept which is originally a combination of one step and two steps which transformed into one step concept. Steps are based on the adaptation of the previous researchers (Waluyo, 2009).

Assessment of Measurement Model

The first step of SEM tool is to test the Measurement Model which from its results can be concluded that the dimensions used by researchers do not yet reflect the analyzed latent variables, but all indicators of value CR> t-table after discharge indicator ($X1.6$, $Y_{4.4}$), so it can be concluded that the indicators were the dimension of the latent variable formed significantly (Waluyo, 2009). The second step is

to test the structural model which from its results can be concluded that the dimensions used by researchers do not yet reflect the analyzed latent variables, as shown by test results of Goodness of Fit Indices which has not shown a great value. The next step is to modify the model so that it turns to be good, this step is next to be reference in arranging business strategies and tactics, this model will be used as the basic capital and benchmark for the coming year measurement. The recapitulation of model modification can be seen on the next page. A complete evaluation can be done as follows is ok:

RESULTS

Testing of research hypotheses

Hipotesis One (H-1)

Ho:The learning process does not significantly influence the internal business processes.
H1: The learning process significantly influences the internal business.

Test results of H1 hypothesis, accepted

The learning process (X1) significantly influences internal business processes, with a regression coefficient of .321, the probability value < .05 (.041) (see Table 3), it means that the learning process increase by 1 unit in the form of activities (employee satisfaction, employee retention, employee productivity, motivation and empowerment and alignment) which will improve internal business processes at .321. One thing to note is once employee's learning process and performance are positive, they must be maintained so that the best workers that will not leave the organization / company, the worker has to have a high artistic soul, this business line involves in the arts (shoe and furniture industry).

Table – 1. Variable and Indicator
Source: The primary data analyzed

Variable	Formed Variable	Indicator	Intervening Variable
Strategic Object tive (Y_4)		Learning & Growth ($Y_{4.1}$), Business Process/ Internal ($Y_{4.2}$) *Customer ($Y_{4.3}$)* Finance ($Y_{4.4}$)	Finance (Y_3)
Finance (Y_3)		Revenue Growth($Y_{3.1}$) Reduced Cost ($Y_{3.2}$)	Customer (Y_2)
Customer (Y_2)	Increasing Customer's Trust ($Y_{2.1}$)	*Time* ($Y_{2.1.1}$), *Quality* ($Y_{2.1.2}$) *Cost* ($Y_{2.1.3}$)	Business Process/ Internal (Y_1)
	Speed of Service ($Y_{2.2}$)	Availability of product images ($Y_{2.2.1}$) On time delivery ($Y_{2.2.2}$).	
Business Process/ Intern (Y_1)	Increasing the quality of customer service process ($Y_{1.1}$)	Inovation Process ($Y_{1.1.1}$) Operational Process ($Y_{1.1.2}$) After Sale Service Process ($Y_{1.1.3}$), Ease of Payment Process($Y_{1.1.4}$) Satisfying maket segment through product or service ($Y_{1.1.5}$).	
	Stateof the art technologi ($Y_{1.2}$)	High level taste of art worker ($Y_{1.2.1}$) Fast Response on art technology ($Y_{1.2.2}$)	
Learning (X_1)		*employee satisfaction* ($X_{1.1}$) *employee retention* ($X_{1.2}$), *employee productivity* ($X_{1.3}$) Motivation $X_{1.4}$ Empowerment and Alignment ($X_{1.5}$) Empowerment and Alignment ($X_{1.6}$)	
Growth (X_2)		Organization Growth increases ($X_{2.1}$) Organization Growth decreases($X_{2.2}$)	

Table – 2. *Goodness Of Fit* and *Cut Off Value* (modification)

Goodness of Fit Indices	Result of Model Testing	*Cut – Off Value*	Remark
X^2 Chi Square	276.550	Small*	Good
Probabilitas	0,058	≥ 0,05	Good
CMIN/DF	1.150	≤ 2,00	Good
RMSEA	0.039	≤ 0,08	Good
GFI	0.836	≥ 0,90	Marginal
AGFI	0.811	≥ 0,90	Marginal
TLI	0,965	≥ 0,95	Good
CFI	0,972	≥ 0,95	Good

Source: The primary data analyzed

Tabel - 3. Regression Weights (λ): (Default Model modification)
Source: The primary data analyzed

Vari able			Esti mate	S.E	C.R.	P	(λ)
y1	<--	x1	-.024	.153	-.158	.041	.321
y1	<--	x2	.576	.248	2.326	.020	.327
y2	<--	y1	.302	.076	3.963	.039	.419
y3	<--	y2	.046	.156	2.294	.036	.311
Y4	<--	Y3	.399	.166	3. 291	***	.275

Second hypothesis (H-2)

Ho: The process of growth does not significantly affect on internal business processes.
H1: The process of growth significantly affects internal business processes.

Test results of H1 hypothesis, accepted

The process of growth significantly influences the internal business processes with a regression coefficient of .327 with a probability value < .05 (.020). In table-3, it means that the process of growth increased by 1 unit, in the form of growth-enhancing activities of the organization and preventing the organization from declining will improve the internal business processes at .327. The test results support the (Bhattacharya et al, 2014; Chen at al, 2011; Geuser at al, 2009; Ken Ogat at al, 2014; Tseng, 2010; Sangjae at al, 2013; Seyed at al, 2011; Stewart, 2001).

Third (H-3) & Fourth Hypothesis (H-4)
Ho: Internal business processes does not significantly affect on the quality elevation of the customer service processes.
H1: Internal business processes significantly affects the elevation quality of the customer service processes, and
Ho: Internal business processes does not significantly affect on state of the art technology
H1: Internal business processes significantly affects on state of the art technology

Hypothesis test results can not be proved because there is a "Warning" in which there are indications of multi-colinearity. Supported by (Waluyo, 2009).

Fifth Hypothesis (H-5)
Ho: Internal business processes does not significantly affect customers.
H1: Internal business processes (Y1) significantly affects customers.

Test results of H1 hypothesis, accepted
Internal business processes significantly influence the customer with a regression coefficient of .419, with the probability value of <.05 (.039). (See Table 3) This means if internal business process increases by 1 unit in the form of activities to improve the coordination of all existing indicators so that it is more synergized, it will increase the number of customers amounted to .419, resulting in products that are of high artistic quality. The test results support the (Chen, at al, 2011; Wu, 2012; Mendes, at al, 2012; Sangjae, at al, 2013).

Sixth (H-6) Seventh Hypothesis (H-7)
Ho: Customer does not significantly affect increasing customer trust.
H1: Customer significantly affects increasing customer trust, and
Ho: Customer does not significantly affect service speed.
H1: Customer significantly affects service speed.

Hypothesis test results can not be proved because there is a "Warning" in which there are indications of multi-colinearity. Supported by (Waluyo, 2009).

Eighth Hypothesis (H-8)

Ho: Customer does not significantly affect finance.
H1: Customer significantly affects finance.

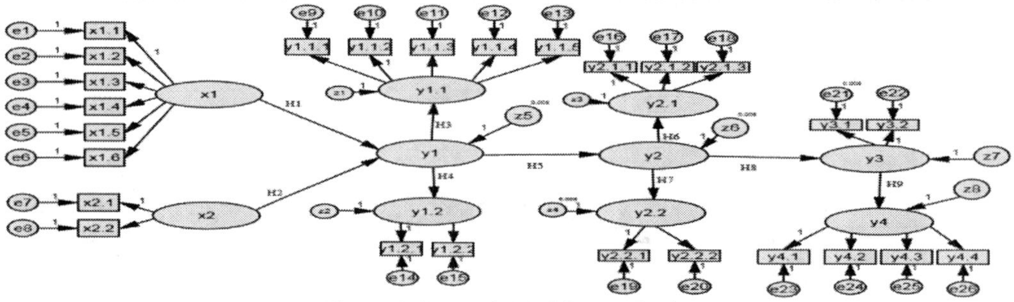

Figure-1. Research Model Hypothesis
Source: The primary data analyzed

Test results of H1 hypothesis, accepted
Customers have a significant effect on the financial with a regression coefficient of .311, and a probability value of < .05 (.049) (see table 2). The score .311 means that if customer increases by 1 unit in the form of negotiation activities that enhance customer trust with three indicators (on time finished order, the quality standards to order, and low cost to improve the speed of service to customers), it will increase the finance by .311. The test results support the (Bhattacharya, at al, 2014; Chen, at al, 2011; Mendes, at al, 2012; Sangjae, at al, 2013).

Ninth Hypothesis (H-9)
Ho : Finance does not significantly affect on strategic objective perspective.
H1 : Finance significantly affects on strategic objective perspective.

Test results of H1 hypothesis, accepted
Finance significantly affects the strategic objective perspective with a regression coefficient of .275 and probability value of < .05 (.050) (see table 2). The score of .275 means that if finance increases by 1 unit in the form of reduction of the cost, the revenue growth will increase the strategic objective of .275 by synergizing strategic objective perspective indicator. However, since financial perspective indicator is not significant, it is disposed of. This is done because the majority of SMEs capital is independent capital. This is due to the mindset of those affected by the parental discourse that bank debt gives a headache. The test results support the (Behn, et al, 2010; Bhattacharya, et al, 2014; Grigoroudis, et al, 2012; Huang, 2009; Mehrdad, et al, 2013; Mendes, et al, 2012; Naranjo, et al, 2009; Sangjae, et al, 2013)

DISCUSSIONS

Implications for researchers

The results of implications of model simulation are as follows: Goodness of Fit modifications DF = 227, Chi-square = 276.550, P = .058, CMIN/DF = 1.150, GFI = .831, AGFI = .791, CFI = .972, TLI = .965 and RMSEA = .039, where all variables have significant and direct effect. BSC model synergized with the SEM is a model of the new discovery combined models between one step and two steps. Because the process occurs in multi-colinearity, it eventually becomes the concept of one step that is supported by (Waluyo, 2009).

Implications for practitioners

In the work pattern of sandals and furniture industry before the monetary crisis (1997-1998), there were not many players in the industry who knew about banking. In running their businesses, they used their own capital added with down payment for each order of products (slippers and furniture). By having this unique method of work, it makes the industry still exists, but obtaining a capital loan should also be considered to accelerate the process of growth when the outlook is promising. This mindset led to the process of learning and growth which significantly affects the internal business processes, customers, and finances.

Based on the primary data, through the interview of mindset and small industry players' behavior at a time of crisis in Indonesia, they had made innovations as suggested in the modified model, in which keeps them exist.

This experience is an important grip in running entrepreneurship so that Indonesia would still exist and is always ready to face the economic crisis. This also proves that the stability of economic growth in Indonesia is number one in the world in this era. In the current era, through its lecturers, universities are approaching in the implementation of its three principles of higher education, namely one of which is teaching on how to easily get a bank loan and they also oversee the process to get it. This makes SMEs grow faster.

The basic model of measurement model of BSC is a comparison measurement of BSC models in the years to come. If the result of measurement model of BSC is better, then the SMEs model exist. The measurement model of BSC model has a cause and effect relationship, through which will oversee the process to get it. This makes SMEs grow faster relationship model, a strategy which can be easily animated as well as criticized at the same time because every relationship and causality can be tested in detail. (Kaplan and Norton, 1996b).

An effective government's decision is not immune from the perspective of the BSC combining four steps focus on systems with behavior / different steps, namely by adding a strategic variable. BSC causes all variables to have a significant effect, which is supported by (Lipe and Salterio, 2000; Malina and selto, 2001; Dess and Shaw, 2001; Ittner, et al, 1998; Amado, et al; 2012, Epstein, et al, 1998; Wu, et al, 2012).

CONCLUSIONS

The BSC model in this paper was proposed to improve the original model. At the start of the three-step analysis and structural equation Measurement and modification of the model, the value of its Goodness of Fit Indices was good, although there were two values that gave test results in marginal status.

The effect of BSC variable is significant and direct the learning perspective for significant influence internal business processes with a regression coefficient of .321 and the Growth Perspective significant effect for internal business processes with a regression coefficient of .327. The next in a chain of internal business processes are customers with a regression coefficient of .419. Customers have a significant effect for Financial with a regression coefficient of .311. A significant effect for Financial Perspective strategic objectives is with a regression coefficient of .275, where the model of an integrated approach produced test results. The goodness of Fit Indices is good after the model was modified.

REFERENCES

Amado, C.A.F, Sérgio P. Santos, Pedro M. Marques, Integrating The Data Envelopment Analysis and The Balanced Scorecard Approaches for Enhanced Performance

Arbuckle J.L, IBM® SPSS® Amos™ 22, Amos Development Corporation,2013

Baggio R, Sainaghi R. Complex and Chaotic Tourism Systems: Towards a Quantitative Approach. International Journal of Contemporary Hospitality Management, 23 (6), 2011, pp. 840–861.

Banker R.D. Chang, H. Pizzini M.J. The Balanced Scorecard: Judgemental Effects of Performance Measures Linked to Strategy. The Accounting Review 79 (1), 2004, pp 1–23.

Bhattacharya A, Priyabrata Mohapatra, Vikas Kumar, Prasanta Kumar Dey, Malcolm Brady, Manoj Kumar Tiwari & Nudurupati Sai S, Green Supply Chain Performance Measurement Using

Butler. A, Letza, S.R, Neale, B. Linking The Balanced Scorecard to Strategy, Long Range Planning 30 (2), 1997, pp. 242–253.

Behn, B.K, Riley R.A, Using Non-Financial Information to Predict Financial Performance: The Case of The U. S. Airline Industry, Journal of Accounting, Auditing and Finance 14, 1999, pp. 29–56. Scorecard Approach, Expert Systems with Applications 37 (8), 2010, pp. 5931–5938.

Chen F-H, Hsu T-S, Tzeng G-H, A Balanced Scorecard Approach To Establish a Performance Evaluation And Relationship Model For Hot Spring Hotels Based On a Hybrid MCDM Model Combining DEMATEL and ANP, International Journal of Hospitality Management, 30 (4), 2011, pp. 908–932

Dess G.G, Shaw D.J, Voluntary Turnover, Social Capital, and Organizational Performance. The Academy of Management Review 26 (3), 2001, pp. 446–456.

Epstein M, Manzoni J.-F, Implementing Corporate Strategy: From Tableaux De Bord to Balanced Scorecards, European Management Journal 16 (2), 1998, pp. 190–203.

Geuser, S. Mooraj F.D. & Oyon D, Does The Balanced Scorecard Add Value Empirical Evidence on Its Effect on Performance, European Accounting Review 18(1), 2009.

Gomes J, Romao M, Mário Caldeira, The Benefits Management and Balanced Scorecard Strategy Map : How They Match 4 (1), 2013, pp. 11.

Gonzalez-Padron.T, Chabowski B, Hult G.T.M, Ketchen D.J, Knowledge Management and Balanced Scorecard Outcomes: Exploring the Importance of Interpretation, Learning and Internationality, British Journal of Management 21 (4), 2010, pp. 967–982.

Grigoroudis. E, Orfanoudaki E, Zopounidis C, Strategic Performance Measurement In a Healthcare Organisation : A multiple Criteria Approach Based on Balanced Scorecard, Omega40 (1), 2012, pp. 104–119.

Harvey M.G., Lusch F.R, Balancing the Intellectual Capital Books : Intangible Liabilities, European Management Journal 17 (1), 1999, pp. 85–92.

Huang. H-C. Designing a Knowledge-Based System for Strategic Planning : A Balanced Scorecard Perspective, Expert Systems with Applications36 (1), 2009, pp. 209–218.

Ittner C.D, Larcker D.F, Are Non-Financial Measures Leading Indicators of Financial Performance? An Analysis of Customer Satisfaction Journal of Accounting Re-search Supplement 36, 1998, pp. 1–35.

Johnson H.T, Kaplan R.S, Relevance Lost, The Rise and Fall of Management Accounting. Harvard Business School Press, Boston, MA.1987

Kaplan R.S, Norton D.P, Linking the Balanced Scorecard to Strategy.California Management Review 39 (1), 1996, pp. 53–79.

Kaplan R.S, Norton R.D, Using The Balanced Scorecard As a Strategic Management System, Harvard Business Review 1996b, pp. 75–85.

Ken Ogat, Gary Spraakman, Beyond Control: Using the Balanced Scorecard to Change Culture and Implement Strategy, *Canadian Academic Accounting Association (CAAA) Annual Conference 2014,* pp. 1-50.

Lee S, Park, S.Y, Do Socially Responsible Activities Help Hotels and Casinos Achieve Their Financial Goals International Journal of Hospitality Management 28 (1), 2009, pp. 105–112.

Lipe M.G, Salterio S.E, The Balanced Scorecard: Judgmental Effects of Common and Unique Performance Measures, Accounting Review 75 (3), 2000, pp. 283– 298.

Malina, Selto .F, Controlling and Communicating Strategy : An Empirical Test of The Effectiveness of The Balanced Scorecard. Journal of Management Accounting Research, 2001, pp. 47–90.

Martinsons M, Davison, R, Tse. D, The Balanced Scorecard: a Foundation For The Strategic Management of Information Systems, Decision Support Systems 25 (1), 1999, pp. 25–36.

Mayo A, The Role of Employee Development in The Growth of Intellectual Capital, Personnel Review 29 (4), 2000, pp. 2000.

Mehrdad N. A, Vahid Majazi Dalfard, Kiarash Poursalik, A Combination Model Using Strategic Alignment Model and Balanced Scorecard and Strategies' Prioritisation Based on TOPSIS Technique, International Journal of Productivity and Quality Management12 (3), 2013, pp. 313-326.

Mendes P, Ana Carina Santos, Fernando Perna, Margarida Ribau Teixeira, The Balanced Scorecard as an Integrated Model Applied to The Portuguese Public Service: a Case Study in The Waste Sector Journal of Cleaner Production24, 2012, pp. 20–29.

Naranjo-Gil, David, Strategic Performance in Hospitals: The Use of The Balanced Scorecard by Nurse Managers, Health Care Management 34 (2), 2009, pp. 161-170.

Nielsen S, Nielsen E.H, Discussing feedback System Thinking in Relation to Scenario Evaluation in A balanced Scorecard Setup, Production Planning & Control 23 (6), 2012, pp. 436–451.

Nikzad Manteghi, Abazar Zohrabi, A Proposed Comprehensive Framework for Formulating Strategy: a Hybrid of Balanced Scorecard, SWOT Analysis, Porter‘s Generic Strategies and Fuzzy Quality Function Deployment. Procedia - Social and Behavioral Sciences, 15, 2011, pp. 2068–2073.

Petty R, Guthrie J, Intellectual Capital Literature Review: Measurement, Reporting, and Management, Journal of Intellectual Capital 1 (July (2)), 2000, pp. 155–176.

Ratnasingam P, The Evolution of Balance Scorecard and its Impact on Web Services Quality, International Journal of Organizational and Collective Intelligence (IJOCI) 4 (1), 2014, pp. 12.

Sabah M. Al-Najjar, Khawla H. Kalaf, Designing a Balanced Scorecard to Measure a Bank's Performance : A Case Study, International Journal of Business Administration 3 (4), (2012), pp. 44.

Sangjae. Lee, Sung Bum Park, Gyoo Gun Lim, Using Balanced Scorecards For the Evauation of "Software-as-a-Service" Information & Management, 2013, pp. 553-561.

Seyed Mohammad Seyedhosseini, Ahmad Ebrahimi Taleghani, Arash Bakhsha, Solmaz Partovi Extracting Leanness Criteria by Employing the Concept of Balanced

Scorecard, Expert Systems with Applications 38 (8), 2011, pp 10454–10461.

Sim K.L, Koh H.C, Balanced scorecard: A Rising Trend in Strategic Performance Measurement, Measuring Business Excellence 5, 2001, pp.18–27.

Stewart W.E., Balanced Scorecard for Projects, Project Management Journal 32 (1), 2001, pp. 38–53.

Taylor J, Baines C, Performance Management in UK Universities: Implementing the Balanced Scorecard, Journal of Higher Education Policy & Management 34 (2), 2012, pp. 111–124.

Tobia, R.D. An Introduction to Least Squares Regression, SAS Institution Inc, 1999.

Tseng. M-L, Implementation and Performance Evaluation Using The Fuzzy Network Balanced Scorecard, Computers & Education55 (1), 2010, pp. 188–201.

Waluyo. M, Object and Applied Structural Equation Modelling, PT. Indeks, 2009, Jakarta.

Wu H.Y, Constructing a Strategy Map for Banking Institutions With Key Performance Indicators of the Balanced Scorecard, Evaluation & Program Planning 35 (3), 2012, pp. 303–320.

Yawson R, Amoa-Awua W, Sutherland A, Smith D, Noamesi S, Developing a Performance Measurement Framework to Enhance the Impact Orientation of the Food Research Institute, Ghana, R&D Management 36 (2), 2006, pp. 161– 172.

Zahirul Hoque, 20 Years of Studies on the Balanced Scorecard: Trends, Accomplishments, Gaps and Opportunities for Future Research, The British Accounting Review46, (1), 2014, pp. 33–59.

COLLABORATION AND SUSTAINABLE AGRI-FOOD SUPLY CHAIN: A LITERATURE REVIEW

Wike Agustin Prima Dania[12], Ke Xing[1], Yousef Amer[1]
[1]University of South Australia, 101 Curie St, Adelaide 5001, South Australia
[2]University of Brawijaya, 1 Veteran St, Malang 65144, Indonesia
E-Mail: wike.dania@mymail.unisa.edu.au

ABSTRACT

Maintaining collaboration among the entire stages in the agri-food supply chain to achieve sustainability is complex. All the stakeholders involved in the activities have to prioritize their financial benefits without putting aside social development and environmental responsibilities. Some scholars have paid attention to this topic. The objective of this paper is to review current research on sustainable supply chain and collaboration model in agri-food industry. Sustainability aspects that consist of economic, environment, and social and the model of sustainable supply chain in agri-food industry are analyzed. Moreover, collaboration in sustainable agri-food supply chain management is also studied thoroughly from vertical and horizontal perspectives. The result shows that there are few studies focusing on the integrated collaboration to achieve sustainable supply chain system. Additionally, not all sustainable aspects are covered thoroughly. The scholars pays more attention to economic and environmental aspects than social aspects. Furthermore, some studies only focus on one type of collaboration in sustainable agri-food supply chain. Often, these studies do not even consider all elements in the triple bottom line.

Keywords: sustainable supply chain, agri-food industry, collaboration

INTRODUCTION

Sustainable supply chain and collaboration is the key concept that essential to be applied in business environment, especially agri-food industry. Those concepts have been discussed in the literature. The purpose of this paper to review the current research in both area and to identify the limitations and provide suggestions for the future research in agri-food industry related to sustainable supply chain collaboration. In specifically, this study covers sustainability aspects in sustainable supply chain management in agri-food industry that consists of economics, environment, and social dimensions. The sustainable supply chain management models in agri-food industry also discussed comprehensively. Moreover, collaboration models in sustainable agri-food supply chain management are investigated by clustered it into vertical collaboration, horizontal collaboration, and the combination of both types of collaboration. Therefore, this review will highlight the area of improvement which lead to suggestion of the future works.

Agri-food supply chain and sustainability

Agri-food industry is one of the most important sectors in the world. It gives impacts in economic, environment, and social aspects in several countries. For example, agri-food sector supports the economic activities in national and international scopes in the US while maintain the human life (Maloni & Brown, 2006). In Europe, this sector has significant impact on the economic, environmental, and social simultaneously (Turi, et.al., 2014). In Scotland, food and drink supply chain become major sector that support economic benefits in Scottish community and deal with large number of employee (Leat, et al., 2011). Moreover, agri-food industry has also supported economic benefits in developing countries. For examples, rapid changing on product and process innovation in agri-food supply chain brings significant impact on socioeconomic development in Malaysia (Mohezar & Nor, 2014). In Indonesia, agri-food product such as sugar, rice, meat, soybean, corn, chili, red onion, and palm become the main commodities that support national long term development (Rusono et al, 2014).

Supporting the development of agri-food product, supply chain has the important role. In agri-food, supply chain encompasses all processes and activities from farmers and suppliers who provide raw material, food manufacturers who process the food products to increase the added value, distributors and retailers who distribute the product into the customers through the systematic business processes. Each stakeholder in agri-food supply chain suffers the cost and collects the benefit even though sometimes it is unfair for some stakeholders. Therefore, supply chain need to be maintained to spread the benefit along and across the supply chain in fair and positive ways.

The more complex a supply chain configuration is, the more challenges the parties will face. Global competitions and unique characteristics of agri-food products demand that the food chain maintains the food quality throughout the processes from farm to the end consumers. Otherwise, it will become waste since it cannot be consumed (Yu & Nagurney, 2013). One of critical factors in the food supply chain is how to ensure fair collaboration among stakeholders and give attention to economic, environmental, social, organizational, marketing, food safety factors, and obligation over the firms, consumers, and society (Fritz & Schiefer, 2008).

Deal with the complexity of agri-food supply chain, sustainability is one of perspectives that can be applied to maintain the competitive strategies in economic, environmental, and social aspects that is called triple bottom line (TBL). The economic dimension includes revenue, cost, and consumer satisfaction, and service level (Varsei, et al., 2014; Wang, et al., 2011). The environmental dimension refers to natural resource consumption, carbon footprint, environmental legislation, waste management, and hazardous chemical and materials (Sanders, 2012; Varsei, et al., 2014). The social dimension covers impacts on a society such as working conditions, community development, consumer health and safety, human rights and child labour (Gosling, et al., 2014; Klassen & Vereecke, 2012; Mota, et al., 2014). Within those three dimensions, optimal competitive advantages can be achieved while support the social development and reduce the environmental impacts.

There are several issues raised in the agri-food industry regarding the sustainability. The requirements of product quality assurance without compromising the price, the availability of natural resources and the continuity of raw material without damaging the environment, climate change, and employee welfare have become the main targets of consumers and societies. The dependence on weather, short shelf life, limited raw materials, food safety regulations, and global competition force the stakeholders to keep abreast of the demographic changes in order to offer high quality, value added, and sophisticated food products in the right time and the right quantity to consumers. On the other hand, financial benefits will remain the main priority to be achieved whilst supporting social development and environmental responsibilities. Another critical issue is material handling processes which require particular treatments and conditions due to the unique food product characteristics (fragile, perishable, and lower yield). Those issues can be addressed if there is strong commitment from all stakeholders involved to achieve the competitive advantages without sacrifice certain parts in supply chain.

Collaboration in sustainable agri-food supply chain

Strong commitment among the stakeholders involved can be realized in the form of collaboration. Supply chain collaboration is a joint partnership within the various stages in supply chain as well as its external environments to optimize their competitive advantage throughout the entire processes (Cao & Zhang, 2011; Liao & Kuo, 2014). Collaboration is important to support the long term partnership and spread the benefits throughout the entire supply chain system, from strategic level into operational level. Within collaboration, stakeholders able to share their assets (materials, labor, infrastructures, facilities and equipment, and machines) and their capabilities (technology, business processes, policy and legislation, and finance). Therefore, they can reduce the uncertainty, share the risk and cost, and able to serve customers in the right time, right quantity, and right quality without disregards the interest of other stakeholders.

There are two types of collaboration that should be viewed as a single entity, namely vertical collaboration and horizontal collaboration (Figure 1). Vertical collaboration is relationship among stakeholders from upstream to downstream along the supply chain. Horizontal collaboration is relationship among stakeholders that play in the same level including competitors and complementary, as well as external parties such as government, NGOs, associations, and universities. Those two types of collaboration have to be considered in order to achieve a better sustainability system for all stakeholders without adversely suffering other stakeholders such as local farmer and SMEs. Without collaboration, the price that comes to consumers will be higher because each stakeholder will increase the price to get the higher benefit and to minimize the risk. Therefore, collaboration is important to increase the product's value and to ensure the flexibility of the food production while following the environmental policy integration (Ernšteins, et al., 2013).

In the last decade, stakeholders in food industry have been started to concern about collaboration in achieving the SSC system targets since it gains optimal benefits in maximising the profit and minimising the risk among the parties involved. However, maintaining supply chain collaboration among the entire stages in the sugar supply chain is complex. The more stakeholders take part in the collaboration system, the more complex the system becomes. The lack of support from stakeholders, insufficient assessment systems, limited information systems, organizational culture, and reluctance to change create obstacles for the application of the collaborative systems in sugar supply chain (Bezuidenhout, et al., 2012). On the other hand, sufficient technology and information are required to support the collaboration. In addition, to build mutual trust among partners, there needs to be eagerness from both parties to commit to a positive collaboration.

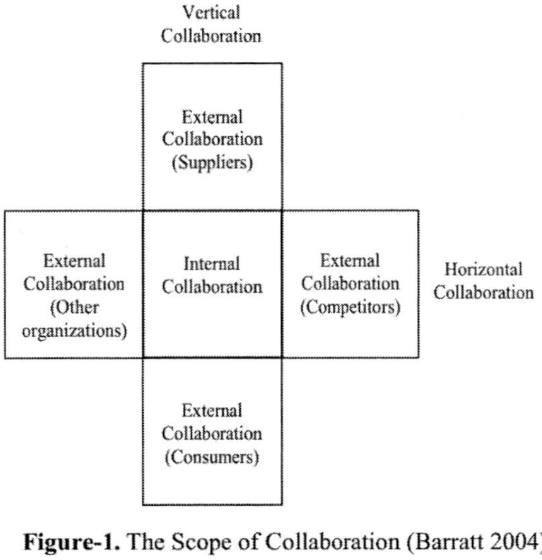

Figure-1. The Scope of Collaboration (Barratt 2004)

LITERATURE REVIEW

This section will be started by reviewing the dimensions of sustainable supply chain management in agri-food industry that consist of economic, environment, and social dimensions. Furthermore, how these dimensions have been used in the current sustainable supply chain model, specifically for agri-food industry, are also analysed. The last, research related to collaboration among stakeholders in agri-food supply chain management is identified. Therefore, it will give overview about the current research in both concepts that can lead to the identifying area for further research.

Sustainable supply chain management in agri-food industry

In agri-food industry, sustainability is a concept that is still continuously improved. Food industry has a main role in sustaining consumers' health, taking part in an economic growth, and assuring an environmental impact reduction (Turi, et al., 2014). Therefore, sustainability is important for maintaining a value, awareness, society, and business reputation as well as enhancing a business environment and cooperation along a supply chain (Lin & Tseng, 2014; Shokri, et al., 2014). Consisting of economic, environment, and social aspects, it is expected to respond to the dynamic environment in the food supply chain that has high food safety expectation, food regulation, and environmental legislation (Validi, et al., 2014). Moreover, the implementation of sustainability in an agri-food supply chain can be seen as an effective way to attract stakeholders to collaborate in achieving common goals (Rota, et al., 2013).

1. Economic Dimensions in Agri-food Industry

Economic dimensions are critical drivers in food business processes which implement sustainability (Shokri, et al., 2014). These dimensions refer to an allocation of resources in an appropriate manner to achieve efficiency and competitiveness to enhance contribution to the society (Leat, et al., 2011). The implementation of sustainability may lead to an incremental cost while adjusting internal and external facilities to create the advantages for all food business partners (Li, et al., 2014). Indeed, economic dimensions are influenced by social and environmental dimensions while they become the most important aspect in the sustainable supply chain (Yakovleva, et al., 2010).

Economic dimensions can be categorized into macroeconomics and microeconomic factors (Brandenburg, et al. 2014). Macroeconomic factors focus on labour productivity, market concentration, and import dependency under economic sustainability to achieve several goals such as promoting the economic growth, enhancing the competitive economy, and altering the customers preferences in food products (Yakovleva, et al., 2012). Microeconomic factors address revenues, production and transportation costs, and overtime costs as economic indicators (Dwi, et al. 2013). In addition, Yan and Ma (2012) assessed the economic dimensions by using inventory level and supply cost (procurement aspect), energy and fuel costs, delivery methods, inefficiency operation processes (internal operation), packaging materials and suitability between packaging and product requirements (product development and stewardship). Based on the survey, they proposed that sustainability in the food industry can be achieved by creating the proper business model, finding the trade-off between quality and cost-effectiveness, and finding the mature sourcing (Gold, et al., 2013).

2. Environmental Dimensions in Agri-food Industry

While running the business, the decision making is not only consider the economic aspects, they also have to pay attention to environmental aspects (Ala-Harja & Helo, 2014). Currently, most countries need to support the reduction of GHG emissions since the declaration of Kyoto Protocol which has a target to achieve it in the next certain years (Ramanathan, et al., 2014). In the low income countries, the contribution of the carbon footprint in food industry is higher compare to developed countries (Li, et al, 2014).

Environmental dimensions encompass input oriented (energy and natural resources) and output oriented indicators (waste and pollution) (Brandenburg, et al., 2014). Yan and Ma (2012) divided the environment supply chain issues into three categories, namely procurement (raw material, long and short term supply, waste and packaging), internal operation (water, air, soil pollution, health impact, and waste management), and product development and stewardship (product impact, substitute product, disposal, and traceability). On the other hand, Leat, et al., (2011) argued that there are three major categories in the environmental issue, namely climate change and eco-efficiency, green production and food safety, and animal welfare.

3. Social Dimensions in Agri-food Industry

Social issues is also needed to be considered in all strategic management processes in supply chain, even though these aspects are the most difficult aspect to be measured compare to others since it is has correlation with intangible aspects such as with culture, social communities, lifestyle, politics, health, human rights, and communities' aspiration (Vachon & Mao, 2008; Wang, et al. 2011) (Cambero & Sowlati, 2014). However, social standards such as ISO 26000 and Social Accountability SA8000 are still beyond the notice of organization management systems in the agri-food supply chain (Gold, et al., 2013).

The social focus in the food industries can be related to the raw material procurement from the local farmer, vitamin added in the local food, and increase the local income by providing the healthy and affordable local products (Gold, et al., 2013). These social dimensions support community development, labour opportunities, and human welfare (Leat, et al., 2011). Yakovleva, et al., (2012) determined wages, employment, and gender ratio to survive in creating fair market share and maintain productivity. Yan and Ma (2012) added labour standards, life balances, working hours, consumer demands, and inflation become the indicator in social dimensions. Furthermore, quantitative indicators have been established to measure the social dimensions such as number of employees trained, management levels with specific environment responsibility, number of improvement suggestions submitted by employees (Turi, et al., 2014).

Models of sustainable supply chain management in agri-food industry

Sustainable supply chain management is important to be applied in agri-food industry. It will influence not only internal organizations but also external relationship with other parties. It will help the development of socio-economic and environmental aspects, as well science and technology (Li, et al., 2014). These sustainability aspects also give a strong influence in the food supply chain particularly with regard to the natural resources and labour issues, health and safety, biotechnology, fair trade, community, and procurement

(Maloni & Brown, 2006). Research in sustainable agri-food supply chain is limited. Some models for assessing sustainable agri-food supply chain from various perspectives. However, not all of the models cover economic, environment, and social dimensions. In this section, the current research on sustainable supply chain management in agri-food industry has been investigated to identify the sustainability dimensions and indicators used that are shown in Table 1.

Table-1. Research in Sustainable Agri-food Supply Chain

Authors	Industries	Sustainability Dimensions	Indicators
Hamprecht et al. (2005)	Nestle food supply chain	Economics	-
		Environmental	Nutrition demands of the soils
		Social	Labour standard
Egilmez et al. (2014)	33 U.S. food manufacturing sectors	Economics	-
		Environmental	The carbon footprint, the water withdrawals impact category, the energy footprint of each sector, the cropland footprint, the grazing land footprint, the forestland footprint, and the fishery land footprint
		Social	-
Validi, Bhattacharya and Byrne (2014)	Irish dairy market supply chain	Economics	Total distribution cost
		Environmental	CO_2 emission from transportation
		Social	-
Zanoni and Zavanella (2012)	Chilled or frozen foods supply chain	Economics	Distribution cost
		Environmental	Energy effort required
		Social	-
Bourlakis et al. (2014)	Greek food supply chain	Economics	Materials cost, storage cost, delivery and distribution cost, financial cost, gross profit margin, flexibility, responsiveness, and quality
		Environmental	energy consumption, and waste
		Social	Firm's perception of its own supply chain performance, and firm's perceptions of market opinion regarding its supply chain performance
Yakovleva (2007)	U.K. food industries	Economics	Productivity, diversity and structure of the industry, and reducing transportation of imported products
		Environmental	Energy consumption, water consumption, and waste disposal
		Social	Employment volumes, quality of employment, and gender balance at workplace
Tajbakhsh and Hassini (2014)	U.K. food industries	Economics	Raw material cost, transportation cost, supplier capability factor, defect-free parts per million, advertisement cost, investment in sustainability design, personnel cost, service diversity, time deliveries, and procurement cost
		Environmental	CO_2 emission, and number of green products
		Social	Average reputation factor, and average customer satisfaction factor
Turi, Goncalves and Mocan (2014)	U.K. food industries	Economics	Transport cost, perfect order percentage, and total validity period in transport
		Environmental	Reverse logistics, reduce energy consumption, and CO_2
		Social	Number of employees trained, management level with specific environment responsibilities, and number of improvement suggestion submitted by employees

Some research only focuses on environmental and economic aspects in sustainable supply chain. They tried to link the environmental impact into economic value in their model. For instance, Hamprecht et al. (2005) developed a model to control the sustainability of upstream material flow in cereal and fresh milk industry. It adapted total quality management (TQM) to support the social development while following the environmental policy by monitoring the labour standards and the nutrition demands of the land. The result indicates the important variable to achieve the high quality fulfil the food safety regulation is traceability, Moreover, the incentives program for high quality farmers is also help to support the quality assurance. Other research also argue that sustainability has strong correlation with quality, temperature and energy (Zanoni & Zavanella, 2012). This model has been proposed to enhance the preservation process in chilled or frozen foods by linking economic and environmental dimensions. Another model that was proposed by Turi, Goncalves and Mocan (2014) considers quality, time, cost, logistics, and productivity to assess the sustainability in food industry.

Sustainability performance can be assessed by several ways. Economic Input-Output Life Cycle Assessment (EIO-LCA) and Data Envelopment Analysis that has been designed by Egilmez et al. (2014) can be applied to analyse the direct and indirect carbon footprints. In this model, Sustainability Performance Index (SPI) has been used as an input for policy makers to propose recommendations. Other model has been designed by Validi, Bhattacharya and Byrne (2014) to analyse the sustainability performance of two layers supply chain from manufacturer to drop off points of one dairy product supply chain. This study only focus on environmental perspectives by emerging trade-off between carbon emissions and distribution costs.

Furthermore, there are some research that considered all triple bottom line aspects in their proposed model. Sustainability can be assessed based on quality, responsiveness, flexibility, consumption, and total supply chain (Bourlakis et al. 2014). Comparing between different food firm sizes (micro, small, and medium enterprises), it shows that medium manufacturers are more mature in maintaining the sustainable supply chain, because they tend to play in local scopes that have fewer risks and under control, and they seem to be more responsive and flexible. However, small firms, except manufacturers, present better sustainability performance.

Yakovleva (2007) proposed a model that can be applied to benchmark different kind of agri-food companies in the UK based on economic, environment, and social dimensions. She recommended several indicator and analysed it by using spider diagram. In the current year, Yakovleva, Sarkis and Sloan (2012) applied another method for benchmarking. They use Analytic Hierarchy Process (AHP) and expert opinion to determine overall index of sustainability. However, this study has limitation since it is assumed that all variables are independent. Therefore, they suggested Analytic Network Process (ANP) and optimisation tool to be applied in the future study. Moreover, efficiency score of sustainable supply chain can be analysed by using Data Envelopment Analysis (DEA) (Tajbakhsh & Hassini, 2014). This model has the ability to analyse the individual and overall efficiency score of stakeholders in beverage industry from farmer to retailer.

In summary, this review highlights that not all of the studies cover all the three aspects of triple bottom line on models as well as all stages in the supply chain. Some studies also limit the case by assuming that all indicators are independent. In fact, there are many potential issues that still need to be investigated comprehensively.

Collaboration in agri-food supply chain management

Collaboration is needed in agri-food supply chain system in order to minimize cost, increase the profit, fulfil the quality assurance, and as the result is gaining the trust from consumers. Collaboration involves all activities such as production processes, sharing information and infrastructure, skills and knowledge among all stakeholders in the agri-food supply chain such as farmers, food manufacturers, distributors, retailers, consumers, government, NGOs, and finance providers. Each stakeholder has limitation that can be solved by conducting collaboration. This collaboration need strong commitment from all organization involved to achieve the common goal (Steele & Feyerherm, 2013).

However, there is a complexity in the application of collaboration, particularly integrated collaborations in a agri-food industry. Trust, Commitment, and willingness to share risks become the main keys in achieving a long term goal to create strong collaboration (Bezuidenhout, et al., 2012). Unfortunately, encouraging stakeholders to address these characteristics is not an easy task, particularly for food industry that has a complex system (Rota, et al., 2013). Moreover, the existence of global regulations, global trading, and emerging consumer preferences also bring another risk to the collaboration system (Matopoulos, et al., 2007).

Various studies on collaboration on agri-food supply chain have been conducted. There are some studies that only focus on vertical collaboration, horizontal collaboration, or considering both types of collaboration that is shown in Table 2.

Table-2. Research on Collaboration in Agri-food Supply Chain Management

Authors	Industries	Vertical Collaboration	Horizontal Collaboration
Masuku et al. (2003)	Swaziland sugar supply chain (growers-millers)	√	
Matopoulos et al. (2007)	Greece food catering (growers-processors)	√	
Proches and Bodhanya (2013)	South Africa sugar supply chain (upstream supply chain)	√	
Rota, Reynolds and Zanasi (2013)	Non-specific food industries (food manufacturers-retailers)	√	
Hildbrand and Bodhanya (2014)	South Africa sugar supply chain (growers-millers)	√	
Laham (2014)	Malaysian pineapple supply chain	√	

Van der Krogt, Nilsson and Høst (2007)	European dairy industries		√
Malaza and Myeni (2009)	Swaziland Sugarcane farmers		√
Ghaderi et al. (2012)	confectionary SMEs		√
McAdam et al. (2014)	artisan bakers' SMEs in the UK		√
Wright, Score & Conner (2007)	Michigan food system economic partnership	√	√
Weaver (2008)	Non-specific food industries	√	√
Gellynck and Kühne (2010)	traditional food supply chain in Belgium, Hungary, and Italy	√	√
Steele and Feyerherm (2013)	Canadian seafood supply chain	√	√
Warsanga (2014)	Tanzania banana supply chain	√	√
Leat and Revoredo-Giha (2013)	Scotland pork supply system	√	√

1. Vertical Collaboration

Various studies on vertical collaboration on agri-food supply chain have been conducted. In dyadic relationship between growers and millers in Swaziland sugar industry supply chain, organizational behaviour such as commitment, trust, and cooperation are important in supporting the contractual relationship (Masuku et al. 2003). Their research investigated that individual trust is more important than authorized relationship. On the other hand, in fresh product and food processing supply chain, it is important to combine macro factors (globalization, consolidation, consumers' attitude, and fixed regulations) and micro factors (industry's structure and product features) (Matopoulos, et al., 2007). These factors are linked with the two pillars of supply chain: design and the government of supply chain activities, and the establishment and the maintenance of supply chain relationship.

A theoretical framework, such as complexity theory can also be applied to analyse the collaboration among stakeholders in supply chain (Proches & Bodhanya, 2013). In the upstream supply chain interaction, that involving sugarcane farmers, hauliers, and millers, it indicates that different goals, interests, power levels, and perspectives become the obstacles to generate good collaboration. Other theoretical frameworks such as supply chain management, transaction cost economics, and resource based view theories can be used to analyse a sustainable relationship and collaboration (Rota, et al., 2013). The results suggest the key driver of sustainability that influences the performance of sustainability is vertical collaboration.

Another model, Viable System Model (VSM) can also be applied to analyse the complexity of supply chain collaboration (Hildbrand & Bodhanya, 2014). In the farmers-millers interaction, inefficiency is tend to be caused by soft issues and organizational behaviour rather than hard issue. Additionally, local autonomy is important in supporting farmer-local mills relationship. Furthermore, qualitative approach also can be applied to analyse supply chain vertical collaboration (Laham, 2014). He divided supply chain into two categories, upstream supply chain and downstream supply chain and analyses their different characteristics based on product characteristics, business relationships, business processes, positions in supply chain, and information sharing. The results indicate that the business relationship is the most important drivers in collaboration. Furthermore, a high level collaboration exists in upstream supply chain.

2. Horizontal Collaboration

Beside vertical collaboration, some studies analyse the relationship among stakeholders that play in the same level including external organizations that is called horizontal collaboration. These studies discuss the collaboration from different perspectives and variables.

In dairy industries, cooperative and non-cooperative organizations have differences preferences in applying horizontal collaboration (Van der Krogt, et al., 2007). Cooperative organizations prefer to join in joint ventures, mergers, general collaboration and licensing agreements. On the other hand, equity share holdings and acquisitions are more suitable for non-cooperative organizations. Cooperative strategies are more appropriate for industries that have limited equity capital, need minimum risk, and during the growth stage.

If horizontal collaboration is applied in sugarcane farmers' collaboration, strategic alliance, farm management, and ratoon management fund are important to overcome the common problems, such as high production costs, poor ratoon management, low income, and high levels of debt (Malaza & Myeni, 2009). It will enhance the productivity and competitiveness among local farmers. Horizontal collaboration can also be applied in reducing outbound transportation among confectionary SMEs participating in a logistic system consortium (Ghaderi, et al., 2012). For SMEs that have small orders can implement the horizontal collaboration by combining the load among the parties.

Horizontal collaboration is also can be applied to support the relationship among the food retailers (Nielsen, et al., 2014). It is identified that collaboration among retailers need enthusiasm, strong commitment, a high level of maturity, and willingness to override individual actions. Companies prefer to cooperate with an organization that has a similar size, structure, capabilities, and resources. However, it is difficult to get strong commitment in the collaboration and supporting the sustainability.

3. Combining Vertical and Horizontal Collaboration

Despite the fact that collaborations have been studied independently, some researchers realized that both vertical and horizontal collaborations are also important to be investigated comprehensively. Wright, Score and Conner (2007) investigated the collaboration among local government, universities, agri-business, and local societies. It shows that motivation from stakeholder to engage in collaboration form is important. The reasons are there is dependency of the community into agricultural aspects to achieve the prosperity and to support the socio economic development. Collaboration strategy also has the implication into pull innovation in agri-food supply chain to increase the value creation (Weaver, 2008). Vertical collaboration helps to gain better coordination within the supply chain stages in term of product design and development. However, the processing time can be reduced significantly by horizontal collaboration since innovation processes work parallel.

Another study discovered that both types of collaborations, vertical and horizontal, exist in the traditional food supply chain in Belgium, Hungary, and Italy (Gellynck & Kühne, 2010). The result indicates that vertical collaboration is well established. However, horizontal collaboration only exists if there are associations among organizations. The biggest barrier in creating collaboration form is lack of understanding, trust, knowledge, financial and physical resources.

The difficulties to enhance the fair collaboration form is also happened in banana supply chain in Tanzania (Warsanga, 2014). Farmers suffer the most cost and get the less benefit compare to others. It indicates that the collaboration is inefficient. To overcome this problem, horizontal collaboration among farmers is important by creating association organization. Furthermore, power sharing and agreement among stakeholders in the within vertical collaboration are essential to support good quality of collaboration.

Another study argues that external collaboration is more important than internal collaboration within organization for controlling performance achievement (Steele & Feyerherm, 2013). In seafood industry, sustainable organization can be achieved by strong commitment and organization's clarity. Furthermore, sustainable collaboration is also influenced by having good management systems and performance tracking. Moreover. risk management is important to be considered in achieving the flexibility through vertical and horizontal collaboration (Leat & Revoredo-Giha, 2013). It shows that the flexibility of the primary supplier has an impact on the supply chain resilience. Therefore, good collaboration on facilitate risk management among the stakeholders is important.

DISCUSSIONS

Based on literature review, it can be seen that even though the agri-food industry has a significant influence on social and economic developments globally, research on sustainable agri-food supply chain is limited. Some studies do not cover all sustainability aspects. Most of them focus on economic and environmental aspects and disregards social aspects that also important to build the strong social development in the communities. Even though, they paid attention in all sustainability aspects, economic indicators are also more dominant rather than others since this aspect are easier to be quantified. Moreover, some studies also limit the case by assuming that all indicators are independent. However, in the more reality, there is possibility that the indicators have the correlation to each other (Kahraman, et al., 2006). Therefore, it can be potential issues that can be explored broadly in the future by analysis the dependency among variable and sub-variable.

Furthermore, some studies only focus on one type of collaboration in sustainable agri-food supply chain. Often, these studies do not even consider all elements in the triple bottom line. They focus on what collaboration strategies that are applicable in different food sectors from different perspectives. However, even though the products are different, the nature of the research is quite similar. Most of the studies show that indicators of sustainability such as food safety, quality, nutrient, natural resources, labour issues, energy, and carbon footprint become the main issues that need to be addressed in several agri-food industries. Moreover, even though the problems are discovered from different perspectives, the collaboration structures among the agri-food industries are similar. Mainly, it consists of suppliers/farmers, food manufacturers, distributors, and retailers. In addition, related to collaboration behaviour communication becomes the main problem in collaboration, despite the fact that there are other slightly different aspects that need to be concerned. Communication will lead to positive relationship (Bezuidenhout, et al., 2012). In brief, in the future studies, integrated collaboration model that can accommodate all sustainable aspects is potential to be investigates comprehensively. Moreover, other organizational theories can be applied to analyse the complexity of the collaboration to propose the integrated collaboration model.

CONCLUSIONS

Sustainability and collaboration are concept that is important in supporting the complex system in agri-food supply chain management. Sustainability is one of perspectives that can be applied to maintain the competitive strategies in economic, environmental, and social aspects that is called triple bottom line (TBL). Several studies discuss all those element, however economic aspects are still become dominant variables that are considered in analyzing the sustainable supply chain. It is important to achieve a better sustainability system for all stakeholders without adversely suffering other stakeholders such as local farmer and SMEs by reducing the uncertainty and sharing the risk and costs among stakeholders involves.

Moreover, only few research that consider both vertical and horizontal collaboration as a network to achieve the positive competitive advantages. Collaboration as a network will support the long term partnership and spread the benefits throughout the entire supply chain system, from strategic level into operational level. Both concepts brings significant impact on agri-food industry in term of achieving socio-economic development as well as supporting the reduction of environmental impacts in the world. From this literature review, it can be concluded that it is important to develop model can accommodate both sustainability and integrated collaboration model is essential. For the further literature review, methods that have been used to develop sustainable supply chain and collaboration in supply chain management need to be analyzed comprehensively. Moreover, review about assessment method to examine the robustness of collaboration among stakeholders also need to be considered

REFERENCES

Ala-Harja, H & Helo, P 2014, 'Green supply chain decisions–case-based performance analysis from the food industry', *Transportation Research Part E: Logistics and Transportation Review,* vol. 69:97-107.

Barratt, M 2004, 'Understanding the meaning of collaboration in the supply chain', *Supply Chain Management: An International Journal,* vol.9(1):30-42.

Bezuidenhout, CN, Bodhanya, S & Brenchley, L 2012, 'An analysis of collaboration in a sugarcane production and processing supply chain', *British Food Journal,* vol.114(6):880-895.

Bourlakis, M, Maglaras, G, Aktas, E, Gallear, D & Fotopoulos, C 2014, 'Firm size and sustainable performance in food supply chains: Insights from Greek SMEs', *International Journal of Production Economics,* vol.152:112-130.

Brandenburg, M, Govindan, K, Sarkis, J & Seuring, S 2014, 'Quantitative models for sustainable supply chain management: Developments and directions', *European Journal of Operational Research,* vol.233(2):299-312.

Cambero, C & Sowlati, T 2014, 'Assessment and optimization of forest biomass supply chains from economic, social and environmental perspectives – A review of literature', *Renewable and Sustainable Energy Reviews,* vol.36:62-73.

Cao, M & Zhang, Q 2011, 'Supply chain collaboration: impact on collaborative advantage and firm performance', *Journal of Operations Management,* vol.29(3):163-180.

Dwi, Y, Nouaouri, I, Allaoui, H & Goncalves, G 2013, 'Sustainable supply chain and collaboration: What is the link and what are the benefits?', *2nd International Conference on Sustainable Intelligent Manufacturing, SIM 2013*, 631-636.

Egilmez, G, Kucukvar, M, Tatari, O & Bhutta, MKS 2014, 'Supply chain sustainability assessment of the US food manufacturing sectors: A life cycle-based frontier approach', *Resources, Conservation and Recycling,* vol.82:8-20.

Ernšteins, R, Lontone, A, Zvirbule, L & Kauliņš, J (eds) 2013, *Sustainable food governance development perspective on the local municipality level*, University of Latvia, Latvia.

Fritz, M & Schiefer, G 2008, 'Food chain management for sustainable food system development: a European research agenda', *Agribusiness,* vol.24(4):440-452.

Gellynck, X & Kühne, B 2010, 'Horizontal and vertical networks for innovation in the traditional food sector', *International Journal on Food System Dynamics,* vol.1(2):123-132.

Ghaderi, H, Darestani, SA, Leman, Z & Ismail, MY 2012, 'Horizontal collaboration in logistics: a feasible task for group purchasing', *International Journal of Procurement Management,* vol.5(1):43-54.

Gold, S, Hahn, R & Seuring, S 2013, 'Sustainable supply chain management in "Base of the pyramid" food projects—A path to triple bottom line approaches for multinationals?', *International Business Review,* vol.22(5):784-799.

Gosling, J, Jia, F, Gong, Y & Brown, S 2014, 'The role of supply chain leadership in the learning of sustainable practice: Toward an integrated framework', *Journal of Cleaner Production*, Available online 16 October 2014.

Hamprecht, J, Corsten, D, Noll, M & Meier, E 2005, 'Controlling the sustainability of food supply chains', *Supply Chain Management: An International Journal,* vol.10(1):7-10.

Hildbrand, S & Bodhanya, S 2014, 'Application of the viable system model in a complex sugarcane supply chain', *British Food Journal,* vol.116(12):2048-2068.

Kahraman, C, Ertay, T & Büyüközkan, G 2006, 'A fuzzy optimization model for QFD planning process using analytic network approach', *European Journal of Operational Research,* vol. 171(2):390-411.

Klassen, RD & Vereecke, A 2012, 'Social issues in supply chains: Capabilities link responsibility, risk (opportunity), and performance', *International Journal of Production Economics,* vol.140(1):103-115.

Laham, J 2014, 'Supply chain collaboration in the Malaysian pineapple chains', PhD thesis, School of Agriculture and Food Sciences, The University of Queensland.

Leat, P, Revoredo-Giha, C & Lamprinopoulou, C 2011, 'Scotland's food and drink policy discussion: Sustainability issues in the food supply chain', *Sustainability,* vol.3(4):605-631.

Leat, P & Revoredo-Giha, C 2013, 'Risk and resilience in agri-food supply chains: the case of the ASDA PorkLink supply chain in Scotland', *Supply Chain Management,* vol.18(2):219-231.

Li, D, Wang, X, Chan, HK & Manzini, R 2014, 'Sustainable food supply chain management', *International Journal of Production Economics,* vol.152:1-8.

Liao, S-H & Kuo, F-I 2014, 'The study of relationships between the collaboration for supply chain, supply chain capabilities and firm performance: A case of the Taiwan's TFT-LCD industry', *International Journal of Production Economics,* vol.156:295-304.

Lin, Y-H & Tseng, M-L 2014, 'Assessing the competitive priorities within sustainable supply chain management under uncertainty', *Journal of Cleaner Production,* Available online 17 July 2014.

Malaza, S & Myeni, D 2009, 'Integrating smallholder growers into competitive sugarcane production', *South African Sugarcane Technologists Association,* vol.82:405-408.

Maloni, MJ & Brown, ME 2006, 'Corporate social responsibility in the supply chain: an application in the food industry', *Journal of business ethics,* vol.68(1):35-52.

Masuku, MB, Kirsten, JF, Van Rooyen, CJ & Perret, S 2003, 'Contractual relationships between small-holder sugarcane growers and millers in the sugar industry supply chain in Swaziland', *Agrekon: Agricultural Economics Research, Policy and Practice in Southern Africa,* vol.42(3):183-199.

Matopoulos, A, Doukidis, GI, Vlachopoulou, M, Manthou, V & Manos, B 2007, 'A conceptual framework for supply chain collaboration: Empirical evidence from the agri-food industry', *Supply Chain Management: An International Journal,* vol.12(3):177-186.

McAdam, M, McAdam, R, Dunn, A & McCall, C 2014, 'Development of small and medium-sized enterprise horizontal innovation networks: UK agri-food sector study', *International Small Business Journal,* vol.32(7):830-853.

Mohezar, S & Nor, MNM 2014, 'Could supply chain technology improve food operators' innovativeness? A developing country's perspective', *Trends in Food Science & Technology,* vol.38(1):75-82.

Mota, B, Gomes, MI, Carvalho, A & Barbosa-Povoa, AP 2014, 'Towards supply chain sustainability: economic, environmental and social design and planning', *Journal of Cleaner Production*, Available online 1 August 2014.

Nielsen, FAG, Angeli, F, Luppe, MR & Giesbrecht da Silveira, JA 2014, 'Horizontal alliances among independents and non-integrated food retailers: A worldwide analysis ', *International Journal of Business & Commerce,* vol.3(10):21-50.

Proches, CNG & Bodhanya, S 2013, 'An analysis of multi-stakeholder interactions in the sugar industry using a social complexity framework', *Problems and Perspectives in Management,* vol.11(4):77-85.

Ramanathan, U, Bentley, Y & Pang, G 2014, 'The role of collaboration in the UK green supply chains: An exploratory study of the perspectives of suppliers, logistics and retailers', *Journal of Cleaner Production,* vol.70:231-241.

Rota, C, Reynolds, N & Zanasi, C 2013, 'Sustainable food supply chains: The role of collaboration and sustainable relationships', *International Journal of Business and Social Science,* vol.4(4):45-53.

Rusono, N, Suanri, A, Candradijaya, A, Muharam, A, Tejaningsih, IM, Hadi, PU, Susilowati, SH & Maulana, M 2014, *Studi pendahuluan: Rencana Pembangunan Jangka Menengah Nasional (RPJMN) bidang pangan dan pertanian 2015-2019,* Direktorat Pangan dan Pertanian,Bappenas, Jakarta, Indonesia.

Sanders, NR 2012, *Supply chain management : A global perspective*, John Wiley & Sons, Hoboken, NJ.
Shokri, A, Oglethorpe, D & Nabhani, F 2014, 'Evaluating sustainability in the UK fast food supply chain', *Journal of Manufacturing Technology Management,* vol.25(8):1224-1244.

Steele, B & Feyerherm, A 2013, 'Loblaw sustainable seafood: Transforming the seafood supply shain through network development and collaboration', *Organizing for Sustainable Effectiveness,* vol.3:101-132.

Tajbakhsh, A & Hassini, E 2014, 'A data envelopment analysis approach to evaluate sustainability in supply chain networks', *Journal of Cleaner Production*, Available online 9 Agustus 2014.

Turi, A, Goncalves, G & Mocan, M 2014, 'Challenges and competitiveness indicators for the sustainable development of the supply chain in food industry', *Procedia - Social and Behavioral Sciences,* vol.124:133-141.

Vachon, S & Mao, Z 2008, 'Linking supply chain strength to sustainable development: A country-level analysis', *Journal of Cleaner Production,* vol.16(15):1552-1560.

Validi, S, Bhattacharya, A & Byrne, PJ 2014, 'A case analysis of a sustainable food supply chain distribution system—A multi-objective approach', *International Journal of Production Economics,* vol.152:71-87.

Van der Krogt, D, Nilsson, J & Høst, V 2007, 'The impact of cooperatives' risk aversion and equity capital constraints on their inter-firm consolidation and collaboration strategies—with an empirical study of the European dairy industry', *Agribusiness,* vol.23(4):453-472.

Varsei, M, Soosay, CA, Fahimnia, B & Sarkis, J 2014, 'Framing sustainability performance of supply chains with multidimensional indicators', *Supply Chain Management: An International Journal,* vol.19(3):242-257.

Wang, PC, Halim, I, Adhitya, A & Srinivasan, R 2011, 'Integrating economic, environmental and social indicators for sustainable supply chains', in E.N. Pistikopoulos, MCG & Kokossis, AC (eds), *Computer Aided Chemical Engineering*, Elsevier, 1220-1224.

DOI: 10.1051/matecconf/20165802004

Warsanga, WB 2014, 'Coordination and structure of agri-food value chains: Analysis of banana value chain strands in Tanzania', *Journal of Economics and Sustainable Development,* vol.5(7):71-78.

Weaver, RD 2008, 'Collaborative pull innovation: Origins and adoption in the new economy', *Agribusiness,* vol.24(3):388-402.

Wright, W, Score, M & Conner, DS 2007, 'Food system makers: Motivational frames for catalyzing agri-food development through multi-stakeholder collaboration', *Community Development,* vol.38(3):39-59.

Yakovleva, N 2007, 'Measuring the sustainability of the food supply chain: A case study of the UK', *Journal of Environmental Policy & Planning,* vol.9(1):75-100.

Yakovleva, N, Sarkis, J & Sloan, T 2010, 'Sustainability indicators for the food supply chain', Woodhead Publishing Limited, 297-329.

Yakovleva, N, Sarkis, J & Sloan, T 2012, 'Sustainable benchmarking of supply chains: The case of the food industry', *International Journal of Production Research,* vol.50(5):1297-1317.

Yan, X-h & Ma, J 2012, 'Development of sustainable food supply chain', *International Journal of Engineering & Technology,* vol.4(2):166-169.

Yu, M & Nagurney, A 2013, 'Competitive food supply chain networks with application to fresh produce', *European Journal of Operational Research,* vol.224(2):273-282.

Zanoni, S & Zavanella, L 2012, 'Chilled or frozen? Decision strategies for sustainable food supply chains', *International Journal of Production Economics,* vol.140(2):731-736.

DISTRIBUTION RELIABILITY OF THE ENERGY SUPPLY CHAIN

Rachmad Hidayat[1], Sabarudin Akhmad[2]

[1,2]Industrial Engineering Department, University of Trunojoyo Madura, Indonesia
E-mail: rachmad.trunojoyo@gmail.com

ABSTRACT

Reliability of an oil fuel distribution system can be achieved by considering multiple attributes. Load-point indicators consisted of the frequency of failures, average duration of an outage and average annual outage time. Indicators of system performance were SAIFI and SAIDI and the method of the study was the Learning Vector Quantization (LVQ). Decisions were made by considering the multi-attribute oil fuel distribution system to determine high reliability, which was simulated on the agent. The use of agents in the simulation to determine the oil fuel distribution system reliability helped visualize the oil fuel distribution in meeting the quality of fuels received by consumers and made it easier to learn the problems faced by oil fuel distribution.

Keywords: oil fuel distribution, LVQ, agent, reliability

INTRODUCTION

Rising world oil prices have tremendous impacts on Indonesia's economy. The burden of oil fuel subsidy borne by the government is becoming increasingly high. This is compounded by the sub-optimal fuel distribution system. The risk of distribution fraud is very likely to occur at the level of distributor or retailer. Among those forms of fraud are misallocation of oil fuels, sales of oil fuels for households or individuals to industries and smuggling of subsidized oil fuels abroad. (Ardiansyah, *et.al.*, 2012). Oil fuels constitute a necessity which concerns the lives of many people. Oil fuel distribution requires transparency, necessitating public disclosure of distribution data as a form of the operator's accountability. A breakthrough should be found for smooth, transparent and traceable process of oil fuel distribution. The process of oil fuel distribution starts from the supply to the end user, the oil fuel consumers. The quality of oil fuels received by consumers is influenced the distribution system; therefore, a reliable oil fuel distribution system is required. Reliability of fuel distribution system is measured by the oil fuel availability and service level from the system to the end-users. (Simões, *et. al.*, 2011). The benchmark of distribution system reliability includes how often the distribution system experiences a delay, how long a delay occurs and how quickly is the distribution system to recover to its previous normal condition. (Portugal and Esteban, 2014). A highly reliable system will be able to distribute oil fuels any time as needed, while a system is said to be poorly reliable when it provides a low level of oil fuel availability.

A highly reliable distribution system is indicated by the load points and system performance, consisting of the frequency *of failures* (λ), *average duration of a failure* (*r*), *average annual outage time* (*U*). (González *et. al.*, 2009). Based on these indications, multiple attributes can be considered to achieve a highly reliable distribution system, such as the load points, SAIFI and SAIDI. Subsequently, the multiple attributes are subjected to learning by the use of Learning Vector Quantization (LVQ) in order to automatically classify the given input vectors. Among the advantages of LVQ is the method's ability to provide learning to the competitive layers so that it can automatically classify the given input vectors (Fujiki and Kazuko, 2009).

The process for determining a highly reliable oil fuel distribution system is further simulated on an agent. An agent can help determine a highly reliable oil fuel distribution system with by considering multiple attributes. An agent is a component or individual capable of responding to the environment by considering many attributes and of adaptation so that they can make independent decisions. Reliability is defined as the ability of a component or system to perform the functions required in a particular environment and operational condition for a predetermined period of time (Ahmad and Kamaruddin, 2012). Thus, reliability constitutes one of the aspects capable of affecting the success of production process. Reliability is of paramount importance since it will affect the cost of maintenance, which in turn will affect the profitability of the company.

Reliability of the distribution system is a measure of oil fuel availability or service level of oil fuel provision from the system to the end-users. The measure of reliability can be expressed as how often the distribution system experiences a delay, how long a delay occurs and how quickly is the distribution system to recover to its previous normal condition. A highly reliable system will be able to provide oil fuel any time when needed, while a poorly reliable system is frequently late to distribute oil fuel. The service continuity level of the means of distribution is structured on the basis of length of the efforts to re-establish the supply after a downtime due to interruptions (Dhillon, 2005; González *et. al*, 2009).

Oil fuel distribution is part of the Energy Supply Chain, since the quality levels of oil fuels distributed from the refinery to the consumer are determined by the distribution system, particularly the reliability of the oil fuel distribution system. Distribution of oil fuels is influenced by multiple aspects, including minimizing losses, minimizing cost, improving reliability of supply, and satisfying operational constraints (Dhillon, 2005). Attribute planning for determining the reliability of an oil fuel distribution system is based on the load point indeices consisting of λ, *r* and *U* and the system performance index consisting of SAIFI and SAIDI can be seen in Figure 1.

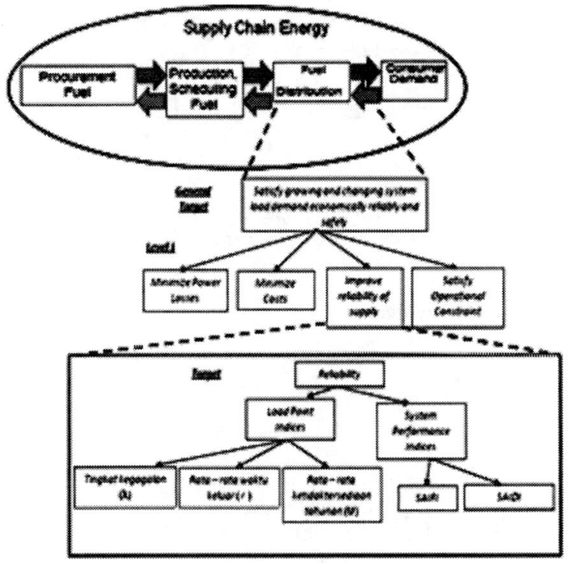

Figure-1. Oil fuel distribution in the Energy Supply Chain

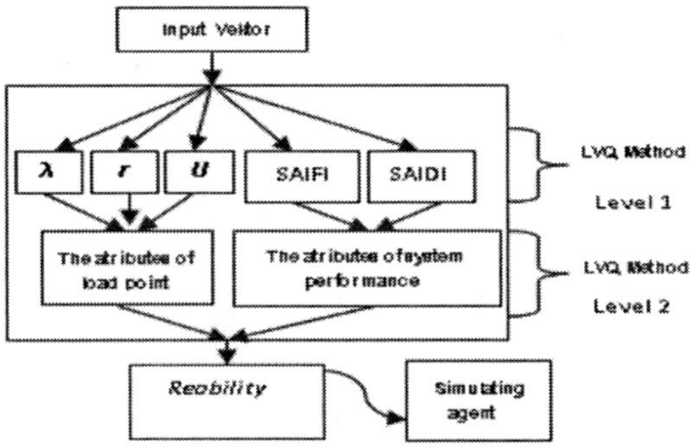

Figure-2. Flowchart of the study

A classification process is required to determine the reliability of the distribution system by the use of the attributes of load points and system performance index since the attributes of load point index has input vectors of λ, r, and U and the attributes of system performance index has input vectors of SAIFI and SAIDI. Classification flowchart such as figure 2.

Classification of multiple attributes to determine the reliability of the distribution system using LVQ can be seen in detail in Figures 3, 4 and 5. The classification steps of the attributes of load point index and system performance index until the shortest distance from the input vectors to the class is obtained.

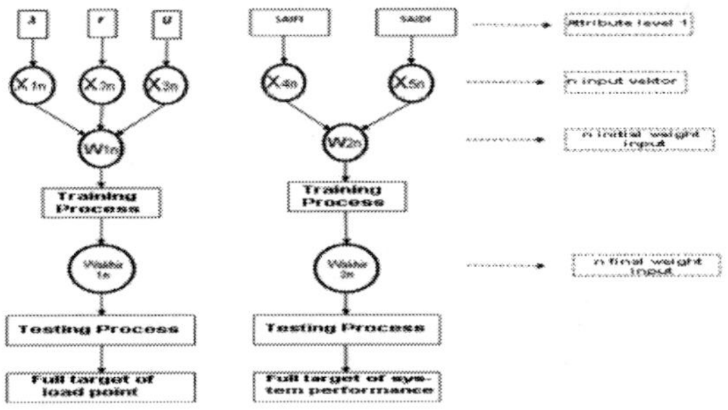

Figure-3. Flowchart of study step for attribute level 1

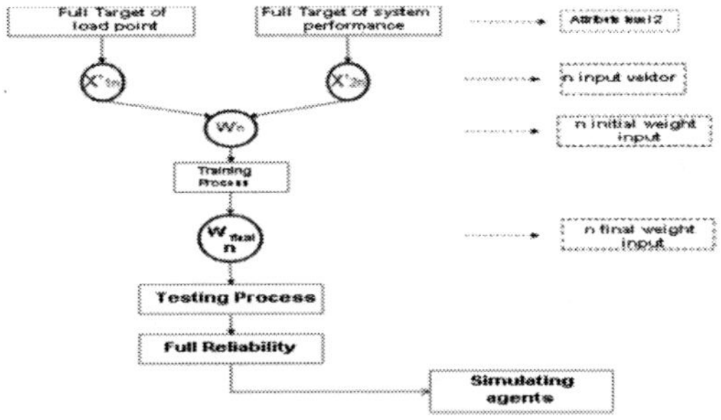

Figure-4. Flowchart of study step for attribute level

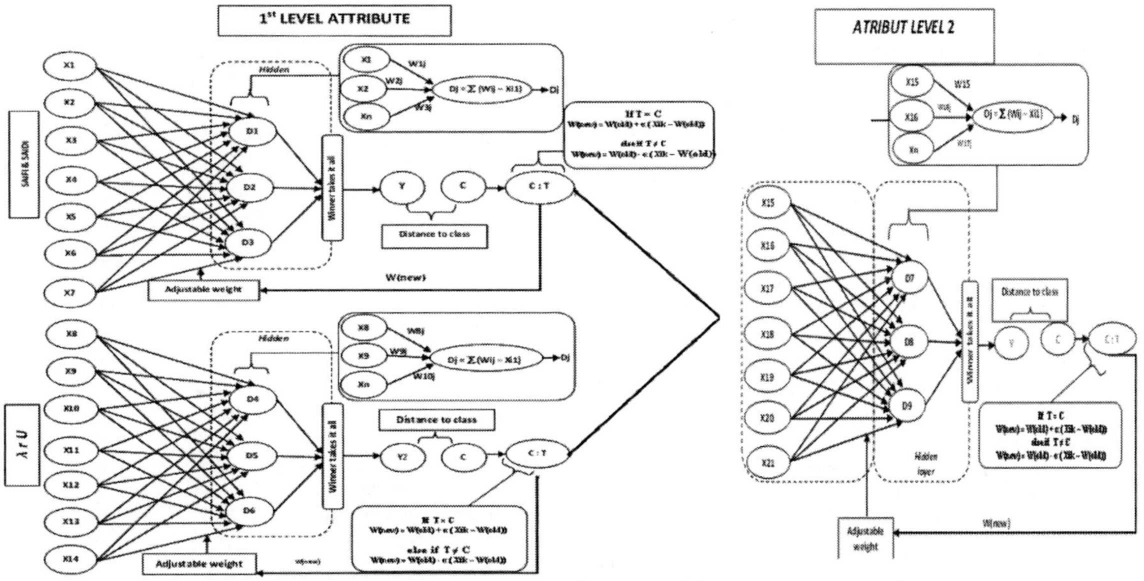

Figure-5. Process for attributes level 1 and attributes level 2

METHODOLOGY

Reliability Index

Reliability index is an indicator of reliability expressed in a probability scale. Load point failure indices typically used include the frequency of failure λ (failure/month), the average duration of an outage r (h/failure) and the average annual outage time U (h/year). The average value of the three basic load point indices for load point j can be calculated from the (up-down) operation history of the load point by using the following equation (Díaz and Márquez, (2014).

$$\lambda = \frac{N_j}{\sum T_{uj}} \quad (1)$$

$$r_j = \frac{\sum T_{dj}}{N_j} \pi r^2 \quad (2)$$

$$U_j = \frac{\sum T_{dj}}{\sum T_{uj} + \sum T_{dj}} \quad (3)$$

$\sum T_{uj}$ = uptime for load point j

$\sum T_{dj}$ = downtime for load point j

N_j = number of failures during the simulation period for load point j

And based on Viktoria Neimane (Crespo and Benoit, 2007) to obtain the frequency of failure λ (failure/year), the average duration of outage r (h/failure) and the average annual outage time U (h/year) as follows:

$$\lambda_s = \lambda_1 + \lambda_2 + \cdots . \lambda_n \quad (4)$$

$$r_s = \frac{\lambda_1 . r_1 + \lambda_2 . r_2 + \cdots . + \lambda_n . r_n}{\lambda_1 + \lambda_2 + \ldots + . \lambda_n} \quad (5)$$

$$U_s = \lambda_s . r_s = \lambda_1 . r_1 + \lambda_2 . r_2 + \cdots + \lambda_n . r_n \quad (6)$$

System Average Interruption Frequency Index (SAIFI) is defined as the average number of failures per customer served by the system per unit of time (usually a year). (Fujiki and Kazuko, 2009).

$$\text{SAIFI} = \frac{\sum \lambda k . Mk}{\sum M} \quad (7)$$

λ_k = outage rate of the component

M_k = number of customers at load point k

M = total customers in the distribution system

System Average Interruption Duration Index (SAIDI) is defined as the average duration of failures for each consumer for one year. This index is determined by dividing the sum of continuous duration of failures for all customers during the specified period of time by the number of customers served during the year. The equation for SAIDI (average duration of interruption for each customer) can be seen in the following equation: (Fujiki and Kazuko, 2009).

$$SAIDI = \frac{\sum U_k . M_k}{\sum M} \quad (8)$$

U_k = component repair time

M_k = number of customers at load point k

M = total customers in the distribution system

The classification steps of the attributes of load point index and system performance index are described as follows: (Kate *et. al.*, 2001; Robi, 2001; Olden *et. al.*, 2004).

1. Determine the input vectors for each attribute;
2. Specify the initial weight randomly;
3. Determine the initial class as the target class in the training process of the input vectors.
4. During the learning process in the hidden layers, find the minimum distance to the input vectors of predetermined initial classes by using the *Euclidean* distance formula. Dj = ∑ (Wij – Xi) (9)
5. Once the minimum distance in the learning process has been obtained, the point is said to be the winner and the output, i.e., the minimum distance of the input vector to the target class is symbolized as the winner class.
6. Compare the winner class to the target class to find using the new weight by using the formula:
   if
   T = C
   W(new) = W(old) + α (Xik – W(old)) (10)
   else if
   T ≠ C
   W(new) = W(old) - α (Xik – W(old) (11)

   Results obtained, i.e., the new weight W, are then used for trials.

Trials are continuously carried out with the term 'epoch' and will stop when the learning rate reaches a minimum value. The output of the process will be a minimum distance of the input vector to the class, in which the minimum distance data is simulated on the agent.

LVQ is a method that is based on an unsupervised learning algorithm. LVQ can also be used as a supervised vector quantizer since LVQ network has class points linked to the input vectors. (Kate *et. al.*, 2001).

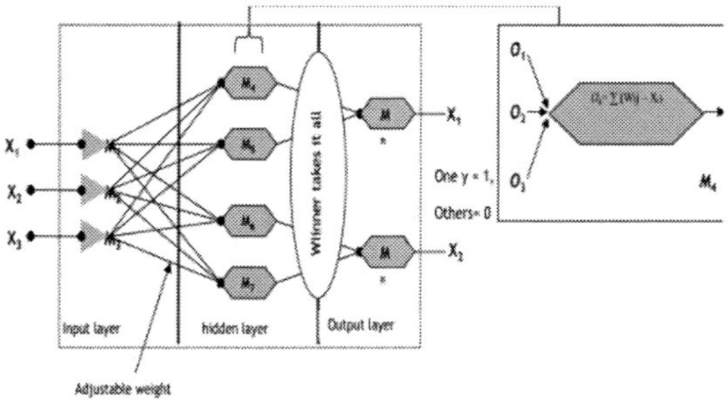

Figure-6. LVQ architecture: One hidden layer with kohonen neurons, adjustable weights between the input layer and the hidden layer (www.neural-forcasting.com)

The LVQ algorithm is as follows: (Sivanandam and Deepa, 2006)

1. Initialize weights and learning rate;
2. It will stop if it is incorrect, do steps 3–7;
3. For training input vector x, do steps 4–5
4. Calculate j using the Euclidean distance formula
   D(j)=∑(wij – xi)²
   find j when D (j) is at its minimum
5. Update wj by follow the following requirements
   If T=Cj, then
   wj(new) = wj(old) + α[x - wj(old)]
   If T ≠ Cj, then
   wj(new) = wj(old) - α[x - wj(old)]
6. Calculate the error of the weighting values
   e(t+1) = ||W(t+1) – W(t)||
7. Reduce the learning rate
8. Perform trials for stop condition; the condition will stop when the iteration rate or learning rate reaches the smallest value.

FSM AND AGENT SCENARIOS

FSM (Finite-State Machine) of the agent's simulated oil fuel distribution reliability is as figure 7. Agent scenarios based on the FSM in the simulation to determine the reliability of oil fuel distribution system is as follows:

1. Agents scan feeder (supply) 1 through feeder 7 to read the data of each feeder;
2. Subsequently, agents compare the data of the distance from the input vector to the class in all feeders;
3. Agents determine the shortest distance from the input vector to class 1 (class with the lowest possible interruption);
4. Then, agents block or mark green the feeder;

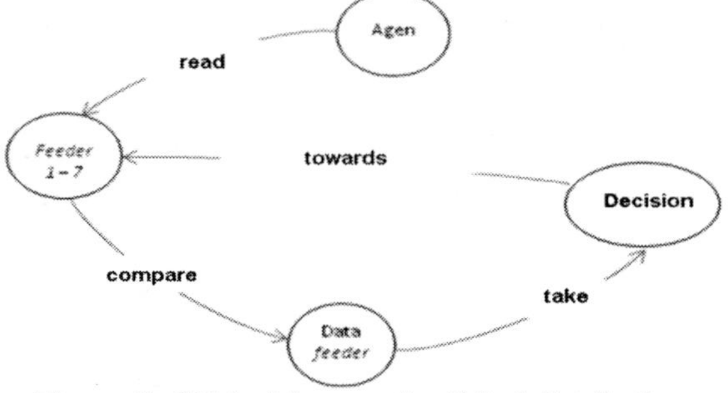

Figure-7. FSM of the agent's oil fuel distribution reliability

TRIAL RESULTS

Trial 1

1. Indication values of load points are derived from the total LP (load point) of each feeder on BUS 4;
2. Values of SAIFI and SAIDI are calculated using equations 7 and 8 based on data;
3. Class inputs for load point indications and SAIFI SAIDI are determined randomly as in Tables 1-2;
4. Total data = 7
5. Number of the initial weight of each attribute = 3
6. Number of class of each attribute = 3
7. Max epoch = 10000000
8. Based on case (A): disconnect – fuses – alternative supply – repair transformer.

Table-1. Data of load point and initial class indications

Feeder	λ	*r*	U	Class
F1	1.343	157.25	30.14	1
F2	0.35	28.85	3.36	2
F3	1.299	161.55	30.01	3
F4	1.558	176.13	34.34	1
F5	0.344	29.19	3.36	2
F6	0.36	26.51	3.18	3
F7	1.313	160.07	30.01	1

Table-2. Data of system performance and initial class indications for the case A

Feeder	SAIFI	SAIDI	Kelas
F1	0.191	4.29	1
F2	0.117	1.12	2
F3	0.186	4.30	3
F4	0.195	4.30	1
F5	0.115	1.12	2
F6	0.120	1.06	3
F7	0.188	4.31	1

Results obtained from the attribute level 1 for SAIFI and SAIDI (chain 1) ;

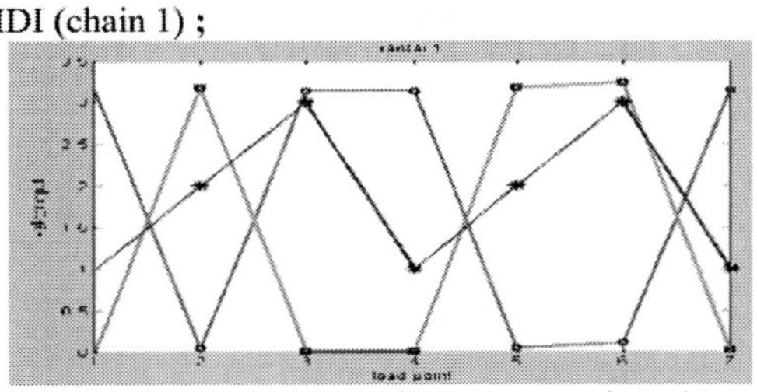

Figure-8. SAIFI and SAIDI input vector chart (case A)

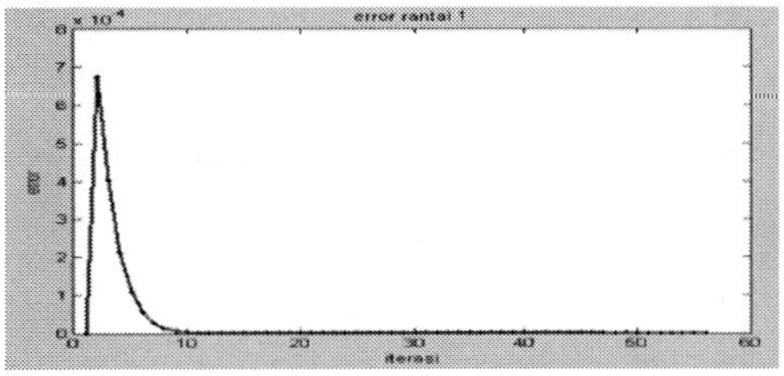

Figure-9. SAIFI and SAIDI input vector error chart (case A)

********** Chain 1 **********
Initial weight matrix
0.1910 0.1170 0.1860
4.2900 1.1200 4.3000
weight matrix after epochs to 56
0.1925 0.1127 0.1762
4.2985 1.1651 4.2940
Distance after epochs to 56
0.0087 3.1258 0.0154
3.1794 0.0453 3.1746
0.0067 3.1357 0.0115
0.0029 3.1359 0.0198
3.1795 0.0452 3.1746
3.2393 0.1054 3.2345
0.0123 3.1458 0.0199
The initial class
1 2 3 1 2 3 1
The bafter Training Class
1 2 1 1 2 2 1

Results obtained from the attribute level 1 for λ, *r* and *U* (chain 2);

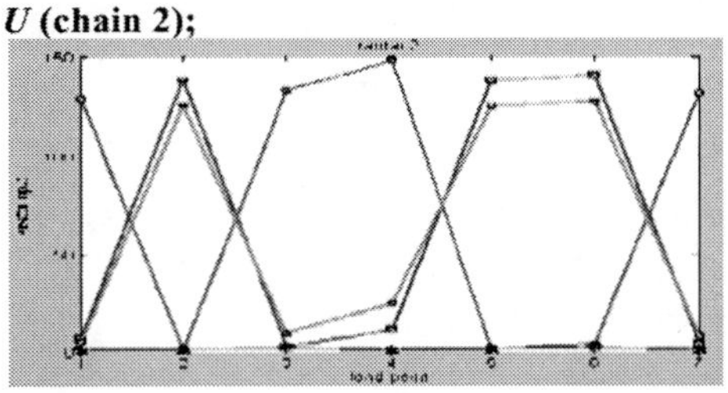

Figure-10. Input vector chart for λ, *r* and *U* (case A)

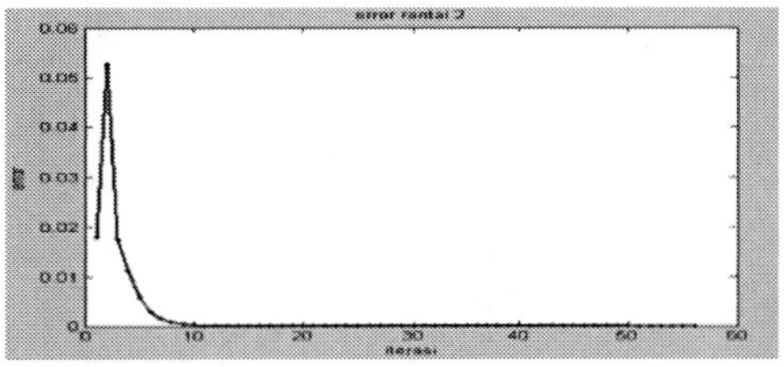

Figure-11. Error input vector chart for λ, *r* and *U* (case A)

Simulation results Chain 2 (case A)
********** Chain 2 **********
Initial weight matrix
1.3430 0.3500 1.2990
157.2500 28.8500 161.5500
30.1400 3.3600 30.0100
weight matrix after epochs to 56
1.4179 0.3364 1.1122
164.6319 30.9523 152.1007
31.6555 3.4954 26.9819
Distance after epochs to 56
7.5363 129.0816 6.0450
138.7030 2.1067 125.4963
3.4958 133.2656 9.9244
11.8081 148.4232 25.1346
138.3702 1.7675 125.1624
141.0307 4.4535 127.8285
4.8508 131.8156 8.5276
The initial class
1 2 3 1 2 3 1
The bafter Training Class
3 2 1 1 2 2 1

Results obtained from the attribute level 2 (chain 3);

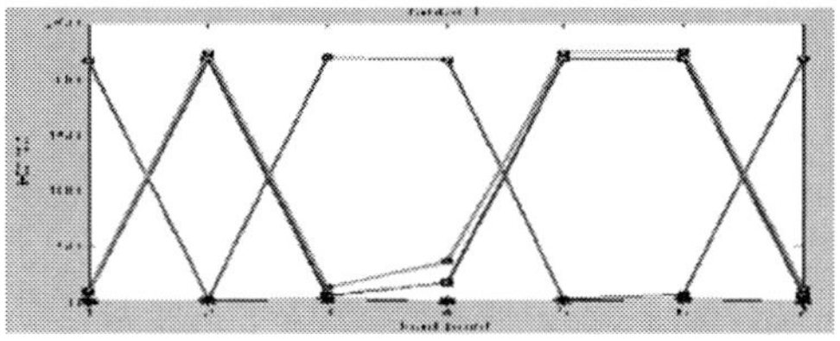

Figure-12. Input vector chart of the results of load point indications & system performance indications (caseA)

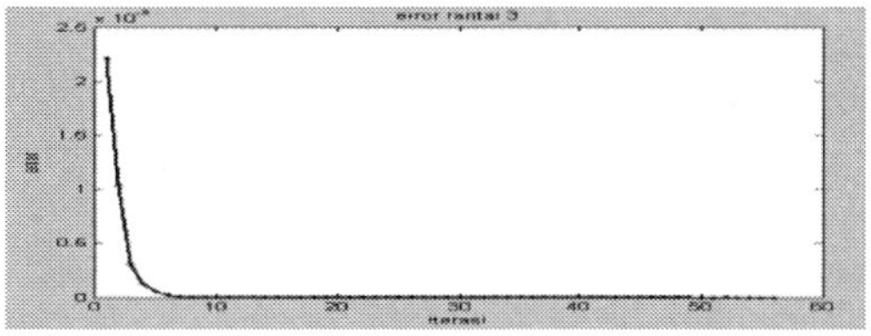

Figure-13. Input vector error chart of the results of load point indications & system performance indications (case A)

Simulation results of all the input vectors (case A)
********** Chain 3 **********
Initial weight matrix
0.0087 3.1794 0.0067
3.1258 0.0453 3.1357
0.0154 3.1746 0.0115
7.5363 138.7030 3.4958
129.0816 2.1067 133.2656
6.0450 125.4963 9.9244
weight matrix after epochs ke 56
0.0093 3.1344 0.0089
3.1371 0.0000 3.1356
0.0203 3.1295 0.0066
8.6258 136.6172 -1.4916
135.7772 0.0001 124.1710
12.6068 123.4060 0.7983
Distance after epochs ke 56
j class 1 j class 2 j class 2
9.4379 217.0870 11.5389
218.0859 3.6281 223.9050
6.3104 219.9737 13.8156
18.0830 217.4703 36.8417
217.9233 3.0476 223.6963
219.2783 7.6748 225.4143
6.8253 219.0071 12.5861
The initial class
1 2 3 1 2 3 1
The bafter Training Class
1 2 1 1 2 2 1

Analysis of Trial 1

Trials for the attributes level 1 of chain 1 with SAIFI and SAIDI input vectors produced a feeder with the shortest distance from the input vector to class 1 or the class with high reliability. The feeder was feeder 4 which had a distance of the input vector to class 1 of 0.0029; in other words, feeder 4 had higher reliability as an oil fuel distribution system than other feeders. The next trials for the attributes level 1 of chain 2 for the input vectors λ, *r* and U produced feeder 3 as one that had high reliability as an oil fuel distribution system. It was due to the fact that feeder 3 had the shortest distance from the input vector to class 1 of 3.4958.

Trials for chain 1 (SAIFI and SAIDI) and chain 2 (λ, *r* and U) were followed by classification of the attributes level 2 of chain 3, or the attributes with the input vectors derived from classification of chains 1 and 2. Classification of the attributes level 2 of chain 3 was the last step of classification for case A, taking into account multiple attributes and using LVQ as the method of classification and obtained feeder 3 of 6.3104 as one with the shortest distance from the input vector to class 1.

Results of the Overall Trials

Trials 1 to 6 for case A to case F on feeder 1 to feeder 7 showed the Table 3. Input vector distance data from all the attributes of all the cases to class 1 of *each feeder*. Multi-attribute classification was conducted on three classes:

Class 1 = good, representing feeders with high system reliability
Class 2 = moderate, representing feeders with medium system reliability
Class 3 = poor, representing feeders with low system reliability

Determination of the distribution system reliability as high, medium, low is shown only for feeders with the input vector distance for all cases approaching class 1 (high system reliability) as shown in figure 13. Reliability was also determined for attribute data consisting of load point indices and system performance indices. Furthermore, the data were simulated on the agents.

Table-3. Input vector distance of all feeders to class 1

Feeders	*Case*					
	A	B	C	D	E	F
1	9.4379	9.6142	9.4381	9.4736	9.4379	10.0074
2	218.0859	222.6989	218.1668	218.2161	218.0188	219.0970
3	6.3104	6.6961	6.3168	6.3643	6.3104	7.3195
4	18.0830	18.4926	18.0834	18.0994	18.0830	18.0878
5	217.9233	222.5497	218.0080	218.0537	217.8560	218.9338
6	219.2783	223.8759	219.3514	219.4085	219.2107	220.2855
7	6.8253	7.1523	6.8277	6.8319	6.8253	6.8354

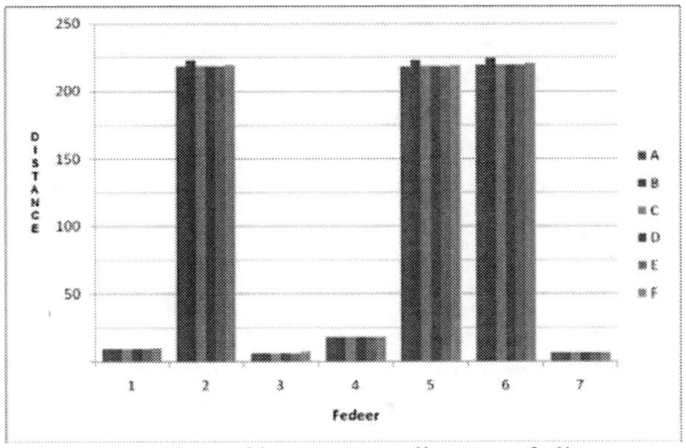

Figure-14. Chart of input vector distance of all cases to class 1

Agent Simulation Flowchart

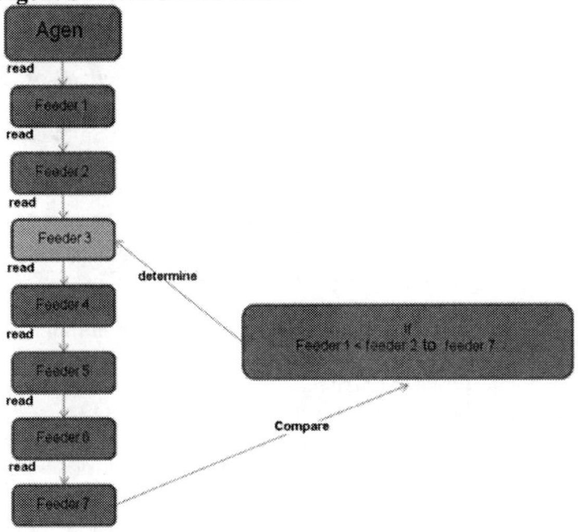

Figure-15. Agent simulation flowchart

CONCLUSION

Trials performed to determine the reliability of the oil fuel distribution system using LVQ (Learning Vector Quantization) showed that an oil fuel distribution system can be considered reliable when the interruption value was low. Utilization of more detailed attributes would be able to determine the reliability of the distribution system more accurately. This was due to the fact that oil fuel distribution system reliability is influenced by multiple aspects that can be used as attributes. The use of simulation in some problems is of paramount importance since it helps understand them visually. In particular, agent-based simulations for oil fuel distribution system reliability can visually help determine a reliable oil fuel distribution system. Future research should emphasize the determination of the initial weights of input vectors from LVQ method since initial weighting is essential to achieve classification accuracy. Future research should consider the use of Multi-Layer LVQ on each hidden layer capable of affecting the degree of accuracy in order to obtain better classification results.

REFERENCES

Ahmad, R., & Kamaruddin, S. (2012). An overview of time-based and condition-based maintenance in industrial application. *Computers & Industrial Engineering*, *63*(1), 135-149.

Ardiansyah, F., Gunningham, N., & Drahos, P. (2012). An environmental perspective on energy development in Indonesia. In *Energy and non-traditional security (NTS) in Asia* (pp. 89-117). Springer Berlin Heidelberg.

Crespo A. Márquez, Benoit Iung. 2007. A Estructured Approach for the Assessment of System Availability and Reliability Using Montecarlo Simulation. Journal of Quality in Maintenance Engineering. 13 (2): 125-136.

Dhillon, B.S. 2005. Reliability, Quality, and Safety for Engineers, CRC Press, USA

Díaz, V. G. P., & Márquez, A. C. (2014). On the Assessment and Control. In *After–sales Service of Engineering Industrial Assets* (pp. 175-211). Springer International Publishing.

Fujiki Morii, Kazuko Kurahashi. 2009. Clustering Based on Multiple Criteria For LVQ and K-Means Algorithm, Journal of Advanced Computational Intelligence and Intelligent Informatics. 13: 360-370.

González Díaz, A. Crespo, P. Moreu, J. Gómez, C. Parra. 2009. Availability and reliability assessment of industrial complex systems: A practical view applied on a bioethanol plant simulation. Safety and Reliability for Managing Risk. Taylor & Francis Group: 687-695.

Kate A. Smith, Frederick Woo, Vic Ciesielski and Remzi Ibrahim. 2001. Modeling the Relationship Between Problem Characteristics and Data Mining Algorithm Performance using Neural Networks. Intelligent Engineering Systems Through Artificial Neural Networks. 11: 356-362.

Olden, Julian D., Michael K. Joy, and Russell G. Death. 2004. An accurate comparison of methods for quantifying variable importance in artificial neural networks using simulated data. *Modelling* 178 (3): 389-397.

Portugal-Pereira, J., & Esteban, M. (2014). Implications of paradigm shift in Japan's electricity security of supply: A multi-dimensional indicator assessment. Applied Energy, 123, 424-434.

Robi Polikar. 2001. Learn++: An Incremental Learning Algorithm for Supervised Neural Networks. IEEE Transaction on System, Man, and Cybertics Aplications and Reviews. 31: 497-508.,

Sivanandam S. N., Sumathi and Deepa. 2006. Introduction to Neural Networks Using Matlab 6.0. Tata McGraw-Hill Education.

Simões, M. G., Roche, R., Kyriakides, E., Miraoui, A., Blunier, B., McBee, K., and Ribeiro, P. (2011, September). Smart-grid technologies and progress in Europe and the USA. In Energy Conversion Congress and Exposition (ECCE), 2011 IEEE (pp. 383-390). IEEE.

POLICY STRATEGIES OF COCOA FOR LEAD UP AGROINDUSTRIAL FOOD AND DRINKS IN JEMBER REGENCY, INDONESIA

Eko Nurhadi [1], Syarif Imam Hidayat[2], Pawana Nur Indah[3], Sri Widayanti [4]
Agribusiness Department, Faculty Of Agriculture, University of Pembangunan Nasional "Veteran", Surabaya, East Java, Indonesia
E-mail:ekonhadi@yahoo.com

ABSTRACT

Cocoa is one of the commodity plantation for foreign exchange, the source of income of farmers, agro-industries, and regional development. Indonesia is currently the third largest country in the world of cocoa production with a 15 percent share of production after Ghana (16%) and Ivory Coast (40%). The research aims to analyze the potential for superior Cocoa products in order to accelerate economic growth and development of the region. This study was conducted in Rambipuji and Wuluhan District in Jember, East Java. The study population is bringing cocoa households (RTP) and sampling using random sampling. Cocoa in the study site has a comparative advantage indicated by the magnitude of the greatly enhanced by DRCR <1 and the competitive advantage represented by the PCR value of 0.5803. The research location is also the location of the cocoa plant base that could potentially seed.

Keywords: Characteristics of cocoa Farmers, Agricultural System, Cocoa beans, food, drink.

INTRODUCTION

Cocoa is one of the mainstay plantation commodities as a major foreign exchange earner, the source of the income of farmers, job creation for farmers, to encourage agribusiness and agro-industry and regional development. Indonesia is now in the production of third-largest cocoa world with a share production of 15% after Ghana (16%) and Ivory Coast (40%) (Brandt, 2007)

The study revealed that, even though the quality is important to all categories of actors in the cocoa sector, interactions among them are hampered by problems of information asymmetry that result especially in farmers evading recommended practices. While cocoa sector policies ensure the export of premium quality cocoa, they have not sufficiently alleviated the information problem especially in the relation between farmers and cocoa purchasing agents. (Quarmine et al.,2012)

That collective marketing has a positive and statistically significant effect on the net price received by farmers. This effect is estimated at 44 FCFA per kilogram of cocoa sold collectively, that means 8% increase on the individual sale price. (Bergaly *et al.*, 2013)

Cultivation cocoa in Indonesia will be done through three forms of community plantation business, namely, large plantations state and private plantations, and there are approximately 965.00 thousands of farmer households directly involved in cacao farming.

The overall environmental impacts resulting from cocoa production and processing activities and improvement options towards the sustain- ability of the system studied are presented and discussed
(Ntiamoah and Afrane, 2008)

According to ICCRI research centers, potential cacao production can reach more than 2 tons/ha/year. Indonesia's cocoa, especially those that produced by the people, is still regarded unpretentious low in the international market because most are not in good quality, even the seeds are dominated with non-fermented seeds, having high offscouring, and contaminated with insects, fungi, or mycotoxin.

However, Indonesia still has a quite big prospect for the development of cocoa both for the local market and international market from high upstream and downstream. This was because of: (a) its natural resources, labor, and technology rolls-off that were spearheaded by the Center for research in cocoa commodities; (b) the government through its strategic plan of the Directorate General of the Department of Agriculture and Plantation to assert that cocoa will be developed as leading export commodities after rubber and palm oil. This policy implementation can be seen from the development in cocoa that has been done since the beginning of the 1980s.

The production of cacao seeds commodities are not only in various milling products but also has been made in the form of food and soft drinks using chocolate raw materials which are quite familiar in the community, among others are chocolate candy (*cocoa candy*), chocolate powder (*cocoa powder*), and the fat chocolate (*cocoa butter*) which is a semi-finished product.

The penchant of consuming chocolate-based food and soft drinks in this millennium era is no longer dominated by certain circles, but it has become a common matter of all layers of society, especially in children and young people. It shows a welcoming development on the market outlook of chocolate-based food and soft drinks. The increase of economic welfare of farmers is one of the factors which turns the community consumption patterns on chocolate food and soft drinks (Faturoti *et al*, 2012).

This condition gives an opportunity that is large enough to possible various parties taking advantage of the opportunities by keeping a priority that there are a basic benefit and cost ratio rational to economic growth and the increase in welfare and regional development community (Brandt, 2007). The study revealed that, even though the quality is important to all categories of actors in the cocoa sector, interactions among them are hampered by problems of information asymmetry that result especially in farmers evading recommended practices. (Quarmin et al,2012) (Kpodo, 2006), (Vigueri, 2008). (Hounkonnou, 2009) (Dormon, 2009) (Fbaah, 2008). Calibration suggests that to ignore margins of pass-through other than price has substantial implications for welfare analysis. (Lorenzo and Tristan, 2013) (Jano and Mainville, 2007)

State of Research Hypotheses

To provide the research hypothesis of the above is not easy. It is undeniable that some cacao farmers still have the knowledge and skills enough to apply ways of managing cocoa plantation. Social and economy matters

 DOI: 10.1051/matecconf/20165802006

of the farmers, local culture, technological innovation, and agro-industry cacao also prompted more and more problems in the field. Responding to the question, it needs to be socialized in the plantation through dissemination by government officials, mentoring establishment, as well as the form of training by various parties, both private and government. The purpose of this research is to analyze the potential superior products in order to accelerate Cocoa economic growth and development areas.

METHODOLOGY

Location

The research focus is developing cacao commodities as superior products of agro-industry food and soft drinks. The place chosen was Rambipuji and Wuluhan Districts in Jember Regency, where one is in ICCRI location and the other is in Renteng Plantation of PTPN XII. The collaboration between farmers and the synergy of government, private, world trade, and higher education is important for the development of cocoa commodities.

Sampling Procedures

The household Objects are cacao farmers (RTP). Random sampling method was used to determine samples. The Data Source is primary data collection of cacao RTP with a structured-administered questionnaire, observation, the Focus Group Discussion (FGD), and field notes while the secondary interview data collection was also done for the government agencies and institutions related to the problems of developing a leading agro-industry cacao food and soft drinks.

Result

A strategy is used to improve the welfare of farmers and the development of cacao commodities as well as to increase the extension program partiality towards the government and private sectors to cacao farmers. It is increasing the system management into an effective and efficient management and also accelerating the policy rule that gives conducive climate for cultivation cacao in the country.

The analysis of matrices policy for divergencies showed a negative value (see table 1), which means it failed to produce an efficient market price. In other words, there has been market distortion that indicate a lack in selling cacao from which should be achieved. This can happen due to lack of effective government policy in post-harvest / price of cacao which began in 2012, where the export cacao seeds must be fermented. Regulation rules are aimed at encouraging cacao processing industry development in the land so that it would give added value for the cacao farmers. The *tradable divergence input* is high-negative, both in Rambipuji and Wuluhan districts. A negative tradable divergence input causes market distortion, where the farmers buy tradable means of production cheaper because of the existence of fertilizer subsidies. Subsidy policy is intended to farmers to be able to buy recommended fertilizer. It can be seen in the table below:

Table-1. Divergencies in Commodity Prices of Cacao

No.	Model	Revenue	Tradable Input	Domestic Factors			Profits
				Labor	Capital	Land	
A	**Rambipuji District**						
1	Private	27,259,893	6,194,509	5,403,208	2,454,108	4,360,000	8,848,066
2.	Social	37,701,266	8,315,480	5,403,208	2,454,108	4,360,000	18,481,122
3.	Divergencies	(10,441,373)	(2,120,970)	-	-	-	(9,633,056)
B.	**Wuluhan District**						
1	Private	27,852,071	6,277,724	5,224,852	2,445,020	4,480,000	9,424,473
2.	Social	38,520,266	8,347,171	5,224,852	2,445,020	4,480,000	19,326,021
3.	Divergencies	(10,668,195)	(2,069,447)	-	-	-	(9,901,547)
C.	**Rambipuji and Wuluhan**						
1	Private	27,482,536	6,455,587	5,336,151	2,445,994	4,420,000	8,824,803
2.	Social	38,009,188	8,660,833	5,336,151	2,445,994	4,420,000	18,428,655
3.	Divergencies	(10,526,652)	(2,205,246)	-	-	-	(9,603,852)

Source: Research Analysis (2013).

Table -2. Divergencies benefits with Price Status and Social in Cacao Farming

No	Location	Private (Rp)	Social (Rp)	Divergencies (Rp)
1.	Rambipuji district	8,848,066	18,481,122	(9,633,056)
2.	Wuluhan district	9,424,474	19,326,022	(9,901,547)
3.	Rambipuji and Wuluhan	8,824,803	18,428,655	(9,603,852)

Source: Research Analysis (2013).

ANALYSIS OF COMPETITIVENESS

Comparative Advantages

The superiority comparative is the size of competitiveness (superiority) potential in the sense of competitiveness will be achieved if it does not have a distortion at all. Commodities that has the comparative advantage also be said to have efficiency in the economy. Comparative Advantages is the relative size that shows the great potential advantage commodities in the trade in free markets (to be able to compete perfectly).

Table-3. Private cost Ratio (PCR) and Domestic Resource Cost Ratio (DRCR) of Cocoa Commodities Development in Jember Regency

No	Location	Domestic Resource cost Ratio (DRCR)	Private cost Ratio (PCR)
1.	Rambipuji	0, 4158	0, 5,800
2.	Wuluhan	0, 4027	0, 5632
3,	Rambipuji and Wuluhan	0.4158	0.5803

Source: Research Analysis (2013).

COMPETITIVE ADVANTAGE

Competitive advantage or often called *"revealed competitive advantage"* is a measuring stick of actual economic condition competitiveness. It appeared in table cacao PCR in the value of 0.5803, while in Rambipuji district is 0.5800 and PCR Wuluhan district is 0.5632 which are <1. It shows that the development cacao in those locations have high competitive advantage. It also means that to produce a value of output in that price status, it is only needed less than one domestic unit-resource cost. This competitive development of cocoa commodities in Jember regency is due to the low usage of tradable input.

Government Policy to the Development of Cocoa Commodities

The size of divergence impact and government policy in The Analysis Matrices policy can be measured by transfer output, transfer input, transfer and other net transfers factors. The Relative Size is voiced by the analysis of protection coefficient output nominal or *nominal protection coefficient on output* (NPCO), coefficient protection input nominal or *nominal protection coefficient on input* (NPCI), and coefficient effective protection or *effective protection coefficient* (EPC). *Profitability coefficient* (PC) and the ratio subsidy for producers or *subsidy ratio to produce* (SRP).

Influence of policy Input to the Development Cocoa Commodities

An incentive policy that was found on tradable input is shown by the transfer value input (IT) and nominal protection coefficient on input (NPCI). Policies to the tradable input factor can be in the form of trade subsidies and taxation while another divergence form can be caused by the market distortion. Transfer input shows the difference between input fees that can be traded at the price of the private input costs which can then be traded at a social price. Nominal Protection Coefficient Input (NPCI) as an indication transfer input which is a tradable input costs ratio is counted using the cost based on the private tradable input price that are counted in the social price. More detailed information on the value of IT on the development and NPCI of cacao commodities in the research location can be seen in table 4.

Table-4. Transfer Input of Cocoa Farmers in Jember Regency

No.	*Location*	*Transfer Input (Rp)*
1.	Rambipuji	(-2.120.970)
2.	Wuluhan	(-2.069.447)
3.	Rambipuji and Wuluhan	(2,205,652)

Source: Research Analysis (2013).

Table-5. Nominal Protection Coefficient Input (NPCI) to the Tradable Input Fertilizer of Cocoa Farmers in Jember Regency

No.	Tradable Input Fertilizer	NPCI
1.	Urea	0.46
2.	ZA	0.69
3.	TSP	0.51
	Average of Tradable Input	0.52

Source: Research Analysis (2013)

Influence of output to the Development cocoa commodities

The existence of government intervention or policy incentive in output can be seen from the amount of transfer output value (OT) and NPCO. A form of Government intervention is the policy in trading, export taxes, import tariffs, and subsidizing policy and taxes.

Transfer output is the difference between the revenue that is counted on private prices with the revenue calculated based on social price. The coefficient protection of nominal output (NPCO) is an indication of the transfer output shown by the ratio between the revenue that is calculated based on status price with the revenue counted based on social price.

Table-6. Value of output Transfer (OT) and NPCO In Cocoa Farmers in Jember Regency

No.	Location	Output Transfer (IDR. in million)	NPCO
1.	Rambipuji	(10,441)	0.723
2.	Wuluhan	(10,668)	0.723
3.	Rambipuji and Wuluhan	(10,526)	0.723

Source: Research Analysis (2013).

Influence of policy of Strengthening Economy Input-output Model

Table-7. NT, PC, EPC and the SRP in Jember Regency.

No.	Sub-district	NT IRD in thousands	PC	EPC	The SRP
1.	Rambipuji	(9,633)	0,479	0,717	-0.256
2.	Wuluhan	(9,901)	0,488	0,715	-0.257
3.	Rambipuji and Wuluhan	(9,603)	0,479	0,716	-0.253

Source: Research Analysis (2013).

DISCUSSION

Cacao Farmers people in the research location are able to gain the profit status and social benefits. Higher profits from the social benefits indicated that the price means of production/input paid by farmers is less, or the output prices received by farmers is lower than the international oil prices. This means that the government subsidizes inorganic fertilizer, especially Urea, TSP, and ZA to the farmers. The implication is that farmers will enjoy more benefits than if they enjoy the international price of cacao. Even when subsidized fertilizers were eliminated, cacao farmers are still able to gain more if the result is fairly sold to international markets or in the domestic market with the world market prices. Meanwhile, divergencies advantage is in the negative value, meaning there is a distorted policy which is the subsidies of input fertilizer.

In Table 3, Rambipuji and Wuluhan districts each has a value of 0. 4158 and 0, 4027, respectively. It means that compared to overseas cacao production, Indonesia's still has comparative advantages. Economically, it will be more advantageous to provide an opportunity for Jember Regency to develop their cacao products rather than outside Jember or even imported from abroad.

In Table 4, the massive transfer input of credit watch is negative, where the transfer input for cacao farmers in the sub-district of Rambipuji is -2,120,970; greater than that of Wuluhan district, which is Rp. - 2,069,447. There was a policy that reflects a distortion, which is fertilizer subsidy policy. The fertilizer subsidy is reflected from the massive transfer input. Thus, cacao farmers in Rambipuji receive fertilizer subsidies more than cacao farmers in Wuluhan. In addition, the ratio to measure tradable input transfer is Nominal Protection Greatly enhanced by on Input (NPCI). The ratio showed how big domestic price of tradable input prices can be with different activities. If NPCI value is more than one, the cost for domestic input costs is more expensive than that of in high world prices. In other words, the system seems to be burdened by tax policy. When NPCI is smaller than one, the price is lower than the domestic world price, and the system seems to be subsidized by existing policy. If there is no transfer input price, the domestic and global prices will not be different, and NPCI will be equal to one.

Based on Table 5, it is showed that the coefficient of NPCI <1, all for Urea, ZA, and TSP, which was 0.46, 0.69, and 0.51, respectively. That means farmers will pay a lower price than it should be. For example, the Urea is paid by farmers to 46% of international oil prices. It was the role of the government to control, in order to give fertilizer price incentives or to protect the farmers so they can develop cacao commodities. To establish the structure protection is by providing an image that farmers still have an incentive in developing cacao commodities which are shown as negative transfer values. Table 6 shows the negative value of output Transfer (OT) and the coefficient NPCO <1. But the high negative value of OT for Wuluhan farmers is still slightly higher than the farmers of Rambipuji. This shows that Rambipuji farmers are more able to compete rather than Wuluhan farmers because Rambipuji farmers are able to sell cacao with higher prices than it should be of Wuluhan' farmers. The NPCO value in a location with the coolness research of 0.723 means that cacao farmers accept the price 72.30% from the cost of goods sold for relatively competing commodities, which reflects the enormous cost production for cacao commodities. This situation made the cacao commodities to be able to compete with the prices of imported cocoa commodities.

Results of the analysis net transfers (NT) for the development of cacao commodities in the research location is NT negative. This means that there is a government policy providing incentives to the input (tradable input and domestic factors) and which in the overall output, benefits cacao farmers. The coefficient PC > 0, which means there is a government's overall policy to give incentive to cacao farmers, in this case in the form of fertilizer subsidies. The coefficient value of EPC < 1, shows a lack of protection to producers or cacao farmers, and means that the government, even though giving subsidy policy to input fertilizer, but the output is still not effective. The coefficient value of SRP in the research locations were relatively the same, which was -0.253 (table 7). This means that net transfer of Rp 9.6 million will happen to the policy of cacao export for 25.30% if there are no other divergencies. The negative SRP value shows, in general, that the policy of the government is distorting for cacao farmers *"Although cacao farmers receive a subsidy from the government as incentives, but the government policy is less effective in output policy, especially in export category"*.(Dormon et al.,2009) (Oluyole, 2010)
(Dongo, 2009).

CONCLUSIONS

A strategy is used to improve the welfare of farmers and the development of cacao commodities through the increase in extension programs partiality toward the government and the private sector to cacao farmers. Analysis of matrixes policy for divergencies gave a negative value, which means that it failed to produce an efficient market price.

Results of the analysis net transfers (NT) for the development commodities cacao in research locations were NT negative. It means that there is government policy providing incentives to the input (*tradable input* and *domestic factors)* and which had the overall output to benefit the cacao farmers. The PC coefficient showed a lack of protection to producers or cacao farmers. It means that even though the government give subsidy policy of input fertilizer, the output is still not effective.

REFERENCES

Adger, W. N. 1996. *An assessment of vulnerability to climate change in Xuan Thuy, Vietnam.* unpublished paper, CSERGE, University of East Anglia and University College London.

Anim-Kwapong, G., Frimpong, E. 2005. Vulnerability of agriculture to climate change impact of climate change on cocoa production. Accra, Ghana.

Ajewole Davies Ojo *and* Iyanda Sadiq. 2010. Effect of Climate Change on Cocoa Yield: A case of Cocoa Research Institute (CRIN) Farm, Oluyole Local Government Ibandan Oyo State. *Journal of Sustainable Development in Africa Volume 12, No.1.*

Atkinson, A.B.; Mogensen, G.V. [eds.] 1993, *Welfare and Work Incentives: A North European Perspect*, Oxford University Press, ISBN: 0198288603
Bergaly, KAMDEM Cyrille,2013. Impact of Collective Marketing.4th International Conference of the African Association of Agricultural Economists.

Blejer, M. I. and I. Guerrero. 1990. '*The Impact of Macroeconomic Policies on Income Distribution: An Empirical Study of Philippines', Review of Economics and Statistics, Vol. 72.*

Brandt, Peter, 2007. Die wirtschaftliche Bedeutung von Kaffee, Tee und Kakao –eine Übersicht für Deutschland, J. Verbr. Lebensm. 2 (2007): 393–398.

Birkhäuser Verlag, Basel.Baah,F and Harnessing, 2008 Farmer associations as channels for enhanced management of cocoa holdings in Ghana, Scientific Research and Essay 3: 395–400.

Cascio, Wayne F., 1991. Applied Psychology in Personel Management. Prentice Hall Ind, Inc
Carter, M.R., 1997. Environment, technology, and the social articulation of risk in West African agriculture. *Economic Development and Cultural Change* 45: 557-590.

Casaburi, Lorenzo and Tristan Reed. 2013.Incomplete Pass-Through and Interlinked Transactions. International Growth Center. Working Paper Harvard University Department of Economics Cambridge.

Deressa,T., Hassan,C. RinglerT., Alemu., and Yesuf,M. 2008 *Analysis of theDeterminants of Farmers' Choice of Adaptation Methods and Perceptions of Climate Change in the Nile Basin of Ethiopia*, IFPRI Discussion Paper No. 798.

Dongo,L.N.,Aigbekaen,C.O.,Jayeola,L.A., Emaku,S.B.,Orisajo.2009 Influence of farmers practices on cocoa bean quality: Nigeria field experience, in: African Crop Sciences Conference, African Crop Sciences Society, Uganda.:299-302.

Dormon, ENA. 2006 From a technology focus to innovation development: the man- agement of cocoa pests and diseases in Ghana, CERES, Wageningen University, Wageningen.

Faturoti B. O. Madukwe M. C., Ogunedojutimi O. and Anyanwu. 2012., Socioeconomic impact of SARO agro allied organic cocoa programme on beneficiary cocoa farmers in Nigeria.*Journal of Agricultural Extension and Rural Development Vol. 4(16):435- 445.*

Hounkonnou,D.,VanHuis,A.,Röling,.Sterk, ,2009. Towards Enhancing Innovation Systems Performance in Smallholder African Agriculture, Elmina, Ghana, pp. 11–14.

Jano,P and Mainville,D. 2007. The cocoa marketing chain in ecuador: analysis of chain constraints to the development of markets for high-quality cacao, Paper Presented at IAMA Conference, Parma, Italy.

Mocan, H. N. 1999, 'Structural Unemployment, Cyclical Unemployment, and Income Inequality', *Review of Economics and Statistics*, Vol. 81, No. 1.

Nhemachena, C.2010. Micro-Level Analysis of Farmers' Adaptation to Climate Change in Southern
Africa. IFPRI Discussion Paper 00714 Nigeria Meteorological Agency (NIMET) Nigeria Climate Review Bulletin. NMA, Abuja.

Ntiamoah, Augustine.,George,Afrane.2008 Environmental impacts of cocoa production and processing in Ghana: life cycle assessment approach.
Journal of Cleaner Production 16:1735-1740.

Oredein O.2011. Nigeria's cocoa output estimated 15%-20% lower. Market Watch – The Wall Street Journal. Internet file downloaded on 10th January-from. http://www.marketwatch.com/story/nigerias-cocoa-output-estimated-15-20-lower-2011-07-29

Oluyole,K.A.2010. The effect of weather on cocoa production in different agro- ecological zones in Nigeria, World Journal of Agricultural Sciences 6 :609–614.

Oyekale A.S., Bolaji M.B, and Olowa O.W.2009. The Effects of Climate Change on Cocoa Production and Vulnerability Assessment in Nigeria. Agricultural Journal. 4: 77-85.

Parry, Martin L., Canziani, Osvaldo F., Palutikof, Jean P., van der Linden, Paul J., and Hanson, Clair E. 2007., Contribution of Working Group II to the Third Assessment Report of the Intergovernmental Panel on Climate Change. Cambridge University Press, Cambridge, United Kingdom.

Quarmine,W.,Haagsma,R.,Sakyi-Dawson,Asante,O .,Van Huis,A ., Obeng-Ofori., D . 2012. . Incentives for cocoa bean production in Ghana: Does quality matter?. NJAS Wagenigen Journal of Life Sciences.: 7 -14.

Seebens, Holger ,.2009. The contribution of female non-farm income to poverty reduction. *Contributed Paper prepared for presentation at the International Association of Agricultural Economists Conference, Beijing, China, August 16-22.*

Selvaraju, R.,Subbiah, S.B and Juergens.I, 2006. Livelihood adaptation to climate variability and change in drought-prone areas of Bangladesh, Food and Agriculture Organization (FAO).

UNEP .,2006. Adaptation and Vulnerability to Climate Change: The Role of the Finance Sector. CEO Briefing. New York and Geneva.

Vigneri,M.and P.Santos,2008. What Does Liberalization Without Price Competition Achieve: The Case of Cocoa in Ghana, Ghana Strategy Support Program (GSSP), International Food Policy Research Institute (IFPRI).

Winchester, Lucy and Raquel Szalachman (2009). The urban poor's vulnerability to the impacts of climate change in latin america and the caribbean.

A policy agenda. The Urban Poor's Vulnerability to the Impacts of Climate Change in Latin America and the Caribbean. A Policy Agenda. Fifth Urban Research Symposium

SUSTAINABLE STRATEGIES SELECTION IN SMEs USING MCDM APPROACH

Ari Basuki
Department of Industrial Engineering, Faculty of Engineering,
University of Trunojoyo Madura, Indonesia Indonesia
E-mail: ari_bsk@yahoo.com

ABSTRACT

Since the nineties, companies have started to recognize the relevance of integrating sustainability in their corporate strategies. However, despite this increasing awareness, integrating sustainability in corporate strategy remains a challenge for many organisations due to the need of dealing with a wide range of issues. This research discusses more about the sustainable strategies selection in a SME and develops a model based on integrated multicriteria decision making (MCDM) methods to solve such problems. The developed model applies Entropy weighting method and Vikor algorithm to effectively analyze the weight of each sustainability criteria and to select the best sustainable strategies while capturing all subjective and objective criteria. A case study is illustrated to test the proposed model in a Batik SME. Results show that profits and facility requirements are the two most important criteria in sustainable strategies selection. This implies that during sustainable strategies selection, it is suggested that emphasis should be placed on revenue generation and facilities needed of the respective strategies.

Keywords: sustainable, strategies selection, MCDM, entropy, vikor

INTRODUCTION

Recently, the conservation of environmental quality has become essential since the scarcity of natural resources and increasing population. In many aspects, environmental problems have affected regional and global cooperation and have even prompted conflicts (Zhao and Guo, 2014). Environmental deterioration and global warming have prompted public concern over sustainability and environmental issues. Such legislation forces manufacturers to decrease pollution during the whole production process. Hence, the practice of green activities has become mandatory to balance these conflicts; even manufacturing processes cannot make an exception. Green issues have gained more importance in contemporary globalization.

External pressure regarding the environmental issues encourages industrial creativity using cleaner technology of greener production process. Regarding with environmental issues that call for more stringent environmental regulations, there are future trends that yield a new impact on the future prosperity of business firms since they have to cover the extra cost for green materials and new designs or new processes. In practice, an environmental management system is seen as a tool to provide a framework for assessing environmental impacts and then to help management to form firm-level environmental strategies.

It is essential for firms to have a clearly identified their strategy to adapt with green policy which should describe the objectives and targets. Before planning and doing any stages, top management needs to refine and review the environmental objectives and policies of green policy to assure effective planning and implementation. More and more countries have regulated public procurements in favor of green products by granting some premiums. Taiwan set up a "governmental procurement law" for public enterprises or administrative offices in 1998. This law allows the purchaser, to buy green products at 10% higher prices than normal products with the same specifications. In the UK, some legal regulations have been imposed on potential pollution from industrial activities. Furthermore, Japan has set up time schedule for guiding the firms to engage in green purchasing.

Many manufacturers have adjusted their manufacturing philosophies and introduced environmental programs into their organizations. Some firms proactively recognize the urgency of environmental protection and have integrated environmental goals into their corporate strategies. Those firms with relatively limited resources must outsource some business functions or operations, or purchase raw materials or components/ sub-components from other suppliers to establish an interrelated supply-chain network. To advance their environmental performance, they must monitor their own operations and coordinate with other partners in their supply networks, including material suppliers, manufacturers, distributors, users, and so on. Supply-chain managers consider both traditional performance criteria and environmental criteria (Deshmukh and Vasudevan, 2014).

Poole and Simon (1997) suggested life-cycle analysis as a method of examining the overall environmental impact of a product. Kurk and Eagan (2008) also extracted environmental attributes from each phase of a product's lifecycle, including raw-material extraction, product manufacturing, packaging and transport, use and service, and final disposal. Most studies emphasize reduction, re-manufacturing, recycling product design, process design, manufacturing practices, procurement, and some mixture of items across managerial levels. Integrating environmental concepts into these business functions ameliorates environmental pollution. Effectively achieving corporate green goals means linking an environmental corporate strategy with every business functional strategy, thus eliminating obstacles to environmental integration.

Sustainable development ties together concern for the carrying capacity of natural systems with the social, political, and economic challenges faced by humanity.

Furthermore, the implementation of sustainable management system has emerged as an important research topic in recent years. An entrepreneurial activity, can be labelled as sustainable if satisfy the three aspects, People, Planet and Profit (Isaksson et. al, 2015).

There are many research on sustainable strategies implementation on larger companies (Elkington, 1994; Harms, et al, 2013; Egels-Zanden and Rosen, 2015). However, the studies of sustainable strategies implementation on SME's are somewhat limited, especially in Indonesia. The number of SME in Indonesia growth rapidly year by year. SMEs are found to have been weathering the crisis better than larger companies, though many have been hit hard too (Berry, et al, 2001). SMEs have historically been the main player in domestic economic . Therefore, the studies of sustainable strategies on SME's are something urgent.

Take for example the SME's of batik producers that have long been existed. This research explores Batik SME in Sidoarjo, East Java – Indonesia. In Batik production process, craftsmen may use synthetic or natural dyes. Synthetic dyes material is relatively easier found in the market and much cheaper compared to natural dyes material. Furthermore, using synthetic dyes makes the coloring process faster than using natural dyes. However, the synthetic dyes, such as napthol, indigosol, etc., are not environmentally friendly. If such synthetic wastes flow into the ground, those would damage the soil ecosystem because the soil bacteria are not able to degrade the chemicals.

Some SMEs with awareness about green issues and sustainability investigate the use of natural dyes and other technology which is considered more eco-friendly. On the other hand, they have a shortage of resources to implement the sustainable strategies into their production process. Therefore, the SME's should select the best sustainable strategies which is very compatible to be adopted to their business.

METHODOLOGY

Many SMEs face challenges to implement the sustainable policy into their organization. Most SMEs have shortage of resources, include money, manpower, and time as well (Fernández-Ortiz, R., et al., 2015). Studies have identified three key barriers to SME adoption of environmental practices. The first barrier is SMEs' perception that they have little individual impact on the environment (Sanchez-Medina et. al., 2014; Wilson et. al., 2012; Williams S., and Schaefer A., 2013). The second barrier is the lack of expertise and understanding of strategies to address environmental issues (Govindan K. et al., 2014; Arena M., 2012; Mathiyazhagan, K., 2013). And finally, cost is a major barrier to more proactive environmental behavior in SMEs, in which the managers perceive that there is only little financial benefit from environmental investments (Brophy, M., 2014).

Obviously, changes in the use of dyes in Batik can not simply be implemented. Many factors must be taken into consideration, such as the sustainability strategy, business strategy, technology used, and also social factors. Decision-makers should appropriately modify the contents and aims of environmental practices to match changes in business development. Many companies have just begun exploring environmental concerns and implemented environmentally-friendly activities, so they have not yet identified many environmentally-related factors. Therefore, rethinking the relationships between each factor of environmental practices is necessary. This research uses hybrid MCDM aproach in this study to select the best sustainability strategy for a SME's as a guideline to aid company decisions.

Multi-Attribute Decision Making is the most well known branch of decision making. It is a branch of a general class of Operations Research (or OR) models which deal with decision problems under the presence of a number of decision criteria. This super class of models is very often called multi-criteria decision making (or MCDM) (Triantaphyllou, et.al, 1998). An MCDM methods as a compromise solution for a problem with conflicting criteria can help the decision makers to reach a final decision.

In MCDM context, the alternative selection is obtained using the evaluation of the alternatives where each alternative is described by its performance on each of a number of criteria. The Multi-criteria Decision Making aims to provide support to the decision maker in the process of making the choice among the available alternatives. Moreover, the MCDM method may include the generation of a proposed "compromise" solution and/or some form of preference ranking.

The SME sustainable strategies selection is modelled in accordance with the criteria set by the company. Multi Criteria Decision Making (MCDM) is used in this research to solve the problem. Among numerous methods of MCDM, outranking methods have a rapid progress because of their flexibility to the most real decision situations. In addition, outranking methods using Vikor algorithm which was proposed by Opricovic (1998) is applied. The proposed research methodology consists of several steps and it is shown as in Figure 1.

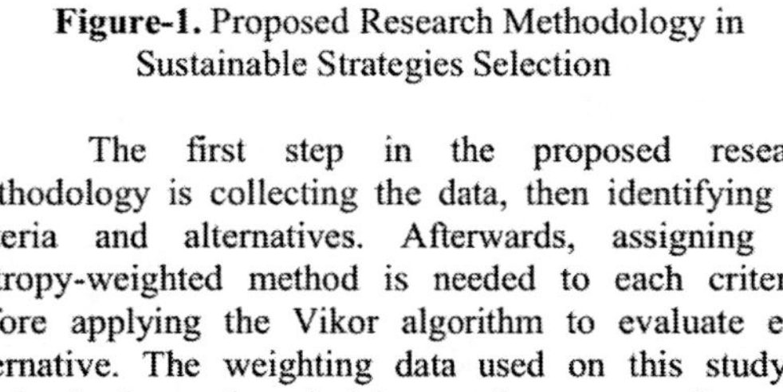

Figure-1. Proposed Research Methodology in Sustainable Strategies Selection

The first step in the proposed research methodology is collecting the data, then identifying the criteria and alternatives. Afterwards, assigning the Entropy-weighted method is needed to each criterion before applying the Vikor algorithm to evaluate each alternative. The weighting data used on this study is obtained from the decision maker score for each alternative. Finally, the last step is obtaining the alternatives of sustainable strategy's rank.

Entropy-weighted method

Entropy was originally a thermodynamic concept. Shannon (2001) is first introduced the entropy application into information theory. Since then, it has been widely used in the engineering, socioeconomic and other fields. This research utilized entropy method to obtain W_j, weight for each criterion (C_j). There are three steps to acquire the weight using entropy-based weighting method.

First of all is normalizing the value of each criterion using Eqn. (1) with $r_{ij}$is the normalized value of each x_{ij}

$$r_{ij} = \frac{x_{ij}}{\sum_{i=1}^{m} x_{ij}} \quad (1)$$

Then, calculate entropy values of each criterion using Eqn. (2).

$$e_j = -K \sum_{i=1}^{m} r_{ij} * \log r_{ij} \quad (2)$$

where

- e_j indicate entropi value of C_j
- $K = \frac{1}{\log m}$ is a constant value to make the entropi (e_j) value remains between $0 \leq e_j \leq 1$

And, the last step is calculating W_j, a weight of C_j criteria using Eqn. (3).

$$W_j = \frac{1-e_j}{\sum_{i=1}^{n}(1-e_j)} \quad (3)$$

The obtained W_j reveals the weight of each C_j criteria. Afterwards, the weight of each criteria is used to calculate the ranking method.

Vikor algorithm method

After performing Entropy weighting method, the next step is to apply Vikor algorithm method to get the rank of sustainable strategies option. Regarding on Opricovic (1998), Vikor algorithm is one of multi-criteria decision methods based on ideal point method to solve a complex system. The basic view of VIKOR is determining positive-ideal solution and negative-ideal solution. The positive-ideal solution is the best value of alternative for each criterion, while the negative-ideal solution is the worst value of alternative for each criterion. Below are the calculation steps of Vikor algorithm.

First, normalize the dimension of each criterion's value. The normalization for benefit-type criteria used Eqn. (4), while for the cost-type criteria used Eqn. (5).

$$v_{ij} = \frac{x_{ij}}{\max_i(x_{ij})} \quad (4)$$

$$v_{ij} = \frac{\min_i(x_{ij})}{x_{ij}} \quad (5)$$

Second, calculate each criteria's positive-ideal solution value f_i^* and negative-ideal solution's value f_1^-, $i = 1,2, \dots, n$ using Eqn. (6) and Eqn. (7), respectively.

$$f_i^* = \max_j f_{ij} \quad (6)$$

$$f_i^- = \min_j f_{ij} \quad (7)$$

Third, Obtain the optimal solution (S_i) and the most inferior solution(R_i), $i = 1,2, \dots, n$ using Eqn. (8) and Eqn. (9). At this point, each criteria's weight (w_i) will be used to determine entropy-weighting method as explained in the previous subsection.

$$S_i = \sum_i^n \frac{w_i(f_i^* - f_{ij})}{(f_i^* - f_i^-)} \quad (8)$$

$$R_i = max[w_i(f_i^* - f_{ij})/(f_i^* - f_i^-)] \quad (9)$$

Fourth, Obtain the value of interest ratio (Q_j) using Eqn. (10).

$$Q_j = \frac{v(S_i - S^*)}{(S^- - S^*)} + \frac{(1-v)(R_j - R^*)}{(R^- - R^*)} \quad (10)$$

where

- $S^* = \min_i S_j$
- $S^- = \max_i S_j$,
- $R^* = \min_i R_j$,
- $R^- = \max_i R_j$,

v is introduced as a weight for the strategy of maximum group utility, whereas 1 - v is the weight of the individual regret.

Fifth, rank the alternatives sorting by values S, R and Q in an ascending order. The minimum value is the best rank regarding the value S, R, and Q.

Sixth, check whether these two conditions are satisfied:

C1: Acceptable Advantage

$$Q(a^{(2)}) - Q(a^{(1)}) \geq 1/(m-1) \quad (11)$$

Where

- $a^{(1)}$ is the alternative with best rank regarding the minimum value of Q.
- $a^{(2)}$ is the alternative with second position in the ranking list by Q.
- m is the number of alternatives

C2: ***Acceptable acceptable stability in decision-making process***

C2 is the condition where $a^{(1)}$ is also the best rank alternative regarding the value of S or and R. This compromise solution is stable in decision-making process, it may differ as varies:

- when $v > 0.5$, decision-making will be according to majority criteria;
- when $v \approx 0.5$, the selection will consider to overall and individual's evaluation;
- when $v < 0.5$, veto the schemes set.

If one of the above two conditions is not satisfied, then the compromise solution should be obtained, including:

a. If the condition $C2$ is not met, then $a^{(1)}$ and $a^{(2)}$ schemes are both compromise solution.
b. If the condition $C1$ is not met, then the schemes is used: $a^{(1)}$, $a^{(2)}, \dots, a^{(r)}$, where $a^{(r)}$ is determined by the relation $Q(a^{(r)}) - Q(a^{(1)}) \geq 1 / (m-1)$ with maximum value.

RESULTS

In this section this study proposes a numerical example to illustrate an application of the proposed method in the previous section. The case study has been conducted in a middle-sized SME of Batik located in Sidoarjo, Indonesia. The SME is in the process of implementing certain sustainable concept to increase awareness and concern over the environmental impact and economic growth of business and trade. This research is then proposed to select the best sustainable strategy. The inputs have been gathered from the decision maker, who is responsible for implementing sustainable strategy in the SME's organisation.

The SME in this case study is still looking at sustainability in terms of cost efficiency rather than as something that can add value to their company. Furthermore, material properties and selection are very important in sustainable manufacturing. Sustainability can also be achieved, if the designers include sustainable development aspects into their design (Kerr, I. R., 2006; Maxwell, D., and Van der Vorst, R., 2003). Innovations are obtained by new production methodologies, so the choice of technology is very important, not only for technical reasons, but also because of the influence of economical, ecological and social factors. Therefore in this case study sustainability orientations are taken as green design, green manufacturing, green purchasing and green methodology, which are designated as (S_1, S_2, S_3, S_4).

There are four sustainable strategies orientation (S_1, S_2, S_3, S_4) selected against five criteria. The evaluation criteria are: facility requirement (C_1), clean technology (C_2), workforce engagement (C_3), profits (C_4) and environmental degradation (C_5). The description and orientation of each criterion is described in Table 1.

Table-1. Criterion with description

Evaluation Criteria	Description	Type
Facility requirement (C1)	Usage of the facilities while using the method	Percent
Adaptability (C2)	Ability to adapt efficiently and fast to changed circumstances	Percent
Workforce engagement (C3)	Measurable degree of an employee's positive or negative emotional attachment to their job	Number of hours
Profits (C4)	Profit obtained by implementing the concepts	Rupiah
Environmental degradation (C5)	The reduction of the capacity of the environment to meet social and ecological objectives, and needs	Percent

After identifying criterion, the decision maker is invited to rate the strategies according to the evaluation criteria. The criteria's values are shown in Table 2.

Table-2. The criteria's values of each alternative

Evaluation Criteria	Type	Sustainable strategy orientation			
		S1	S2	S3	S4
Facility requirement (C1)	Cost	75	65	70	60
Adaptability (C2)	Benefit	70	85	70	80
Workforce engagement (C3)	Benefit	60	75	75	80
Profits (C4)	Benefit	60	80	75	80
Environmental degradation (C5)	Cost	70	65	70	50

The next step, the entropy-weighted method is applied using Eqn. (1), Eqn. (2) and Eqn. (3). As the result, the weight of each criteria, $\mathrm{w_i} = (0.382, 0.031, 0.119, 0.376, 0.041)$. Before applying the weight into Vikor algorithm, the normalization process should be done to normalize each criterion's value using Eqn. (4) for benefit-type criteria and Eqn. (5) for cost-type criteria. The transformation result is shown in Table 3.

Table-3. The transformated criteria's value

Evaluation Criteria	Sustainable strategy orientation			
	S1	S2	S3	S4
Facility requirement (C1)	0,800	0,923	0,857	1,000
Adaptability (C2)	0,824	1,000	0,824	0,941
Workforce engagement (C3)	0,750	0,938	0,938	1,000
Profits (C4)	0,750	1,000	0,938	1,000
Environmental degradation (C5)	0,714	0,769	0,714	1,000

At this point, the positive-ideal solution (f_i^*) and the negative-ideal solution (f_i^-) are obtained, which is presented in Table 4.

Table-4. The f_i^* and f_i^- value

	The positive-ideal solution (f_i^*)	The negative-ideal solution (f_i^-)
C1	1,0000	0,8000
C2	1,0000	0,8235
C3	1,0000	0,7500
C4	1,0000	0,7500
C5	1,0000	0,7143

Calculate each strategy's S, R, Q value according to Eqn. (8), Eqn. (9) and Eqn. (10). In this calculation process, the weight of the decision-making strategy is set as "consensus" (v = 0.5). The result is presented in Table 5.

Table-5. S, R, Q values of each strategy orientation

	Sustainable strategy orientation			
	S1	**S2**	**S3**	**S4**
S	1,000	0,379	0,682	0,042
R	0,302	0,244	0,302	0,042
Q	1,000	0,564	0,834	0,000

Rank the S, R, Q value of each strategy orientation in ascending order. The result is shown in Table 6.

Table-6. Rank of S, R, Q value of each strategy orientation

	Sustainable strategy orientation			
	S1	**S2**	**S3**	**S4**
S	4	2	3	1
R	4	2	3	1
Q	4	2	3	1

DISCUSSION

After obtaining the value of S, R, and Q, the next procedure is to check whether this scheme meet the condition of C1. Regarding rank list in Q value, it is found that $a^{(1)}$ is S4 and $a^{(2)}$ is S2. Therefore, according Eqn. (11), it is obtained that:

$$Q(S_2) - Q(S_4) \geq 1/(4-1)$$
$$0.564 - 0 \geq 1/3 \quad \text{(satisfy)}$$

Subsequently, the next step is to check whether this scheme met the condition of C2 using the Eqn. (11) applied to S and R list. To satisfy the C2, then $a^{(1)}$ should also the best rank alternative in S or / and R list.

$$S(S_2) - S(S_4) \geq 1/(4-1)$$
$$0.379 - 0.042 \geq 1/3$$
$$0.337 \geq 0.333 \quad \text{(satisfy)}$$

$$R(S_2) - R(S_4) \geq 1/(4-1)$$
$$0.244 - 0.042 \geq 1/3$$
$$0.202 \geq 0.333 \quad \text{(does not satisfy)}$$

From the computation above, it is found that S4 is the optimal solution which met the condition of C1 and C2. Therefore, the final ranking result using entropy-weighting method and Vikor algorithm is S4 > S2 > S3 > S1.

CONCLUSION

The implementation of green concept in the manufacturing field has been the theme of intense research in recent years. This study utilizes hybrid MCDM to find the best sustainable strategy orientation. It is essential to the required information for the analysis from the case organisation, and detailed procedure of analysis has been carried out. This analysis is carried out using four evaluation criteria. From the discussion, it is concluded that the production methodology is the best strategy orientation to improve the sustainability in the case organisation under the given evaluation criteria. This result implies that there should be a change of production methodology, such as designing the workplaces and production procedure which is continuously reduced or eliminated physical, chemical, biological, and technologies that present hazards to human health or the environment.

ACKNOWLEDGMENT

The author would like to express thanks the students and academics for their involvement in this research. This research was supported and funded by Faculty of Engineering, University of Trunojoyo Madura – Indonesia.

REFERENCES

Arena, M., & Azzone, G. 2012. A Process-Based Operational Framework For Sustainability Reporting In SMEs. Journal of Small Business and Enterprise Development, 19(4): 669-686.

Berry, A., Rodriguez, E., & Sandee, H. 2001. Small And Medium Enterprise Dynamics In Indonesia. Bulletin of Indonesian Economic Studies, 37(3): 363-384.

Brophy, M. 2014. Environmental Policies. Corporate Environmental Management 1: Systems and Strategies, 90.

Deshmukh, A. J., & Vasudevan, H. 2014. Emerging Supplier Selection Criteria In The Context Of Traditional Vs Green Supply Chain Management. International Journal of Managing Value and Supply Chains, 5(1): 19-33.

Egels-Zandén, N., & Rosén, M. 2015. Sustainable Strategy Formation at A Swedish Industrial Company: Bridging The Strategy-As-Practice And Sustainability Gap. Journal of Cleaner Production, 96: 139-147.

Elkington, J. 1994. Towards The Sustainable Corporation: Win-Win-Win Business Strategies For Sustainable Development. California management review, 36(2), 90.

E. Triantaphyllou, B. Shu, S. Nieto Sanchez, and T. Ray. 1998. Multi-Criteria Decision Making: An Operations Research Approach. Encyclopedia of Electrical and Electronics Engineering, (J.G. Webster, Ed.), John Wiley & Sons, New York, NY, Vol. 15: 175-186.

Fernández-Ortiz, R., Arteaga-Ortiz, J., & Clavel San Emeterio, M. 2015. Factors That Foster Export Commitment: An Empirical Study In SMEs. Engineering Economics, 26(3): 272-283.

Govindan, K., Kaliyan, M., Kannan, D., & Haq, A. N. 2014. Barriers Analysis For Green Supply Chain Management Implementation In Indian Industries Using Analytic Hierarchy Process. International Journal of Production Economics, 147: 555-568.

Harms, D., Hansen, E. G., & Schaltegger, S. 2013. Strategies In Sustainable Supply Chain Management: An Empirical Investigation Of Large German Companies. Corporate Social Responsibility And Environmental Management, 20(4): 205-218.

Isaksson, R. B., Garvare, R., & Johnson, M. 2015. The Crippled Bottom Line–Measuring And Managing Sustainability. International Journal of Productivity and Performance Management, 64(3): 334-355.

Kerr, I. R. 2006. Leadership Strategies For Sustainable SME Operation. Business Strategy And The Environment, 15(1): 30-39.

Kurk, F., & Eagan, P. 2008. The Value of Adding Design-For-The-Environment To Pollution Prevention Assistance Options. Journal of Cleaner Production 16 (6): 722–726.

Mathiyazhagan, K., Govindan, K., NoorulHaq, A., & Geng, Y. 2013. An ISM Approach For The Barrier Analysis In Implementing Green Supply Chain Management. Journal of Cleaner Production, 47: 283-297.

Maxwell, D., & Van der Vorst, R. 2003. Developing Sustainable Products And Services. Journal of Cleaner Production, 11(8): 883-895.

Opricovic. 1998. Multi-Criteria Optimization Of Civil Engineering Systems, Faculty of Civil Engineering, Belgrade.

Poole, S., & Simon, M. 1997. Technological Trends, Product Design And The Environment. Design Studies, 18 (3): 237–248.

Sánchez-Medina, A. J., Romero-Quintero, L., & Sosa-Cabrera, S. 2014. Environmental Management In Small and Medium-Sized Companies: An Analysis From The Perspective Of The Theory Of Planned Behavior. PluS one, 9(2): e88504.

Shannon, C. E. 2001. A Mathematical Theory Of Communication. ACM SIGMOBILE Mobile Computing and Communications Review, 5(1): 3-55.

Williams, S., & Schaefer, A. 2013. Small And Medium-Sized Enterprises And Sustainability: Managers' Values And Engagement With Environmental And Climate Change Issues. Business Strategy And The Environment, 22(3): 173-186.

Wilson, C. D., Williams, I. D., & Kemp, S. 2012. An Evaluation Of The Impact And Effectiveness Of Environmental Legislation In Small And Medium-Sized Enterprises: Experiences from the UK. Business Strategy and the Environment, 21(3): 141-156.

Zhao, H., & Guo, S. 2014. Selecting Green Supplier Of Thermal Power Equipment By Using A Hybrid MCDM Method For Sustainability. Sustainability, 6(1): 217-235.

THE IMPLEMENTATION OF VENDOR MANAGED INVENTORY IN THE SUPPLY CHAIN WITH SIMPLE PROBABILISTIC INVENTORY MODEL

Ika Deefi Anna
Departement of Industrial Engineering, Faculty of Engineering, University of Trunojoyo Madura, Indonesia
E-mail: deefi_fian@yahoo.com

ABSTRACT

Numerous studies show that the implementation of Vendor Managed Inventory (VMI) benefits all members of the supply chain. This research develops model to prove the benefits obtained from implementing VMI to supplier-buyer partnership analytically. The model considers a two-level supply chain which consists of a single supplier and a single buyer. The analytical model is developed to supply chain inventory with probabilistic demand which follows normal distribution. The model also incorporates lead time as a decision variable and investigates the impacts of inventory management before and after the implementation of the VMI. The result shows that the analytical model has the ability to reduce the supply chain expected cost, improve the service level and increase the inventory replenishment. Numerical examples are given to prove them.

Keywords: inventory, supply chain, economic, probabilistic

INTRODUCTION

Industry players are increasingly aware that to provide a cheap, qualified and quick products they need the role of all players. The players are suppliers, manufacturers, distributors, retails and corporate supporters such as logistics. The companies establish a network that work together to create and deliver products to the end user. The formed network is called supply chain (Pujawan, 2005). Supply chain is a dynamic system that includes all activities like inventory control, manufacturing, distribution, warehousing, and customer service to deliver product from raw material supplier to the end customer (Pasandideh et al., 2010). All activities need to be coordinated and integrated by supply chain management (SCM). SCM gives profitability and cost reduction for whole supply chain member in facing global competition (Tyana and Wee, 2003). SCM coordinates the enterprises along the logistic network and focus in finding the best strategy (Simchi-Levi et al.,2003).

In the traditional supply chain, the buyers determine the timing and size of orders itself based on the information they have. Supplier will respond the buyer's request without finding out more why the buyer ordered that volume. To meet the buyer's demand, the supplier predicts the raw materials and products needed in inventory based on the buyer past orders. This condition forces the supplier to store more inventory to overcome the buyer uncertainty orders. One possible case that often occurs in the traditional supply chain is the buyer's demand changes suddenly. Something that supplier did not anticipate. This condition causes the supplier to change the production schedule. Frequent changes in production schedules lead to schedule nervousness (Pujawan, 2005).

Many companies use the vendor managed inventory (VMI) to solve this problem. In the VMI model, the buyer no longer decides what, when, and how the goods will be purchased from the supplier. The buyer only gives the information about the customer demand and the remaining inventory. They are useful for suppliers to decide the timing and amounts of inventory replenishment (Disney and Towell, 2003). And of course the buyer should also provide information about the minimum and maximum inventory they expect (Pujawan, 2005).

Numerous studies show that the implementation of VMI benefits all members of the supply chain. VMI gives the competitive advantages to the retailers such as the higher product availability and service level, lower ordering cost and inventory monitoring (Waller et al., 1999). The benefit of applying VMI for the supplier are reducing the bullwhip effect (Disney and Towell, 2003) and synchronizing the inventory replenishment planning (Cetinkaya and Lee, 2000).

This research develops a model to prove the benefits obtained from implementing WMI to supplier-buyer parthership analytically. The model is developed for two level supply chain which consist of a single supplier and a single buyer. This research is the extension of Pasandideh et al.'s research which investigated the application of VMI model in supply chain in the EOQ model with shortage (Pasandideh et al., 2010). This research changes the demand assumption from deterministic to probabilistic.

In the probabilistic model, lead time becomes an important issue. Its management can prevent the stock out, reduce the safety stock, improves the customer service level and the competitive advantage of business (Ouyang et al, 2007). Reducing the lead time usually causes an added cost namely the crashing cost. Research of inventory model which consider lead time as decision variable have developed by many researchers. The probabilistic inventory model which incorporate lead time as decision variable was firstly conducted by (Liao and Shyu, 1991). The lead time usually consist of several component, such as order preparation, order transit, set up time, delivery time and the supplier lead time (Tersine, 1994).

This research develops the analytical model of the VMI implementation to supply chain inventory with probabilistic demand which follows normal distribution. The model incorporates lead time as decision variable and investigates the impacts of inventory management before and after the implementation of VMI.

METHODOLOGY

The developed model is used to prove the benefits obtained by implementing VMI to the supply chain inventory with probabilistic demand. In the traditional supply chain (non VMI), the supplier can not obtain the information of consumer demand directly but through the buyer's ordering policy. On the contrary, in the supply chain with VMI, the supplier obtains the information of consumer demand directly by its information system. As a result, the supplier has a combination of two inventory cost, that is order set up and holding cost (Dong and Xu, 2002).

To prove the benefits obtained from VMI implementation on supply chain with probablistic demand, mathematical models will be established for two supply chain models. They are mathematical model for traditional supply chain (without VMI) and mathematical model for integrated (VMI) supply chain. The mathematical models are established by using the following assumptions and notations. The next step is modeling the total cost of both the traditional (Non VMI) and the integrated (VMI) supply chain.

The mathematical models are developed based on the following assumptions :

a. The supply chain consists of a single supplier and a single buyer with a single item which is considered.
b. Demand during planning horizon is a simple probabilistic with average demand *D*, deviation standart *S* and normal distribution.
c. The ordering lot size Q^* is constant for every ordering time, the products will come simultaneusly with leadtime *L*, and the order is done when the inventory level attains to the reorder point *r*.
d. The price of product *p* is constant.
e. The service level η or the probability of inventory shortage is known and determined by the company.
f. The crashing cost is not allowed in this model.

The following notations will be used to develop the model :

TC_{noVMI}	Total expected cost of the traditional (non VMI) supply chain
TC_{VMI}	Total expected cost of the integrated (VMI) supply chain
Q_{noVMI}	The order quantity of the traditional (non VMI) supply chain
Q_{VMI}	The order quantity of the integrated (VMI) supply chain
A_S	The supplier's ordering cost per unit
A_B	The buyer's ordering cost per unit
D	The buyer's demand rate
S	The standard deviation of demand
h_B	Product holding cost per unit held in buyer's store in a period
α_{noVMI}	The probability of inventory shortage in the traditional (non VMI) supply chain
α_{VMI}	The probability of inventory shortage in the integrated (VMI) supply chain
BI_{noVMI}	The buyer's expected inventory cost of the traditional (non VMI) supply chain
BI_{VMI}	The buyer's expected inventory cost of the integrated (VMI) supply chain
SI_{noVMI}	The supplier's expected inventory cost of the traditional (non VMI) supply chain
SI_{VMI}	The supplier's expected inventory cost of the integrated (VMI) supply chain
L_{noVMI}	Lead time of the traditional (non VMI) supply chain
L_{VMI}	Lead time of the integrated (VMI) supply chain
$\hat{\pi}$	The shortage cost per unit per time
Z_α	The safety factor
$f(Z_\alpha)$	The standard normal probability density function
$\psi(Z_\alpha)$	The cumulative distribution function

The case of traditional (Non VMI) Supply Chain

In the traditional supply chain with probabilistic demand, the buyer places an every Q number of demands. The average cycle time is $\frac{Q}{D}$ respectively. The safety stock (SS) is the expected net inventory level which formulated by $SS = Z_\alpha S\sqrt{L}$. The expected net inventory level is $Q + SS$ immediately after the arrival of procurement. The average inventory is about $\frac{Q}{2} + SS$ over the cycle. At the end cycle, the expected demand shortage is given by N with $N = S\sqrt{L}\beta$ and $\beta = f(Z_\alpha) - Z_\alpha\psi(Z_\alpha)$ (Hadley and Within, 1963).

The buyer's inventory cost in the non VMI supply chain comprises of the ordering cost, holding cost and the shortage cost which can be expressed as Eqn. (1).

$$BI_{noVMI} = \frac{A_B D}{Q_{noVMI}} + h_B\left(\frac{Q_{noVMI}}{2} + Z_\alpha\, S\sqrt{L_{noVMI}}\right) + \frac{\hat{\pi} D\, S\sqrt{L_{noVMI}}\,\beta}{Q_{noVMI}} \tag{1}$$

The supplier's inventory cost in the non VMI supply chain comprises of the ordering cost which can be expressed as Eqn. (2).

$$SI_{noVMI} = \frac{A_S D}{Q_{noVMI}} \tag{2}$$

The total inventory cost of traditional supply chain is the summation of the buyer's inventory cost and the supplier's inventory cost indicated by the following Eqn. (3).

$$TC_{noVMI} = \frac{A_B D}{Q_{noVMI}} + h_B\left(\frac{Q_{noVMI}}{2} + Z_\alpha\, S\sqrt{L_{noVMI}}\right) + \frac{\hat{\pi} D\, S\sqrt{L_{noVMI}}\,\beta}{Q_{noVMI}} + \frac{A_S D}{Q_{noVMI}} \tag{3}$$

The buyer's inventory cost in Eqn. (1) is a function of both Q_{noVMI} and L_{noVMI}. The optimal values of both Q_{noVMI} and L_{noVMI} can be obtained by partially differentiating of Eqn. (1). The result from differentiating Eqn. (1) with respect to Q_{noVMI} and setting it zero is as follows.

$$\frac{\partial BI_{noVMI}}{\partial Q_{noVMI}} = 0 \tag{4}$$

$$\frac{A_B D}{Q_{noVMI}} + h_B\left(\frac{Q_{noVMI}}{2} + Z_\alpha\, S\sqrt{L_{noVMI}}\right) + \frac{\hat{\pi} D\, S\sqrt{L_{noVMI}}\,\beta}{Q_{noVMI}} = 0 \tag{5}$$

$$Q^2{}_{noVMI} = \frac{2D}{h_B}\left([A_B + \hat{\pi} DS\,\beta\sqrt{L_{noVMI}}\,]\right) \tag{6}$$

$$Q_{noVMI} = \sqrt{\frac{2D}{h_B}\left([A_B + \bar{\pi} DS\,\beta\sqrt{L_{noVMI}}\,]\right)} \quad (7)$$

Differentiating Eqn. (1) with respect to L_{noVMI} and setting it zero is as follows.

$$\frac{\partial BI_{noVMI}}{\partial L_{noVMI}} = 0 \quad (8)$$

$$\frac{h_B Z_\alpha S}{2\sqrt{L_{noVMI}}} + \frac{\bar{\pi} DS\,\beta}{2\sqrt{L_{noVMI}}\,Q_{noVMI}} = 0 \quad (9)$$

$$h_B Z_\alpha S Q_{noVMI} = -\bar{\pi} DS\beta \quad (10)$$

$$Q_{noVMI} = -\frac{\bar{\pi} D\beta}{h_B Z_\alpha} \quad (11)$$

By substituting Eqn. (11) into Eqn. (7), we can obtain:

$$\frac{2D[A_B + \bar{\pi} S\sqrt{L_{noVMI}}\,\beta]}{h_B} = \left[-\frac{\bar{\pi} D\beta}{h_B Z_\alpha}\right]^2 \quad (12)$$

$$2D\bar{\pi} S\sqrt{L_{noVMI}}\,\beta = \frac{1}{h_B Z_\alpha^2}(\bar{\pi} D\beta)^2 - 2DA_B \quad (13)$$

$$\sqrt{L^*_{noVMI}} = \left(\frac{\bar{\pi} D\beta}{2Sh_B Z_\alpha^2}\right) - \left(\frac{A_B}{\bar{\pi} S\beta}\right) \quad (14)$$

Substituting Eqn. (14) into Eqn. (7), we can obtain the optimal order quantity of the non VMI supply chain as follows:

$$Q^*_{noVMI} = \sqrt{\frac{2D}{h_B}[A_B + \bar{\pi} S\beta\left(\frac{\bar{\pi} D\beta}{2Sh_B Z_\alpha^2} - \frac{A_B}{\bar{\pi} S\beta}\right)} \quad (15)$$

$$Q^*_{noVMI} = \sqrt{\frac{2D}{h_B}\left[A_B + \left(\frac{\bar{\pi}^2 D\beta^2}{2h_B Z_\alpha^2} - A_B\right)\right]} \quad (16)$$

$$Q^*_{noVMI} = \frac{\bar{\pi} D\beta}{h_B Z_\alpha} \quad (17)$$

Furthermore, by substituting Eqn.(14) and Eqn. (17) into Eqn. (3), the expected total inventory cost of the traditional supply chain becomes as Eqn. (18):

$$TC^*_{noVMI} = \frac{3D\bar{\pi}\beta}{2Z_\alpha} + \frac{h_B Z_\alpha}{\bar{\pi}\beta}(A_S - A_B) \quad (18)$$

The case of the integrated (VMI) supply chain

As a result of the implementation of VMI, the buyer's inventory cost switches to supplier. Thus, the buyer's cost becomes zero. The inventory cost of both supplier and buyer also the total inventory cost of the integrated (VMI) supply chain are calculated as follows.

$$BI_{VMI} = 0 \quad (19)$$

$$SI_{VMI} = \frac{A_B D}{Q_{VMI}} + h_B\left(\frac{Q_{VMI}}{2} + Z_\alpha S\sqrt{L_{VMI}}\right) + \frac{\bar{\pi} DS\sqrt{L_{VMI}}\,\beta}{Q_{VMI}} + \frac{A_S D}{Q_{VMI}} \quad (20)$$

$$TC_{VMI} = \frac{A_B D}{Q_{VMI}} + h_B\left(\frac{Q Q_{VMI}}{2} + Z_\alpha S\sqrt{L_{VMI}}\right) + \frac{\bar{\pi} D(S\sqrt{L_{VMI}}\,\beta)}{Q_{VMI}} + \frac{A_S D}{Q_{VMI}} \quad (21)$$

The optimal values of both Q^*_{VMI} and L^*_{VMI} can be obtained by partially differentiating of Eqn. (21). The result from differentiating Eqn. (21) with respect to Q_{VMI} and setting it zero is as follows.

$$\frac{\partial TC_{VMI}}{\partial Q_{VMI}} = 0 \quad (22)$$

$$Q^2{}_{VMI} = \frac{2D}{h_B}\left(A_B + A_S + \bar{\pi} S\,\beta\sqrt{L_{VMI}}\,\right) \quad (23)$$

$$Q_{VMI} = \sqrt{\frac{2D}{h_B}\left(A_B + A_S + \bar{\pi} S\,\beta\sqrt{L_{VMI}}\,\right)} \quad (24)$$

The optimal value of L^*_{VMI} can be obtained by partially differentiating of Eqn. (21) with respect to L_{VMI}, and the result is as follows.

$$\frac{\partial TC_{VMI}}{\partial L_{VMI}} = 0 \quad (25)$$

$$\frac{h_B Z_\alpha S Q_{VMI} + \bar{\pi} DS\,\beta}{2\sqrt{L_{VMI}}\,Q_{VMI}} = 0 \quad (26)$$

$$Q_{VMI} = -\frac{\bar{\pi} D\beta}{h_B Z_\alpha} \quad (27)$$

Inserting Eqn. (27) to Eqn. (24), we will obtain :

$$\frac{2D[A_B + A_S + \bar{\pi} S\sqrt{L_{VMI}}\,\beta]}{h_B} = \left[\frac{\bar{\pi} D\beta}{h_B Z_\alpha}\right]^2 \quad (28)$$

$$2D\bar{\pi} S\sqrt{L_{VMI}}\,\beta = \frac{1}{h_B Z_\alpha^2}(\bar{\pi} D\beta)^2 - 2D(A_B + A_S) \quad (29)$$

$$\sqrt{L^*_{VMI}} = \left(\frac{\bar{\pi} D\beta}{2Sh_B Z_\alpha^2}\right) - \left(\frac{A_B + A_S}{\bar{\pi} S\beta}\right) \quad (30)$$

The optimal order quantity of the VMI supply chain (Q^*_{VMI}) is obtained by inserting Eqn. (30) to Eqn. (27), as follows.

$$Q^*_{VMI} = \sqrt{\frac{2D}{h_B}\left(A_B + A_S + \frac{\bar{\pi} D\beta}{2Sh_B Z_\alpha^2} - A_B - A_S\right)} \quad (31)$$

$$Q^*_{VMI} = \frac{\bar{\pi} D\beta}{h_B Z_\alpha} \quad (32)$$

The expected total inventory cost of the VMI supply chain is obtained by inserting Eqn. (30) and Eqn. (32) to Eqn. (21) and the result is as Eqn.(33).

$$TC^*_{VMI} = \frac{3D\bar{\pi}\beta}{2Z_\alpha} - \frac{h_B Z_\alpha}{\bar{\pi}\beta}(A_S + A_B) \quad (33)$$

RESULTS AND DISCUSSION

The comparison of two supply chain models is carried out to variables such as the optimal order quantity, lead time, service level, and the expected total cost of supply chain.

The comparison between the order quantity Q^*_{noVMI} with Q^*_{VMI}

Developing inventory models on probabilistic demand and lead time as decision variable result in the same value of order quantity between both of supply chain. We can conclude as Eqn. (34).

$$Q^*_{noVMI} = Q^*_{VMI} = \frac{D\hat{\pi}\beta}{h_B Z_\alpha} = Q^* \quad (34)$$

The comparison between the lead time L^*_{noVMI} with L^*_{VMI}

The developed assumption is L^*_{noVMI} greater equal than L^*_{VMI} . Let $L^*_{noVMI} \geq L^*_{VMI}$.

$$L^*_{noVMI} \geq L^*_{VMI} \quad (35)$$

$$\left[\left(\frac{\hat{\pi}D\beta}{2Sh_B Z_\alpha^2}\right) - \left(\frac{A_B}{\hat{\pi}S\,\beta}\right)\right]^2 \geq \left[\left(\frac{\hat{\pi}D\,\beta}{2Sh_B Z_\alpha^2}\right) - \left(\frac{A_B+A_S}{\hat{\pi}S\,\beta}\right)\right]^2 \quad (36)$$

$$\left[\left(\frac{\hat{\pi}D\beta}{2Sh_B Z_\alpha^2}\right) - \left(\frac{A_B}{\hat{\pi}S\,\beta}\right)\right]^2 \geq \left[\left(\frac{\hat{\pi}D\,\beta}{2Sh_B Z_\alpha^2}\right) - \left(\frac{A_B}{\hat{\pi}S\,\beta}\right) - \left(\frac{A_S}{\hat{\pi}S\,\beta}\right)\right]^2 \quad (37)$$

let $\left(\frac{\hat{\pi}D\beta}{2Sh_B Z_\alpha^2}\right) - \left(\frac{A_B}{\hat{\pi}S\,\beta}\right) = \theta$, therefore Eqn. (37) becomes Eqn. (38).

$$\theta^2 \geq \left[\theta - \left(\frac{A_S}{\hat{\pi}S\,\beta}\right)\right]^2 \quad (38)$$

Eqn. (38) shows that the value of L^*_{noVMI} is greater than L^*_{VMI} which proves that the above developed assumption is fulfilled. This condition shows that the implementation VMI can reduce lead time. The shorter lead time impact to reducing the value of safety stock SS^*_{VMI} and the reorder point r^*_{VMI}, so we can generalize that $SS^*_{noVMI} \geq SS^*_{VMI}$ and $r^*_{noVMI} \geq r^*_{VMI}$. The short lead time of VMI supply chain will make the inventory replenishment more frequently. This brings benefits to the buyer in anticipation of uncertain demand and improves customer service level.

The comparison between the expected total inventory TC^*_{noVMI} with TC^*_{VMI}

let $TC^*_{noVMI} \geq TC^*_{VMI}$ to compare the expected total inventory costs of two supply chains.

$$TC^*_{noVMI} \geq TC^*_{VMI} \quad (39)$$

$$\frac{3D\hat{\pi}\beta}{2Z_\alpha} + \frac{h_B Z_\alpha A_S}{\hat{\pi}\beta} - \frac{h_B Z_\alpha A_B}{\hat{\pi}\beta} \geq \frac{3D\hat{\pi}\beta}{2Z_\alpha} - \frac{h_B Z_\alpha A_S}{\hat{\pi}\beta} - \frac{h_B Z_\alpha A_B}{\hat{\pi}\beta}$$
(40)

$$\text{Let } \lambda = \frac{3D\hat{\pi}\beta}{2Z_\alpha} - \frac{h_B Z_\alpha A_B}{\hat{\pi}\beta} \quad (41)$$

by inserting Eqn. (41) to Eqn. (40), the Eqn. (40) becomes Eqn. (42).

$$\lambda + \frac{2h_B Z_\alpha A_S}{\hat{\pi}\beta} \geq \lambda \quad (42)$$

The Eqn. (42) indicates that value of TC^*_{noVMI} is greater than value of TC^*_{VMI}. The difference value of both is about $\frac{2h_B Z_\alpha A_S}{\hat{\pi}\beta}$. This indicates that the assumption above is met analytically and proves that VMI supply chain can reduce the expected total cost.

The comparison between the probability of inventory shortage α^*_{noVMI} with α^*_{VMI}

The comparison of the probability of stock out between two supply chain is used to determine the service level provided by them. Based on (Jha and Shanker, 2009), the equation of service level constrain is as Eqn. (43).

$$\frac{S\sqrt{L}\beta}{Q} = \alpha \quad (43)$$

The developed assumption is α^*_{noVMI} greater equal than nilai α^*_{VMI}.

$$\alpha^*_{noVMI} \geq \alpha^*_{VMI} \quad (44)$$

$$\frac{S\sqrt{L^*_{noVMI}}\,\beta}{Q^*_{noVMI}} \geq \frac{S\sqrt{L^*_{VMI}}\,\beta}{Q^*_{VMI}} \quad (45)$$

Since $Q^*_{noVMI} = Q^*_{VMI} = Q^*$, so Eqn. (45) can be wrote in Eqn. (46).

$$\frac{S\beta}{Q^*}\sqrt{L^*_{noVMI}} \geq \frac{S\beta}{Q^*}\sqrt{L^*_{VMI}} \quad (46)$$

$$\frac{\hat{\pi}D\beta^2}{2Q^* h_B Z_\alpha^2} - \frac{A_B}{Q^*\hat{\pi}} \geq \frac{\hat{\pi}D\beta^2}{2Q^* h_B Z_\alpha^2} - \frac{A_B}{Q^*\hat{\pi}} - \frac{A_S}{Q^*\hat{\pi}} \quad (47)$$

$$\text{Let } \gamma = \frac{\hat{\pi}D\beta^2}{2Q^* h_B Z_\alpha^2} - \frac{A_B}{Q^*\hat{\pi}} \quad (48)$$

Eqn. (48) can be simplified into Eqn. (49).

$$\gamma \geq \gamma - \frac{A_S}{Q^*\hat{\pi}} \quad (49)$$

The value of α^*_{noVMI} is greater than α^*_{VMI} in Eqn. (49). The above assumption in Eqn. (44) is proved. This condition shows that the VMI supply chain gives the value of probability of inventory shortage smaller than traditional supply chain. It means that the VMI supply chain provides the service level better than the Non VMI supply chain. The difference of service level value of both is about $\frac{A_S}{Q^*\hat{\pi}}$.

NUMERICAL EXAMPLES

This section has three numerical examples to justify and demonstrate the proposed method.

Table-1. The initial data

The Parameters	Example 1	Example 2	Example 3
A_B	30	21	45
A_S	40	75	150
h_B	85	85	85
D	9000	9000	9000
S	900	900	900
$\hat{\pi}$	90	90	90

Table-2. The result obtained for example 1

The variables	Non VMI supply chain	VMI supply chain
Q^*	395	395
L^*	0.016234	0.014444
α^*	0.002527	0.002248
TC^*	50,529.13	48,704.11

Table-3. The result obtained for example 2

The variables	Non VMI supply chain	VMI supply chain
Q^*	395	395
L^*	0.016651	0.013337
α^*	0.002592	0.002076
TC^*	51,532.88	48,110.98

Table-4. The result obtained for example 3

The variables	Non VMI supply chain	VMI supply chain
Q^*	395	395
L^*	0.01555	0.009524
α^*	0.00242	0.001482
TC^*	52,696.33	45,852.53

Table 2 shows that the value of non VMI supply chain variables with α=0.05 are $Q^*_{noVMI} = 395$, $L^*_{noVMI} = 0.016234$, the actual $\alpha^*_{noVMI} = 0.002527$, and the total cost $TC^*_{noVMI} = 50{,}529.13$. While the value of VMI supply chain variables are $Q^*_{VMI} = 395$, $L^*_{VMI} = 0.014444$, the actual $\alpha^*_{VMI} = 0.002248$, and the total expected cost $TC^*_{VMI} = 48{,}704.11$. Both models have the same value of optimal order quantity. But for the other three variables such as lead time, the actual probability of inventory shortage, and the total cost, VMI supply chain has a better value. By implementing the VMI program, the total expected cost can be reduced as 3.75%. This savings is also obtained for the example 2 and 3. The value of total expected cost in table-3 is decreased from 51,532,88 to 48,110.98 with the decreasing value as 7.11%. While the decreasing value of total expected cost is 12.99% for example 3 in table-4.

We can conclude from all table result, the optimal order quantity of non VMI supply chain is equal to the optimal order quantity of VMI one ($Q^*_{noVMI} = Q^*_{VMI}$). This condition occurs in the probabilistic inventory model which considering lead time as a decision variable. The optimal lead time of non VMI supply chain is longer than the optimal lead time of integrated (VMI) one ($L^*_{noVMI} > L^*_{VMI}$). While the actual of inventory shortage probability of Non VMI supply chain is greater than the optimal lead time of integrated (VMI) one ($\alpha^*_{noVMI} > \alpha^*_{VMI}$).

The decreasing of the total expected cost in the VMI supply chain is caused by the decreasing buyer's holding cost and transferring buyer's ordering cost to supplier. Subsequently, the decreasing lead time reduces safety stock and reorder point in the integrated (VMI) supply chain. The shorter lead time can increase inventory replenishment which can ensure the product availability in buyer's store. Finally, it can improve the product service level.

The application of VMI program in the probabilistic inventory model whereas lead time is treated as a decision variable is able to reduce the actual value of α. Both of the supply chain has the smaller actual value of α than the initial one ($\alpha = 0.05$). In brief, implementation of VMI program in supply chain which has condition such as probabilistic demand, a single supplier and a single buyer and lead time as a decision variable has ability reducing the total expected cost, lead time and improving the service level.

CONCLUSIONS

This paper contributes an analytical model that proves benefits obtained from implementing VMI in the supply chain. The model considers a two-level supply chain which consists of a single supplier and a single buyer. The analytical model investigates the impacts of inventory management before and after the implementation of the VMI.

This model is extension of Pasandideh et al's model by changing the deterministic assumption into probabilistic assumption with normal distribution. In this model, lead time is treated as decision variable together with the order quantity. This model has a broader application of the real system compared with the previous one since the demand in the real system is uncertain.

The model results show that the application VMI to the supply chain can reduce the total expected cost of supply chain, lead time and improve the service level. The study of changing the above assumptions will be a possible topic for further research.

ACKNOWLEDGEMENT

The author thanks the referees for the constructive comments and suggestions. This research was supported by International Journal Grant of Engineering Faculty, University of Trunojoyo Madura.

REFERENCES

Cetinkaya S. and Lee C.Y. 2000. Stock Replenishment and Shipment for Vendor Managed Inventory Systems. Management Science. 46 (2) : 217 – 232.

Disney S.M. and Towill D.R. 2003. The Effect of Vendor Managed Inventory Dynamics on the Bullwhip Effect in Supply Chains. International Journal of Operation and Production Economics. 85(2) : 199-215.

Dong Y. and Xu K. 2002. A Supply Chain Model for Vendor Managed Inventory. Transp Res Part E Logist Trans Rev 38: 75 – 95.

Hadley G. and Within T.1963. Analysis of Inventory System. Eaglewood Cliffs. New Jersey. Prentice Hall.

Jha J.K. and Shanker K. 2009. Two-Echelon Supply Chain Inventory Model with Controllable Lead Time and Service Level Constrain. Computers & Industrial Engineering. 57 : 1096-1104.

Liao C.J. and Shyu C.H.1991. An Analytical Determination of Lead Time with Normal Demand. International Journal of Operations and Productions Management. 11(90) : 72-78.

Ouyang L.Y, Wu K.S., and Wo C.H. 2007. An Integrated Vendor-Buyer Inventory Model with Quality Improvement and Lead Time Reduction. International Journal of Production Economics. 108(1-2) : 349 -358.

Pasandideh S.H.R., Niaki S.T.A., and Nia A.R.. 2010. An Investigation of Vendor Managed Inventory Application in Supply Chain : The EOQ Model with Shortage. International Journal of Advanced Manufacturing Technology. 49 : 329-339.

Pujawan, I.N. 2005. Supply Chain Management. Penerbit Guna Widya. Surabaya, Indonesia.

Simchi-Levi D., Kaminsky P., and Simchi-Levi E. 2003. Designing and Managing the Supply Chain : Concepts, Strategy, and Case Studies. Mc Graw Hill. New York, USA.

Tersine R. J. 1994. Principles of Inventory and Materials Management. Eaglewood Cliffs. New Jersey. Prentice Hall.

Tyana J. and Wee H.M. 2003. Vendor Managed Inventory : A Survey of the Taiwanese Grocery Industry. Journal of Purch Supply Management. 9 : 11-18.

Waller M.A., Johnson M.E., and Davis T.1999. Vendor Managed Inventory in the Retail Supply Chain. Journal of Business Logistics. 20 (1) : 183 – 203.

ANALYSIS OF THE EFFECT OF INFORMATION SYSTEM QUALITY TO INTENTION TO REUSE OF EMPLOYEE MANAGEMENT INFORMATION SYSTEM (SIMPEG) BASED ON INFORMATION SYSTEMS SUCCESS MODEL

Tri Lathif Mardi Suryanto[1*], Djoko Budiyanto Setyohadi [2], Asif Faroqi [3]
[1,3]University of Pembangunan Nasional "Veteran" Jawa Timur, Surabaya, East Java, Indonesia
[2]Universitas Atma Jaya Yogyakarta, Department of Informatics Engineering, Yogyakarta, Indonesia
E-Mail: trilathif.si@upnjatim.ac.id

ABSTRACT

This study examines the effect of Information Quality, Systems Quality and Service Quality on the user intention to reuse Employee Management Information System (SIMPEG) in University in the city of Surabaya, based on the theoretical foundation of DeLone and McLane Information Systems Success (ISS) Model. The distribution of questionnaire was conducted to 120 employees of different universities by means of stratified random sampling. The results showed that: (1) there is a significant positive effect of the System Quality on the Quality of Information, (2) there is a significant positive effect of the Information Quality on the Intention to Reuse, information related to the fulfillment of the user's needs; (3) there is a significant positive effect of the Quality of the Intention on system re-use, the system related to the fulfillment of the needs of users; (4) there is no effect of the Quality of Service to the Intention to Reuse. In the end, the results of this study provide an analysis and advice to The University officials that can be used as a consideration for Information Technology/Information System investment and development in accordance with the Success of Information System and Intention to Reuse model.

Keywords: information system success model, intentions to reuse, information technology/information system.

INTRODUCTION

In recent years, due to the rapid advancement of Information Systems, business information systems have been implemented for management operations with the aim of creating a profit for the organization and, without exception, this also applies to the teaching and learning activities. To support the learning process and well organize employees an information system which is related to employees in University is made, commonly referred to as Employee Management Information System (SIMPEG). In order for information system to be utilized effectively and contribute to the performance of organization, the members of the organization should be able to use these technologies properly Lucas and Spitler (1999).

The learning process accomplishment in University involving all relevant aspects. The process, have a correlation with the student, employee, administrative clerks, financial officers, student support officer, up to campus official, and many more. University are in charge for an excellent service and special treatment to be stand out among others. Methods of integrating facility management also need to be considered in order to get a measure of performance required by the organization's core Hinks (2002)

By considering the good impact on teaching and learning activities, the University is expected to treat their employees well, such as: provide a good facilities, fulfill their employee's rights, provide a comfortable atmosphere, and the staff will give the University all the best they have in return. In other words, a good treatment by the University will have cause an implications for students. Providing a good facilities for employees need a special attention from management, this is because the university has a high complexity of needs compared to other educational institutions. The easier way to get information related to this reciprocal relationship is necessary to be respond seriously. Facilities create an environments for occupants to work effectively within organisations and the performance of these environments influence the activities that are carried out (Amaratunga, et. al, 2002).

In regard of that special needs, it is necessary for University to find a solution. One solution is the application of Information Systems/Information Technology (IT/IS) to provide a good information for University and to reduce the gap between employees. To simplify, accelerate, and adding quality of university services for students. Thus, the achievement that makes competition among University has developed in the era of Information System/Information Technology competition at internal and external services, Atkin define Facilities Management as an integrated approach to the operation, maintenance, repair and adapting buildings and infrastructure of an organization in order to create an environment that strongly supports the main objective of the organization Atkin and Brooks (2000)

At first the discussion related to the ability of the Information System/Information Technology with a positive impact on University can be tested using the Information System Success (ISS) DeLone and McLean Model (1992), relevant study is comprised of six dimensions: system quality, information quality, use, user satisfaction, individual impact and organizational impact. However, other researchers Pitt, Watson and Kavan (1995) adds that the quality of service becomes an additional test dimension in achieving organizational impact. So DeLone and McLean Model (2003) adds quality of service to be independent factors other than the quality of information and system quality that can have an impact on the organization, reinforced by Tseng and Li (2008) ISS i.e. including system quality, information quality, service quality, the use of system , user satisfaction, and benefits to the organization. Furthermore, the three factors are develop to integrity, firmness, accuracy, precision, relativeness, facilitation, and the timeliness of any report produced, and assessment quality system refers to the accessibility, functionality, operation, responsiveness, and system reliability, while the quality of service refers to how often users use the system and the willingness to use.

The attempt to measure the relationship between the desire to use, usage, user satisfaction and benefit to the organization and this contest is an important indicator for the success of information systems Liu, et. al (2010). The discussion above of the ISS shows that the quality of information, systems quality, and quality of service can be used to assess information to allow the quality of information systems to meet user expectations, improve user satisfaction, and increase profits for the organization systems Wixom (2001); DeLone & McLean (2004)

The research related to educational activities that are so complex to manage the employees and increase of University standard with the Information Technology /Information System, can be a limited applied research. The development of the IT/IS has a significant way to the University to explore and develop significant ways for educational programs, therefore, the intention of returning the system uses is important for the success of the role of a system Li (2011). Thus, this study is using the ISS as a theoretical foundation to examine the influence of information quality, system quality, and service quality of Employee Information Systems to the intention of re-use SIMPEG on University in Indonesia especially in the city of Surabaya.

METHODOLOGY

Research subjects

The core of this research is the user of the Employee Management Information System in the State of Indonesia especially in the city of Surabaya that was surveyed using questionnaires given to respondents who are employee at different universities in Surabaya. This activity of sampling was held from 2nd to14^{th} February 2015. The questionnaires distributed by the researchers directly (offline). In total, 120 copies of questionnaires were distributed and collected, with a response rate of 100%. After removing invalid questionnaire, there are 100 copies, with an effective response rate of 97.1%.

Research Tools

The research tool consists of two parts. The first is the study of literature, gathering all the information that is contained in each relevant SIMPEG University. Secondly, ISS using Delone Mclane Scale Model (2003) developed by Li (2011). Thus it is supported by a Likert scale which is divided into five points including a "very poor", "less", "pretty", "disagree", "strongly agree" that corresponds to 1 to 5 points respectively.

Research Stucture

In accordance with the purpose of the research and literature review, this study developed a structure (shown in Fig.1) by ISS to examine the effect of information systems, systems quality, and quality of the employee's intention to reuse SIMPEG on University.

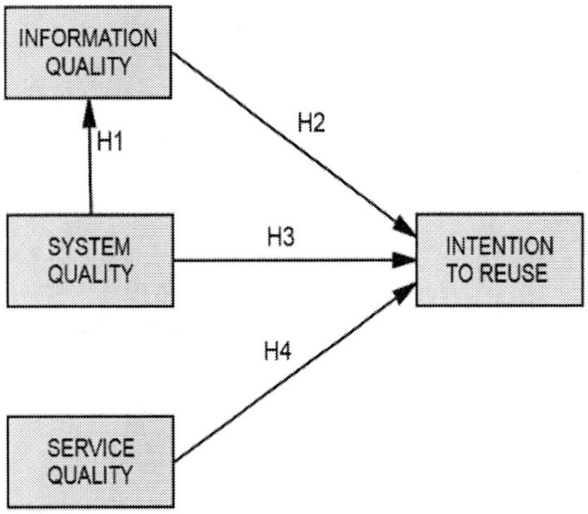

Figure-1. Research Hypothesis

Research hypotesis

This research uses the ISS proposed by DeLone and McLean (1992) as the basis for the theory. There are three factors forming the intention to reuse a system of information quality, system quality, and quality of service. All these three are the most important base factors for measuring the quality of information systems, on the other research Saeed et. al. (2003) found that the quality of information is an important influential factor in determining the willingness to buy back and customer satisfaction.

Statement that supports and adds that the quality of information is an influential factor important in determining the willingness to reuse and satisfaction emerges from Swan and Trawick (1981) which states that during the process of forming the intention to use the system, the level of service quality determine the intentions of the user's back. Thus, this study proposes two formulation of the problem as follows:

System quality has a significant influence on the information quality to support the establishment of information quality factors which significantly influence the intention to reuse SIMPEG.

Data analysis

This study uses the software, SPSS 21 and AMOS 16.0 for Windows 7 for the needs of data analysis.
Statistical methods adopted by this study are described as follows:

Descriptive statistics were used as the data analysis support to illustrate SIMPEG user data has been collected, make the results of the demographic distribution at the University SIMPEG users who respond to this study.

Path analysis is used to provide the results of the analysis relating to the significance of the relationship and the influence of information quality, significance and influence of the quality system, and the significance and impact of service quality to the intention of reusing SIMPEG University

RESULTS AND DISCUSSION

Sample characteristics

Table-1. Summary of Background Variables of Research

Background variable	Name of group	Frequency (time)	%	Accumulated %
Sex	Male	59	59	59
	Female	41	41	100
Age	20–30	49	49	49
	31–40	20	20	69
	41–50	14	14	83
	More than fifty	17	17	100
Education Level	Senior High School	4	4	4
	Undergraduate	39	39	43
	Postgraduate	50	50	93
	Doctoral	7	7	100
Position	Staff Employees	42	42	42
	Lecturer	48	48	90
	Campus Officials	10	10	100
Ability to use a Computer	Below Average	11	11	11
	Average Ability	61	61	72
	Above Average	28	28	100

This study used descriptive statistics to analyze the distribution of basic information (see Table 1). The effective sample size is 100 respondents, 59 were male (59%) and 41 were female (49%), showed a majority of employees University in Surabaya are men, but at least the difference between genders is can be interpreted comparison male employees and women are still equal. In terms of age, the largest age group is between the age of 20-30 with a percentage (49%), followed by the age of 31-40 with percentage (20%), then at the age of more than 50 with percentage (17%), and the smallest group at the age of 41-50 with percentage (14%).

Thus we can say that the use SIMPEG competence can be applied as expected given the level of the users are in the productive age. In terms of educational background may also affect the likely position at the University, there are (4%) SIMPEG users who have a high school degree, 39% of users have a college degree, 50% have a master's degree and there is a 7% users have a doctorate. Subjects also received questions about the ability of a computer, and as a result, there are 61% of the employees who have the ability average, and 28% of employees have an above-average ability.

Testing for offending estimates

Table-2. Test of offending estimates

	Question	Standardized Regression Coefficient	Error Variance
X1	SIMPEG provide actual information to the users	79	30
X2	SIMPEG has a data records to give user an easy way to look for some information	76	31
X3	SIMPEG provide a correct information which fits the needs of users and institution	79	27
X4	SIMPEG provide a concise information which help your work	73	37
X5	SIMPEG provide a Help Menu with which users can easily operate it without the assistance from colleagues	64	48
X6	SIMPEG has a direct benefit, hence users is not need to contact the person in charge to SIMPEG	70	43
X7	SIMPEG has a user guide with which help the users to operate it	69	43
X8	SIMPEG is easy to use and helpful both for users and institution	75	48
X9	SIMPEG provide a valid and complete data which suitable with user's needs according to his job/responsibility	60	56
X10	SIMPEG can be accessed by multiple users, which can gives a sense of security for users	68	48
X11	SIMPEG can be used and understood easily so that it can make the user's jobs done easier	78	40
X12	SIMPEG can be customized by the user to fits their needs and based on the level of their work	76	40
X13	SIMPEG can give a fast response of user input, so that users doesn't need help	69	41
X14	I intend to spend more time to use SIMPEG in order to share knowledge with colleagues	62	31
X15	I intend to use SIMPEG consistently in order to support my work	57	27
X16	I intend to learn the SIMPEG thoroughly in order to help the success of the institution work program	53	30
X17	I intend to use SIMPEG regularly to help institution to develop its assets	53	25

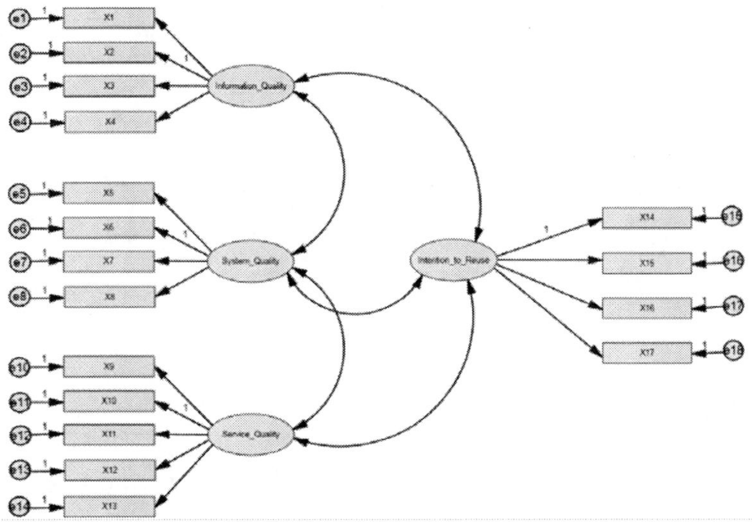

Figure-2. Structurre of the Confirmatory Factor Analysis of the Informationn System Successs Model

This study uses (1) The negative error variance and (2) the proximity of standardized regression coefficients are useful as a basis for testing estimates offensive. As shown in Table 2, there is no negative error variance, and standard coefficient falls between 0.68 and 0.86, both lower than 0.95, it indicates no offensive estimates in the model, and the overall model fit.

Test for convergent validity

This study conduct confirmatory factor analysis (CFA) for the IST Scale and found four dimensional model (system quality, information quality, service quality and willingness to use), the load is between 0.65 ~ 0.86, component reliability between 0.89 ~ 0.90 , and the average variance extracted (AVE) is between 0.58 ~ 0.66, which are all meet the standards, as shown in Table 3. as a result, each dimension of the ISS reaches convergent validity.

Table-3: Reliability and AVE Construc cted by Observable Variable and Latent Variable

Latent Variable	Observable Variable	Factor Loading	Component Reliability	AVE
IQ	X1	79	1,33	,51
	X2	76		
	X3	79		
	X4	73		
SQ	X5	64	1,33	,51
	X6	70		
	X7	69		
	X8	75		
SEQ	X9	60	1,25	,51
	X10	68		
	X11	78		
	X12	76		
	X13	69		
ITR	X14	62	1,33	,51
	X15	57		
	X16	53		
	X17	53		

Testing discriminant validity

This study adopts Bootstrap Distribution Efect proposed by Torkzadeh, Koufteros and Pflughoeft (2003) to test the discriminant validity by building confidence interval for the correlation coefficient of each dimension. Discriminant validity test for the ISS is shown in Table 4, show that the correlation coefficient has a standard 95% confidence intervals were not included. Thus, ISS has discriminant validity between dimensions.

Table-4. 95% of Confidence Intervals of Bootstrap Correlation Coefficient

Parameter	Estimate	Lower	Upper	P
Information Quality <--> System Quality	,957	,942	,969	,000
Information Quality <--> Service Quality	,609	,456	,724	,000
System Quality <--> Service Quality	,545	,370	,672	,000
Service Quality <--> Intention to Use	,627	,463	,749	,000
System Quality <--> Intention to Use	,945	,927	,961	,000
Information Quality <--> Intention to Use	,975	,966	,986	,000

Structural Model Analysis

This study uses six indices for evaluating the overall fit model ie, the ratio of χ2Chi-Square, root mean square error of approach (RMSEA), goodness of fit index (GFI), adjusted 3 goodness of fit index (AGFI), the comparative fit index (CFI), and the comparative fit index parsimony-customized (PCFI). Results from this analysis are reproduced in Table 5, and shows the overall fit standard and acceptable to the model used in this study.

Table-5. Fit Analysis of Research Mode

Fit Indices	Range of Allowance	Research Structural Model	Determination of Model fit
χ^2Chi-Square	Minimized Valued	301,980	Good
RMSEA	$\leq$ 0,08	0,13	Less
GFI	$\geq$ 0,8	,80	Good
AGFI	$\geq$ 0,8	,73	Less
CFI	$\geq$ 0,9	,86	Less
PCFI	$\geq$ 0,5	,73	Good

Test Results of Research Hypothesis

As shown in Table 6, the system quality has a proven positive impact on information quality, then the first hypothesis proved the existence of a positive relationship between system quality and information quality. The 2nd hypothesis can be seen in Table 6, that information quality can directly influence positively the intention of re-use. Something similar happened to hypothesis 3, wherein the system quality has positive influence on the intention to reuse, but the hypothesis 4, where service quality can be a positive influence on the intention of re-use is not proven.

Table-6. Empirical Results of Research Hypothesis

H	Path Relationship	Path Value	Support of Hypothesis
H1	System Quality --> Information Quality	1,17*	Yes
H2	Information System --> Intention to Reuse	0,64*	Yes
H3	System Quality --> Intention to Reuse	0,56*	Yes
H4	Service Quality --> Intention to Reuse	0,09	No

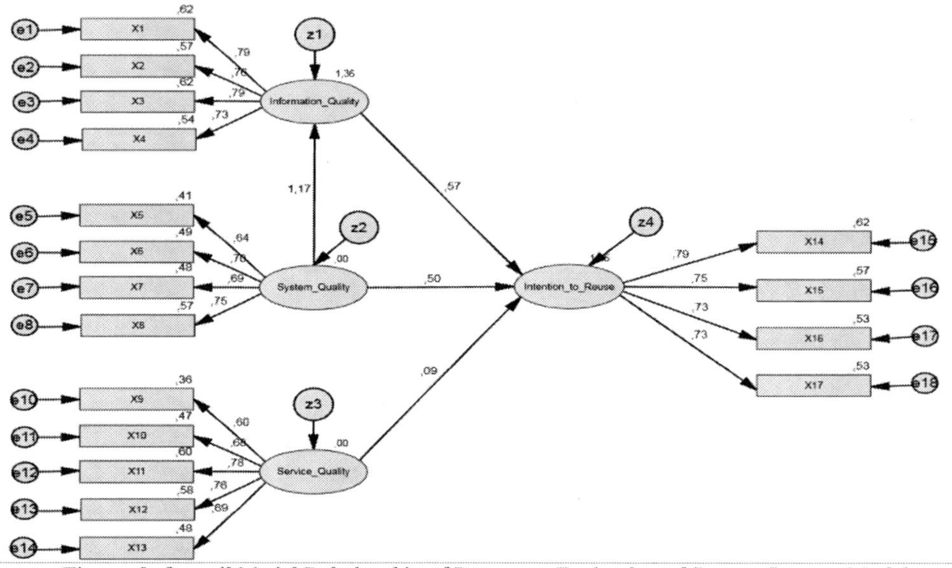

Figure-3. Overall Model Relationship of Parameter Evaluation of System Success Model

CONCLUSIONS

This study uses IIS as a theoretical basis, to examine the three-dimensional model (system quality, information quality, and quality of service) as an independent variable to explain the intention to reuse SIMPEG, research results are described as follows:

The positive effects of significant quality on the quality of the information system appears to have contributed in encouraging the formation of the quality of information of significant value to the intention of re-use SIMPEG on University. It can be assumed that there is a correlation between the capabilities and robustness of the system on the quality of information that will be presented to the user. Therefore, to create a good quality of information it can be started by developing a system which is strong, reliable, trustworthy, and capable of providing more flexibility in accessing information to both general and special users of the system at university.

REFERENCES

Amaratunga, D. & Baldry, D. (2002). Sample View of Current Performance Measurement Practices in Facilities Management. Proceedings of International Postgraduate Conference, The University of Salford.

Atkin, B. & Brooks, A. (2000). Total Facilities Management. London: Blackwell Science.

Cheng, Mu. Wu., A Study on the Willingness to Use Information System of Sport Event Based on Information System Success Model, The Journal of Human Resource and Adult Learning, Vol. 9, Num. 2, December 2013 issue

DeLone, W. H., & McLean, E. R. (1992). Information system success: the quest for the dependent variable. Information Systems Research, 3(1), 60-95.

DeLone, W. H., & McLean, E. R. (2003). Measuring e-commerce success: applying the DeLone & McLean Information Systems

Hinks, J. (2002). Lies, Damned Lies and KPIs: Moving Towards More Meaningful Performance Measurement for FM. In the Proceedings of Euro FM Symposium. The University of Salford, UK.

Li, Yan , Yanqing Duan, Zetian Fu, and Philip Alford. "An empirical study on behavioural intention to reuse e-learning ." British Journal of Educational Technology, 2011.

Liu, N. C., Huang, H. C., & Liao, C. C. (2010). An empirical study on the use effectiveness of decision-making supportive system: combining task-technology fit and information success model. Journal of E-Business, 12(3), 407-430.

Lucas, Henrt C, dan Spitler, V.K.. (1999)."Tecnology Use and Performance: A Field study of Broken Workstasions." Decision sciences.

Pitt, L. F., Watson, R. T., & Kavan C. B. (1995). Service quality: a measure of information systems effectiveness. MIS Quarterly, 19(2), 173-187.

Saeed, K. A., Hwang, Y., & Yi, M. Y. (2003). Toward an integrative framework for online consumer behavior research: a meta-analysis approach. Journal of End User Computing, 15(4), 1-26.

Swan, J. E., & Trawick, I. F. (1981). Disconfirmation of expectations and satisfaction with aretail service. Journal of Retailing, 57, 40-67.

Tseng, F. C. & Li, T. H. (2008). A study on evaluation of library E-Information integrated search system. Journal of Library and Information Studies, 6 (2), 111-142.

Wixom, B. H. (2001). An empirical investigation of the factors affecting data warehousing success, MIS Quarterly 25(1), 17-41. Yeh, H. C. (1990). Management of schools ports event. National Sports Quarterly, 19(1), 22-35.

DETERMINATION OF SELECTION METHOD IN GENETIC ALGORITHM FOR LAND SUITABILITY

Asti Dwi Irfianti[1], Retantyo Wardoyo[2], Sri Hartati[2], EndangSulistyoningsih[3]

[1]Department of Information System, Faculty of Industrial Technology
University of Pembangunan Nasional "Veteran"Jawa Timur, Surabaya, Indonesia, [2]Department ofComputer Science and Electronic
Faculty of Mathematic and Natural Science, Gadjah Mada University, Yogyakarta,Indonesia
[3]Faculty of Agriculture Gadjah Mada University, Yogyakarta Indonesia
E-Mail: asti.ilkom2000@gmail.com

ABSTRACT

Genetic Algoirthm is one alternative solution in the field of modeling optimization, automatic programming and machine learning. The purpose of the study was to compare some type of selection methods in Genetic Algorithm for land suitability. Contribution of this research applies the best method to develop region based horticultural commodities. This testing is done by comparing the three methods on the method of selection, the Roulette Wheel, Tournament Selection and Stochastic Universal Sampling. Parameters of the locations used in the test scenarios include Temperature = 27°C, Rainfall = 1200 mm, hummidity = 30%, Cluster fruit = 4, Crossover Probabiitiy (Pc) = 0.6, Mutation Probabilty (Pm) = 0.2 and Epoch = 10. The second test epoch incluides location parameters consist of Temperature = 30°C, Rainfall = 2000 mm, Humidity = 35%, Cluster fruit = 5, Crossover Probability (Pc) = 0.7, Mutation Probability (Pm) = 0.3 and Epoch 10. The conclusion of this study shows that the Roulette Wheel is the best method because it produces more stable and fitness value than the other two methods.

Keywords: Method Selection, Genetic Algoritm, Roulette Wheel, Tournament Selection, Stocastic Universal Sampling

INTRODUCTION

Genetic Algorithm (GA) is a stochastic optimization algorithms block adopting Darwin's theory. GA has been applied to solve problems such as optimization and modeling, automatic programming and machine learning (Goldbeg, 1999). There are six basic GA need to be considered which are the issue of the chromosome (genome) representation, the selection function, genetic operators such as mutation and crossover, the initial population creation, termination criteria and the evaluation function (Zadeh and Ghazalian, 2010).

Search on GA begins with a set of the population (Blickle and Thiele, 1995). The population consists of chromosomes. Each chromosome is a problem-solving solution. Populations will produce new offsprings which are expected to be better than the previous population. Good natural population will have an opportunity to continue to be developed in order to produce good offspring.

In order for GA to run, first, fitness function stating the success rate of a population needs to be defined. By calculating the fitness function, then it will be determined population to be set to produce the next generation. This process is commonly referred to as the selection process. This process is one of the stages set out in an iterative process (Blickle and Thiele, 1995).

The selection process is a process that plays an important role in GA (Huang and Wang, 2011). Examine the importance of the selection process will be used as a reference for determining the chromosomes quality. The best generation is the generation that can move on to the next selection method (Mitchell, 1997).

Selection process GA is defined as a fitness function which states the success rate of the population, calculation of fitness function will get the population to maintain dan produce the next generation. This process is a stages series of the iteration process.

There are various techniques that can be used in the selection process on the issue. Selection techniques used depends on the problem to be solved. There are several methods for parent selection, such as the rank-based Centre Task, Roulette Wheel Selection, Stochastic Universal Sampling, Selection of local, cutting Selection, Selection Tournament (Mohebi *et al.*, 2009). In this paper, the selection process used Roulette Wheel Selection (RWS), Selection Tournament and Stochastic Universal Sampling (SUS). Parameters used in the process selection called fitness. Fitness is used to determine how well the chromosome will survive (Jebari and Madiafi, 2013).

METHODOLOGY

Block Diagram used in this research is as follows:

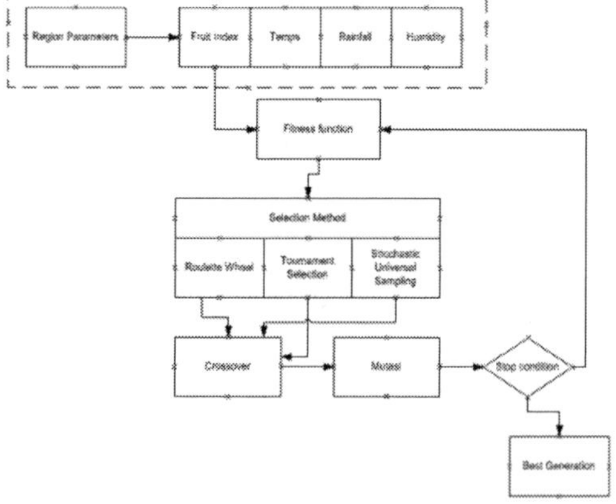

Figure-1. Block Diagram Selection Method

Genetic Algorithm

Genetic Algoritm was proposed by John Holland in early 1970s, it applies some of natural evolution mechanism such as *crossover, mutation,* and survival of the fitness to optimization and machine learning. GA provides search method efficiently working on population, and has been applied to many problems of optimization and classification (Goldbeg, 1999; Mohebi, 2009*)*. General GA processes are as follows:

1) Initiate the population of genes
2) Calculate the fitness of each individual in the population
3) Reproduce the individual selected to form a new population according to each individual's fitness
4) Perform crossover and mutation on the population
5) Repeat step (2) through (4) until some condition is statisfied.

```
Algorithm: GA(n, χ, μ)
// Initialise generation 0:
k := 0;
P_k := a population of n randomly-generated individuals;
// Evaluate P_k:
Compute fitness(i) for each i ∈ P_k;
do
{   // Create generation k + 1:
    // 1. Copy:
    Select (1 − χ) × n members of P_k and insert into P_{k+1};
    // 2. Crossover:
    Select χ × n members of P_k; pair them up; produce offspring; insert the offspring into P_{k+1};
    // 3. Mutate:
    Select μ × n members of P_{k+1}; invert a randomly-selected bit in each;
    // Evaluate P_{k+1}:
    Compute fitness(i) for each i ∈ P_k;
    // Increment:
    k := k + 1;
}
while fitness of fittest individual in P_k is not high enough;
return the fittest individual from P_k;
```

Figure-2 Pseudocode GA (Chusadama et al., 2011)

Crossover operation swap part of genetic bit string within parents. It emulates just as crossover of genes in the real world that descendants are inherited characteristics from both parents.

Mutation operation inverts some bits from whole bit string at very low rate. In real world we can see that some mutants come out rarely. Fig 1 shows the way of applying crossover and mutation operation to genetic algorithm. Each individual in the population evolves to getting higher fitness generation by generations.

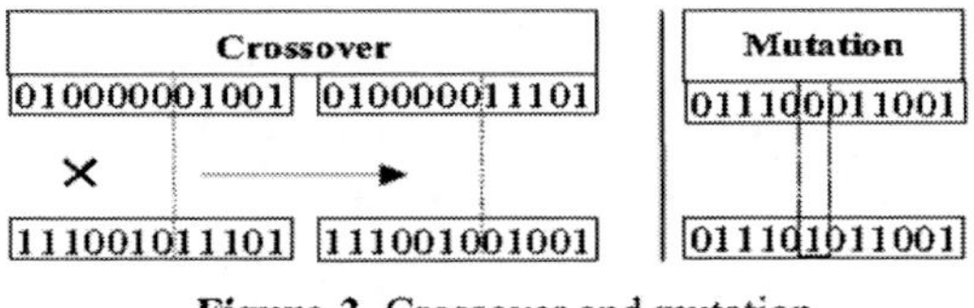

Figure-3. Crossover and mutation

Roullete Wheel

a. Choose a solution that is useful for recombination.

b. The Fitness function provides fitness for a solution chromosomes. The fitnesss level to connect with the probability of selection of each individual chromosome.

c. If Fi is fitness for the individual i in the population, the possibility of elected if pi = fiΣNj = 1fj, where N is the number of individuals in the population.

d. The proportion is assigned to each of the wheels based on the choice of fitness. By dividing the total Gym fitness of all so that the normalization of the 1st choice is made randomly.

e. Candidate solutions with higher fitness is less likely to be eliminated, the selected fitness will eliminate the presentation of the candidates who are weak. Fitness proportionate makes it possible for weak solutions to survive. There is a benefit if the solution is weak since it may include several components that may be useful to follow the process of recombination.

Tournament Selection

```
Algorithm 1: (Tournament Selection)

Input: The population P(τ) the tournament size t ∈ {1, 2, ..., N}
Output: The population after selection P(τ)'

tournament(t, J_1, ..., J_N):
    for i ← 1 to N do
        J'_i ← best fit individual out of t randomly picked
               individuals from {J_1, ..., J_N};
    od
    return {J'_1, ..., J'_N}
```

Figure-4. Tournament Selection (Tobias et al., 1995)

Often tournament held between two individuals (binary tournament) but in general it can change group size t which is also called the tournament size. Outline of the algorithm can be seen from the tournament selection which is implemented very efficiently.

Stochastic Universal Sampling

Parent chosen based on the fitness value, the greater the value of fitness, the greater the likelihood is selected. Stochastic Universal Sampling uses a single random value. Stochastic Universal Sampling assessed from a small random number and select the next candidate from the remaining population. Here below is the algorithm of selection methods Stochastic Universal Sampling:

```
Algorithm 6: (Stochastic Universal Sampling)

Input: The population P(τ) and the reproduction rate for each
       fitness value R_i ∈ [0, N]
Output: The population after selection P(τ)'

SUS(R_1, ..., R_n, J_1, ..., J_N):
    sum ← 0
    j ← 1
    ptr ← random[0,1)
    for i ← 1 to N do
            sum ← sum + R_i where R_i is the reproduction rate
                  of individual J_i
            while (sum > ptr) do
                  J'_j ← J_i
                  j ← j + 1
                  ptr ← ptr + 1
            od
    od
    return {J'_1, ..., J'_N}
```

Figure-5. Stochastic Universal Sample(Tobiaset.al., 1995)

RESULTS

In this paper, we have compared cluster horticulture with selection method in Genetic Algorithm. The selection method analysis are Roulette Whell, Stochastic Universal Sampling and Tournament Selection.

Parameters used in figure 6 Temperature = 27°C, Rainfall = 1200 mm, humidity = 30%, Cluster fruit = 4, Crossover Probability (Pc) = 0.6, Mutation Probability (Pm) = 0.2 dan Epoch = 10.

The graph in figure. 6 shows fitness value from Roulette Wheel method selection. This graph below shows the difference appearance of horticulture. It can also be seen that Carrots appears more than others at 60%. That means Carrots are the most suitable plants in this location.

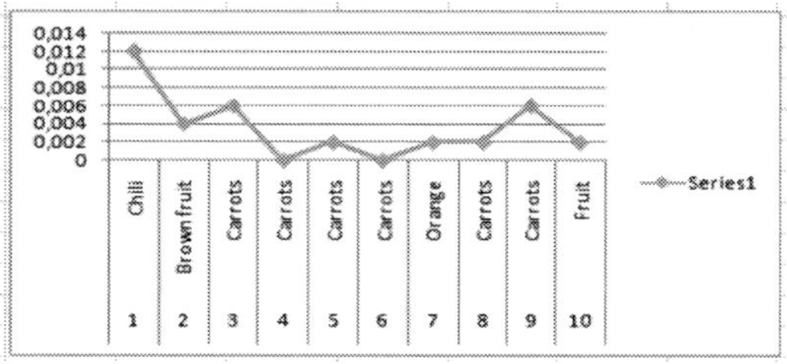

Figure-6. Roulette wheel (a)

The graph in figure 7 show fitness value from Tournament Selection.

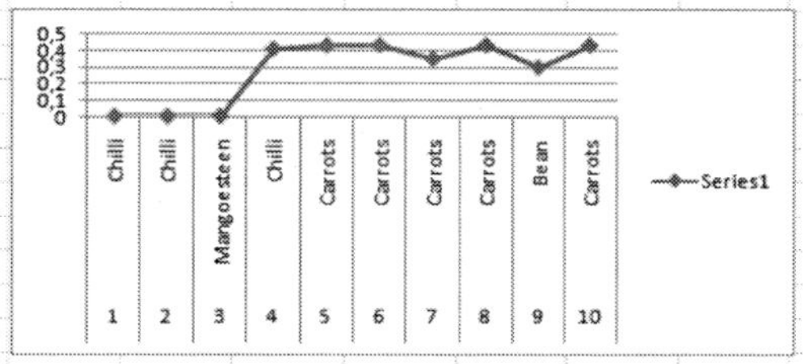

Figure-7. Tournament Selection (a)

In figure 7, percentages of Carrots is 50% and chilli is30%. That means Carrots are the most suitable plants in this location.

In the figure 8 it can be seen in the graph that Orange has a frequency of occurence appears 5 times, followed by Chilli appears 3 times.

Parameters used in the experiments 2 is Temperature = 30°C, Rainfall = 2000 mm, humidity = 35%, Cluster fruit = 5, Crossover Probability (Pc) = 0.7, Mutation Probability (Pm) = 0.3 and Epoch 10. The study produced several experiments as follows: The fitness value generated Roulette Wheel method on iteration 1 to 10 to produce the value of fitness and commodities in sequence as follows:

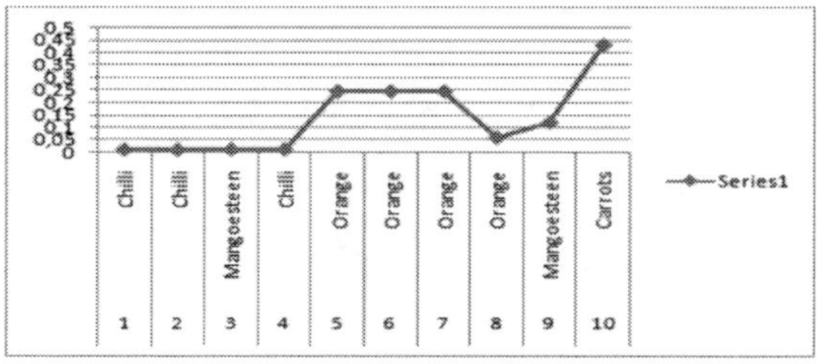

Figure- 8. Stochastic Universal Sampling (a)

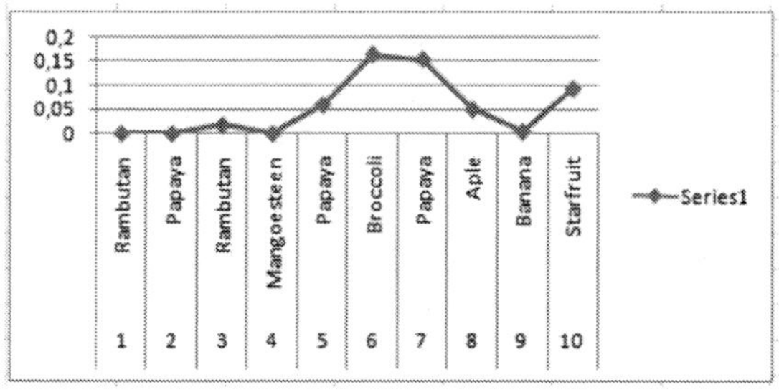

Figure-9. Roulette Wheel (b)

The graph in figure 9 shows fitness value from Roulette Wheel (b) method selection. However, this figure was considerably percentages from Papaya 30% and Rambutan 20%. That means Papaya are the most suitable plants in this location.

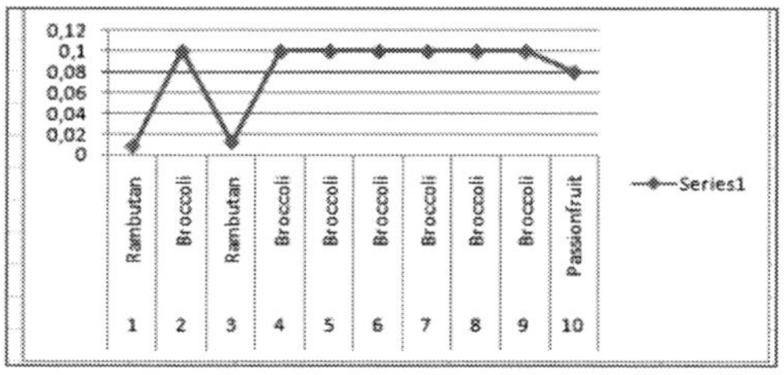

Figure-10. Tournament Selection (b)

In the figure 10 seen in the graph that the frequency of occurence Broccoli appears 7 times, followed by Rambutan appears 2 times.

The graph in figure 10 shows fitness value from Tournament Selection (b) method selection. From the figure, it is worth noting the high percentages of Broccoli at 70% and Rambutan at 20%.. That means Brocolli are the most suitable plants in this location.

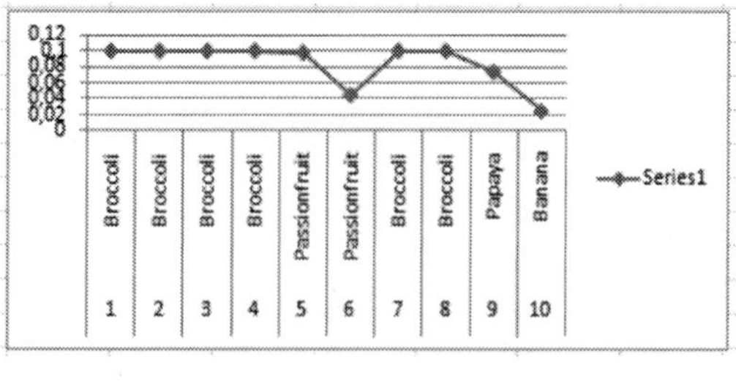

Figure-11. Stochastic Universal Sampling (b)

 DOI: 10.1051/matecconf/20165803002

The graph in figure 11 shows fitness value from Stochastic Universal Sampling (b) method selection. Among others, Brocoli is better at 50% followed by Passion Fruit at 20%. That means Brocolli are the most suitable plants in this location

DISCUSSIONS

The Roulette Wheel (a) shows that after 10 iterations, Carrots and Orange often appear frequently. It means Carrots and Orange are the most suitable plant parameters. In the Roulette Wheel (b), it can be analyzed that after 10 iterations plants that have a higher frequency of appearance are Orange and Chilli. In the Stochastic Universal Sampling (a), it can be seen that after 10 iterations, plants that appears often are Orange and Chilli. Of three trials on crop cluster 4, it can be determined that the priority sequence is Carrots, Chili, and Lemon.

Result of the comparison of the three selection methods shown in table 1 and figure 12.

Table-1. Comparison of three methods Selection in Cluster 4.

No	Roulette Wheel	Tournament Selection	Stochastic Universal Sampling
1	0.012	0.012	0.012
2	0.004	0.012	0.012
3	0.006	0.014	0.014
4	0	0.014	0.014
5	0.002	0.434	0.248
6	0	0,434	0.248
7	0.002	0.355	0.248
8	0.002	0.434	0.061
9	0.006	0.303	0.121
10	0.002	0.434	0.434

In table 1, there seems to be a little resemblance between the fitness value of Tournament Selection with Stochastic Universal Sampling selection.

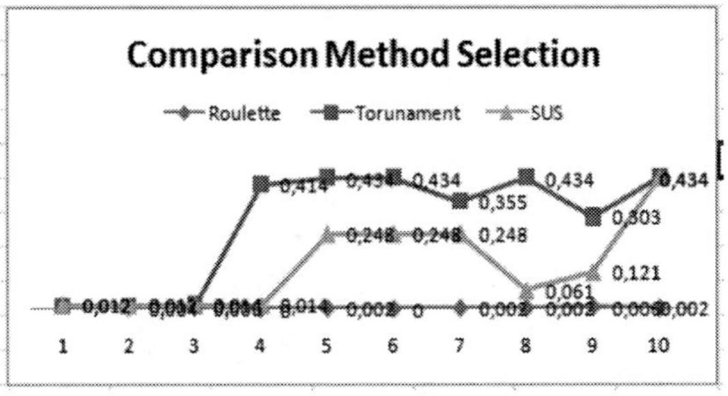

Figure-12. Comparison Fruit Cluster 4

The graph in Figure 12, illustrates comparison of three selection methods in Genetic Algorithm.. The smallest fitness value is generated by the Roulette Wheel selection method.

Table-2. Comparison of three methods selection in Cluster 5

No	Roulette Wheel	Tournament Selelction	Stochastic Universal Sampling
1	0	0,008	0,008
2	0	0,100	0,100
3	0,008	0,002	0,022
4	0	0,100	0,100
5	0,061	0,100	0,100
6	0,164	0,100	0,100
7	0,154	0,100	0,100
8	0,051	0,100	0,100
9	0,004	0,100	0,100
10	0,094	0,080	0,080

The Table 2 illustrates are comparison fitness values from methods selection.

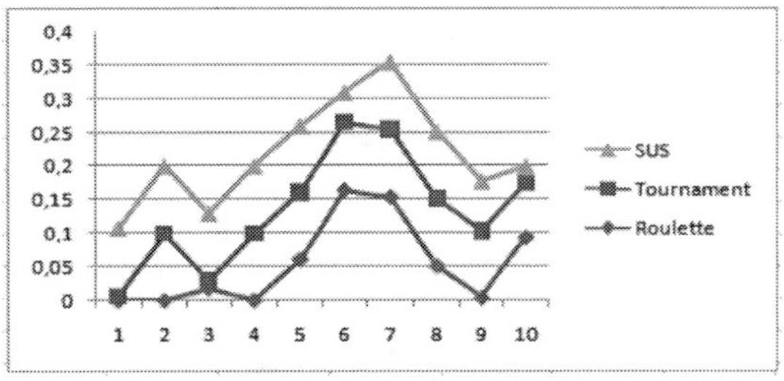

Figure-13. Comparation Fruit Cluster 5

The graphs illustrates the comparison of the three method of selection. The smalllest fitness value is generated by the Roulette wheel selection method.

CONCLUSIONS

1. The best selection method in this experiment is Roulette Whell because this selection method has small fitness values and is stable.
2. Fitness value which has been generated in each process in the genetic algorithm shows the index value of fruit.
3. Fruit index which is appeared on each iteration has demontrated suitability with the location.

ACKNOWLEDGMENTS

The author would like to thank UPN Veteran Jatim on funding the Hibah Doctor and funding the publication of this journal. Department of Computer Science and Electronic Gadjah Mada University (http://mkom.ugm.ac.id) for providing technical support for the research.

REFERENCES

Bolboaca S.D., Jantschi L., Balan M.C., Diudea M.V., Sestras R.E.,2010. State of Art in Genetic Algorithms for Agriculture System. Notula Botanica Horti Agrobotanic. Cluc Napoca. 51-63. ISSN 0255-965X. Electronic 1842-4309.

Blickle T. and Thiele L. 1995. A Comparison of Selection Schemes used in Genetic Algorithms.

Chudasama C., Shah S.M., Panchal,M..2011. Comparison of Parents Selection Methods of Genetic Algorithm for TSP. ,Proceeding published by International Journal of Computer Application (IJCA). International Conference on Computer Communication and Network CSI-COMNET-2011.

Goldberg D.E. and Kalyanmoy D. 1999. A Comparative Analysis of Selection Schemes Used in Genetic Algorithms.

Goldbeg D.E. 1999. Genetic Algoritm in Search Optimization and Machine Learning. Addison-Wesley Publishing Co.inc.

Mohebi E., Bouyer A., Karimi M. 2009. An Effecient Clustering of the SOM using Rough set and Genetic Algoritm. GCC Conference & Exhibition, 5th IEEE. 17-19 March 2009. Page 1-5. ISBN 978-1-4244-3885-3.

Mitchell T. 1997. Machine Learning. Genetic Algorithm Handout. Mc Graw-Hill

Jebari K. and Madiafi M. 2013. Selection Methods for Genetic Algorithms. International Journal Emerg Science. IJES, 3(4). 333-344. Desember 2013. ISSN 2222-4254.

Huang G and Wang L. 2011. Hybrid Neural Network Models for Hydrologic Time Series Forecasting based on Genetic Algorithm. Fourth International Joint conference on Computational Sciences and Optimization, 97-0-7695-4335-2/11. DOI 10 1109/CSO 2011.146. IEEE Computer Society.

Zadeh E. A. and Ghazalian R. 2010. Modulation Classification Using Genetic Algorithm and Radial Basis Neural Network Based on the HOS. Digital Content Multimedia Technology and its Aplication (IDC). 6th International Conference, 16-18 Aug 2010, Page 375-378.. ISBN: 978-1-4244-7607-7.

FAKE REVIEW DETECTION FROM A PRODUCT REVIEW USING MODIFIED METHOD OF ITERATIVE COMPUTATION FRAMEWORK

Eka Dyar Wahyuni [1] and Arif Djunaidy [2]
[1]Major In Information System, Dept. Of Informatics Engineering, ITS, Surabaya, East Java, Indonesia.
University of Pembangunan Nasional "Veteran" Jawa Timur, Surabaya, East java, Indonesia.
[2]Dept. of Information Systems, ITS, Surabaya, East Java, Indonesia.
E-Mail: [1]ekadyar.w@gmail.com,[2]adjunaidy@gmail.com

ABSTRACT

The rapid growth of the Internet influenced many of our daily activities. One of the very rapid growth area is ecommerce. Generally e-commerce provide facility for customers to write reviews related with its service. The existence of these reviews can be used as a source of information. For examples, companies can use it to make design decisions of their products or services, while potential customers can use it to decide either to buy or to use a product. Unfortunately, the importance of the review is misused by certain parties who tried to create fake reviews, both aimed at raising the popularity or to discredit the product. This research aims to detect fake reviews for a product by using the text and rating property from a review. In short, the proposed system (ICF++) will measure the honesty value of a review, the trustiness value of the reviewers and the reliability value of a product. The honesty value of a review will be measured by utilizing the text mining and opinion mining techniques. The result from the experiment shows that the proposed system has a better accuracy compared with the result from iterative computation framework (ICF) method.

Keywords: fake reviews, fake reviews detection, opinion mining, sentiment analysis, text mining, ICF

INTRODUCTION

Reading product reviews before buying the product becomes a habit, especially for potential customers. If they want to buy a product, they usually read reviews from some customers about the current product. If the review is mostly positive, there is a big chance to buy the product, otherwise if it's mostly negative, they tend to buy other products. While, for a company, the positive reviews from customers can generate significant financial benefits for businesses, it can be used as input for decisions related to product design and what services are provided to customers (Heidary et al., 2015).

Related with the financial benefits gained as a result of the positive reviews about the product / service from the customer, the fraudsters tried to play the existing system by writing fake reviews and provide an assessment that is not fair to promote or discredit a product or service (Heidary et al., 2015). Individuals like this are called spammers and their activity called opinion spamming (Liu, 2012).

Automatic detection of spammers is a very important work but still lacks in research. Unlike other types of spam, such as web spam or email spam, spam on a review is far more difficult to detect. The main reason is that spammers can easily disguise themselves. Thus it is difficult, for users to recognize, while web spam or email spam, one can determine spam or not without much difficulty (Heidary et al., 2015). The other reason is difficult to find good data (gold standard) related with fake and genuine reviews to build a model, because it is very difficult, to manually identify/labeling whether the review is fake or not just by reading them (Liu, 2012).

In general, the detection of fake reviews can be seen as a classification problem with two classes, fake and genuine. This classification problems are generally resolved with supervised method. However, as explained earlier, the main problem is the difficulty to identify fake reviews manually (from reading those reviews), because spammers can create a fake review as similar as possible to the original one (Liu, 2012). Because of these difficulties, there is no reliable data related to fake and genuine reviews that can be used to conduct training. Regardless of these difficulties, some supervised algorithms have been developed and evaluated.

Due to the absence of reliable data that has been labeled fake or genuine for training, Jindal and Liu (2008) utilized duplication in the review. In their study of 5.8 million reviews and 2.14 million reviewers of amazon.com, they found a large number of duplicate and near duplicate review, which showed that the presence of fake reviews is widespread. This is likely due to the difficulty of writing some new review when the spammers never buy the product or use the service. They tend to used the same review or slightly revised for different products. This duplication can be divided into four categories, that is (1) duplicates from the same customer id on the same product, (2) duplicates from different customer id on the same product, (3) duplicates from the same customer id on different products, (4) duplicate from different customers id on different products. The first type of duplication can be generated from reviewer's mistakes in clicking the submit button several times (which can easily be checked by the date of entry of the data). However, the last three types of duplication is most likely fake review. Thus, the training data used the last three types of duplication as the fake review and the rest of the other reviews as the original one.

In line with the growing popularity of sentiment analysis and opinion mining, and the increasing number of publications related to the fake reviews detection. Sites that provide product reviews facilities are implementing a wide range of algorithms that utilize a wide range of indicators to detect spammers. But on the other hand, spammers are able to learn from their mistakes and change their tactics and eventually adapt to avoid antispam techniques that are already equipped with these indicators, so that in the end, the algorithm that utilized duplication approach no longer can be used to detect fake reviews anymore. Motivated by the difficulties in manually labeling training data, Wang et al.,(2011) developed an unsupervised method. Computational methods were developed to measure the trustines of reviewers, the honesty of reviews, and the reliability of stores. Properties of review used in this study is its rating. They examined how the interaction between nodes (reviews, reviewers and shop) in a graph can reveal the cause of spam. They also created an algorithm called iterative computation framework (ICF) for calculating the trustiness of reviewers, the honesty of reviews, and the reliability of stores by exploring interdependencies between the three of it. This approach successfully detected fake reviews that are not duplicate with the level of precision (the ratio between the correct result of identification of spam obtained from the algorithm with the results from manually identified by humans) by 49%. The low value of the accuracy of the algorithm is likely influenced by the use of the rating property in measuring the honesty of a review, without looking at the semantic (did not utilizing the text property).

This study attempt to modify ICF by utilizing semantic aspects of a review by using opinion mining (in this research, we used sentiment polarity). Aspects extraction process needs to be done before executing opinion mining on a review. In this research, aspects that want to be axtracted are an attribute from the product. Extracted attributes are an attribute from the review that is explicitly mentioned in the review. This attributes extraction performed by using the FP-Growth. The next process is to determine the sentiment from the attributes by using opinion mining.

The fake review detection will compare the attributes and sentiments from the attributes of the reviews with the attributes and sentiments from the attributes of the other reviews and take into account the value of the similarity between the two reviews. To calculate the value of similarity, this research used Jaccard Similarity because euclidean distance is a measure of the distance, whereas the cosine similarity, the Jaccard coefficient and the Pearson coefficient is a measure of similarity (Sandhya, et al., 2008) and also because Jaccard Similarity is suitable for a wide range of applications, such as textual similarity or similarity of the buying habits of customers (Leskovec, Rajaraman, & Ullman, 2011). The end result of this detection application is the trustines value of reviewers, the honesty value of reviews, and the reliability value of a product.

Opinion mining and sentiment analysis

Textual information generally classified into two main categories, ie the fact and opinion. Facts are objective statements about entities and events in the world. Opinion are subjective statements that reflect the sentiment or perception of entities and events (Liu, 2009). Sentiment analysis, also called opinion mining is a study that analyze opinions, sentiments, evaluation, assessment, stance and emotions towards an entity such as products, services, organizations, individuals, issues and events, topics and attributes.

Opinions are very important in decision making. This applies both to individuals and organizations. The impact, the technology application of opinion mining has wide coverage. For individual consumers, if she/he want to buy the product, reads the opinion summary from the previous customer was very helpful in making a decision to buy or not. The individual can also compare the opinion summary from competing products. This is way better than reading from a various review and create the big picture of the advantages and disadvantages of a product. For organizations and businesses, opinion mining can be used to see how the costumer's views related its products and the products of its competitors. This information is not only useful for marketing and benchmarking of the products, but also can be used for design and product development (Liu, 2009).

Opinion mining starts with identifying the words that contain an opinion, ie, amazing, beautiful, ugly, etc. There are many studies that have been mine the words and identify its semantics orientation (as positive or negative). Hatzivassiloglou, V. and McKeown, K. in Liu (2009) identifies the linguistic rules that will be used for the identification of opinion words in a large corpus. Bootstrap approach carried out by Hu, M and Liu, B; Kim, S. and Hovy, E. in Liu (2009) by using a small set that contains the opinion words and then search for its synonyms and antonyms in WordNet. The major development of sentiment classification in a document level is done by Dave, D., Lawrence, A., and Pennock, D; Pang, B., Lee, L. and Vaithyanathan, S; Turney, P in Liu (2009) which classify each document reviews about an object (eg, movies, camera, or car) into positive or negative sentiment.

Association rule mining

In data mining the task of finding relationship among data is very important. Many algorithm developed for this task, ie apriori, fp growth, eclat, relim etc. Among other algorithm, apriori and fp growth have been studied in large scale in the past few years.

The Apriori algorithm was proposed by Agarwal and Srikant in 1994. Apriori uses a "bottom up" approach, where frequent subsets are extended one item at a time (a step known as candidate generation), and groups of candidates are tested against the data. Apriori uses breadthfirst search and a hash tree structure to count candidate item sets efficiently. It generates candidate item sets of length k from item sets of length k-1. Then it prunes the candidates which have an infrequent sub pattern. After that, it scans the transaction database to determine frequent item sets among the candidates. Candidate generation and support counting is expensive (in space and time).

Whereas The FP-Growth Algorithm, proposed by Han, is more efficient. This algorithm used an extended prefix-tree structure for storing compressed and crucial information about frequent patterns named frequent-pattern tree (FP-tree). Then it extract frequent itemsets directly from the FP tree. In his study, Han proved that his method outperforms other popular methods for mining frequent patterns, e.g. the Apriori Algorithm and the TreeProjection (Han, et al., 2004). Unfortunately, this task is computationally expensive, especially when a large number of patterns exist.

Jaccard simmilarity

Jaccard Similarity is often called Jaccard Coefficient or Tanimoto Coefficient, measured similarity as an intersection of an object that is divided by union of the object (Sandhya, et al, 2008; Leskovec, Rajaraman, & Ullman, 2011). Jaccard Similarity of the set S and T are JS (S, T), with the following equation

$$JS(S,T) = \frac{|S \cap T|}{|S \cup T|} \quad (1)$$

METHODOLOGY

Data preprocessing

Dataset used in this research was dataset from another research conducted by McAuley and Leskovec (2013), which was a product review from Amazon.com from June 1995 - March 2013, data retrieved from https://snap.stanford.edu/data/web-Amazon.html. Examples of data as shown in the figure-1

```
product/productId: B000FTPOMK
product/title: 14k Yellow Gold Butterfly Pendant, 16"
product/price: unknown
review/userId: A3T4KKUM1JATUX
review/profileName: Disappointed Sony customer
review/helpfulness: 1/1
review/score: 4.0
review/time: 1178150400
review/summary: pretty necklace
review/text: It is a nice made necklace, and the butterfly pendant looks beautiful. I love it.
```

Figure-1. Product review

Product / productId : an identification number (standard Amazon).
product / title : describes the product name.
product / price : the price of the product.
review / userId : user identification number.
review / profileName : user name.
review / helpfulness : the number of users who assess whether this review quite helpful.
review / score : the rating of the product.
review / time : the time when the review submitted (unix time).
review / summary : a summary from reviews and
review / text : is the content of the review.

In the preprocessing stage, the data parsed token by token and put on the review tables. Tokens can be any word or character. For this process the applications developed using PHP and AJAX. Once the application executed, the database will be filled, but the existing data still can not be used for the next process, its still need to be filtering. In this research, filtering (in the database) are held in 4 stages. The first stage is the removal of duplicate review. The second stage is the removal of 'anonymous' and 'unknown' users. The third stage is the removal of unpopular product, ie a product that has less than 3 reviews. The fourth stage is the removal of inactive users, ie users that has less than 3 review. After the filtering stage finished, the data generated is clean and are ready to be analyzed.

ICF ++ method

ICF ++ method took the name of a product, the name of a reviewer, the content of a review, the number of iteration and minsup as parameter. After taking those value, this method intialize the counter to 0, and honesty value and reliability value to 1. After that, the next process are pos tagging, creation of transaction file, FP Growth, polarity generation, calculation of agreement value. The next step is iteration. Inside in an iteration, the method continues by calculating the honesty value, calculating the trustiness value, calculating the reliability value, updating agreement value using updated honesty value, updating reliability value and increement counter. Iteration in this research is 3, because the algorithm converges quite fast in practice. The process are described in the figure-2.

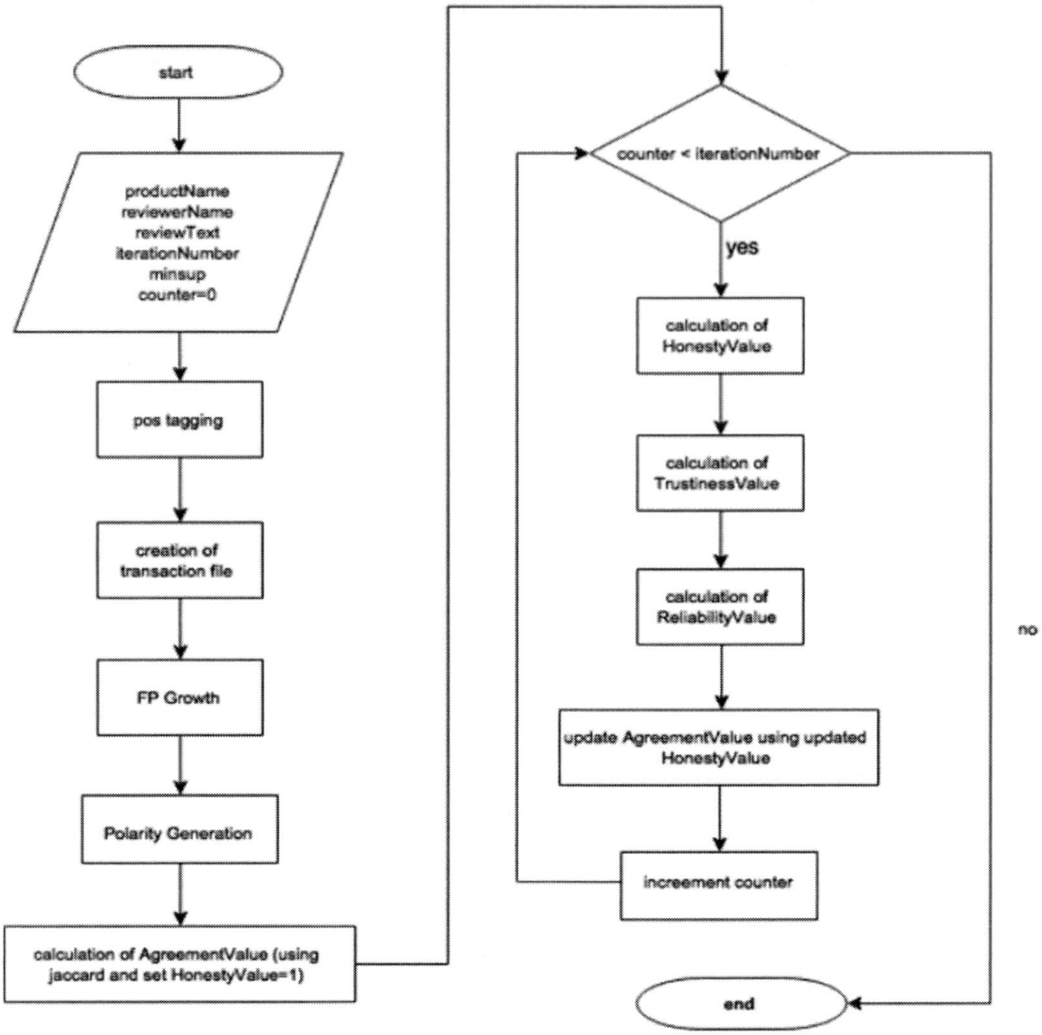

Figure-2. ICF++ flowchart

POS tagging

POS tagging was conducted by utilizing the Stanford Part-Of-Speech (POS) tagger from stanford-corenlp-3.5.1. . Stanford Pos Tagger is a software that reads text and determine the part of speech for each token. This process produced a tag in the form of abbreviations, for english tagger, the resulted abbreviations used Penn Treebank standard.

For this process, a desktop application was developed using Java. This process started with training the tagger model. Input from this process is a sentence from review and produce POS tag for each token that is stored into the database.

Creation of transaction file

Transaction file in this research contains all tokens that are stored in the database from a products that it's tag value is noun, either NN, NNP, NNPS or NNS. Each row in the transaction file was a noun from every sentence for a product. This transaction file became the input for the FPGrowth process.

FPGrowth

This process aimed to identify the features that has the most comment. In this study, the features are properties or attributes from a product. For example, for camera, the feature can be its battery, memory card etc. This information actually available on amazon, but, datasets related to this information were not publicly available, so to get information about product's feature, this research used fp-growth algorithm from association rule mining technique. This study used spmf library developed by (Fournier-Viger, et al., 2014).

The minsup used in this research was 0,1 and the confidence was 0,8. The minsup and confidence values ranging from 0 to 1. The minsup (minimum support) used for the comparison against the support that was owned by an item that compose a rule. Rule produced by this algorithm should consist of items that it's support value are exceed the minsup value. Support value was calculated from the number of item appearance in a transaction. For example, there was an item that appears two times in 5 transactions, then the support was 2/5 which is 0.4. If the minsup value entered was 0.1, then this item would appear in the output, but if the minsup value entered is 0.5 then this item will not appear in the output.

The FP Growth process in this research described as follow, input for this process are the transaction file that contains nouns. The next stage, are FP-Tree generation. FP tree is data structure that contains specific information about item, number of occurance of the item and link to other node. The next stage, this fp tree will be scanned to generate frequent pattern in the form of rule. The resulting rule will be broken down to obtain each phrase. Each of these phrases is an attribute of a product. The last stage is, the attribute will be saved to the transaction file and to the database.

Polarity generation

The feature or attribute resulted from the previous process will be identified its opinion orientation. For example, for a camera, the feature or attribute that can be identified is the battery, the comment related to the battery from userId X is "this battery drain pretty fast, and you do not want to be stuck somewhere with a dead battery". The purpose of this process is to identify the opinion orientation from a sentence that contains the attributes that were identified from the previous process. Whether its orientation positive, negative or neutral. From the example above, its opinion orientation is negative.

Sentiment prediction/opinion orientation prediction can use variety of techniques, from the words level, sentence level and others. According to Socher, et al., (2013), sentiment prediction only from words often misses, because the order of each word was ignored. As in the illustration below, the words funny and witty each are positively oriented, but in the sentence "This movie was actually neither that funny, nor super witty", causing the orientation of the overall sentence to be negative.

The input for this process are two parameters, ie a list of all tokens from a review of a current product (sentences that have been POS tagged and stored into the database), named list i. The second parameter is a list of all the features from a current product, named list f, which is obtained from the database. This process will check whether index i-th from list i contains a token from list f. If this condition is met, it will look for the orientation by using stanford.nlp.sentiment from stanford-corenlp-3.5.1. library. The output of this process is a list of features contained in a single sentence along with the orientation of the feature, as the following examples "necklace pos pendant neg".

Calculation of agreement value

The next process was calculate the agreement value that illustrates how similar the feature and its opinion orientation resulted from the polarity generation process with the other reviews. This process uses Jaccard similarity. Results from Jaccard similarity was in the range of (0,1), 0 means that the second review didn't have the same opinion orientation for any feature, 1 means both reviews had the same opinion orientation for all the features. In this research, both of the reviews agreed each other if their jaccard value exceed 0.8.

The surrounding set of review v (S_v) was the set of v's surrounding reviews about a product.

$$S_v = S_{v,a} \cap S_{v,d} \tag{2}$$

where $S_{v,a}$ is set of the reviews that agree with each other, $S_{v,d}$ is a set of reviews that disagree with each other. From the definition above, $S_{v,a}$ is in the form of a list of a reviewID that contains the other review which it's jaccard value exceed 0.8. Whereas, $S_{v,d}$ contains other review which it's jaccard value less than 0.8.

The agreement value from a review v within a period Δt is described in the following equation

$$A_n(v,\Delta t) = \frac{2}{1+e^{-A(v,\Delta t)}} \tag{3}$$

$$A(v,\Delta t) = \sum_{i \in S_{v,a}} T(\kappa_i) - \sum_{j \in S_{v,d}} T(\kappa_j) \tag{4}$$

where A is the agreement value of the target review. In this research, we used 3 months before and after the posting time of the target review as the time window Δt.

Calculation of honesty value

Review's honesty value is described in the following equation

$$H(v) = |R(\Gamma_v)| A_n(v,\Delta t) \tag{5}$$

Where $R(\Gamma_v)$ is product's reliability value that is being reviewed.

Calculation of trustiness value

Reviewer's trustiness value is defined in the following equation

$$T(r) = \frac{2}{1+e^{-H_r}} - 1 \tag{6}$$

Where T (r) is the reviewer's trustiness value.

Calculation of reliability value

Product's reliability value (R (p)) is described in the following equation

$$R(p) = \frac{2}{1+e^{-\theta}} - 1 \tag{7}$$

Where

$$\theta = \sum_{v \in U_x, T(\kappa_v) > 0} T(\kappa_v)(\Psi_v - \mu) \tag{8}$$

Where Ψ_v is the rating from review v, while is the median value of the rating, which is 3, in the range of 1-5 rating.

RESULTS AND DISCUSSIONS

There are two types of tests in this research, which is verification of results test and performance test. The purpose of the test of verification of the results is to check whether the calculations performed by the application and manualy calculated has the same process and the same result. This test also intend to check if the given input has the correct output. While the aim of the performance test is to compare results from the ICF ++ method with another fake review detection methods. Method from Wang, et al., (2011) is reimplemented using java to see the performance. Then the fake review detection methods in Wang's research used as a reference comparison with method ICF ++.

Result from verification test

Applications that has developed take 4 input, ie, product name, reviewer name, review (text) and minsup. The application will perform three iterations. Each iteration will be shown in the system and then compared with the results from manual calculations which also iterate 3 times. Summary of the results of the test can be seen in the following table.

Table-1. Summary result of manual calculation and ICF++

no	process	Manualresults			ICF++ results			conclusion
		I 1	I 2	I 3	I 1	I 2	I 3	
1	agreement	0	-1,227	-1,6531	0	-1,227	-1,6531	**verified**
2	agreementVal	0	-0,5466	-0,6786	0	-0,5466	-0,6786	**verified**
3	honesty	0	-0,4163	-0,5168	0	-0,4163	-0,5168	**verified**
4	sumHonesty	-0,4621	-1,5615	-1,8197	-0,4621	-1,5615	-1,8197	**verified**
5	trustiness	-0,227	-0,6531	-0,7211	-0,227	-0,6531	-0,7211	**verified**
6	reliability	-0,7616	-0,7616	-0,7616	-0,7616	-0,7616	-0,7616	**verified**
7	theta	-2	-2	-2	-2	-2	-2	**verified**

From table 1, it can be concluded that for every process performed by the application, all the process and result are verified, because for a given input, the output generated by applications has been in line with the expected result, which is derived from manual calculations.

Result from performance test

The Application calculated the honesty value of a review, the trustiness value of the reviewers and the reliability value of a product. The algorithm developed by Wang, et al., (2011) will converges very fast. Therefore, from 855928 review data, not all of it being processed with these two models. Systematic sampling technique carried out, that will select elements number n-th as a sample. In this study, the interval used (n) is 800, so every reviews multiple of 800 taken as sample. The number of samples to be tested for both models is 1070. The steps of this test is described as follows

1. Fake review detection by ICF
2. Fake review detection by ICF ++

Result evaluation

This research used IR-based evaluation strategy (Wang, et al., 2011) with the following steps

1. both models identified fake reviews.
2. human evaluator decide whether the review fake or genuine, in order to calculate the precision values which illustrates the performance of the model.

A method is said to be effective, if different juries have the same assessment related a current review, so in this research, consistency of assessment used as one of the evaluation criteria. The evaluator used in this study were 3 graduates student from computer science and has considerable experience in online shopping.

Judging suspicious spammer was a complex task, which requires intuition and also ability to search for additional information. To decide whether this reviewer is a review spammer or not, the judge must read the reviews and collect evidence about relationship between these reviews to the other reviews, also how the relationship between reviews of all products that has reviewed by this reviewer with another review by other reviewers. To standardize the judgement process, the three judges agreed on two conditions as our evidence to claim that a reviewer is a potential spammer.

1. A reviewer is suspicious if she/he has a significant number of reviews that gives opposite opinion to other reviews about the same product.
2. A reviewer is suspicious if she/he has a significant number of reviews that gives opposite opinion to other reviews about some product as compared to evidences presented by general web search results (eg, the google search results).

This research took 100 suspicious reviews. In evaluation stage, if more than one judges identify that the targetted review was fake, it was labeled as fake. To study the agreement between each jury, this research used Fleiss kappa (Wang, et al., 2011), which is an inter-evaluator agreement measure for any number of evaluators. Kappa values for each model are summarized in the following table.

Table-2. Kappa values of ICF and ICF++

Model	**Kappa Value**	**agreement Interpretation**
ICF	0.648823	substantial agreement
ICF ++	0.697581	substantial agreement

From table 2 can be concluded that the agreement of the three judges for all methods is more than 0.6 and less than 0, 8 which represents substantial agreement.

Precision values of the model ICF is summarized in the following table.

Table-3. Precision value of ICF

evaluator	**1**	**2**	**3**
1	54	46	46
2		57	49
3			56

For the ICF method, our judges identified 57 suspicious reviews, so the precision is 57%. Table 3 shows the agreement of evaluator. Evaluator 2 identified 57 suspicious reviews, out of which 49 were recognized by Evaluator 3 and 46 were caught by Evaluator 1.

Precision values of the model ICF++ is summarized in table 4.

Table-4. Precision value of ICF++

evaluator	1	2	3
	54	51	48
2		63	55
3			59

For ICF++ method, three judges managed to identify 63 out of 100 reviews were suspicious, so the precision is 63%. Table 4 shows the agreement of evaluator. Evaluator 2 identified 63 suspicious reviews, out of which 55 were recognized by Evaluator 3 and 51 were caught by Evaluator 1.

CONCLUSION

From Table 3 and 4 can be seen that the precision value of ICF ++ method is higher than ICF method. It can be concluded that the use of ratings alone to assess whether the review is fake or genuine is inadequate, because the information that can be processed is very limited. By using text properties, as implemented in the ICF ++ method proven to improve the precision value by 6%.

The drawback of this method is, some process need to be optimized, so it can detect a fake review in a short amount of time.

REFERENCES

Coenen, F., Goulbourne, G. & Leng, P., 2003. Tree Structures for Mining association Rules. Journal of Data Mining and Knowledge Discovery, Vol 8, No 1, pp.25-51.

Forman, E.H., 1993. Facts and fictions about the analytic hierarchy process. Mathematical and Computer Modelling, Volume 17, Issues 4–5, pp.19-26.

Fournier-Viger, P., Gomariz, Gueniche, T., A., Soltani, A., Wu., C., Tseng, V. S. 2014. SPMF: a Java Open-Source Pattern Mining Library. Journal of Machine Learning Research (JMLR) 15, pp. 3389-3393.

Han, J., Pei, J., Yin, Y. & Mao, R., 2004. Mining Frequent Patterns without Candidate Generation: A Frequent-Pattern Tree Approach. Data Mining and Knowledge Discovery 8, pp.53–87.

Heydari, A., Tavakoli, M.A., Salim, N., & Heydari, Z., 2015. Detection of review spam: A survey. Expert Systems with Applications, 42(7), pp.3634–42.

Jindal, N., & Liu, B. 2008. Opinion Spam and Analysis. Proceedings WSDM '08 Proceedings of the 2008 International Conference on Web Search and Data Mining,219-230.

Khan, K., Baharudin, B., Khan, A., & Ullah, A. , 2014. Mining opinion components from unstructured reviews: A review. Journal of King Saud University - Computer and Information Sciences, 26(3), pp.258–75.

Leskovec, J., Rajaraman, A. & Ullman, J.D., 2011. Mining of Massive Data Sets. Cambridge University Press.

Liu, B., 2009. Opinion Mining. Encyclopedia of Database System, pp.1986-90.

Liu, B., 2012. Sentiment Analysis and Opinion Mining. Morgan & Claypool Publishers.

McAuley, J., & Leskovec, J. 2013. Hidden Factors and Hidden Topics : Understanding Rating Dimensions with Review Text. Proceeding RecSys '13 Proceeding of the 7th ACM conference on Recommender Systems , 165-172.

Pang, B. & Lee, L., 2008. Opinion Mining and Sentiment Analysis. Foundations and Trends in Information Retrieval Vol. 2, No 1-2, pp.1-135.

Ravi, K. & Ravi, V., 2015. A survey on opinion mining and sentiment analysis: Tasks, approaches and applications. Knowledge-Based Systems, 89, pp.14-46.

Sandhya, N., Lalitha, Y.S., Govardhan, A. & Anuradha, K., 2008. Analysis of Similarity Measures for Text Clustering. CSC Journals 2.

Savage, D., Zhang, X., Yu, X., Chou, P., Wang, Q., 2015. Detection of opinion spam based on anomalous rating deviation. Expert Systems with Applications, 42(22), pp.8650–57.

Socher, R., Perelygin, A., Wu, J.Y., Chuang, J., Manning, C.D., Ng, A.Y., Potts, C., 2013. Recursive Deep Models for Semantic Compositionality Over a Sentiment Treebank. Proceedings of the conference on empirical methods in natural language processing (EMNLP). Vol. 1631 , 1642.

Wang, G., Xie, S., Liu, B., & Yu, P.S 2011. Review Graph Based Online Store Review Spammer Detection. Proceeding ICDM '11 Proceedings of the 2011 IEEE 11th International Conference on Data Mining , 1242-1247.

FORECASTING OF COMPUTER NETWORK PERFORMANCE WITH QUADRATIC EQUATION

[1]Achmad Ubaidillah, [2] S. Ida Kholida
[1] Telecommunication of Electrical Engineering, Madura Trunojoyo University, Indonesia
[2] Physics Education, Madura Islamic University, Indonesia
Email : ca_achmed@yahoo.com

ABSTRACT

The computer network service provider has to observe, evaluate and plan continually its services in order to provide reliable systems. One of the efforts in planning process is forecasting. There are many methods in it, and this paper proposes to use Quadratic Equation Formula as a forecasting method. It is used because of the simplicity and it is suitable to solve problems with parabolic characteristic that can not be solved by linier computational method. It also compares characteristic of Quadratic Equation Formula with Linear Regresion forecasting.The result of this research shows that the computer network performance can be forecasted with Quadratic Equation Formula which has similar case characteristic with it. The standard deviation error of forecast value of the Quadratic Equation Formula is 370 of 3080 forecasted throughput maximum value. While the standard deviation error of forecast value of the linear regresion method is 282 of 3518 forecasted throughput maximum value. But in the forecasted throughput of linear regresion methhod, the output value will allways increase as the increasing of the input variable value. As the result, the error distribution calculation is not interesting anymore because there is no peak value for the forecasted result.

Keywords : Forecasting, Computer Network Performance, Quadratic Equation

INTRODUCTION

There are many advantages of computer network system like resource sharing, reliability, economical side, scalability and communication medium. Madura Trunojoyo University that apply computer network system has to takehold and maintain its system in order to gain the advantages especially with good monitoring, evaluation and planning continually. It is very important to give quality of service of network system to the users in university, especially when the users of computer network in campus increase (Achmad Ubaidillah et al, 2014).

The observation, evaluation and planning of computer network performance are very important to be implemented. There are many research that concern about it. For example, the research that make network planning to define QoS Metric based on total delay and packet loss with Calculus Theory approach on chip modeling architecture (salem Nasri, 2011). Dhobale J.V., Kalyankar N.V., and Khamitkar S.D., observe and evaluate the computer network performance using OMNET++ simulation environment with different input variable (Dhobale J.V. et al, 2014). Other research that concerns in computer network performance are the research that makes remaster of Ubuntu to observe the computer network performance (Andi Pebriananta et al, 2011), the research that measures the computer network performance to manage the availibility (Sri Wulandari et al, 2011), and the research that measures and analizes the distributed computer network performance with different properties for administration (Amit Kumar Sahu et al, 2012).

Beside that, the management of computer network performance can be observed by optimizing and improving the performance. For exmple the research that proposes some recomendations that can improve the network performance based on QoS changes (Winarno Sugeng et al, 2015), The research that combines the MBAC and PBAC to improve admission control and network utilization efficiency (Suleiman Y. Yerima, 2011). Other researchs that develop the improvement of network performance are paper that proposes to improve end to end delay of network management system using network coding (El Miloud Ar Reyouchi et al, 2013), the research that proposes to minimize trannsmission cycle delay between mobiles moving by changing the BEB Algorithm (Ibrahim Syed Ahmad et al, 2013), and the research that proposes to improve TCP Performance over wireless network with frequent disconnections (Purvang Dalal et al, 2011).

The observation, evaluation and planning of computer network performance also can be implemented by developing the performance forcasting system with parameters that influence network performance. Paulo Cortez proposes to use Neural Network, ARIMA and Holt Winter to forecast multi-scale internet traffic (Paulo Cortez et al, 2012). While Poo Kuan Hoong proposes to forecast the network computer traffic using ARMA (Poo Kuan Hoong et al, 2012).

Basically, this paper begins from other research especially (Achmad Ubaidillah et al, 2014). It proposes to use linear and multi linear regresion as a forecasting method of computer network performance. But it is just linear computation. It assumes that the output value will increase as the increasing ofinput variable value. It is contradictive compared with characteristic of computer network performance that can increase and decrease after gain the peak value. So this research proposes to use Quadratic Equation Formula to forecast the computer network performance because of the similarity of the output data characteristic.

Network performance can be measured by turn arround time, response time, throughput, goodput, capacity, availability, reliability, delay, jitter, and blocking probability (Mohammad Iqbal, 2012), (Hendrawan, 2006). This research uses sent packets as the input variable and throughput as the output variable.

Linear Regresion As Forecasting Method

Regression is one of statistical development model that is used to predict a value of response variable based on the variable explanatory (Hendrawan, 2006). Linear regression is the most simple model of regression as linear equation. Basically, linear regression equation is,

$$y = a + bx + e \quad (1)$$

with, x: input variable
a : intercept
b : slope
e : random error

the correlation value between input variable and output variable must be known before determining the linear regression equation in order to know the relation and the influence of input variable to output variable. The correlation value is shown as,

$$r = \frac{1}{n-1}\sum_{i=1}^{n}\left(\frac{x_i-\bar{x}}{S_x}\right)\left(\frac{y_i-\bar{y}}{S_y}\right) \quad (2)$$

with, r : Correlation value between input and output variable
n : event
x_i: input variabel
$\bar{x}$: mean of input variable
S_x : standard deviasion of input variable
y_i : output variable
$\bar{y}$: mean of output variable
S_y : standard deviasion of output variable

The next step after calculating the correlation value is calculating the slope or gradient value of equation with,

$$b = r\left(\frac{S_y}{S_x}\right) \quad (3)$$

Afterwards, the intercept value can be calculated after knowing the slope value with,

$$a = \bar{y} - b\bar{x} \quad (4)$$

Quadratic Equation Formula As Forcasting Method

The formula of Quadratic Equationproduces parabolic graph, and it is the main contemplation of this research. According to (Frank Ayres, JR., 2006), The main formula is,

$$y = ax^2 + bx + c \quad (5)$$

The steps to make (5) to become forecast formula:

1. begin from input variable and output data
2. take 3 samples of its coordinate, 1 sample before peak value, 1 sample at the peak and 1 sample after the peak
3. obtain 3 equations in a, b and c by making substitution the 3 samples to (5)
4. compute the 3 equations with elimination and substiution method to obatain the values of a, b and c
5. obtain a new Quadratic Equation whit them
6. make computational forecasting with it

Error calculations of this paper are RMSE (Root Mean Square Error) that is implemented as standard deviation of the error.

$$MSE = \left[\frac{\sum_{i=1}^{n} e_i^2}{n-2}\right] \quad (6)$$

$$S_e = \sqrt{MSE} \quad (7)$$

with MSE : *Mean Square Error*
S_e : *Unbiased of error standard Deviation*

METHODOLOGY

First, this research simulates a computer network system (figure 1) with Network Simulator 2 (NS2). The system is built as a bottle neck network, all nodes (0, 1, 4, 6, 7) send packets to node 3 over node 2. The traffic types are mixed between TCP and UDP. The variable input is total sent packets, while terminal throughput in node 3 is set as network system performance output. Then, the input and output variable values are computed into Linear Regresion Forecasting method and Quadratic Equation Forecasting Formula. The performance computation result of them are compared each other to be concluded which the better is.

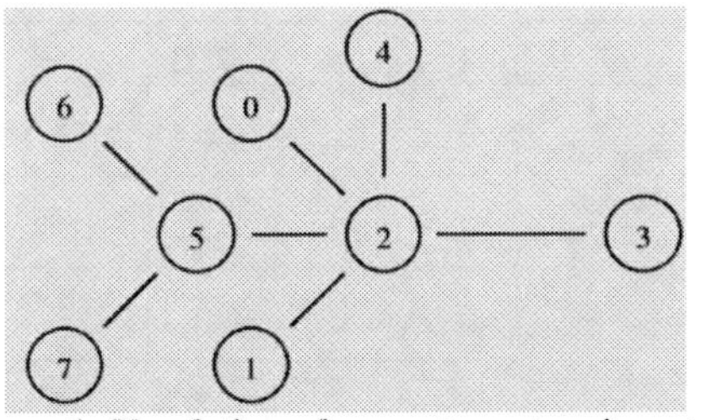

Figure-1. Simulation of computer network system observed

RESULTS

The simulation results are shown in figure 2 and figure 3.

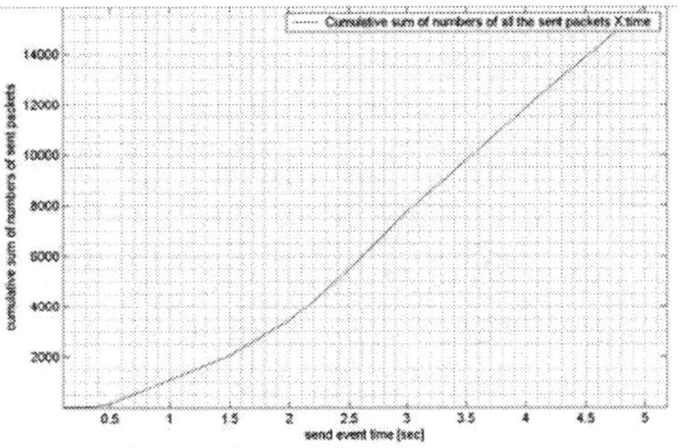

Figure-2. Graph of event time to total sent packets of simulation result

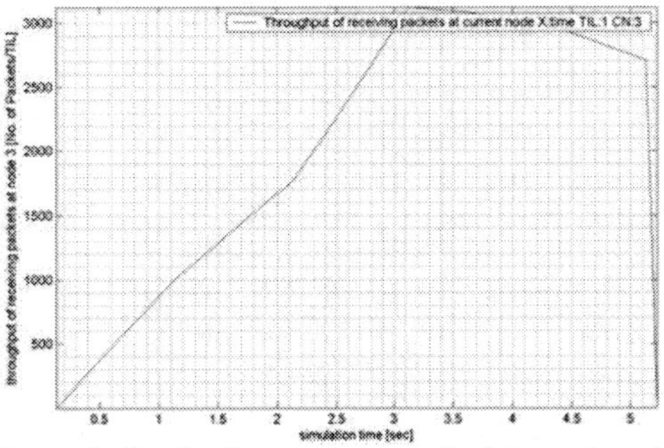

Figure-3. Graph of event time terminal throughput in node 3 of simulation result

Then, the performance resultsare sampled in some node samples shown in table 1. It shows the correlation between total packets sent and terminal throughput values in node 3.

Table-1. Samples of network performance of the simulation and the forecasted value with Linear Regresion method

Sent_Pckt (Packets)X	**Trhoughput (Packets/second)Y**	**Forecasted Throughput Y'**
305	370	1047
1230	890	1195
2215	1297	1352
3490	1680	1556
5510	2295	1880
7750	2944	2238
9726	3102	2554
11981	3005	2915
13833	2928	3211
15752	2720	3518

From X and Y oftable 1, some parameters that can be calculated :

- $n = 10$ (8)
- $\bar{x} = 7179$ (9)
- $S_x = 5490$ (10)
- $\bar{y} = 2123$ (11)
- $S_y = 995$ (12)

Using (2),

$r = 0{,}86$ (13)

Using (3),

$b = 0{,}16$ (14)

Using (4),

$a = 998$ (15)

So, the Linear Regrasion equation that will be used as forecasting equation is,

$y = 998 + 0{,}16x$ (16)

The Forecasting values (Y') is obtained by using (16). Then the next step is to calculate error distribution with equation (6) and (7). The MSE value is 79.664 and the standard deviation of error (S_e) is 282.

Then, this research proposes to use Quadratic Equation formula as forecast method. Three nodes (X,Y) are taken as samples of the network performance, (5510, 2295), (9726, 3102) and (13833, 2928). By inserting them to Quadratic Equation then it obtains,

$30360100a + 5510b + c = 2295$ (17)

$94595076a + 9726b + c = 3102$ (18)

$191351889a + 13833b + c = 2928$ (19)

From (17), (18) and (19), value of a, b and c can be obtained by using Elimination and Substitution computational method. Then,

$a = -0{,}000031$ (20)

$b = 0{,}66$ (21)

$c = -442{,}5$ (22)

So, the forecasting Quadratic Equation is,

$y = -0{,}000031x^2 + 0{,}66x - 442{,}5$ (23)

The result of forecasted throughput Y' is shown in table 2.

Table-2. Samples of network performance of the simulation and the forecasted Throughput with Quadratic Equation Formula

Sent_Pckt (Packets)X	**Trhoughput (Packets/second)Y**	**Forecasted Throughput Y'**
305	370	-243
1230	890	327
2215	1297	876
3490	1680	1496
5510	2295	2273
7750	2944	2839
9726	3102	3080
11981	3005	3060
13833	2928	2807
15752	2720	2321

DISCUSSIONS

It is interesting to watch forecasted throughput Y' in table 1, that the value will always increase as the increasing of the input variable value. It becomes the main problem and the main weakness of forecasting with linear regresion method. It is not appropriate to be implemented to computer network performance that has parabolic characteristic. Beside that, the error distribution like MSE and S_e is not important anymore. While the real throughput value will decrease after peak value, but the forecasted throughput value will always increase. It will cause the error valeu is getting higher and higher as the increasing of input variable.

On the opposite, the result of forecasted throughput Y' of Quadratic Equation shown in table 2, shows that the forecasted throughput Y' values increase, obtain the peak value then decrease like parabolic graph. It has similar characteristic with values of Y. As the result, error distribution becomes important to be observed. The MSE value is 136.622 and the standard deviation of error (S_e) is 370 of 3080 maximum forecasted throughput value.

CONCLUSION

From the analysis of the computational result before, it is clearly proven and concluded that Linear Regresion method is not appropriate to be used as forecasting method for computer network performance because of the linearity problem, while network performance has parabolic characteristic. Even, calculation of error distribution is not important anymore because of it. In the other side, Quadratic Equation Formula can be used as forecasting method for computer network performance because of the characteristic similarity. So, for the case like this, Quadratic Equation Formula is better than Linear Regresion method.It can be said the research of this paper

The error standard deviation of forecasted throughput Y' with Quadratic Equation Formula is not to bad, 370 of 3080 maximum forecasted value or about 12%. It means, there is wide possibility to make it better. So, the suggestion for next research is how to make it better.

REFERENCES

Achmad Ubaidillah Ms., Koko Joni, Miftahul Ulum, Haryanto, Diana Rahmawati, and Riza Alfita, 2014. Forecasting of Computer Network Performance with Linear and Multi-Linear Regresion Method, FORTEI, ISBN:978-602-9056-70-9.

Amit Kumar Sahu and Naveen Hemrajani, 2012, An Analysis of Distributed Computer Network Administration, International Journal Computer Technology and Applicaton. 3(2) : 660-667.

Andi Pebriananta, Indrastanti R.W., T. Indra Bayu, 2011, Development of Computer Network performannce observer Through Remastering Ubuntu. JUTI 9(1) : 35-40.

Dhobale J.V., Kalyankar N.V., and Khamitkar S.D., 2014, Computer Network Performance Evaluation Based on Different Data Packet Size Using OMNET++ Simulation Environment, International Journal of Advanced Research in Computer Science and Technology vol. 2, issue 2, ver. 1 : 91-95

Dhobale J.V., Kalyankar N.V., and Khamitkar S.D., 2014, Computer Network Performance Evaluation Based on Network Scalability Using OMNET++ Simulation Environment, International Journal Advanced Networking and Application. 5(5) : 2041-2045.

Dhobale J.V., Kalyankar N.V., and Khamitkar S.D., 2014, Computer Network Performance Evaluation Based on Data rate and Number of Clients Per Server Using OMNET++ Simulation Environment, Global Journal Computer Science and Technology : E Network, Web and Security. vol. 14, issue 3, ver. 1.0.

El Miloud Ar Reyouchi, Kamal Ghoumid, Koutaiba Ameziane, and Otman El Mrabet, 2013, The Improvement of End to End Delays in Network Management System Using Network Coding, International Journal Computer Network and Communication. 5(6) : 65-84.

Frank Ayres JR., 2006, Schaum : University Mathematics (Third Edision), Erlangga

Hendrawan, Dr., 2006, Handout : Introduction of Telecommunication and Computer Network Performance, Electrical Engineering of Bandung Technology Institute

Hendrawan, Dr., 2006, Handout : System Comparison and Forecasting, Electrical Engineering of Bandung Technology Institute

Ibrahim Sayed Ahmad, Ali Kalakech, and Seifedine Kadry, 2013, Minimizing Mobiles Communication Time Using Modified Binary Exponential Backoff Algorithm, International Journal Computer Network and Communication. 5(6) : 85-102.

Mohammad Iqbal, Dr., 2012, Handout : System Performance Analysis, Gunadarma University

Paulo Cortez, Miguel Rio, Miguel Rocha, and Pedro Sousa, 2012, Multi-Scale Internet Traffic Forecasting Using Neural Network and Time Series Methods, Expert System. 29(2) : 143-155.

Poo Kuan Hoong, Ian K.T. Tan, and Chee Yik Keong, 2012, Bit Torrent Network Traffic Forecasting with ARMA, International Journal Computer Network and Communication. 4(4) : 143-156.

Purvang Dalal, Nikhil Kothari, and K.S. Dasgupta, 2011, Improving TCP Performance Over Wireless Network with Frequent Disconnections, International Journal Computer Network and Communication. 3(6) : 169-184.

Salem Nasri, 2011, Network on Chip : A New Approach of QoS Metric Modeling Based on Calculus Theory, International Journal Computer Network and Communication. 3(5) : 53-60.

Sri Wulandari and Achmad Affandi, 2011, Measuring of Computer Network Service Performance for Management of availability, SESINDO, Department of Information System of ITS

Suleiman Y. Yerima, 2011, Implemetation and Evaluation of Measurement Based Admission Control Schemes within a Converged Networks QoS Management Framework, International Journal Computer Network and Communication. 3(4) : 137-152.

Winarno Sugeng, Jazi Eko Istiyanto, Khabib Mustofa, and Ahmad Ashari, 2015, The Impact of QoS Changes Towards Network Performance, International Journal Computer Networks and Communications Security. 3(2) : 48-53.

FORECASTING THE NUMBER OF PATIENTS DISEASES USING BACKPROPAGATION

Aeri Rachmad, Devie Rosa Anamisa
Faculty of Engineering, University of Trunojoyo Madura, Indonesia
Email: aery_r@yahoo.com

ABSTRACT

Forecasting with various types of disease is important for health centers, because it can be used to help the health center management in conducting strategic planning and decision making. Health Care Center Torjun in Indonesia has made estimationabout the number of patients with various types of diseases, such as Acute Respiratory Infections(ISPA), RA (Rheumatoid Arthritis), diarrhea, HT (Hypertension), Skin Allergies, Conjunctivitis, Asthma, Febrile, TB(Tuberculosis). Lung, scabies, Gastritis, typus and scarlet fever with reports the number of patients with certain diseases in the coming period and prepare the necessary needs both medical services and as well as drugs for use later. In this study, Artificial Neural Network (ANN) is one model that is used to identify patterns of images of people with various kinds of diseases. Backpropagation is one of the popular models of Neural Networkwhich is used for forecasting, prediction, and decision makers based on the input of data entry that has been studied in advance. The resultsis HT (0.35 %)with parameters for forecasting system using Neural Network Backpropagation is the best of the trial results that shows disease HT which are obtained from the experiments. They predict the number of patients with a disease that needs to be watched for in the coming period and prepare all the needs of both medical and medication needed to handle the number of people with the disease.

Keywords: Forecasting, Health Care Center, Diseases, Backpropagation

INTRODUCTION

Forecasting is a process to predict the needs of the future which includes the need to measure the quantity, quality, time and location needed in order to meet the demand for goods and services. Forecasting demand is the level of demand for the products that are expected to be realized for a certain period in the future. Basically forecasting approaches can be classified into two approaches: qualitative approach and quantitative approach (Memmedli, 2012)

There are some solutions that are needed such as a good quality of initial data and in sufficient quantity, because principally, the method of prediction is to identify the pattern of data which is based on the existing data, so mathematically predicted results will be relatively accurate (Chang, 2012). One technique that can be considered in pattern recognition neural network for prediction is a clone. Neural Network Backpropagation is the development of least mean square algorithm that can be used to train the network with multiple layers which is based on steepest descent algorithm approach and uses its performance index as the mean square error (Kara, 2011).

The outcome of a forecasting considers that forecastingmust contain errors, meaning that forecasters could only reduce the uncertainty that will happen, but it can not eliminate this uncertainty. Forecasting should provide information about how large the size of the erroris, meaning that forecasting certainly contains errors.It is important for forecasters to inform how large the errors that may occur, or even short-term forecasting is more accurate than long-term forecasting (Santoso,2009).

Along with the development of technology, there are more and more emerging methods that can be used for forecasting such as the classic method and fuzzy time series. Time series methods for modeling and forecasting monthly number of patients at a health clinic which has produced levels of forecasting accuracy of the existing methods are not good enough. They can only predict seasonal problem, but they failed to foresee problems with linguistic values. Moreover the model also requires a number great historical data to generate accurate forecasting results (Setiyoutami, 2012).

In a previous study, several approaches were made to obtain the results of the best forecasting such as Artificial Neural Network, Ensemble Neural Network, Backpropagation Networks, Radial Basis Function Network, General Regression Neural Network, Genetic Algorithm, Multilayer Perceptron, clustering Fuzzy, etc. To determine the daily weather forecast, it can use back propagation technique with 28 input parameters to foresee some parameters such as temperature, rainfall, humidity, clouds, and weather days (Meera 2015). Instead of predicting the weather phenomenon through traditional ways such as the estimated amount of rainfall, barometric pressure, wind speed, temperature etc, we can use a neural network. Back Propagation neural network has high accuracy and efficiency to predict the weather (Pooja 2014; Meera 2015).

By considering the results of the above studies, this research uses back propagation neural network to predict the number of patients with various types of 13 diseases at the health center. The predictive model is used to estimate the number of patients with a disease that needs to be watched in the coming period.

METHODOLOGY

Neural networks are information processing systems that have certain characteristics by adapting performance of biological neural network(Deepthi 2015;

Eka, 2014;Santhosh 2010). An information processing systems is interconnected in a network that work together to produce an output. The output of a neural network has the size on the relationship of each individual node in the network operates. Information processing in neural network is often implemented in parallel and not sequentially. Therefore, the members of neurons which have the same size work together to carry out its function. It is unique from a function of the overall neural network that can still be carried out even if the majority of the neurons are not functioning. That is because the artificial neural networks are very tolerant to faults or failures (Meera 2015). Backpropagation Neural Network may be the reason behind the popularity of artificial intelligence. Because this metode does not restrict the types of algorithms which are used in the learning process. This method spread enthusiasm by the developers of artificial intelligence though it is still much controversy whether the model can be applied as the learning steps of the brain or not. Partly because the mechanism for reversing the signal or error, it can not be accepted. Moreover, because there is no reasonable source for learning signal or target output.Backpropagation uses a multilayer architecture with supervised training methods training. Behind Propagation models, they have a couple of units that exists in one or more hidden layers. Behind Propagation model architecture with n inputs plus a bias, a hidden layer consisting of p units plus a bias, and m pieces of unit output, as shown in Figure 1.

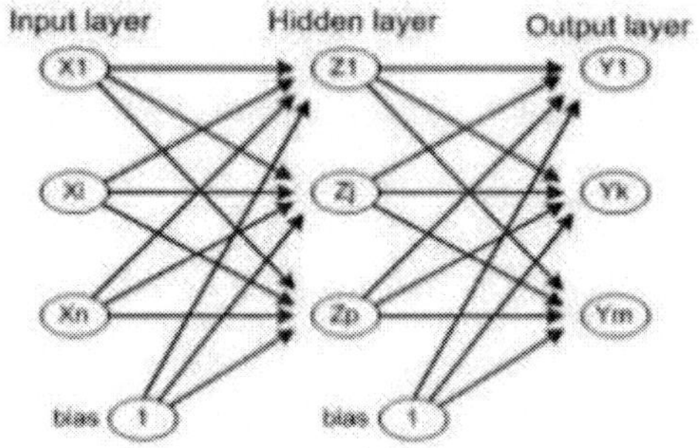

Figurer-1. Backpropagation architecture (MePooja, 2014; Arti 2012)

ANN consists of many elements called neurons, unit, cell or science node connected to one another via the synaptic link / communication links that have weight-related (Kadar 2011). Weights represent information to be used by the network to resolve the issue. They can be applied to various types of problems, for example for pattern classification, making the mapping of input patterns, grouping with many constraints. Each neuron has an internal state called activity level, where the function of the input has been received. Neurons transmit a signal at a given moment; can be seen in Figure 2.

In the training, ANN Backpropagation initialized the weights to determine the numbers of learning (α), determine the value of the error tolerance or threshold values (when using a threshold value as a condition to stop) or set maximum iterations (when using many iterations as the stop condition), then performed step next-step for the stop condition is worth FALSE.

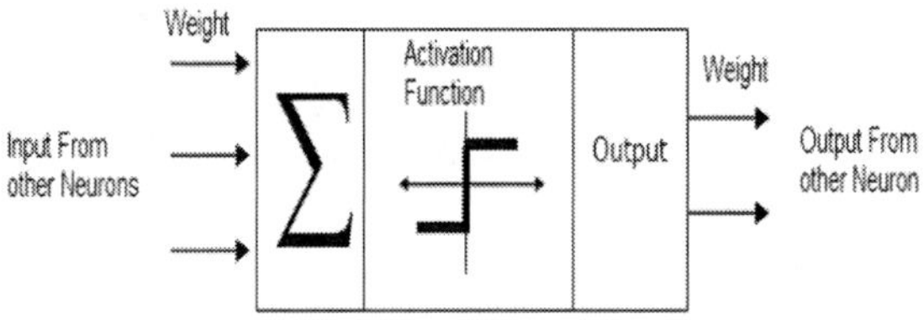

Figure-2. Neuron structure

Every unit of output (y_k, k = 1,2,3, ..., m) receives the target pattern associated with the input pattern learning arithmetic error information and then calculate the correction weights (which will be used to improve the value w_{jk}), as in equation 2 and compute their bias correction (which will be used to improve the value w_{0k}), as in equation 1 (Noviani,2010).

$$\delta_k = (t_k - y_k) f'(y\_in_k)$$
$$\Delta w_{jk} = \alpha \delta_k z_j$$
$$\Delta w_{0k} = \alpha \delta_k \qquad (1)$$

Backpropagation ensures the completion of the minimum mean square error during the learning rate which is not too big. The drawback is when the learning rate is small, then the achievement of convergentvalue is slow, while if the learning rate is large, the achievement of rapid convergence gives a good value, but there is a danger that it can lead to global oscillations when the minimum value is not reached. To overcome this, it used backpropagation variations, such as momentum, where it works with the aim to smooth oscillations that may occur. Filter momentum will be added to the equation weight matrix and bias. Variable Learning Rateworks by trying to raise the learning rate when it encounters a flat surface and then lowering the learning rate in the event of an increase in slope.

RESULTS

The system of forecasting the number of people with the disease at the health center with Backpropagation ANN method has the design development as in Figure 3.Model neural network with back propagation algorithm has three layers with an input layer consists of several unit cells (x_1-x_n) which contains the number of patients disease defined earlier. It can determine the number of people with the disease further, one hidden layer with the number of unit cells is determined at random (y_1-y_n), and one layer of output is amounted to one unit cell (z) as the target of the number of patients with disease outcome prediction next period

Figure-3. Propose Model for number of patients

$$X_i = y(X_{max} - X_{min}) + X_{min} \quad (3)$$

where:

X_i = number of patients with the disease
Y = the output network
X_{min} = Data with a minimum value
X_{max} = Data with a maximum value

Table-1. The best parameter configuration

Disease Name	Parameter					MSE	MAPE (%)
	Hidden	Epoh	Mom	*Lrate*	*Error*		
Acute Respiratory Infections(ISPA)	2	75	0.05	0.008	0.001	1567	0.24
Rheumatoid Arthritis (RA)	10	50	0.02	0.008	0.001	5519	32.65
Diarrhea	4	50	0.02	0.008	0.000001	295.4	4.68
Hypertension (HT)	2	50	0.02	0.008	0.000001	57.94	0.35
Skin Allergies	4	50	0.02	0.008	0.0001	67.63	3.91
Conjunctivitis	2	75	0.02	0.008	0.001	401.3	3.57
Asthma	2	1000	0.02	0.008	0.001	289.1	15.86
Febrile	6	50	0.02	0.008	0.001	422.8	9.78
Tuberculosis (TB). Lung	2	50	0.05	0.008	0.001	16.76	3.14
Scabies	10	1000	0.02	0.008	0.00001	0.001	0
Gastritis	6	50	0.02	0.5	0.001	0.01	0
Typus	8	50	0.02	0.008	0.001	11.95	13.03
Scarlet Fever	2	100	0.02	0.008	0.001	1.05	0.57

The training process of data patterns is normalized first so that the system can determine the weights that can map between data input with the desired output target data. Normalization is done by dividing the value of data with the value of the data range (maximum data value-minimum data value). Input data normalization is aimed to adjust the value range of data with activation function in the system backpropagation. This means that the value of the square of the input should be in the range of 0 to 1. Thus, input range that qualifies is the input data values from 0 to 1 or from -1 to 1. Therefore, the resulting output will be in the range of 0 to 1, such as in equation 2 (Frank 1997).

$$f(x) = \frac{X_i - X_{min}}{X_{max} - X_{min}} \quad (2)$$

where:

X_i =Data to i
X_{min} = Data with a minimum value
X_{max} = Data with a maximum value

In the testing process, the output is produced by the network ranges between 0 and 1 so we need denormalization to convert the output result of network back into the normal number of people with the disease, as in Equation 3 (Noviani,2010). After that, it will be a comparison between the actual data with the data of the prediction, so the error can be calculated.

to calculate the value of the error of the calculation process by the system to the initial value in accordance with the original data using the mean square error (MSE) because MSE is more sensitive to the value of the error. MSE does not have the magnitude of the value. A system is said to have good performance when the MSE value is close to zero (MSE $\approx$ 0) or can be said to be an error in the system. MSE is calculated by summing the squares of all the forecasting error in each period and dividing by the number of forecast period (Upadhyay 2011).

The trial conducted Backpropagation ANN forecasting model to determine the accuracy of the system in the process of disease predicted number. There are 13 types of disease testing process based on the parameters that are used with real data scenarios comparison with predicted results. Test results of each disease prediction produce the best parameters of different configurations, such as in Table 1.

Each disease produces MSE and MAPE with the best parameter configuration. Configuration parameters generate value system accuracy,which is small.It shows that the difference between the data which predicts results with real data is not large, ascan be seen in Table 2. While the accuracy of the predictions for each disease can be seen in Table 3.

Table-2. Accuracy with real data and prediction

Disease Name	2015									
	January		February		March		April		May	
	Real	Prediction	Real	Prediction	Real	Prediction	Real	Prediction	Real	Prediction
Acute Respiratory Infections(ISPA)	29	208	201	201	209	207	200	198	199	195
Rheumatoid Arthritis (RA)	179	211	180	210	189	213	201	220	187	217
Diarrhea	49	45	57	61	61	57	65	60	65	60
Hypertension (HT)	30	29	30	31	31	30	36	35	25	24
Skin Allergies	30	25	31	28	37	35	27	30	39	35
Conjunctivitis	38	35	39	35	41	36	42	36	33	35
Asthma	34	26	32	24	37	25	39	25	32	26
Febrile	33	29	26	31	24	30	36	31	40	32
Tuberculosis (TB). Lung	8	6	17	16	14	15	18	16	19	16
Scabies	7	7	8	9	8	8	9	8	10	8
Gastritis	2	3	1	2	1	1	1	2	1	2
Typus	6	1	6	2	2	1	5	1	4	1
Scarlet Fever	2	2	0	1	0	2	1	2	0	2

Tabel-3.Prediction accuracy each disease

Disease Name	Acute Respiratory Infections (ISPA)	Rheumatoid Arthritis (RA)	Diarrhea	Hypertension (HT)	Skin Allergies	Conjunctivitis	Asthma	Febrile	Tuberculosis (TB). Lung	Skabies	Gastritis	Typus	Scarlet Fever
MAPE (%)	0.24	32.65	4.68	0.35	3.91	3.57	15.86	9.78	3.14	0	0	13.03	0.57

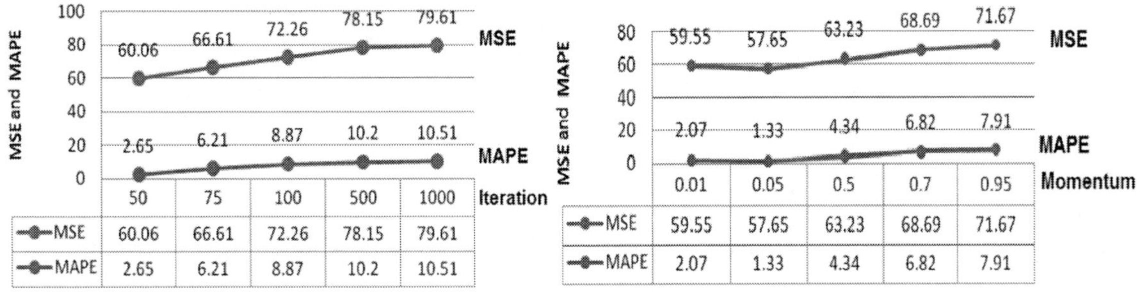

Figure-4. Forecasting evaluation results with iteration and momentum

DISCUSSION

To evaluate the results of forecasting in this research, it used MAPE. MAPE is the percentage difference between the values of the error of the original data with the results of forecasting. MAPE measures the absolute error as a percentage instead of each period rather than the average absolute error in the number of periods of actual data. It can avoid the problems in the interpretation of measurement accuracy relative to the magnitude of the actual value and the predicted value. From 50 to 1000 iterations MSE and MAPE proportional same for any changes momentum, where the smaller iterations or momentum, the value of MSE and MAPE are getting smaller and vice versa, it is shown in Figure 4. When the comparison of accuracy and the learning rate changes, the smaller the value of learning the higher the rate the accuracy or otherwise. And the value of learning rate 0.008 is resulted in the MSE and MAPE smallest,as shown in Figure 5).

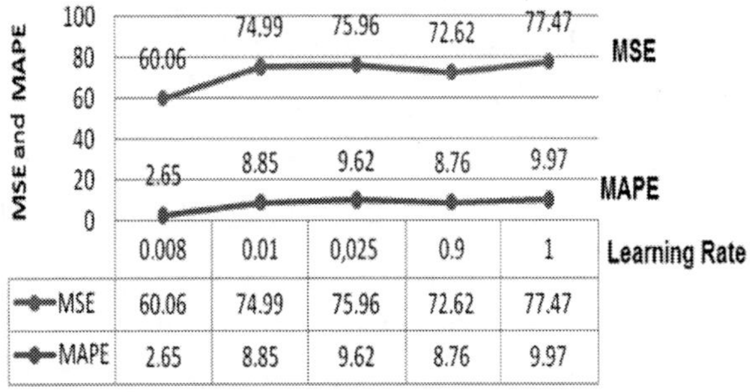

Figure-5. Forecasting evaluation results with learning rate shown in Figure 5).

Artificial neural network architecture consists of one input layer, one hidden layer, and one output layer. The number of input and output node in accordance with the number of lags is obtained from the autocorrelation, while the number of hidden layer is obtained from twice the number of input layer nodes. The condition of the system stops if the system generated error value is less than or equal to the value specified error tolerances or the number of iterations performed has reached the maximum number of iterations specified. From some of the experiments that have been done, it can be concluded that the predicted number of patients using the model ANN Backpropagation disease depends on several parameters. In addition to the number of hidden layer, there are other parameters that need to be regulated, namely learning rate and momentum. Learning rate affects the speed and the accuracy of ANN in the learning process. The greater the learning rate, the faster ANN learns but the results are less accurate. The smaller the learning rate, the slower ANN studies but the results are more accurate (Tanu 2015). Momentum effects on the changes in the chart error. The larger the value of momentum, the shape of the curve is getting steeper. The smaller the value of momentum, then the error is getting sloping curve shape. The parameters are chosen to be the best parameter because it produces the number of iterations that has a value reasonably good accuracy error during the system test that is equal to 0.35%.

CONCLUSION

Configuration parameters for forecasting system using Neural Network Backpropagation is the best of the trial results that shows disease HT which are obtained from the experiments yield learning rate configuration at 0.008, 0.02 momentum, 0.000001 error tolerance, a maximum of 50 iterations and the number of hidden layer 2 as well as a percentage value error 0.35%. At scabies disease, the best parameters were 0.008 learning rate, momentum 0.02, 0.00001 of error tolerance, 1000 maximum iteration, 10 hidden layers, and 0.001 MSE. This shows that the neural network Backpropagation method is suitable to be used for prediction of the number of people with the disease. The smaller the value of learning rate, the better the accuracy of the system. Momentum affects the changes in the chart error. The larger the value of the momentum, the steeper the shape of the error curve. On the opposite, the smaller the value of the momentum, then the flatter the shape of the error curve.

REFERENCES

Arti R. Naik, Prof. S.K.Pathan. 2012. Weather Classification And Forecasting Using Back Propagation Feed-Forward Neural Network. International Journal of Scientific and Research Publications. 2(12): 1 - 3.

Chang, P. C., Wang, D. d., & Zhou, C. l. 2012. A Novel Model By Evolving Partially Connected Neural Network For Stock Price. Expert Systems with Applications. 39: 611–620.

Deepthi.I.Gopinath, G.S. Dwarakish. 2015. Wave Prediction Using Neural Networks At New Mangalore Port Along West Coast Of India, International Conference On Water Resources. Coastal And Ocean Engineering. Aquatic Procedia. 4 : 143 – 150.

Eka Malasari R, 2014. Stock Price Predictions Were Considering External Factors Using Artificial Neural Network, Scientific Journals NERO. 1(2): 5 – 11.

Frank M. Thiesing and Oliver Vornberger. 1997. Sales Forecasting Using Neural Networks. IEEE Proceeding ICNN'97 Houston – Texas. 4: 2125 – 2128.

Kara, Y., Boyacioglu, M. A., & Baykan, O. K. 2011. Predicting Direction Of Stock Price Index Movement Using Artificial Neural Networks And Support Vector Machines: The Sample Of The Istanbul Stock Exchange. Expert Systems with Applications 38: 5311–5319.

K. G. Upadhyay, A. K. Choudhary, M. M. Tripathi. 2011. Short-Term Wind Speed Forecasting Using Feed-Forward Back-Propagation Neural Network. International Journal of Engineering, Science and Technology. 3(5): 107-112.

Meera Narvekar, Priyanca Fargose. 2015. Daily Weather Forecasting Using Artificial Neural Network. International Journal of Computer Applications. 121(22): 9 – 13.

Memmedli, M., & Ozdemir, O. 2012. An Empirical Study of Fuzy Aproach with Artificial Neural Network Models. International Journal of Mathematical Models and Methods in Aplied Sciences. 1(4): 321-326.

Noviani, IR. 2010. Decision Support System Housing Sales Price Determination Method Using Backpropagation Neural Network. Thesis: University of Trunojoyo Madura.

Santoso, S. 2009. Business Forecasting: Methods of Business Forecasting Present with MINITAB and SPSS. Jakarta: PT. Elex Media Komputindo.

Setiyoutami, A. 2012. Prediction Poli Surgical Patient Visits in Surabaya Oncology Hospital Fuzy Using Time Series. Thesis: Teknologi Sepuluh Nopember Institute of Surabaya.

S. Santhosh Baboo and I.Kadar Shereef. 2010. An Efficient Weather Forecasting System Using Artificial Neural Network. International Journal Of Environmental Science And Development. 1(4): 136-142.

Pooja Malik, Prof. Saranjeet Singh, Binni Arora. 2014. An Effective Weather Forecasting Using Neural Network. International Journal of Emerging Engineering Research and Technology. 2(2): 209 - 212.

Tanu Verma, Dr R.K. Srivastava. 2015. Artificial Neural Networks Based Heart Disease Predictive Approach. International Journal of Application or Innovation in Engineering & Management. 4(3): 029-032.

GAMIFICATION APPROACH TO ENHANCE STUDENTS ENGAGEMENT IN STUDYING LANGUAGE COURSE

Andharini Dwi Cahyani
Informatics Department., Faculty of Engineering,
University of Trunojoyo Madura
E-Mail: andharini.dwi.cahyani@gmail.com

ABSTRACT

Many researchers have attempted to utilize gamification to increase student engagement, motivation and achievement in the classroom with varying degrees of accomplishment. This research attempts to review of existing literature on the subject as well as the implementation of gamification on Language course. It aims to get better understanding of how gamification can efficiently be used in education. This analysis reveals that the underlying fact that make games engaging are already utilized in pedagogical practices. There are two important recommendations from this research. First, use gamified learning scenario as an activity, to encourage students trying new things and avoiding fear to make a mistakes. Secondly, the gamification in education setting needs students to participate deliberately to ensure the gamification retains the game-like nature.

Keywords: gamification, enjoyment, motivation, engagement

INTRODUCTION

For students, Language is a course that can be very challenging. This work is aimed to use gamification to help students understand the Madurese language more thoroughly, as well as to make the learning of the topic more fun and engaging. Successful and efficient educational applications have a similar relation to instruction, goals, feedback and interaction. Unfortunately, many educators find it difficult to implement strategies in their curriculum to compete with the engagement of computer games.

Background

Gamification is currently adopting the game elements to improve user experience and user engagement, specifically in non-game contexts (Groh, 2012). Anderson and Rainie's (2012) report defined gamification as the online interactive system design that occupy on people's competitive instincts and rewards to motivate the player. Rewards may be virtual rewards, such as payments, points, badges, or free gifts. According to (McGonigal, 2011), people worldwide spend 3 billion hours a week playing video and computer games. Therefore, it would be potential to make the students' practice for their language skill and knowledge instead of perfecting their skill in "Clash of Clans" strategies.

Many researchers studied the relationship between gamification with in-class participation, students' motivation and student's engagement. Thamvichai et.al. (2012) implemented a points-based system in an attempt to integrate game mechanics into an engineering curriculum. Through the use of a quick questionnaire, the researchers found students motivated to engage in the given tasks. At the same year, Cronk (2012) also adopted gamification concept as a reward-based system in the form of a virtual tree that would get bigger relates to students' points/score. This study also reported an increase in an in-class students' participation. Sheldon (2011) also implemented gamification concept in the classroom. In his experiments, the students are divided into guilds and compete to gain experience points (XP). At the end of the semester, the XP is translated into letter grades. This study reported that the students are pleasing to the class participation.

Educational Games

It is believed that nature of games may facilitate students' engagement and involvement, motivation and interest, and the retention of learned skills. More recently, it has also been suggested that players of commercial games are developing problem solving and literacy skills (Urh et al, 2015). It means that the good commercial games have good learning principles that provide opportunities for gamers to engage and improve their skills during gameplay. Furthermore, Barab and Dede (2007) stated that the "immersive" nature of educational games have the potential to transform the school content standards knowledge into "just-in-me" knowledge to solve meaningful problems actively and interactively.

Lee and Hammer (2011) stated that there are three reasons why the game-like elements have a high engagement in high schools and institutions of higher education. First, the players who usually use different avatars, such as the act as a fantasy character or a secret agent, is proven to be successful in schools (Nasir and Saxe, 2003). This fact leads to experiences of social credibility and recognition.

Second, good game designs perfectly match the player's cognitive abilities with the difficulty level. For example, good games present personalize tutorials while keeping players' engaged and motivated. Moreover, good games usually have incremental sub-goals and also alternative ways to solve its problem that is can be very challenging and motivating. This situation answers the challenges of the educational system nowadays that is student motivation and engagement (Bridgeland et al., 2006).

Third, games especially simulations, are perfectly-fitting sandboxes that allow students to learn from mistakes in quick recovery. Games and simulations enable students to see failures as an opportunity to learn along with frequent feedback cycles. As a result, the students cover an important emotional problem faced in traditional institutions, such as the excessive demand to pass a final exam, or to fail the course.

METHODOLOGY

In this study, the effect of the gamified learning activity was analyzed along three dimensions: perceived easy to use, engagement, and motivation. There are totally 30 students participating in this study, of which 18 are male and 12 female, with the average age of 6-8 years.

Gamification in Education : enjoyment, engagement, motivation

Gamification is a strategy to design the engagement power of games in a non-gaming context (e.g., industry, education, etc.). Others have emphasized the fact that gamification is implemented through technology. For example, Domínguez et al. (2013) defined gamification as "incorporating game elements into a non-gaming software application to increase user engagement and experience". This definition appears more specific and also includes the purpose of gamification as fostering engagement in studying. Prior scholars have defined the educational participation as a bound in learning process (Filsecker and Hickey, 2014), committed learning (diSessa, 2000), and intentional learning (Scardamalia et al, 1989). Therefore, we may conclude that gamification is brought up the enjoyment while studying by fostering students' engagement and students' motivation.

In the gamification literature, usually we found game elements, for example, rewards systems and the use of badges for displaying achievement (Landers and Callan, 2011). This badge is an "informal way" to offer a learning certificate (Abramovich et al., 2013). The badges on the learners' profile are viewable by other players and are awarded for either incidental activity or mastery of skills/demonstration of knowledge. In other words, badges may symbolize a learner's participation and performance (Simões, et al,, 2013).

Some researchers have adopted gamification concept to their prior e-learning system. For example, Schoooools.com for K-6 (Simões et al., 2013), casual social games for undergraduate education and employee training (Landers and Callan, 2011) and, intelligent-tutor-systems (Abramovich et al., 2013). In Landers and Callan (2011) experiments, there is a social experiment where the college students have to answer small quiz and based on their score, the students would be ranked and provided with badges indicating their performance level. There are five performance levels, namely: "Newbie", "Novice", "Intermediate", "Expert" and "Master". These badges are viewable by others on the students' profile. A preliminary evaluation of this project regarding student's motivation presented that a considerable number of students used the gamification features and that by doing so taking the quiz seemed to the more enjoyable and rewarding activity.

Abramovich et al. (2013) introduced gamification in the context of an intelligent tutor system. In gamification learning scenario, the badges were awarded based both on individuals mastery of a skill and on their participation in the system. After using the intelligent tutor system, which use gamification learning scenario – they found that the students showed less concerned about poor performance as compared to their peers and more interest in learning language. Therefore, in summary, research on gamification has just begun, and questions related to their impact on learning, students' motivation and engagement needs a systematic exploration. This study represents an empirical exploration of the relationship between enjoyment, motivation, engagement while studying in gamification learning scenario.

Application Design

Gamification has been used in many contexts, from military training (Halter, 2006) and education (Kapp, 2012) to service marketing (Huotari and Hamari, 2012). In this research, we build gamification in Language course for 6th grades with five learning modules. This user leveling in this application is based on the students knowledge. There are three user levels for each module, namely "beginner", "intermediate" and "advanced". Each level contains three core steps, which are: doing pre-test, learning the material and doing exercise. The pre-tests' result presents the user level at the current module. The detailed of this gamification learning scenario is as Figure 1.

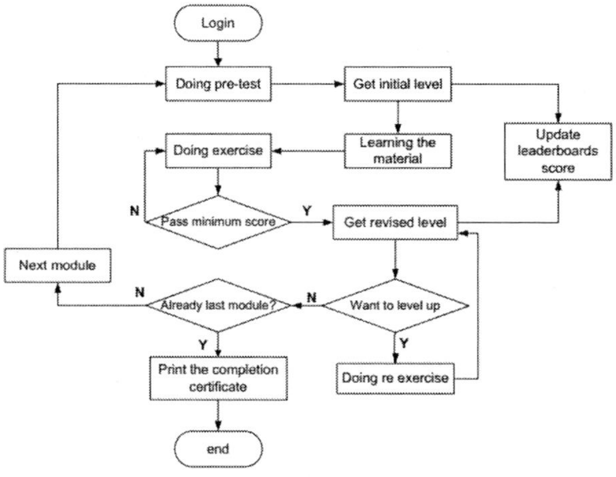

Figure-1. Application Learning Scenario

After doing pre-test, the application calculates the score and obtain the a user initial level. In the meantime, the leaderboards score for that specific level is also updated. User level represents the "students' badges". Furthermore, to be specific, there is a score for each "students' badges" within leaderboards score.

After getting to the initial level, the user should follow the learning material instruction and do some exercise. If he/she passes the minimum score, then he/she gets revised level and also the system update the leaderboards score. Also, the user may level up by doing re exercise and get the revised level. This scenario applies to all modules, so when a user is already in the last module, the application made a completion certificate.

RESULTS

In this study, the effect of the gamified learning activity was analyzed along three dimensions: perceived easy to use, engagement, and motivation. There are totally 30 students participating in this study, of which 18 are male and 12 female, with the average age of 6-8 years. For each of these dimensions, participants indicated their agreement or disagreement on a 5-point Likert scale (strongly disagree, disagree, neither agree or disagree, agree, strongly agree) to a number of related statements. The results for each dimension are discussed next in details.

Enjoyment

An important effect of gamification is that the application should be enjoyable. Furthermore, the enjoyment of the application can lead to improved learning. Therefore, it was important to know how the students enjoy the gamified learning activity. Statements related to enjoyment are presented in Table-1.

Table-1. Statements related to students enjoyment

No	Description
1	I feel genuinely fun
2	I feel happy when playing the game `
3	I feel that it is great for killing time productively
4	I feel exhausted when playing it
5	I feel miserable when playing it

Engagement

An important effect of gamification is that the application can engages the student to the learning material. The student were not only learning what their schools offer but also trying to understand and also internalizing the material towards their life (Newmann, 1992). Gamification within learning process allows students to fail and not feel rejected, so they are willing to try more and more. Statements related to engagement are presented in Table-2.

Table-2. Statements related to students engagement

No	**Description**
1	I wanted to explore all the options because it was very challenging
2	I felt that time passed quickly
3	I wanted to complete the game
4	I did not care how the game ended
5	I feel bored when playing it

Motivation

Students' motivation plays an important role towards conceptual change processes within the affective components, critical thinking, learning strategies and learning achievement. When students have determined their achievement goal, they are intrinsically motivated. Therefore, the students intend to accomplish something to satisfy their innate needs for improving their own performance (Ryan et al., 1991). We suggest that gamified learning interventions may enhance student motivation in their learning process.. Statements related to engagement are presented in Table-3.

Table-3. Statements related to students motivation

No	**Description**
1	It was important to me to do well at this task
2	I would describe this activity as very interesting
3	I tried very hard on this activity
4	I didn't try very hard to do well at this activity
5	I didn't put much energy into this

The response of students is summarized in Table-4. It appears that the majority of students felt enjoyed, engaged and motivated through this gamification learning scenario.

Table-4. Summarized result of the questionnaire

No	Question	Strongly Disagree	Disagree	Neither	Agree	Strongly Agree
Statement related to enjoyment						
1	I feel genuinely fun	0%	10,000%	20,000%	33,333%	36,667%
2	I feel happy when playing the game	3,333%	10,000%	16,667%	36,667%	33,333%
3	I feel that it is great for killing time productively	6,667%	10,000%	16,667%	30,000%	36,667%
4	I feel exhausted when playing it	20,000%	36,667%	20,000%	13,333%	10,000%
5	I feel miserable when playing it	43,333%	30,000%	20,000%	6,667%	0%
Statement related to engagement						
6	I wanted to explore all the options because it was very challenging	0%	6,667%	10,000%	50,000%	33,333%
7	I felt that time passed quickly	6,667%	20,000%	26,667%	16,667%	30,000%
8	I wanted to complete the game	3,333%	6,667%	16,667%	46,667%	26,667%
9	I did not care how the game ended	33,333%	36,667%	13,333%	10,000%	6,667%
10	I feel bored when playing it	6,667%	10,000%	10,000%	40,000%	33,333%
Statement related to motivation						
11	It was important to me to do well at this task	3,333%	6,667%	13,333%	36,667%	40,000%
12	I would describe this activity as very interesting	3,333%	6,667%	16,667%	30,000%	43,333%
13	I tried very hard on this activity	40,000%	26,667%	13,333%	10,000%	10,000%
14	I didn't try very hard to do well at this activity	33,333%	23,333%	10,000%	10,000%	23,333%
15	I didn't put much energy into this	30,000%	26,667%	13,333%	10,000%	20,000%

Regarding to enjoyment questions, 33.33% of participants agreed and 36.67% strongly agreed (a total of 70%) that they had fun while studying in this gamification learning scenario. When asked aboutthe opposite, i.e., if they were exhausted while studying in this gamification learning scenario, 36.67% disagreed and 20% strongly disagreed (a total of 56.67%). While there is some ambivalence about the participants' feelings of enjoyment while studying in this gamification learning scenario, it seems that most felt some degree of enjoyment while a small number of participants did not specifically enjoy the activity.

When asked if the students were feel challenged in this gamified learning activity, the majority of participants answered that they were engaged enough to explore all the options (83.33%) and only 16.67% felt bored when playing it. More than half of the participants (73.33%) wanted to complete this gamified learning activity and slightly less than half found that time passed quickly during the gamification (46.67%). Only 16.67 % stated they did not care how this gamification is ended.

This research also revealed if gamified learning scenario may increase students' motivation. The overall responses were positive: 76.67% of participants felt that it is important for them to do well at this task, and

73.33% believed that this activity is interesting. When asked about the opposite, i.e., if they didn't very hard to do well at this activity, 23.33% disagreed and 33.33% strongly disagreed (a total of 56.67%).

DISCUSSION

The results of the study have a positive impact, particularly on enhancing learning, which is the primary objective of the work. The majority of participants felt that the gamified learning activity improved their learning. Furthermore, the activity did engage the participants and result in some degree of enjoyment.

There are two important recommendations to ensure that the gamified activity retains its game-like feel. The recommendations are: (1) use gamification as an activity, and (2) students' participation should be optional. If we use this as an assessment or students are forced to undertake it, then it will reduce students' motivation to complete it. Students will see it likely as another learning task and takes away from its game-like nature.

One of our intentions in the development of the gamified learning activity was to develop an approach and supporting tool that is not discipline-specific. However, so far, it has only been trialled in Language subject. Trials in courses from other disciplines would be useful to ascertain that the approach and tool are non-discipline specific.

Our proposed work is aimed to investigate how gamified learning approach influence students engagement and motivation during the learning process. This approach was trialed in 6th grade of Elementary School. To evaluate this approach, 30 participants have completed the questionnaire survey that has three dimensions, namely: enjoyment, engagement, and motivation. We found that 70% members believe that they feel happy while doing gamified learning activity. Furthermore, 73.33% believe that this activity is very interesting, and 83.33% perceived that doing gamified learning activity makes them feel challenged and want to complete and master all level.

These results present that this implementation of gamified learning scenario has substantial influence and deserve further researched. Moreover, the students' participation should be voluntary to ensure that the gamification retain as a "game-like" activity. Therefore, students will be encouraged to try new things and overcome the fear of making mistakes. Students who are enthusiastic to participate in this gamification also will be engaged in enjoyable experiences in the learning activity.

CONCLUSION

A gamification is an innovative approach to education that is continuously emerging. Future studies should investigate the new mechanics and new applications associated with emerging gamification technologies.

The results of this survey presented that the gamification can have a positive effect on learning. However, it should be remarked that these findings are about one particular implementation of gamification in education. Further research is needed to discover if gamification improves learning to use a variety of methods and to determine other ways gamified activity in school.

ACKNOWLEDGMENT

We would like to thank the students and academics for their involvement in this study. This study was supported and funded by Engineering Faculty of Universitas Trunojoyo Madura – Indonesia.

REFERENCES

Abramovich, S., Schunn, C., & Higashi, R.M. 2013. Are Badges Useful in Education?: It Depends Upon The Type Of Badge And Expertise Of Learner. Educational Technology Research and Development, 61, 217-232.

Anderson, J., & Rainie, L. 2012. Gamification And The Internet: Experts Expect Game Layers To Expand In The Future, With Positive And Negative Results. Games For Health : Research, Development, and Clinical Applications, 1(4), 299-302.

Barab, S., & Dede, C. 2007. Games And Immersive Participatory Simulations For Science Education: An Emerging Type Of Curricula. Journal of Science Education and Technology, 16(1), 1-3.

Bridgeland, J. M., DiIulio Jr, J. J., & Morison, K. B. 2006. The Silent Epidemic: Perspectives Of High School Dropouts. Civic Enterprises.

Cronk, M. 2012. Using Gamification To Increase Student Engagement And Participation In Class Discussion. In World Conference on Educational Multimedia, Hypermedia and Telecommunications 2012 (1), 311-315.

diSessa, A. A. 2000. Changing Minds: Computers, Learning and Literacy. Cambridge, MA: MIT Press.

Domínguez, A., Saenz-de-Navarrete, J., De-Marcos, L., Fernández-Sanz, L., Pagés, C., & Martínez-Herráiz,J. 2013. Gamifying Learning Experiences: Practical Implications and Outcomes. Computers & Education,

63, 380-392.

Filsecker, M., & Hickey, D. T. 2014. A Multilevel Analysis Of The Effects Of External Rewards On Elementary Students' Motivation, Engagement And Learning In An Educational Game. Computers & Education, 75, 136-148.

Groh, F. 2012. Gamification: State of The Art Definition And Utilization. Institute of Media Informatics Ulm University, 39.

Halter, E. 2006. From Sun Tzu to Xbox: War and Videogames. Thunder's Mouth Press, New York.

Huotari, K., & Hamari, J. 2012. Defining Gamification: A Service Marketing Perspective. In Proceeding of the 16th International Academic MindTrek Conference. 17-22.

Kapp, K.M. 2012. The Gamification Of Learning And Instruction: Game-Based Methods And Strategies For Training And Education. John Wiley & Sons.

Landers, R. N., & Callan, R. C. 2011. Casual Social Games As Serious Games: The Psychology Of Gamification In Undergraduate Education And Employee Training. In Serious Games And Edutainment Applications. 399-423. Springer London.

Lee, J. J., & Hammer, J. 2011. Gamification in education: What, How, Why Bother?. Academic Exchange Quarterly, 15(2), 146.

McGonigal, J. 2011. Reality is broken: Why Games Make Us Better And How They Can Change The World. Penguin.

Newmann, F.M. 1992. Student Engagement and Achievement in American Secondary Schools. Teachers College Press.

Ryan, R. M., Edward L., and Deci. 1991. A Motivational Approach to Self: Integration in Personality. Perspectives On Motivation, 38, 237.

Scardamalia, M., Bereiter, C., McLean, R. S., Swallow, J., & Woodruff, E. 1989. Computer-Supported Intentional Learning Environments. Journal of Educational Computing Research, 5(1), 51-68.

Sheldon, L. 2011. The Multiplayer Classroom: Designing Coursework As A Game. Boston, MA: Course Technology Press.

Simões, J., Díaz, R., & Fernández, A. 2013. A social gamification framework for a K-6 learning platform. Computers in Human Behavior, 29, 345–353.

Thamvichai, R., & Supanakorn-Davila, S. 2012. A pilot study: Motivating students to engage in programming using game-like instruction. Proceedings of Active Learning in Engineering Education. St. Cloud University.

Urh, M., Vukovic, G., & Jereb, E. 2015. The model for introduction of gamification into e-learning in higher education. Procedia-Social and Behavioral Sciences, 197, 388-397.

INFORMATION TECHNOLOGY IMPLEMENTATION ON SOCIALIZATION OF HARMONIOUS CULTURE IN BALI

I Putu Astawa, Ni Putu Wiwiek Ary Susyarini and Ni Nyoman Triyuni.
Tourism Business Management, Bali State Polytechnic, Bali, Indonesia
E-Mail: putuastawa1@pnb.ac.id

ABSTRACT

The research objective is to build a model of information delivery harmonious culture that is run by Village Credit Institutions (LPD) through information technology to rural communities so that financial performance can be improved. Thirty-five rural credit institutions which meet the criteria were used empirically as a sample. Data such as the practice of cultural values harmony collected through in-depth interviews were then analyzed using descriptive method and ethnomethodology. The model is built starting with the collection of data from each row, perform data classification into three categories of cultural harmony, making a website that contains information on the activities of cultural harmony, then mounting internet in each row to facilitate public access to information on the implementation of cultural harmony. Results of model simulations for three months showed that the level of community participation increased by an average of five percent through credit lending and credit issues decreased by 3.5 percent. Cultural harmony is practiced by 55% in touch with the God, 35% related to humans, and 10% related natural premises. Results of this study implied that IT is no longer expensive but able to create a better performance.

Keywords: information technology, harmonious culture, financial performance.

INTRODUCTION

The success of microfinance institutions is not located in lending but the range of credit services provided to the public (Yaron *et al.*, 1997). Credit services to rural communities is one of the tasks village credit institutions and encourage the creation of an increase in the rural economy, creating jobs, reducing the urbanization of rural areas to cities, and reducing crime. Services such as microfinance institutions are now widely found in Bangladesh, Philippines, Malaysia or in other ASEAN countries. However, the services provided by microfinance institutions in rural areas in Bali are very different because in addition to credit services, it also provides services relating to the maintenance of a culture that developed in each village (Astawa *et al.*, 2013). Maintenance culture is done by microfinance institutions summarized in the concept of harmony culture that emphasizes harmony with God's relationship between the company, employees, communities and the natural environment. Applications of cultural harmony in microfinance institutions have an impact on the efficiency, credit risk and financial performance (Astawa *et al.*, 2012, 2013). Companies that implement a culture of harmony as well have little credit risk (Astawa *et al.*, 2013). The client believes that not paying on time is a denial of the law of karma belief (Astawa, 2013) which means that if done well, it will get the goodness in return. Credit payments on time is part prostrate devotion to the God, because the money in pay will be used by others means helping the embodiment of God.

Application of cultural harmony is one of the tools to encourage people to adhere to an agreement they credit with the aim of not more than five percent (Regional Devt Bank, 2013). These conditions must be addressed by providing an understanding of cultural harmony to all customers or rural communities in order to avoid misunderstandings or different interpretations of harmony cultural practices performed by microfinance institutions. The process of providing an understanding can be done by using information technology.

Today information technology (IT) has been able to assist companies in improving operations and more productive (Brynjolfsson and Hitt, 1996, 2003; Bloom et al., 2012). On the other hand some researchers have found that IT it weakens the natural value of an IT that are innovative (Hobijn and Jovanovic, 2001; Gârleanu *et al.*, 2012a, b; Bartelsman, 2013; Kogan and Papanikolaou, 2013, 2014). Research Chun *et al.*, 2015 confirms that the use of IT to contribute positively to the manufacturing and non-manufacturing. The use of IT in the company has an impact on employment (Sniukienea and Sarkane, 2014), which is associated with increased skills and provide motivation to work better and more productive. Another impact caused by the use of IT is the quality of service and decreasing costs (Bardhan and Thoin, 2013). Quality service through the use of IT fosters consumer confidence, because it is served with fast, accurate, and inexpensive. This concept is no longer dominated by the manufacturing company but in the non-manufacturing companies (Chun *et al.*, 2015) healthcare, hospitality, education, and banking.

IT provides benefits for the company and also social life (Habermas (1984); Zaidi *et al.* (2015). The use of IT in this study is very different from previous studies where IT is used for socialization program socialization culture of harmony not tangible products such as the company in general. This research builds a model of information delivery harmonious culture run by the Village Credit Institutions (LPD) through information technology to rural communities in order to increase the financial performance can be achieved. Data were collected through in-depth interviews to 35 managers of companies that are selected through purposive sampling with criteria have IT staff and there is telephone access to community (banjar). The data relate to cultural activities that emphasize harmony harmonious relationship with God, man, and nature analyzed by ethnomethodology. Results of analysis of the cultural practices of harmony as the content will be disseminated through IT to the public or customers through a model built.

LPD is one of the non-bank microfinance institutions in which the owner is a traditional village institutions or Pakraman (institution) and is the only non-bank microfinance institutions that are managed by custom in Indonesia. The existence of LPD in Bali is set in Bali Provincial Laws No. 2 in 1988, and has been updated with local regulations No. 8 of 2002, with the following characteristics : (1) is a business entity; (2) belonging village (*Desa Pakraman*); (3) established and managed by *Desa Pakraman*; (4) organizing functions *Desa Pakraman* community financial institutions, such as: accept / collect funds from village manners, lend only to the village manners, and manage limited to financial institutions *Desa Pakraman* environment; and (5) perform the functions of the business as a financial institution's internal efforts *Desa Pakraman* or as far as inter- *Desa Pakraman*.

Based on data from the Provincial Government of Bali, LPD's population in the province of Bali in eight districts and one municipality in 2010 is as much as 1,405 units of LPD with employment of 7,396 people and total assets of Rp. 5.18 trillion. LPD condition has exceeded the rural banks (BPR) is there, where the number of assets only totaled 3.142 trillion in the same year, but has a problem of uncollected loans of 9.76% (Kontan, 2010).

LPD has a unique view on governance by using the traditional rules, including norms, social sanctions, and engagement of leader Pakraman in the selection process and the enforcement of the loan agreement which is not directly made by the Central Bank (BPD). Models such as these tend to exploit the comparative advantage in monitoring and enforcement capacity of sanctions (Stiglitz, 1990). Brinkerhoff (1992) tried to explain the role of institutions in an effort to improve the performance of microfinance institutions (such as: legal, formal regulations, customs, conventions, social norms and codes of conduct) that is held as a driving force of growth companies. Institutional environment is a set of basic rules of social and political fundamental law that became the foundation for the production, trade, and distribution (Davis and North, 1971). Therefore, institutions have an important role because it can create a structure in everyday life by defining and limiting a set of options that are owned by individuals and organizations.

Astawa and Sudika (2014) examined cultures such as praying, honest, hard work, mutual cooperation, and *tatwamasi* in associate with the financial performance of the company property in Bali. Results of the study showed that culture such as praying, honest, hard work, mutual cooperation, and tatwamasi have a positive effect on improvement of financial performance. Cultural harmony that runs on LPD is the development of an organizational culture Scein (2004) described the study Astawa (2013).

Tabel-1. *The Comparison of Schein's Cultural Theory and Harmonious Culture*

Schein	Harmonious Culture	Harmonious Sub-Culture
Basic assumption	*Parahyangan*	*Sradha* dan *bhakti*, *karmaphala*
Value system	*Pawongan*	*Tatwam asi, rwa bhineda*
Artifact	*Palemahan*	*Olas asih*

Source : Astawa (2013)

Cultural Comparison harmony with the culture Schein (2004) as the harmonization of the relationship with God (parahyangan) which is equal to Basic assumption in practice in LPD reflected on the culture Sradha and bhakti, karmaphala. Where the owners and managers believe through closeness to God means God's attributes will lead in acting like honesty, sincerity, and sincerity. Service is a sacrifice to God and also believed that good deeds will produce good results and vice versa. Through the closeness of God and will no longer lie with the conditions experienced honest. People are afraid to not pay the loan because the services that touch the heart will bring shame to not pay the debt. Harmonization among humans (pawongan) equal to the value system in practice is found respect for the human being with respect for differences. Through respect for human tantamount to honor God because in man there is a soul that is a part of God. Through human treatment equal with God means God's character will be united that will come a sense of mutual respect for each other. There is a harmonization relationship between LPD with the surrounding natural environment (palemahan) together with the artifact. Practice reflected that LPD has a sense of care about the environment. Sense of caring is reflected in the culture olas asih towards the people in the village by taking into account the needs of indigenous and arrangement of the physical environment.

Today the use of IT is important because it has an impact on productivity, creativity, employees, and economic growth (Katz (2009). Lipaj & Davidaviciene (2013) argues that the information system is constituted by IT can improve the achievement of corporate objectives, targets and strategies owned. The use of IT can add value to the company so it can compete in world global terms. At small companies with good IT governance can compete in global terms and is able to obtain consumer information quickly (Серикова, 2003). Small company that controls IT can play a powerful role as mediating between large companies with consumers. Some researchers link between IT with employee performance such as Oulton (2001), Pilat, Lee and van Ark (2002), Khan and Santos (2002), Van Ark, Inklaar, McGuckin (2003), and Van Ark, Piatkowski (2004). Kogan and Papanikolaou (2013, 2014) analyzed the impact on the value of investments in technology companies. The result is that the sensitivity of each company is different to the performance and profit shares this opinion and gain support from Hobijn and

Jovanovic (2001) and Gârleanu et al. (2012b). Special technology investments including the computers, the steamengine, and the dynamo (David, 1990; Greenwood and Jovanovic, 1999) provides an important role in improving corporate performance

METHODOLOGY

This research was done in three stages of the first to perform in-depth interviews to thirty-five LPD manager to get data harmonies cultural practices performed. The sampling process used purposive sampling with criteria that have IT staff and there is telephone access to banjar. thirty-five LPD qualified as research sample of 1406 LPD in Bali (Bali Development Bank, 2013). Analysis tools in view of cultural practices that run on LPD through ethnomethodology approach. Ethnomethodology are attempts to describe and understand the people in everyday life, such as how the interaction patterns, ways of thinking, feelings and how to talk (Bungin, 2007). The term ethnomethodology issued by Garfinkel in 1967, refers to the study of the ways in which people make sense of Reviews their social world (Poore, 2000). Results of this analysis were grouped into three parts, namely cultural activities related to the Lord (parahyangan), cultural activities associated with humans (pawongan), and cultural activities related to nature. The second stage of modeling to convey information by utilizing cultural practices harmony IT owned by LPD and the final step is to carry out the implementation of the model in the village of Kuta Badung. Kuta village is elected as a test of the model because it has telephone access to banjar and LPD has the largest asset in Bali. Test models for three month of May until July 2014) and the results will be analyzed descriptively.

RESULTS AND DISCUSSION

Practical Harmonious Culture

The results of in-depth interviews that reflect the harmony of cultural activities related to the God, human beings and the natural environment.

Table- 2. Practical Harmonius Culture

Reflexivity study Cultural Harmony	Assessing Indices
1. The development temple (Tempat Ibadah) 2. Temple ceremony (upacara di Pura) 3. Donation for ceremony 4. religious trip	Parahyangan (God)
5. Education and training 6. Health 7. *Ngaben* and *Nyekah* 8. Donation for death 9. Arts Group 10. Labor Union	Pawongan (Human)
11. Built building tradisional market 12. Development of Village Hall 13. Development of Sports Facilities 14. Clean beach 15. Greening 16. Road repair 17. Repair of drains 18. Making the trash 19. Help plant seeds and fish 20. Parking area	Palemahan (Environment)

Source: primary data

Rural credit institutions in Bali have run activities associated with God through the creation of a place of worship of God, the teachings through ceremonies and religious activities (Table-2). Belief in God encourages managers to prostrate devotion through ceremonial activities and keep close to God (Astawa, 2013). The closeness to God reflected in the attitude that puts the LPD managers' harmonization in providing credit services. Sincere service without imposed will encourage closeness to customers and the added confidence that if we do good will obtain good results (karmapala) Astawa and Sudika (2014). Through sincere service raises heart closeness between customers with LPD that will encourage honesty to occupy all the agreements made so that the performance can be increased (Yaron *et al.*, 1997; Astawa, *et al.*, 2013).

Human relations problems with LPD managers have helped the villagers to improve education and health through several programs such as scholarships, medical benefits. Other aid in the form of funds deaths and conduct a joint ceremony on a tribute to the ancestors (cremation and nyekah). This activity provides a touch hearts for villagers who feel respected human ancestors even well respected by LPD. Touch services out of the modern management rules that are rigid create LPD closeness with the community difficult to separate. It can be seen LPD has an important role in driving the economy of the village in Bali (Nurcahya, 2006). Pawongan dimension is the harmonious relationship between human beings (Surpha, 2001; Wiana, 2004; Ashrama, 2005), in the context of the business necessary to build a harmonious relationship of internal employees, between employees and management of the company and between the company's management with communities (Ashrama, 2005). In details Ashrama (2005) says that the harmonious relationship can be indicated on several things, such as: the existence of groups in organizations such as trade unions, sekeha-sekeha and companies maintain a harmonious relationship with the community through poverty alleviation programs, utilizing local labor and skills training.

Palemahan, the last dimension, is associated with the physical aspects of the environment around us or the company (Surpha, 2001; Wiana, 2004; Ashrama, 2005). Ashrama (2005) explains that the layout of the company and the existing building should be adapted to religious beliefs and cultures that exist. LPD managers have supported previous research that physical aspect has been considered properly through the construction of facilities and infrastructure (Table 2) as well as providing assistance plant seeds and fingerlings. Keeping the environment in which the company was founded in harmonious order LPD managers have understood well. The concept of harmonious relationship with environments provides benefit to the value of the company and supports the government's green program. This concept is already a famous Balinese culture with compassion Olas (Astawa and Sudika, 2014) meaning that damage the environment means damage to the company, thus safeguarding the environment is a must.

Submission of Harmony Culture Model-Based IT

Researchers had previously explained that IT has a role in improving the performance or productivity of employees. In this study makes the model to convey the harmony of cultural reflexivity of the activities carried out by the LPD. Qualitative analysis results obtained twenty reflexivity of harmony culture (Table- .2) which were divided into three indices namely parahyangan, pawongan, and palemahan. The third dimension of cultural harmony created a model in the form of a website.

The function of the website is to convey information on the plans and activities of cultural harmony to the villagers. To maintain information on each row is given the right to inform the harmony of cultural activities that have been planned or are already underway. Submission of this information is done by the head of the banjo by using the ID and password provided by admin at LPD. Each community can access and view the information culture of harmony in the website by using the ID and password provided by admin. All information is stored in a database and in each row is provided along with wi-fi computer that is used as a tool by the head row to enter the data from the public. This model can be described in Figure-1.

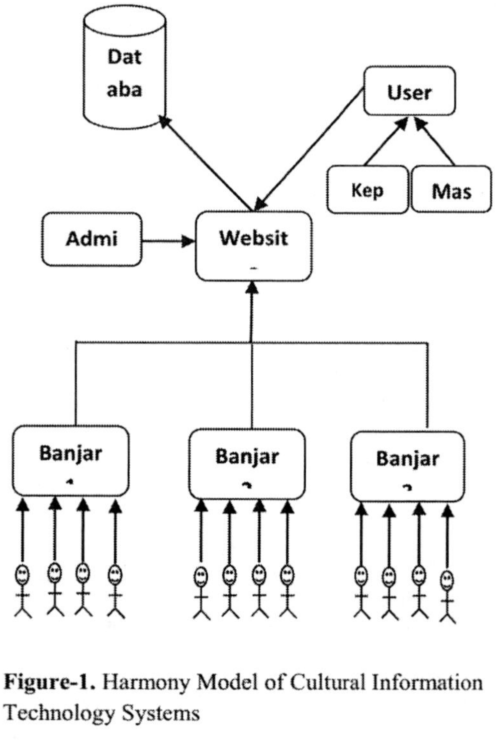

Figure-1. Harmony Model of Cultural Information Technology Systems

IMPLEMENTATION MODEL RESULT

Once the model is created, it was followed by implementation for three months (March - May 2014) and the results are described in Table- 3.

Table- 3. Result l Implementation Model

Reflexivity study Cultural Harmony	Assessing Indeksikalitas	No. Activities
Parahyangan	The development temple (Tempat Ibadah)	0
	Temple ceremony (upacara di Pura)	4
	Donation for ceremony	6
	religious trip	1
Pawongan	Education and training	1
	Health	2
	Ngaben dan *Nyekah*	0
	Donation for death	3
	Arts Group	1
	Labor Union	0
Palemahan	Built building traditional market	0
	Development Balai Desa	0
	Development of Sports Facilities	0
	Clean beach	0
	Greening	0
	Road repair	1
	Repair of drains	0
	Making the trash	1
	Help plant seeds and fish	0
	Parking area	0

Source: Study result model (2014)

The test results obtained by the model during the three months of activity related to the donation for the biggest ceremony then continued with piodalan in the temple and death donation number three. The third major activity lies in parahyangan and pawongan. Another said that the twenty existing activities 55% is related to the activity of God (parahyangan), 35% relate to human beings (pawongan), and 10% related to the environment (palemahan). These results have been demonstrated in previous studies Astawa *et al.*, (2013) the cultural practice of harmony which contains the trust God has been run regularly through religious activities. Balinese people are confident and believe that prosperity is achieved when realized harmonious relationship between man and God creator (Surpha, 2001; Wiana, 2004; Ashrama, 2005). Business activity is an offering which therefore does not escape from the control of the Lord (Widia and Goddess, 2007). So the company is expected to seek as much as possible the emergence of shades relegius (divinity) both inside and outside the corporate environment (Ashrama, 2005, Widia and Dewi, 2007). The observation of model testing also found that the number of clients served in April 2014 amounted to 734 people; the month of May 2014 totaled 801, in June 2014 amounted to 841, and in July 2014 amounted to 883. The increase in loan servicing of observation has increased an average of 5% per month. Number of bad credit problems in the early months of May 2014 amounted to Rp 167 million and at the end of the month of July 2015 amounted to 161 million decreased by 3.5%. Program clarity and evidence submitted through the website activities provide benefits to society. IT has an important role in accelerating the delivery of information, so the impact will be more credit services. This condition will drive corporate performance increase (Lipaj & Davidaviciene, 2013). One measure of the performance of financial institutions is credit problems. The trial period decreased credit crunch models suggesting IT has a role in the achievement of corporate performance (David, 1990; Greenwood and Jovanovic, 1999; Katz, 2009, Liu S. and Wang L., 2014). The results also support the research Chun *et al.*, 2015 confirms that the use of IT to contribute positively to the manufacturing and non-manufacturing companies. The result also support the research (Foster *et al.* 2001; 2006; Acemoglu *et al.;* 2012; Kogan *et al.*; 2012, 2013; 2014; Marinagi *et al.*, 2014) use IT can improve productivity.

CONCLUSION

Cultural socialization of harmony using IT accelerates the communication between the customer and LPD, causing a sense of belonging to the LPD higher because the service is based on a harmonious relationship. Harmonious relations based on the culture of keeping harmony between LPD harmony with the community, environment and God as creator of all that exists on earth.

REFERENCES

Ashrama, Berata, 2005. "The Essence of THK and Its Referential Teaching " *Tri Hita Karana –Tourism Awards and Accreditations.*

Astawa P. I, Sudarma M., Aisjah S., Djumahir, 2012. Credit Risk and Harmonious Values Practice (Study at Village Credit Institution (Lembaga Perkreditan Desa) Of Bali Province), *Journal of Business and Management, Volume 6, Issue 4, PP 16-20*

Astawa P. I., 2013. Ownership in the Perspective of Ethnomethodology at the Village Credit Institutional in Bali, *Research Journal of Finance and Accounting*, Vol. 4, No.8, pp 55-62

Astawa P. I., Sudarma M., Aisjah S., Djumahir, 2013. Institutional Ownership and Harmonious Values in Increasing Financial Performance of Village Credit Institution (Lembaga Perkreditan Rakyat/ LPD) in Bali Province, *Journal of Basic and Applied Scientific Research,* Vol. 3, No. 6, pp 813-824.

Astawa P. I. and Sudika P., 2014. The Impact Local Culture On Financial Performance In Property Firms In Bali, *Asia Pacific Management and Business Application*, Vol 3, No. 2, pp 106-115

Bank Pembangunan Daerah (BPD). 2013. *Laporan Tahunan* Lembaga Perkreditan Desa Provinsi Bali.

Bardhan Indranil R and Thouin Mark F. 2013. Health information technology and its impact on the quality and cost of healthcare delivery. *Decision Support Systems 55*, pp 438–449

Bartelsman, E.J., 2013. ICT, reallocation and productivity. In: *European Economy–Economic Papers* No. 486.

Bloom, N., Sadun, R., Van Reenen, J., 2012. Americans do I.T. better: US multinationals and the productivity miracle. *American Economic Review* No. 102, pp 167–201.

Brinkerhoff, 1992. Promoting the Sustainability of Development Institutions : A Framework for Strategy. *World Development*, Vol. 20 (3), pp 369-383.

Brynjolfsson, E., Hitt, L., 1996. Paradox lost? Firm-level evidence on the returns to information systems spending. *Management Science* Vol. 42, pp 541–558.

Brynjolfsson, E., Hitt, L., 2003. Computing productivity: firm-level evidence. *Reviewof Economics and Statistics* Vo. 85, pp 793–808.

Bungin, Burhan. 2007. *Metodologi Penelitian Kualitatif: Aktualisasi Metodologis ke arah Ragam Varian Kontemporer*. Penerbit. Jakarta.

Chun H., Kim J.W., Lee J., 2015. How does information technology improve aggregate productivity? A new channel of productivity dispersion and reallocation. *Research Policy* Vol. 44, pp 999–1016

Серикова, Н. К. 2003. Оценка влияния ИКТ на экономический рост. Available from internet: http://wiasite.com/elektronnoy-kommertsiiosnovy/otsenka-vliyaniya-ikt-ekonomicheskiy.html

David, P.A., 1990. The dynamo and the computer: a historical perspec-tive on the modern productivity paradox. *American Economic Review* Vol. 80, pp 355–361.

Davis, Lance E. & Douglas C. North, 1971. *Institutional Change and American Economic Growth*. Cambridge: Cambridge University Press.

Foster, L., Haltiwanger, J., Krizan, C.J., 2001. Aggregate productivity growth: lessons from microeconomic evidence. In: Hulten, C.R., Dean, E.R., Harper, M.J. (Eds.),New Developments in Productivity *Analysis. University of Chicago Press*, Chicago, pp. 303–372.

Foster, L., Haltiwanger, J., Krizan, C.J., 2006. Market selection, reallocation, and restructuring in the U.S. retail trade sector in the 1990. *Review of Economics and Statistics* Vol. 88, pp 748–758.

Gârleanu, N., Kogan, L., Panageas, S., 2012a. Displacement risk and asset returns.*Journal of Financial Economics* Vol. 105, pp 491–510.

Gârleanu, N., Panageas, S., Yu, J., 2012b. Technological growth and asset pricing. *Journal of Finance* Vol. 67, pp 1265–1292.

Greenwood, J., Jovanovic, B., 1999. The information-technology revolution and thestock market. *American Economic Review* Vol. 89, pp 116–122.

Habermas J. 1984. The theory of communicative action (T.McCarthy, Trans.). Boston: *Beacon Press*; 1984 (Original work published 1981).

Hobijn, B., Jovanovic, B., 2001. The information technology revolution and the stock market*: evidence. American Economic Review* 91, 1203–1220

Katz, R.L. 2009. The Economic and Social Impact of Telecommunications Output. *Intereconomics*, pp 41–48. http://dx.doi.org/10.1007/s10272-009-0276-0.

Khan, H. & Santos, M. 2002. Contribution of ICT Use to Output and Labour Productivity Growth in Canada. *Bank of Canada Working Paper*

Kogan, L., Papanikolaou, D., 2013. Firm characteristics and stock returns: the role of investment-specific shocks. *Review of Financial Studies* Vol. 26, pp 2718–2759.

Kogan, L., Papanikolaou, D., 2013. Firm characteristics and stock returns: the role of investment-specific shocks. *Review of Financial Studies* Vol. 26, pp 2718–2759.

Kogan, L., Papanikolaou, D., Seru, A., Stoffman, N., 2012. Technological innovation, resource allocation,and growth.In:*NBER Working Paper* No.17769.

Kogan, L., Papanikolaou, D., 2014. Growth opportunities, technology shocks, and asset prices. *Journal of Finance* Vol. 69, pp 675–718.

Kogan, L., Papanikolaou, D., Seru, A., Stoffman, N., 2012. Technological innovation, resource allocation, and growth. In: *NBER Working Paper* No. 17769.

Kontan, 2010. Lembaga Kredit Desa Beraset Raksasa, *Kontan* 29 Desember 2010 Halaman 11

Lipaj, D. & Davidavičienė, V. 2013. Influence of Information System of Business Performance. Science – Future of Lithuania. Vol. 5(1), pp 38–45, http://dx.doi.org/10.3846/mla.2013.06

Liu S. and Wang L., 2014. Understanding the impact of risks on performance in internal and outsourced information technology projects: The role of strategic importance. *International Journal of Project Management* Vol. 32 pp 1494–1510

Marinagi C., Trivellas P.,Sakas D.P., 2014. The impact of Information Technology on the development of Supply Chain Competitive Advantage. *Procedia Social and Behavioral Sciences* Vol. 147 pp 586 – 591

Nurcahya Ketut I. 2006. LPD di Bali Suatu Contoh dari Lembaga Keuangan Mikro Yang Bertahan.*Buletin Studi Ekonomi* . Vol. 11. No. 3. Pp 210-217

Oulton, N. (2001). ICT and Productivity Growth in the UK. *Bank of England*: London.

Pilat, D., Lee, F. & van Ark, B. (2002). Production and Use of ICT: a Sectoral perspective on Productivity Growth in OECD Area. *OECD Economic Studies*. No 35, pp 48–76.

Poore, Simon. 2000. An ethnomethodology- An introduction. www.hewett.norfolk.sch.uk. Diunduh 3 April 2008.

Schein, Edgar, H. 2004. *Organizational Culture and Leadership*,John Wiley and Sont.Inc

Šniukienėa Alma Mačiulytė and Sarkane Elina Gaile. 2014. Impact of information and telecommunication technologies development on labour productivity. *Procedia - Social and Behavioral Sciences* Vol. 110, pp 1271 – 1282.

Stiglitz, Joseph E. 1990. Peer Monitoring and Credit Markets. *The World Bank Economic Review*,Vol. 4(3), pp 351-66

Surpha Wayan Gede. 2001. *Budaya Organisasi, Teori Praktis, Kasus, dan Aplikasi Penelitian*. Penerbit Udayana University Press, Denpasar

Van Ark, B., Inklaar, R. & McGuckin, R.H. 2003. ICT and Productivity in Europe and the United States. Where Do the Differences Come From? *CESifo Economic Studies*, Vol. 49(3), pp 295–318http://dx.doi.org/10.1093/cesifo/49.3.295

Van Ark, B. & Piatkowski, M. 2004. Productivity, Innovation and ICT in Old and New Europe. University of Groningen. *Research Memorandum*.

Wiana, I Ketut, 2004. *Mengapa Bali disebut Bali*. Penerbit Paramita Surabaya.

Wiana, I Ketut, 2007. Tri Hita Karana Menurut Konsep Hindu. Penerbit Paramita Surabaya

Windia, W., dan Ratna K. Dewi. 2007. *Analisis Bisnis yang Berlandaskan Tri Hita Karana*, Penerbit Universitas, Udayana, Denpasar.

Yaron, Yacob, McDonald Benjamin & Gerda Piprek.1977 Rural Finance: Issue, Design, and the Best Practice, *Environmentally and Socially Sustainable Development Studies and Monographs Series* 14. Washington D.C.: The World Bank.

Zaidi A U., Shanti F., Nawal A., 2015 . An exploratory study of the impact of information communication technology (ICT) or computer mediated communication (CMC) on the level of violence and access to service among intimate partner violence (IPV) survivors in Canada. *Technology in Society* Vol. 41, pp 91-97

MANAGEMENT OF VIRTUAL MACHINE AS AN ENERGY CONSERVATION IN PRIVATE CLOUD COMPUTING SYSTEM

Akhmad Fauzi[1], Edy Mulyadi[2], Abdullah Fadil[3], Mohammad Idhom[4]
UPN "Veteran" Jatim
Email: akhmadfauziupn@gmail.com[1],ed1mulyad1@gmail.com[2],
ziyanza@yahoo.com[3], moh.idhom@gmail.com[4]

ABSTRACT

Cloud computing is a service model that is packaged in a base computing resources that can be accessed through the Internet on demand and placed in the data center. Data center architecture in cloud computing environments are heterogeneous and distributed, composed of a cluster of network servers with different capacity computing resources in different physical servers. The problems on the demand and availability of cloud services can be solved by fluctuating data center cloud through abstraction with virtualization technology. Virtual machine (VM) is a representation of the availability of computing resources that can be dynamically allocated and reallocated on demand. In this study the consolidation of VM as energy conservation in Private Cloud Computing Systems with the target of process optimization selection policy and migration of the VM on the procedure consolidation. VM environment cloud data center to consider hosting a type of service a particular application at the instance VM requires a different level of computing resources. The results of the use of computing resources on a VM that is not balanced in physical servers can be reduced by using a live VM migration to achieve workload balancing. A practical approach used in developing OpenStack-based cloud computing environment by integrating Cloud VM and VM Placement selection procedure using OpenStack Neat VM consolidation. Following the value of CPU Time used as a fill to get the average value in MHz CPU utilization within a specific time period. The average value of a VM's CPU utilization in getting from the current CPU_time reduced by CPU_time from the previous data retrieval multiplied by the maximum frequency of the CPU. The calculation result is divided by the making time CPU_time when it is reduced to the previous taking time CPU_time multiplied by milliseconds.

Keywords: cloud computing, virtual machine, energy conservation.

INTRODUCTION

Cloud computing is a service model that provides convenience to the end-users and brokers. These services in the pack in the base of computing resources that can be accessed through the Internet on demand. These resources are presented to the customer self-service through the efforts of management and interaction with minimal service providers (Mell & Grance, 2011). Cloud Service Provider puts the computing resources in the data center.

In terms of management, data center architecture in cloud computing environments are heterogeneous and distributed, composed of a cluster of network servers with different capacity computing resources in different physical server. Cloud computing service providers still continue to look for a proper solution in order to optimize the cloud data center infrastructure. (Zhang, et al., 2010). The challenge is that if the capacity of the computing resources provided by the data center cloud more than the demand then the service provider would suffer losses due to the overhead of operating costs and processes occurring in the data center will experience a shortage utilization (under utilization). Conversely, if the capacity of computing resources provided by the data center experienced a shortfall due to popular demand, the service provider will lose revenue as a result in the stay by the customer and the processes that occur in the cloud data center will have excess usage (over utilization) (Armbrust, et al., 2009).

To overcome the problem of demand and the availability of the cloud services which are fluctuating, then the cloud data center must be made elastic. The technology that makes it possible to realize this is virtualization. Virtual Machine (VM) is a representation of the availability of computing resources that can be dynamically allocated according to demand.

VM consolidation strategy can be done by considering when, where and where VM must migrate. The strategy is broken down into sub decisions namely overloading host detection, underloading host detection, and VM and VM placement selection (Beloglazov & Buyya, 2012). The host overloading detection and the host underloading detection are used to determine when a host is considered experiencing overloaded or underloaded.

In the consolidation procedure of VM, host overloading detection is often a major focus of research in order to balance the workload in associate with the aim of saving energy consumption in the data center cloud (Masoumzadeh & Hlavacs, 2013).

VM migration decision is based on a defined policy in the capture of resource utilization and server computing in a deal offense defined through a Service Level Agreement (SLA).

Cloud computing environments that require dynamic and distributed decision-making procedures and strategies within the VM consolidation will be made dynamically as well. So in this study the proposed electoral policy optimization and placement of VM on VM consolidation procedures in cloud data center environments by considering hosting certain types of service applications on a VM instance that require various computing resources levels. Accumulated use of computing resources on a VM that is not balanced in physical servers can be reduced by using a live VM migration to achieve workload balancing. The practical approach will be used in developing OpenStack-based

DOI: 10.1051/matecconf/20165803008

cloud computing environment by integrating Cloud VM and VM Placement selection procedure using OpenStack Neat VM consolidation (Beloglazov & Buyya, 2012).

The term cloud computing or cloud computing was inspired by the image of cloud (cloud) that are commonly used to describe the current Internet network diagram. In fact there is a link between cloud computing and the Internet. Cloud computing is a client-server model, where resources (resources) such as servers, storage, network and software can be viewed as a service that can be accessed by users remotely and at any time. Users can enjoy a wide range of services provided by the provider of cloud computing without the need for too much ask for technical assistance or support from the provider. Cloud computing service model can be illustrated by (Mell & Grance, 2011).

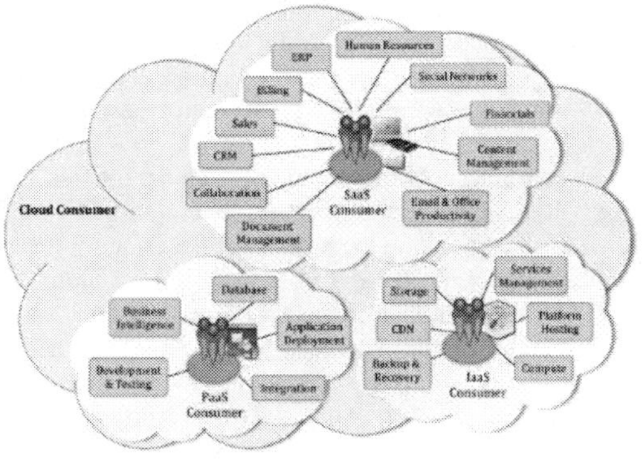

Figure-1. Model of cloud computing services (Zhang, et al., 2010)

Cloud computing infrastructure such as servers, storage, network and various software known as cloud and placed in the data center.

Virtual Machine Migration

The migration process demonstrate the functionality that removes the process from one computer to another. However, there are differences in the operating concept of virtual machine migration and migration processes (Khiyaita, et al., 2012). Migration of the entire OS and all applications as one unit to avoid the many difficulties faced by the process approach the level of migration. Narrow interface between the virtual OS and the virtual machine monitor (VMM) to avoid residual dependency issues. Migration at the level of the entire virtual machine in the state means that the memory can be transferred consistently and efficiently. Migration of virtual machines allows separation of interests between users and data center operators.Virtual Machine (VM) migration can be used for consolidation VM dynamically based on analysis of historical data on the use of resources by the VM. The problem is divided into four parts: (1) determine when the host is considered underloaded which leads to the need to migrate all of the VM from the host and switch hosts into sleep mode, (2) determine when the host is considered overloaded leading to the need of migration from one or more VM from the host to reduce the burden, (3) selecting the VM to be migrated from host overloaded, and (4) find a new placement of the selected VM to migrate from host overloaded (Beloglazov & Buyaa, 2013).

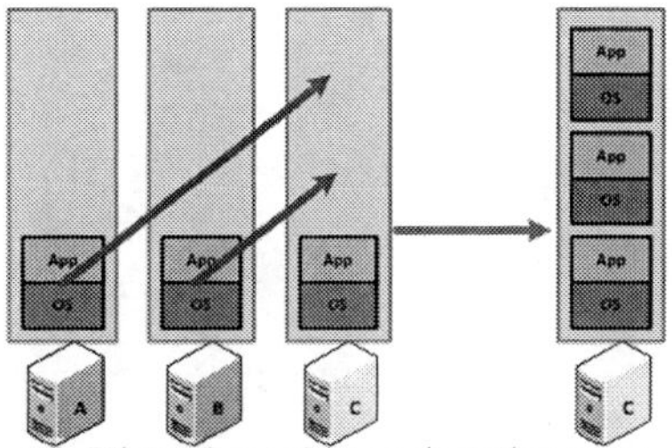

Figure-2. Virtual machine migration to consolidation

Cloud Computing-based OpenStack CloudOpenStack project is a cloud computing platform based on open source that is aimed at public and private cloud. The project initially focused on the presentation of Infrastructure as a Service (IaaS). The main objective of the project is to produce a ubiquitous OpenStack Open Source Cloud Computing platform that will meet the needs of public and private cloud regardless of size, with simple implementation and mass meet scalability.

Openness is a core value behind the project with open standards and open source. OpenStack has been released under the Apache 2.0 license. OpenStack also supports standard through the OpenStack API. Rackspace and NASA originally OpenStack project has commenced in September 2012. Companies such as AT & T, Rackspace, HP, IBM, Canonical, Red Hat, SUSE, Cisco, Dell, and Intel are examples of the 1,505 companies that support and the community very active in contributing to developing OpenStack.

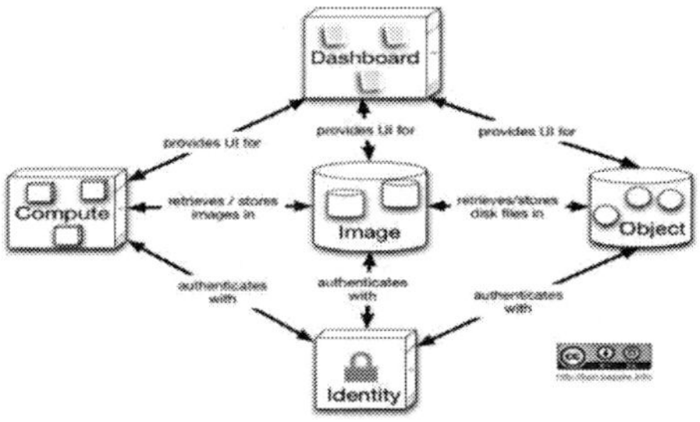

Figure-3: Architecture OpenStack Cloud (Khiyaita, et al., 2012)

OpenStack Neat is a project intended to provide an extension to the OpenStack Cloud implementing dynamic consolidation of Virtual Machines (VM) using live migration. The main objective is the dynamic VM consolidation to improve the utilization of physical resources and reduce energy consumption to reallocate the VM using live migration of resources according to demand real-time and divert hosts idle into sleep mode.

For example, it is assumed that the two VM placed on two different hosts, but the capacity of the combined resources needed by the VM to serve the load current can be supplied by one of the hosts. Then, one of the VM can be migrated to a host of support services other VM on the destination host, and the host that is idle can be diverted into a low power mode to save energy.Neat OpenStack consists of a number of components and data storage units, some of which are used to compute hosts, and some on the Management of hosts

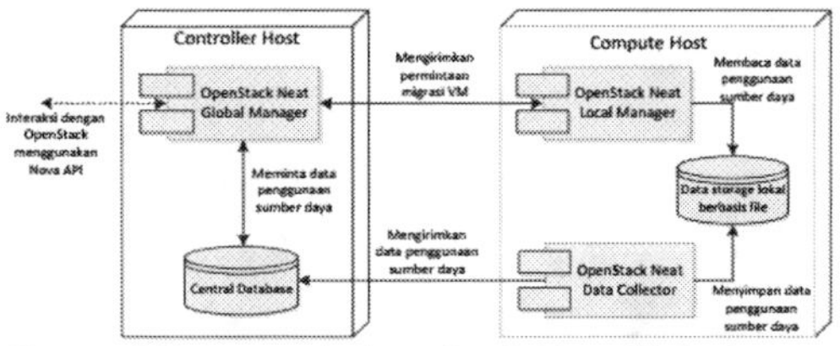

Figure-4. Framework OpenStack Neat (Beloglazov, 2014)

Design of system implementation

A series of tests will be done in the cloud-based data center using the Framework OpenStack OpenStack Cloud Neat. Environmental tests are used as in Figure 5.

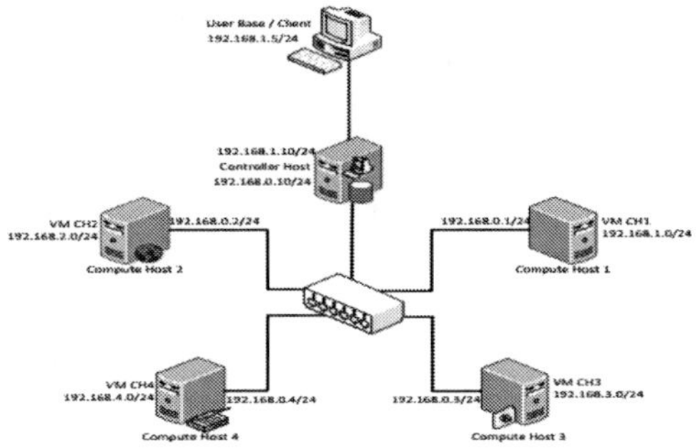

Figure-5. Testbed Network Topology (Beloglazov & Buyya, 2013)

RESEARCH METHODOLOGY

This research was conducted in testbed environments or testing in real terms in cloud computing systems by integrating a dynamic procedure VM consolidation. Stages of implementation of the research method used is as follows:

1) Development of testing in real environment by implementing and integrating OpenStack cloud platform OpenStack framework -Neat as VM dynamic consolidation procedures. At this stage the environment resulting cloud computing systems that can perform the function VM migration between the originating host to a destination host through a dynamic mechanism VM consolidation.
2) Design and implementation of procedures VM consolidation dynamic environment system testing OpenStack-based cloud computing cloud with the methods that have been proposed. At this stage would result VM module selection, placement VM, host overload detection and underload detection host.
3) The design of test scenarios that will be used to test the performance of the system. Designed test scenarios include variations based resource intensive workloads that include CPU-intesive, Memory-Intensive, DiskIO-intesive and Network-Intensive. At this stage that is used to generate scenarios for system testing process.
4) Testing of the entire system. At this stage of testing for the selection mechanism VM that have resulted from the second phase and test scenarios resulting from the third stage.

The development of the election program and the migration of VM through a procedure of consolidation VM dynamically performed by modifying the modules available in the framework OpenStack-Neat (data modules collector, local manager, underload, overload, VM selection, global manager and VM placement as well as modules other support).

RESULT AND DISCUSSION

Models of cloud computing virtualization technology implemented in cloud computing systems development model based private cloud that will build and manage cloud infrastructure independently using the open source platform OpenStack-based cloud. The OpenStack cloud can be implemented on multiple compute nodes to build distributed real testbed environment.

Host or Node used in testbed environments will use CentOS-based Linux distribution. CentOS installation stage following the standard process is described in detail in the Red Hat Enterprise Linux Installation Guide ().
HDD Partition Scheme In Host OpenStack Cloud

To support the needs VM live migration, the distributed file system (DFS) must be provided in order to share storage are also replicated between multiple servers via ethernet. The implementation will be supported by providing a different disk partition on the host used in OpenStack cloud. Disk partitioning scheme results in compute host consists of standar partition for the operating system lv_root, lv_home and lv_swap are grouped in vg_base partition. Special partition lv_gluster prepared for distributed file system (DFS) based GlusterFS contained in vg_gluster group.

Network Gateway OpenStack Cloud

Internet access is required by the host OpenStack cloud in order to facilitate the installation process and a package of modules required to form a testbed environment. One compute hosts (compute1) was Chosen as the gateway to the IP address configuration as in the table () to enable the network address translation (NAT) using iptables as in the following description:

#iptables -t nat -A POSTROUTING -o eth0 -j MASQUERADE
#iptables -A FORWARD -i eth1 -j ACCEPT
#iptables -A FORWARD -o eth1 -j ACCEPT

So that data packets can be forwarded from another host to the internet via compute1, then IP forwarding facility can be activated with the command:

sed -i 's/net.ipv4.ip_forward = 0/net.ipv4.ip_forward = 1/g' /etc/sysctl.conf

Using DFS implementation GlusterFS

The main package and supporting GlusterFS taken from the official website with the command:

#yum –y install wget \
*http://download.gluster.org/pub/gluster/glusterfs/LATEST/CentOS/glusterfs-3.3.0-1.el6.x86_64.rpm *
*http://download.gluster.org/pub/gluster/glusterfs/LATEST/CentOS/glusterfs-server-3.3.0-1.el6.x86_64.rpm *

In the host controller in the host to do initialization is involved in the formation of DFS to perform the command:

#gluster peer probe compute1
#gluster peer probe compute2
#gluster peer probe compute3
#gluster peer probe compute4

Volume gluster for replication between four gluster host created with the command:

*#gluster volume create VM-instances replica 4 *
*compute1:/export/gluster compute2:/export/gluster *
compute3:/export/gluster compute4:/export/gluster

Results of the implementation of DFS using GlusterFS is :

Volume Name: VM-instances
Type: Replicate
Volume ID: 7b0e3467-33bd-42df-962d-288d567a7a11
Status: Started
Number of Bricks: 1 x 4 = 4
Transport-type: tcp
Bricks:
Brick1: compute1:/export/gluster
Brick2: compute2:/export/gluster
Brick3: compute3:/export/gluster
Brick4: compute4:/export/gluster

Implementation Using KVM Hypervisor

Virtualization run only on compute hosts using KVM. KVM host controller does not need because it is not running VM instance. To enable KVM on compute hosts, it must be ensured that the machine supports Intel VT or AMD-V for virtualization in hardware. The feature can be activated via the basic input output system (BIOS). Implementation KVM on CentOS Linux can be done via the command:

yum -y install kVM qemu-kVM qemu-kVM-tools

Abstraction and Hypervisor API Implementation using libvirt Communication between the KVM with OpenStack Cloud bridged by a user application programming interface (API) using libvirt. OpenStack using libvirt as it supports a wide range of hypervisors including KVM and Xen. Libvirt package can be installed using the command:

#yum -y install libvirt libvirt-python python-virtinst avahi dmidecode

To support live migration capabilities on OpenStack Cloud VM, then libvirt configuration needs to be modified to enable communication via TCP without authentication.

#sed -i 's/#listen_tls = 0/listen_tls = 0/g' /etc/libvirt/libvirtd.conf
#sed -i 's/#listen_tcp = 1/listen_tcp = 1/g' /etc/libvirt/libvirtd.conf
#sed -i 's/#auth_tcp = "sasl"/auth_tcp = "none"/g' /etc/libvirt/libvirtd.conf
#sed -i 's/#LIBVIRTD_ARGS="--listen"/LIBVIRTD_ARGS="--listen"/g'
/etc/sysconfig/libvirtd

Implementation of cloud computing using OpenStack Platform Node or host that is used for the implementation of cloud computing using the OpenStack platform is divided into the controller and compute hosts. On each host needed an update repository to retrieve the packages needed to build OpenStack Cloud with the command:

#yum install -y http://dl.fedoraproject.org/pub/epel/6/i386/epel-release-6-8.noarch.rpm

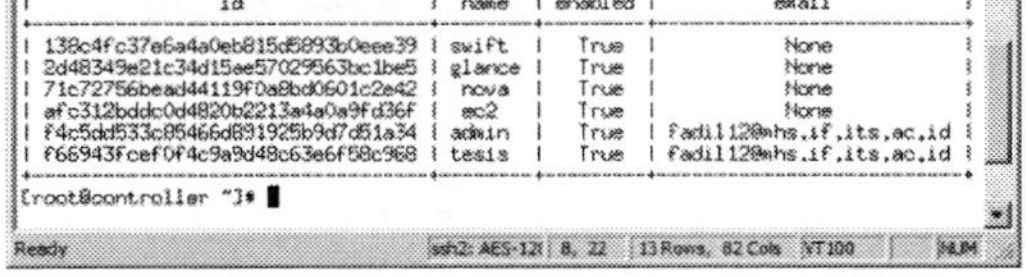

Figure-6. At the OpenStack User List

Glance component package is a VM image management service that is used in OpenStack Cloud. Glance requires MySQL to store the VM image metadata. To check service Glance running properly can be done by downloading Ubuntu Cloud Image and put it into the command Glance:

#wget http://uec-images.ubuntu.com/precise/current/precise-server-cloudimg-amd64-disk1.img
#glance add name="ubuntu" is_public=true disk_format=qcow2 container_format=bare < precise-server-cloudimg-amd64-disk1.img

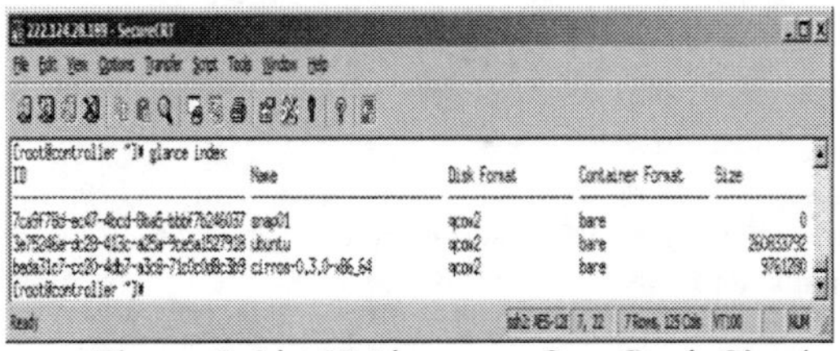

Figure-7. List VM image on OpenStack Cloud

Integration Framework Neat on OpenStack OpenStack CloudOpenStack-Neat is a module used by OpenStack Cloud in implementing the consolidation of a VM dynamically using a live VM migration to improve the performance of cloud services, maximizing the utilization of physical resources and reduce energy consumption by reallocating VM using live VM migration based on demand of resources in real terms and disable the host is not busy into power saving mode.In research conducted, can be integrated into Neat OpenStack OpenStack Cloud environment without changing the configuration of existing and already running. The installation procedure OpenStack OpenStack Cloud Neat on the environment, namely:

1) Duplication OpenStack neat on each host or node that is used with the command:#git clone *https://github.com/beloglazov/openstack-neat.git*
2) Adjust the configuration file neat.conf on environmental conditions OpenStack Cloud.
3) Package OpenStack Neat installed on each node of the repo directory by running the command:#python setup.py install
4) Run the service of the host controller with the command: #. / All-start.sh

CONCLUSIONS

Consolidation VM can dynamically react to the increasing demand for resources to reallocate the VM back when necessary to avoid degradation in the performance of cloud computing system as a whole. VM consolidation procedures that were implemented into OpenStack-Neat can be divided into several components, namely the data collector, the local manager and global manager.

The components used to handle the consolidation procedure is to decide when the host is considered as host underload and overload, the selection of VM to migrate from overload host and placement of the selected VM on the selected host. Value CPU Time used as fill to get the average value in MHz CPU utilization within a specific time period.

The average value of a VM's CPU utilization in getting from the current CPU_TIME is reduced by CPU_TIME from the previous data retrieval multiplied by the maximum frequency of the CPU. The calculation result is divided by the time making CPU_TIME when it is reduced to the time taking previous CPU_TIME multiplied by milliseconds.

REFERENCES

Armbrust, M. et al., 2009. Above the Clouds: A Berkeley View of Cloud Computing. *Electrical Engineering and Compute Sciences Unversity of California at Berkely.*

Beloglazov, A., 2014. *OpenStack Neat.* [Online] Available at: http://www.openstack-neat.org

Beloglazov, A. & Buyaa, R., 2013. Managing Overloaded Hosts for Dynamic Consolidation of Virtual Machines in Cloud Data Centers Under Quality of Service Constraints. *IEE Transactions on parallel and distributed system.*

Beloglazov, A. & Buyya, R., 2012. Optimal Online Deterministic Algorithms and Adaptive Heuristics. *Concurrency and Computation: Practice and Exprelence,* pp. 1397-1420.

Beloglazov, A. & Buyya, R., 2013. *OpenStack Neat: A Framework for Dynamic and Energy-Efficient Consolidation of Virtual Machines in OpenStack.* s.l.:Cloud Computing and Distributed Systems (CLOUDS) Laboratory.

Corradi, A., Fanelli, M. & Foschini, L., 2014. VM Consolidation: A real case based on OpenStack Cloud. *Future Generation Computer Systems 32,* pp. 118-127.

Khiyaita, A., Zbakh, M., Bakkali, H. & Kettani, D. E., 2012. Load Balancing Cloud Computing:state of art. *Network Security and Systems (JNS2),* pp. 106-109.

Masoumzadeh, S. & Hlavacs, H., 2013. s.l., IEEE, pp. 332-338.

Mell, P. & Grance, T., 2011. *The NIST Definition of Cloud Computing,* s.l.: National Institute of Standard and Technology.

OpenStack, 2013. *OpenStack Cloud Administrator Guide,* s.l.: docs.openstack.org.

Zhang, Q., Cheng, L. & Boutaba, R., 2010. Cloud Computing: state-of-the-art and Research Challenges. *Journal of internet services and applications,* pp. 7-18.

MATURITY LEVEL AT UNIVERSITY ACADEMIC INFORMATION SYSTEM LINKING IT GOALS AND BUSINESS GOAL BASED ON COBIT 4.1

Siti Mukaromah[1], Agung Brastama Putra[2]
[1, 2] University of Pembangunan Nasional "Veteran" Jawa Timur. Surabaya, East Java, Indonesia
E-mail: [1]sitimukaromah.si@upnjatim.ac.id, [2]agungbp.si@upnjatim.ac.id

ABSTRACT

The development of Information Technology (IT) has been mainly discussed nowadays, from the top to the lowest level of society. The application of IT help companies to solve problems and even more, the application of IT has been able to provide business strategic decisions support. Many enterprises decided to allocate large budget on IT implementation where this is in line with increasingly sophisticated expectations from IT. With the assumption that the larger budget on enterprises spent on IT application, the greater the benefits they will receive. Unfortunately, IT implementation does not always give an advantage to the company. There are times when the IT implementation does not give any benefit. This situation is called IT Productivity Paradox. The question is then how IT Productivity Paradox can be prevented. The analysis we will get the significance of the IT processes which is linked to its IT Goal. By knowing the significance of the IT processes, it can be seen which one is the significant process and which one is not to the IT Goal. If the IT process are not significant to the IT Goal, the process does not need to be improved because it has no effect to the IT Goal. This research was conducted to obtain the Maturity Level from each IT process and IT process's significance to IT Goal. The result of this research is that IT Processes are significant to IT Goal. It can be concluded that IT Productivity Paradox was not occurred.

Keywords: COBIT, maturity level, IT goal, business goal, significance.

INTRODUCTION

The reason why the alignment of business and IT is discussed is because there is a need to know the role of IT to the business success of the organization. IT is a tool that is included in the business process which is able to help companies to success, the application of IT in a company/organization should provide benefits, instead merely as an expense only and does not give an advantage to companies/organizations. (Brynjolfsson, 1993)

IT from the perspective of function is that the adoption of IT give benefits to company (Chan and Reich, 2007). Application of IT on business processes will add accuracy and efficiency of working time. (Henderson, 1992)

The importance of alignment between Business and IT originally proposed by Mc. Lean & Soden (1977) and followed by Henderson & Sifonis (1988). In the year of 1981, IBM provide a motivation of the perspective that the emphasis of the alignment process is to focus on the Strategic Plan for Business and IT long-term plan of the organization. Empirically in (Chan 1997; Irani 2002; Kearns & Lederer 2003), found the findings that support the hypothesis, that people who succeed to align their business strategy with IT strategy will be superior to those who do not apply.

In this case it can be concluded that in order to achieve Good Corporate Governance, the role of IT Governance is very important. Control Objective for Information and Related Technology (COBIT) can be used as a tools to streamline the implementation of IT Governance, i.e., as a management guideline by applying the entire domain contained in COBIT 4.1. Information System Audit and Control Association (ISACA) introduced a framework for managing IT Governance in a company which is known as COBIT (Control Objectives for Information and Related Technologies) (Indrajit, 2006).

The main contents are subdivided according to IT processes. There are 34 IT processes which give a perfect picture of how to control, manage and measure each process. (ITGI, 2007)

COBIT Maturity Model can be used as a measurement tool to measure the maturity of a company's position in terms of IT application. The Maturity Model have levels, namely Maturity Level. There are 6 Levels of Maturity: 0 (Non-existent), 1 (Initial/Ad Hoc), 2 (Repeatable but Intuitive), 3 (Defined Process), 4 (Managed and Measurable), 5 (Optimized).

Auditors should develop a questions for each Level of Maturity. Furthermore an Auditors could collect evidence by interviewing employees and observing documents. It can be seen at RACI (Responsible, Accountable, Consulted, Informed) chart (ITGI, 2007).

The most complete guide of best practices for IT management is COBIT 4.1 standard. COBIT is used because it has a fairly good compromise in the management scope and detail processes. (Mukaromah, 2010).

COBIT 4.1 has 17 Business Goals and 28 IT Goals. Business Goal is the goal desired by the company, whereas IT Goal is the purpose of the implementation of IT applied in the company. (ITGI, 2007). With the framework mapped by COBIT 4.1, a companies can focus on Business Goals and IT Objectives to be managed, which is lead to further efficiency of IT management processes.

As it has been described previously, Cobit has 28 IT Goals. This research will be focus on the IT Goals 19

(ITG19) *"ensuring that important and confidential information is hidden from the parties who are not interested"*. The result of this research is that all of the IT Process from ITG19 is significant. So, IT Productivity Paradox does not occur in this case study.

METHODOLOGY

This section describes the steps used to solve the research problem. There are two steps that must be done, they are, determine maturity level and calculate the significance with SEM using GSCA.

There are five steps that must be done to determine maturity level: defining the scope, data collection, due diligence, determine the level of security, and determining the results of the audit of information systems. (Revinggar, et al. 2012)

Defining the Scope

According to Sarno (2009), by defining the scope and purpose of the audit and also by taking into account the direction of the management, an Auditor will have a most high-risk IT process. Studies conducted earlier stated that the Higher Education Academic data has a high degree of risk in terms of data criticality and confidentiality (Tanuwijaya and Sarno, 2010). Academic regulations which are applied will affect the academic data. IT which is applied in the university is expected to ensure that important and confidential information can only be accessed by the right people. The need for the process must meet "Tri Darma Perguruan Tinggi" (Three Mission of Higher Learning Institution) of the university, must comply with government regulations, the quality of education that was promised to the people must be achieved, and it takes a certainty that provides the fulfillment of the compliance to the government laws and regulations.

The study begins by determining Business Goals perspective. The Internal Perspective that will be discussed in this research is Business Goal 12: "Provide compliance with external laws, regulations and contracts".

After determining the Business Goal, the next step is to determine IT Goal. The most appropriate with the case study of Academic regulation is IT Goal 19, as can be seen in Figure - 1.

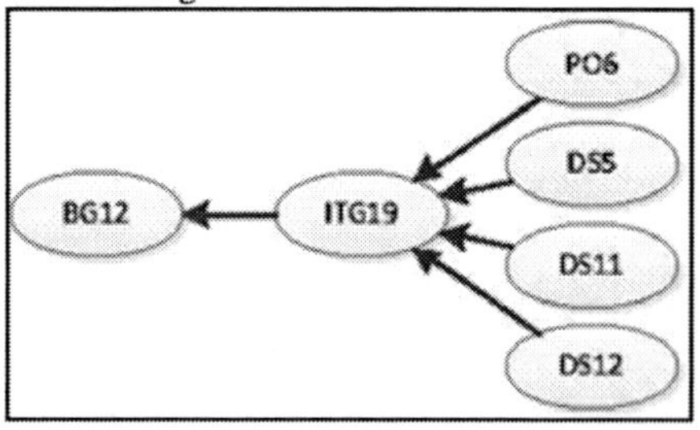

Figure - 1 The processes Impact on IT Goal 19

Data collection

Data collection is done by performing interview with employees to obtain business goals and IT goals data of Research Object. A list of employees who will be interviewed are in accordance with the criteria in RACI Table provided by COBIT 4.1. The data are used as a comparison literature that may affect research.

The question posed is based on Maturity Model Statement contained in COBIT 4.1. Data from the interviews will be used to calculate the level of maturity of IT implementation.

Maturity Level Determination

Data from the interviews will be used to calculate the level of maturity of IT implementation. From the measurement results of the COBIT paper work, the contribution of each level of maturity will be measured to obtain the value of IT process maturity. There are five levels of maturity (0 to 5) which values are needed to be obtained. Each level of maturity value from level 0 to level 5 gives an organization an idea about the overall of IT process maturity.

"Normalize" value (see table 1) give the organization an idea about the compliance influence of the overall IT process maturity. "Normalize" value is derived from the compliance value from each level divided by total compliance. While the "Contribution" value is obtained by multiplying the value of compliance to the "Normalize" value, and the total of the "Contribution" is the level of maturity value.

Determining the Information Systems Audit Results

The results of the evaluation of the implementation of information systems audit will contain findings, based on the due diligence that carried out, as well as recommendations to improve the existing processes. The format of the report will vary in each organization, so there is no standard format. The final report of the audit should present an overview of today's condition from other similar organization, by which allows the management to take the necessary steps to evolve.

Calculating The Significance Value

Structural Equation Modeling (SEM) is a structural equation model which is a combination of the procedures developed in econometrics (Wijayanto, 2008). SEM was originally known as path analysis developed by Sewal Wright in 1984. Structural Equation Modeling (SEM) was used to examine and justify a model and to explain the relationship between variables that exist in the model.

By using SEM, researchers can do things like check the validity and reliability, test the relationship between variables in the model and get the models used for prediction. If the SEM input is in the form of the covariance matrix, SEM will produce a structure that can be used to proof the model. Whereas if the input is in the form of correlation matrix, SEM can be useful for checking the size of the independent variables. By doing so, SEM can be used to determine the dominant variable. Steps to perform SEM according Ghozali on Mukaromah & Subriadi (2015):

1. Development of theoretical models.
   The topic of research in-depth analysis and start doing a hypothesis.

2. Develop a path diagram.
   Arrangement of structural models to connect between latent constructs (endogenous and exogenous) and arrange the measurement models.
3. Change the path diagram into a structural equation.
   The conversion path diagram into the equation.
4. Select the input matrix for data analysis.
   Enter the input data in the form of variants or covariance matrix or correlation matrix.
5. Assessing the model identification.
   The estimation of the structural equation models. The Estimation of structural equations can use maximum likelihood estimation.
6. Assessing Goodness of Fit criteria.
   There are three kinds of goodness of fit: absolute fit measure, incremental fit measure and parsimonious fit measure. Absolute fit measure is used to measure the overall model fit, while incremental fit of the proposed measure is to compare the model with another model. Parsimonious fit measure is to perform adjustment to fit measurements to be compared between models with a number of different coefficient.
7. Interpretation of the model.
   The model modification improvements to improve theoretical explanation.

GSCA is a new method of SEM-based components that can be used to perform score calculation and can also be applied to small sample (Tenenhaus, 2008). GSCA can also be used on the relationships between variables which are complex, involving higher order and Multi group comparison. In addition, GSCA can also be applied to the structural model which has a strong theoretical basis.

Before performing the test GSCA, research data must first pass the linearity test. This is because of the assumption that the required relationship in the equation is linear GSCA (Subriadi, 2013).

RESULTS

This part explain the Results of the interviews conducted. The obtained results is maturity level of each process. Here are the results of Maturity Level assessment:

a). PO6 Maturity Level

Table - 1. PO6 Maturity Level

Level	Compliance	Normalize	Contribution
0	0	0	0
1	0.580	0.221	0.221
2	0.442	0.168	0.337
3	0.661	0.252	0.757
4	0.663	0.253	1.011
5	0.277	0.106	0.528
Sum	2.622		
		Maturity Level =	**2.853**

As it can be seen at Table - 1, PO6 Maturity Level is 2.853. In representation of Maturity Model, it is Repeatable but Intuitive. It means that:

- The needs and requirements of an effective information control environment are implicitly understood by management, but practices are largely informal.
- The need for control policies, plans and procedures is communicated by management, but development is left to the discretion of individual managers and business areas.
- Quality is recognized as a desirable philosophy to be followed, but practices are left to the discretion of individual managers.
- Training is carried out on an individual, as-required basis.

b). DS5 Maturity Level

Table - 2. DS5 Maturity Level

Level	Compliance	Normalize	Contribution
0	0.264	0.077	0.000
1	0.498	0.146	0.146
2	0.248	0.072	0.145
3	0.619	0.181	0.543
4	0.972	0.284	1.138
5	0.816	0.239	1.195
sum	3.416		
		Maturity Level =	**3.166**

As seen at Table - 2, DS5 Maturity Level is 3.431. In representation of Maturity Model, it is Defined Process. It means that:

- Security awareness exists and is promoted by management.
- IT security procedures are defined and aligned with IT security policy.
- Responsibilities for IT security are assigned and understood, but not consistently enforced.
- An IT security plan and security solutions exist as driven by risk analysis.
- Reporting on security does not contain a clear business focus.
- Ad hoc security testing (e.g., intrusion testing) is performed.
- Security training is available for IT and the business, but is only informally scheduled and managed.

c). DS11 Maturity Level

Table - 3. DS11 Maturity Level

Level	Compliance	Normalize	Contribution
0	0.000	0.000	0.000
1	0.466	0.110	0.051
2	0.932	0.220	0.205
3	1.000	0.236	0.236
4	1.000	0.236	0.236
5	0.833	0.197	0.164
Sum	4.231		
		Maturity Level =	**0.893**

As can be seen at Table - 3, DS11 Maturity Level is 0.893. In representation of Maturity Model, it is Non-Existent. It means that:

- Data are not recognized as corporate resources and assets.
- There is no assigned data ownership or individual accountability for data management.
- Data quality and security are poor or non-existent.

d). DS12 Maturity Level

Table - 4. DS12 Maturity Level

Level	Compliance	Normalize	Contribution
0	0.000	0.000	0.000
1	0.333	0.148	0.049
2	0.165	0.074	0.012
3	0.854	0.381	0.326
4	0.889	0.397	0.353
5	0.000	0.000	0.000
Sum	2.241		
		Maturity Level =	0.740

As it can be seen in Table - 4, DS12 Maturity Level is 0.740. In representation of Maturity Model, it is Non-Existent. It means that:

- There is no awareness of the need to protect the facilities or the investment in computing resources.
- Environmental factors, including fire protection, dust, power, and excessive heat and humidity, are neither monitored nor controlled.

DISCUSSIONS

IT process Maturity Level of IT Goal 19 (Ensure that critical and confidential information is withheld from those who should not have access to it) is illustrated in Table - 5.

The two lowest Maturity Level assessment value is on IT process DS11- Manage Data (0.89) and DS12 - Manage the Physical Environment (0.74), it means that those IT processes have to be improved. The improvement of IT process can be done by fulfilling the detail item on each control objectives (each IT process has control objective). IT control objectives provide a complete set of high-level requirements to be considered by management for effective control of each IT process. They are statements of managerial actions to increase value or reduce risk. They consist of policies, procedures, practices and organizational structures designed to provide reasonable assurance that business objectives will be achieved and undesired events will be prevented or detected and corrected.

Table - 5. Result of ITG19 Maturity Level

Domain	Maturity Level
PO6 Communicate Management Aims and Direction	2.85
DS5 Ensure Systems Security	3.17
DS11 Manage Data	0.89
DS12 Manage the Physical Environment	0.74

Inferential Analysis

The result of model FIT from GSCA is explained below as can be seen at table - 6.

FIT

FIT value shows the total value of all the variables that can be explained by the particular model. FIT values ranged numbers from 0 to 1. If the value of the FIT is close to 1, it can be said that the better the model explains the phenomenon under study. This is because the value of 1 describes the proportion of variable that can be explained by the greater structural model. Based on Table 5.2 it can be seen that the model which is formed, can explain all the variables that exist, at 0.460. This means PO6, DS5, DS11, DS12, ITG19 and BG12 can be explained by the model by 46% and the 59% can be explained by other variables.

AFIT

AFIT value is used for comparison of the value of FIT. The difference with the FIT value, is AFIT also taking into account the complexity of the model. Based on the obtained values from Table 5.2 the AFIT has value 0,445. The AFIT value is not much different from the FIT value so it can support the conclusions of FIT value. Based on the AFIT value, PO6, DS5, DS11, DS12, ITG19 and BG12 can be explained 44.5% by the model.

GFI

The GFI value indicates whether the model is appropriate or not. GFI value range is from 0 to 1. From Table 5.2 it can be seen that the value of GFI is 0.995, which means that the overall model is very appropriate because GFI value close to 1.

SRMR

SRMR is the addition of FIT model size. According to Imam Ghozali, SRMR value <0.08 indicates the suitability of the overall model. Based on Table 5.2 it can be seen that the value SRMR produced is 0.228 which indicates that the model does not fit based on SRMR.

NPAR

NPAR value indicates the number of free parameters used in the calculation of GSCA. Based on the FIT model table above, it can be seen that the value NPAR is equal to 77.

Table - 6. Model Fit

Model Fit	
FIT	0.460
AFIT	0.445
GFI	0.995
SRMR	0.008
NPAR	77

R Square Identification

Based on Table - 7 it can be seen that the value of R square on ITG19 is 0.947, it is indicating when there is an increase towards PO6, DS5, DS11, DS12, it can affect ITG19 by 94.7% while the rest is the influence of other variables that are not included in research models. Furthermore, the value of R square on BG12 is equal to 0.312, indicating when there is an increase to ITG19, it can affect the BG12 by 31.2%, while 68.8% is the influence of other variables which is not included in the research model.

Table - 7. R square of Latent Variable

R square of Latent Variable	
ITG19	0.947
BG12	0.312
PO6	0
DS5	0
DS11	0
DS12	0

Based on table - 8. Path Coefficient indicates that the entire hypothesis is accepted.

Table - 8. Path Coefficients

Path Coefficients			
	Estimate	**SE**	**CR**
ITG19->BG12	0.558	0.079	7.09*
PO6->ITG19	0.532	0.152	3.49*
DS5->ITG19	0.669	0.276	2.43*
DS11->ITG19	0.087	0.117	2.74*
DS12->ITG19	-0.406	0.143	2.84*

CONCLUSION

Cobit Mapping of IT Processes (PO6, DS5, DS11, and DS12) and ITG19 were significant in this case study. These relevance mapping, linking with maturity level, show that IT Process of DS11 and IT Process 12 have to be increased by following Control Objectives on Cobit 4.1 framework.

FUTURE RESEARCH

There are a lots of variables on COBIT framework that can be developed as a model to follow the development of case study. This research model is also can be tested on different case study in order to know if there is any difference result.

REFERENCES

Brynjolfsson, Erik. (1993). The productivity paradox of information technology. Communications of the ACM. Volume 36 (12), page 66–77.

Chan, Y.E., Huff, S.L., Barclay, D.W. and Copeland, D.G. (1997). Business Strategic Orientation, Information Systems Strategic Orientation, and Strategic Alignment, Information Systems Research 8(2): 125–150.

Chan. Y.E, Reich. B.H, (2007). IT alignment: what have we learned?. Journal of Information Technology. No. 22, pp. 297–315.

Henderson, J.C. and Sifonis, J.G. (1988). The Value of Strategic IS Planning: Understanding consistency, validity, and IS markets, MIS Quarterly 12(2): 187–200.

Henderson. J. dan Venkatraman. N. (1992). Strategic Alignment: A model for organizational transformation through information technology. T. Kochan & M. Unseem, eds, Transforming Organizations, Oxford University Press, NY.

Indrajit, Richardus Eko. 2006. Maturity Level Measuring Utilization of Information Technology for Educational Institutions an Approach Readiness Stakeholders. EII Journal: 116-120.

Information Systems Audit and Control Association. IS Standards, Guidelines and Procedures for Auditing and Control Professionals, ISACA, 2007.

Irani, Z. (2002). Information Systems Evaluation: Navigating through the problem domain, Information Management 40(1): 11–24.

ITGI. 2007. COBIT 4.1: Control Objective, Management Guidelines, Maturity Models. United States of America: IT Governance Institute.

Kearns, G.S. and Lederer, A.L. (2003). A Resource-Based View of Strategic IT Alignment: How knowledge sharing creates competitive advantage, Decision Sciences 34(1): 1–29.

McLean, E.R. and Soden, J.V. (1977). Strategic Planning for MIS, New York: Wiley.

Mukaromah, Siti. 2010. Information Systems Auditing Academic Information System Using Standards COBIT 4.0 Domain Acquire and Implement (Case Study on Stikomp Surabaya). Stikom Digital Library: Surabaya.

Mukaromah, S., & Subriadi, A.P. 2015. The Significant of Cobit Mapping Business Goal 12 and IT Goal 19 (Case Study: Stikom Surabaya). ISST 2015.

Revinggar, B. N., Tanuwijaya, H., & Mastan, I. A. (2012). Audit Information System Outpatient Installation Based on Customer Perspective Balanced Scorecard and COBIT 4.1 Standard (Case Study: General Hospital Haji Surabaya). Jurnal JSIKA, 1(1).

Sarno, R. 2009. Business Success Strategy with Information Technology, Based Balanced Scorecard & CobiT. Surabaya: ITS Press.

Subriadi A.P., Hadiwidjojo D., Djumahir, Rahayu M., Sarno R. 2013. Information Technology Productivity Paradox: A Resource-Based View and Information Technology Strategic Alignment Perspective For Measuring Information Technology Contribution On Performance. Journal of Theoretical and Applied Information Technology. Vol. 54 No.3 (541-552).

Tanuwijaya H, Sarno R. Comparation of CobiT Maturity Model and Structural Equation Model for Measuring the Alignment between University Academic Regulations and Information Technology Goals. IJCSNS

International Journal of Computer Science and Network Security 10 (2010) 80-92.

Tenenhaus, M. (2008). Structural Equation Modelling for small samples. Working paper no 885.

Wijayanto, S. H. (2008). Structure Equation Modelling, Consep and Tutorial with Lisrel 8.80. Jakarta: Graha Ilmu.

 DOI: 10.1051/matecconf/20165803010

MODIFICATION OF HIDDEN LAYER WEIGHT IN EXTREME LEARNING MACHINE USING GAIN RATIO

Fetty Tri Anggraeny and Intan Yuniar Purbasari
Department of Informatics Engineering, University of Pembangunan Nasional "Veteran" East Java, Surabaya, Indonesia
e-Mail: fetty.ta@gmail.com

ABSTRACT

Extreme Learning Machine (ELM) is a method of learning feed forward neural network quickly and has a fairly good accuracy. This method is devoted to a feed forward neural network with one hidden layer where the parameters (i.e. weight and bias) are adjusted one time randomly at the beginning of the learning process. In neural network, the input layer is connected to all characteristics/features, and the output layer is connected to all classes of species. This research used three datasets from UCI database, which were Iris, Breast Wisconsin, and Dermatology, with each dataset having several features. Each characteristic/feature of the data has a role in the process of classification levels, starting from the most influencing role to non-influencing at all. Gain ratio was used to extract each feature role on each datasets. Gain ratio is a method to extract feature role in order to develop a decision tree structure. In this study, ELM structure has been modified, where the random weights of the hidden layer were adjusted to the level of each feature role in determining the species class, so as to improve the level of training and testing accuracy. The proposed method has higher classification accuracy rate than basic ELM on all three datasets, which were 99%, 96%, and 82%, respectively.

Keywords: extreme learning machine, feature weight, information gain.

INTRODUCTION

Extreme Learning Machine (ELM) is a learning method for a single hidden layer feed-forward neural network (SLFN) that resolves concerns raised by the use of back propagation methods. The learning stages in backpropagation take much longer than ELM despite using the same neural network configuration, i.e. one hidden layer. This is because the Feed Forward Neural Network uses a learning algorithm based on gradient that works slowly, and all the adjustable parameters are repeatedly adjusted in the learning process until an iteration stopping criteria is reached (Huang et al., 2004). Meanwhile, the learning stage in ELM requires only one iteration and the weight parameters are set once randomly, despite having only one hidden layer (Huang et al., 2006). Although ELM gives quickly learning process, it has a fairly good accuracy too.

ELM has been gaining high attention from researchers since its announcement (Huang et al., 2015). It is not only researched within the scope of classification problems, but also in the scope of regression and clustering problems. With its advantages, ELM is considered appropriate to resolve various problems with big data and real-time applications, such as in medical field, image processing, computer vision, etc. They paper show that ELM and its variants are efficient, accurate and easy to implement, also in hardware needs (robot).

ELM research result by Huang et al. (2004) showed that ELM has a higher accuracy than other methods, such as SVM, AdaBoost, C4.5, and RBF. Two datasets used were real diabetes medical diagnosis dataset and forest type dataset.

Data in classification problems consists of several features that represent an object of a specific class species. Each feature has a role level (i.e. a weight feature) which can be categorized as high and low role levels. Weighting is generally performed in feature selection stage and aims to analyze the data and generate the level of a feature role in the classification process. There are two approaches in feature selection process: filter approach and wrapper approach (Karegowda et al., 2010). The filter approach is carried out separately from the classification engine, and is an important preprocessing stage. Since it is separated from the classification engine, the outputs of this feature selection approach can be used by different classification engines. The wrapper feature selection approach uses a classification engine to determine the role levels of each feature in the classification process. In other words, the filter approach is simpler and faster than the wrapper approach. Some of the feature selection methods using filter approach, among others, are gain ratio (Karegowda et al., 2010, Priyadarsini et al., 2011, and Anggraeny et al., 2013), particle swarm intelligence (PSO) (Yang et al., 2007 and Huang et al., 2008), and differential evolution (Khushaba et al., 2011). Some feature selection methods using wrapper approach are ant colony optimization (ACO) (Kanan et al., 2008) and sequential floating forward selection (SFFS) (Liao et al., 2010).

Karegowda et al. (2010) used gain ratio (GR) as feature selection technique, and Radial Basis Function Network (RBF) and Back Propagation Neural Network (BPN) as classifier. The research result showed that classification accuracy for BPN is about 72.88%, GR-BPN is 78.21%, RBF is 81.20%, and GR-RBF is 86.46%.

Anggraeny et al. (2013) used gain ratio (GR) as feature selection technique and Voting of Neural Network Particle Swarm Optimization (VNNPSO) as classifier. The result showed that GR-VNNPSO did not improve all datasets accuracy rate. The method improved classification accuracy on dermatology dataset about 13.28% and reduced accuracy rate about 0.45% on iris dataset and 3.49% on breast cancer Wisconsin. Although only one dataset showed better accuracy, the increasing value was much higher than the decreasing value.

Priyadarsini et al. (2011) used gain ratio as feature subset selection, combined by Naïve Bayes as classifier and K-Means for clustering. Ranking method is used in adult dataset to select a subset of 7 attributes

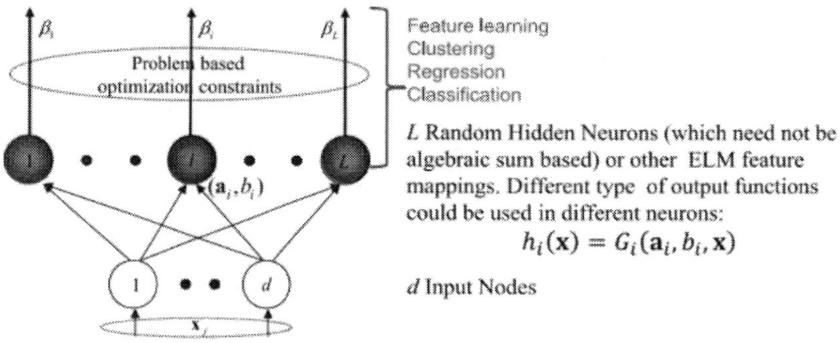

Figure-1. ELM Architecture (Huang et al., 2015).

from the original dataset of 10 attributes. Both on classification and clustering, the utility of the dataset is unaffected by the attribute reduction.

In this research, we will add feature weight using gain ratio method as a multiplier factor of hidden layer random weight in Extreme Learning Machine, in the hope that this modification of ELM weighting will increase the accuracy of training and testing data.

METHODOLOGY

Extreme Learning Machine

Extreme Learning Machine (ELM) is a learning method in single hidden layer feed-forward neural network which is faster and generally has a higher accuracy than backpropagation (Huang et al., 2004). The configuration of the neural network consists of d input nodes in accordance with the number of features, L hidden nodes, and m output nodes in accordance with the number of classes (Figure 1). Unlike backpropagation, ELM is not only aims to achieve a minimum error learning, but also the smallest norm of weight. The function of the output of ELM is (Huang et al., 2015):

$$f_L(x) = \sum_{i=1}^{L} \beta_i h_i(x) = h(x)\beta \quad (1)$$

where $\beta = [\beta_1, \cdots, \beta_L]^T$ is the output weight between L hidden layer nodes and $m \geq 1$ output nodes, and $h(x) = [h_1(x), \cdots, h_L(x)]$ is the ELM nonlinear feature mapping (Figure 2), where $h_i(x)$ is the output of the i-th node to the hidden layer. The output function in hidden nodes may use a different activation function for each node. The output function in hidden nodes is notated as follows:

$$h_i(x) = G(a_i, b_i, x), \quad a_i \epsilon R^d, \ b_i \epsilon R \quad (2)$$

where $G(a_i, b_i, x)$, with hidden nodes parameter (a, b) as a non-linear activation function, x_j is the j-th input value, a_i is random weight of i-th input layer, and b_i is bias of the i-th hidden node.

$$G(a_i, b_i, x) = G(a_i . x_j + b_i) \quad (3)$$

ELM consists of two main stages: random mapping feature and linear completion of parameters. In the first phase, ELM randomly initializes the weights of hidden layer nodes to map input data into ELM feature space. The hidden node's parameters (a, b), are randomly initialized by a probability distribution. In the second phase, the weights connecting the hidden layer and output layer (β) are resolved by minimizing the error output:

$$\min_{\beta \epsilon R^{L \times m}} \|H\beta - T\|^2 \quad (4)$$

where H is the matrix of hidden layer output, T is target data training matrix, and $\|.\|$ denotes Frobenius norm.

Gain Ratio

Gain ratio is an improvement of information gain. Information gain is used to form the induction of a decision tree (ID3), while the gain ratio is used on C4.5 algorithm, which is an improvement algorithm of ID3 (Asha et al., 2010). Information gain produces bias; it prefers features with many variations of values rather than features which has little variation despite being more informative. For example, let us look at a unique feature of a student data, such as its ID, in student table of a database. A separation using student ID creates a lot of partitions, as each data record has a unique value, which is student ID (Asha et al., 2010).

Let S is a data sample set and m is the number of classes. The entropy or information approximation to classify a sample is:

$$I(S) = -\sum_{i=1}^{m} p_i \log_2(p_i) \quad (5)$$

where p_i is the sample probability with a $class_i$ conclusion.

Let feature/attribute A has a value variation of v. Let s_{ij} is the number sample class C_i in subset S_j. S_j consists of samples in S having a value a_j from A. The entropy based on the division of the subset of attribute A is:

$$E(A) = -\sum_{i=1}^{m} I(S) \frac{s_{1i}+s_{2i}+\cdots s_{mi}}{s} \quad (6)$$

Gain information to branch attribute A is:

$$Gain(A) = I(S) - E(A) \quad (7)$$

C4.5 uses the gain ratio by applying a normalization of the gain information obtained from:

$$SplitInfo(S) = -\sum_{i=1}^{v} (|S_i|/|S|) \log_2(|S_i|/|S|) \quad (8)$$

Gain ratio is then computed using the following formula:

$$GainRatio(A) = Gain(A)/SplitInfo(S) \quad (9)$$

Attribute with the highest gain ratio is selected as a splitting attribute.

Feature Weight Adjusted on Extreme Learning Machine

The proposed approach is completely described in the following FW-ELM algorithm in Figure 2. ELM preferred to use only random weights as parameter for hidden layer nodes. In the proposed method, first we calculate the feature role using gain ratio method for all features in dataset. These features role will be used as multipliers factor together with ELM random weighting as the input of neural network structure.

This research added one weight, i.e. feature role, which was obtained using gain ratio method. With the additional feature weight, the formulation in hidden layer becomes:

$$h_i(x) = G\left(a_i, b_i, x, f_j\right), \quad a_i \epsilon R^d, \ b_i \epsilon R \tag{10}$$

$$G\left(a_i, b_i, x, f_j\right) = G(f_j \cdot a_i \cdot x_j + b_i) \tag{11}$$

Where f_j is the feature weight of j-th input (feature) using gain ratio method.

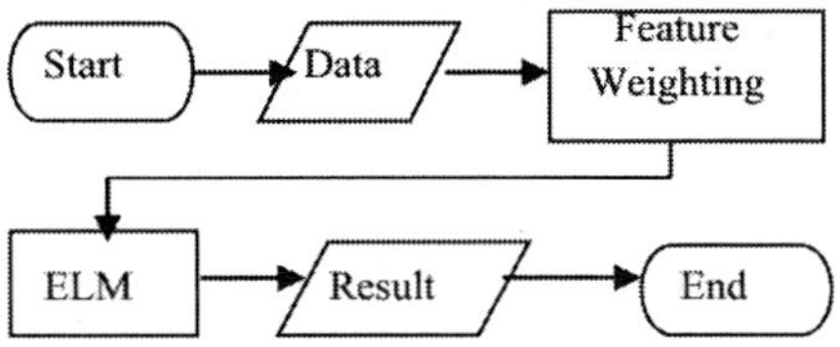

Figure-2. Methodology

RESULTS and DISCUSSION

Several datasets from UCI database were used for testing purpose as listed in Table 1, which were Iris, Breast Wisconsin, and Dermatology. The phase of feature weight computation using gain ratio gave results in the form of an order of importance of features from the largest to the smallest, as listed in Table 2, along with their respective gain ratio values.

Trials were performed 10 times for each dataset, and all data were used both as training and testing. The parameter assessed was the level of accuracy. Table 3 shows a comparison of SLFN tests between Feature Weighted-ELM (FW-ELM), ELM, and GR-VNNPSO (Anggraeny et al, 2013), in terms of classification accuracy (%) and computing time (s). For the original ELM method by Huang et al (2004), the datasets used in their published paper were different from ours. However, we were able to obtain their source code from Huang's website[1] and run it on our datasets. The comparison between ELM and FW-ELM aimed to investigate the effect of variable addition, i.e. feature weight, in ELM architecture. In the case of comparison between FW-ELM and GR-VNNPSO, the latter applied gain ratio before classification process but using different classifier method. This trial was performed using all features in each classifier.

Based on trial results shown in Table 3, the FW-ELM gave a relatively better accuracy than ELM. This showed that the addition of feature weight parameters in the ELM configuration was able to improve the accuracy on all three datasets significantly. In terms of training time, FW-ELM was faster than ELM, provided that feature weight computation process is carried out outside the classification engine. Compared with GR-VNNPSO, FW-ELM has less accuracy on all dataset but faster in computing time.

Table-1. Dataset characteristics

Dataset	feature	class	∑ data
Iris	4	3	150
Breast Wisconsin	9	2	699
Dermatology	34	15	366

Source: (Blake et al., 1998)

Table-2. Order of features according to gain ratio value

Dataset	#feature	Gain Ratio
Iris	4	0.871: 4, 0.734: 3, 0.381: 1, 0.242: 2
Breast Wisconsin	9	0.675: 2, 0.66: 3, 0.564: 6, 0.543: 7, 0.505: 5, 0.466: 8, 0.459: 1, 0.443: 4, 0.198: 9
Dermatology	34	0.7715: 31, 0.7254: 27, 0.7237: 33, 0.7221: 6, 0.7111: 29, 0.7094: 12, 0.7019: 15, 0.6829: 25, 0.6741: 8, 0.628: 30, 0.6012: 22, 0.5919: 20, 0.5303: 21, 0.5297: 34, 0.527: 7, 0.438: 9, 0.4291: 24, 0.3993: 10, 0.3707: 28, 0.3251: 14, 0.325: 16, 0.3171: 26, 0.2941: 23, 0.2911: 11, 0.2674: 5, 0.1978: 3, 0.1769: 2, 0.1687: 19, 0.1599: 13, 0.1491: 4, 0.098: 1, 0.0959: 18, 0.0833: 17, 0.0598: 32

Table-3. Average on accuracy (%) and computing time (s) of SLFN in FW-ELM, ELM, and GR-VNNPSO

Dataset	FW-ELM		ELM		GR-VNNPSO (Anggraeny et al, 2013)	
	accuracy (%)	**computing time (s)**	**accuracy (%)**	**computing time (s)**	**accuracy (%)**	**computing time (s)**
Iris	99.56	0.019	99.89	0.05	100	130.25
Dermatology	81.77	0.013	63.71	0.093	100	445.49
Breast Wisconsin	96.16	0.033	93.49	0.042	97.71	145.73

[1]http://www.ntu.edu.sg/eee/icis/cv/egbhuang.htm

There are features that have a big enough role in helping classification process and those that weaken of all the features possessed by a dataset. For the next improvement, this research can be expanded by adding a feature selection method, so that the features used in classification are those that have major roles in the classification process.

CONCLUSION

In this research, we have modified ELM weighting training of neural network. This approach implemented gain ratio in order to calculate each feature weight in classification process. These weights were used as input to the neural network, and used as weight factor on ELM training process. Moreover, compared with two other methods, FW-ELM appeared to be a promising method as a classifier. In general, FW-ELM has successfully combines high accuracy classification and effectiveness on computing time into becoming a better classifier.

REFERENCES

Anggraeny F.T., Widiasri M. 2013. Voting of Artificial Neural Network Particle Swarm Optimization Biclassifier using Gain Ratio Feature Selection. KURSOR Journal: Research on Computing and Its Application, Vol. 7, No. 2, pp. 69-74, ISSN: 0216-0544.

Blake C.L., Merz C.J. 1998. University of California at Irvine Repository of Machine Learning Databases, University of California, Irvine, 1998. http://www.ics.uci.edu/mlearn/MLRepository.html.

Huang C.L., & Dun J.F. 2008. A distributed PSO–SVM Hybrid System with Feature Selection and Parameter Optimization. Journal of Applied Soft Computing, Vol. 8, Issue 4, pp. 1381–1391.

Huang G.B., Zhu Q.Y., Siew C.K. 2004. Extreme Learning Machine: A New Learning Scheme of Feedforward Neural Networks. Proceedings of IEEE International Joint Conference on Neural Network (IJCNN), Vol. 2, pp.985-990.

Huang G.B., Zhu Q.Y., Siew C.K. 2006. Extreme Learning Machine: Theory and Applications. Journal of Neurocomputing, Vol. 70, Issues 1-3, pp. 489-501.

Huang G., Huang G.B., Song S., You K. 2015. Trends in Extreme Learning Machine: A review. Journal of Neural Network, Vol. 61, pp. 32-48.

Kanan H.R., Faez K. 2008. An improved feature selection method based on ant colony optimization (ACO) evaluated on face recognition system. Journal of Applied Mathematics and Computation, Vol. 205, Issue 2, pp. 716–725.

Karegowda A.G., Manjunath A.S., Jayaram M.A. 2010. Comparative Study of Attribute Selection Using Gain Ratio Correlation based feature selection. International Journal of Information Technology and Knowledge Management, Vol. 2, No. 2, pp. 271-277.

Khushaba R.N., Al-Ani A., Al-Jumaily A. 2011. Feature Subset Selection using Differential Evolution and a Statistical Repair Mechanism. Journal of Expert Systems with Applications, Vol. 38, Issue 9, pp. 11515–11526.

Liao T.W. 2010. Feature extraction and selection from acoustic emission signals with an application in grinding wheel condition monitoring. Journal of Engineering Applications of Artificial Intelligence, Vol. 23, Issue 1, pp. 74–84.

Priyadarsini, R.P., Valarmathi, M.L., Sivakumari, S. 2011. Gain Ratio Based Feature Selection Method For Privacy Preservation. ICTACT Journal On Soft Computing, Vol. 01, Issue. 04, pp 201-205.

Yang H.C., Zhang S.B., Deng K.Z., DU P.J. 2007. Research into a Feature Selection Method for Hyperspectral Imagery Using PSO and SVM. Journal of China University of Mining & Technology, Vol. 17, Issue 4, pp. 473–478.

PERFORMANCE EVALUATION OF DSR WITH MOBILITY MODELS

Henni Endah Wahanani[1], I Made Suartana[2], Yudhi Nur Hasyim[3]
University of Pembangunan Nasional "Veteran" Jawa Timur, Surabaya, East Java, Indonesia
[1]E-Mail: henni_endah@yahoo.com

ABSTRACT

Topology and mobility in a Mobile Ad Hoc Network (MANET) is a key factor that affects the performance of the protocol. Research in the field of MANET is usually done by using simulation. Among other simulation parameters, the model of mobility plays a very important role in determining the performance of the protocol in MANET. This research present Manhattan mobility model in reactive routing protocols DSR with Random Waypoint than that already exists in the network simulator NS2. Performance comparison was done with a variety of nodes, different simulation speed and area. QOS parameters used and analyzed included packet delivery ratio and average end-to-end delay. The research result shows that the performance of DSR routing protocol with mobility was affected two models vary across different mobility models showed with Packet Delivery Ratio fairly stable (between 90 to 100) and there were some improvements in certain simulations, and the value of Average End To End Delay increases and decreases in accordance with certain simulation models used mobility.

Keywords: MANET, DSR, mobility models, random waypoint, manhattan

INTRODUCTION

In mobile ad hoc network, nodes do not rely of any existing infrastructure.instead, the nodes themselves form the network and communicate through means of wireless communications. Mobility causes frequent topology changes and may break existing paths (M. Sreerama Murty, et al., 2011).

The mobility model is designed to describe the movement pattern of mobile users, and how their location, velocity and acceleration change over time. Since mobility patterns may play a significant role in determining the protocol performance, it is desirable for mobility models to emulate the movement pattern of targeted real life applications in a reasonable way. Otherwise, the observations made and the conclusions drawn from the simulation studies may be misleading. Thus, when evaluating MANET protocols, it is necessary to choose the proper underlying mobility model (F. Bai, et al., 2004).

This research discussed how to analyze the performance of DSR routing protocol to the effect of node mobility models that exist in mobile ad hoc network. Previously had much to discuss protocols The DSR routing but still does not give clear results on the subject DSR routing protocol discussion of the influence of node mobility model. In this case the parameters used are: packet delivery ratio and average end to end delay.

Mobility Model

Different mobility models can be differentiated according to their spatial and temporal dependencies.

Spatial dependency: It is a measure of how two nodes are dependent in their motion. If two nodes are moving in same direction then they have high spatial dependency.

Temporal dependency: It is a measure of how current velocity (magnitude and direction) are related to previous velocity. Nodes having same velocity have high temporal dependency (Bhavyesh Divecha, et al., 2007).

To evaluate the performance of a protocol for an adhoc network, it is necessary to test the protocol under realistic conditions, especially including the movement of the mobile nodes. Since not many MANETs have been deployed, most of this research is simulation based. These simulations have several parameters including the mobility models and the communicating traffic pattern. MANET protocol performance may vary drastically across different mobility models. In the literature, there are a lot of models used, mostly in simulations. Among the common one is the Random Waypoint Model, which is a simple model that may be applicable to some scenarios. However, there are other mobility models that may be used to capture the more important mobility characteristics of scenarios that MANETs may develop (Rajesh Deshmukh, et al., 2010).

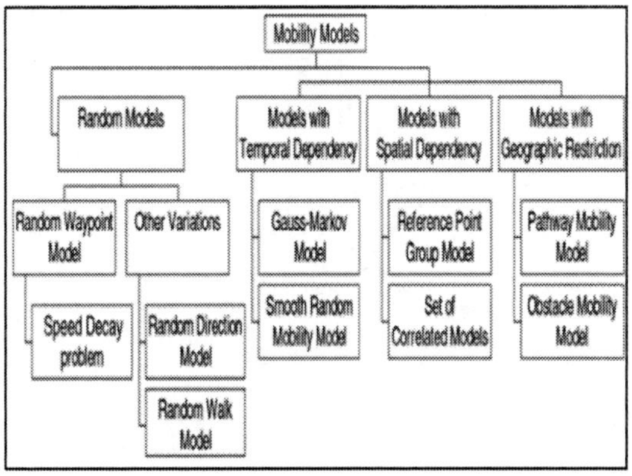

Figure-1. Category of mobility models

However, in the Figure-1, MANETs maybe used in different applications where complex mobility patterns exist. Hence, recent research has started to focus on the alternative mobility models with different mobility characteristics.

Random Waypoint

In Random Waypoint each mobile node moves independently of others. Due to physical constraints of the mobile entity itself, the velocity of mobile node will change continuously and gently instead of abruptly, i.e. the current velocity is dependent on the previous velocity. However, intuitively, the velocities at two different time's slots are independent in the Random Waypoint model. In many cases the movement of a mobile node may be restricted along the street or afreeway. A geographic map may define these boundaries (Sunil Kumar Singh, et.al.,2013).

Manhattan

In the Figure-2. Manhattan models including mobility model Pathway or track. Force mobility model lines each mobile node to move along the shortest path to that destination. Similar behavior is also modeled in mobility models Freeway and the Manhattan mobility model in (F. Bai, et al., 2003). Manhattan mobility model is proposed to model the movement in urban areas. In the model of Manhattan, a mobile node is allowed to move along the horizontal or vertical on a city map. At the intersection of the horizontal and vertical road, the mobile node can turn left, right or straight. The probability of moving on the same road is 0.5, the probability of turning left is 0.25 and the probability of a right turn is 0.25. Speed mobile node at a time slot depends on the speed at the previous time slot. Also, the speed of the node is limited by the speed of the previous node in the same lane of the road.

Manhattan mobility model focuses on the node moves along horizontally or vertically, which is not enough to model the nodes move along the path of non-horizontal and non-vertical. In addition, Manhattan model is not suitable to model the movement occurred at the intersection of the highway system, is much more complex than a local road junction. Thus, Manhattan mobility model that is expected to have a high spatial dependence and high temporal dependence. On top of that, it also imposes restrictions on the mobility of nodes geographically, although giving flexibility to the node to change direction (Rajesh Deshmukh, et al., 2010)

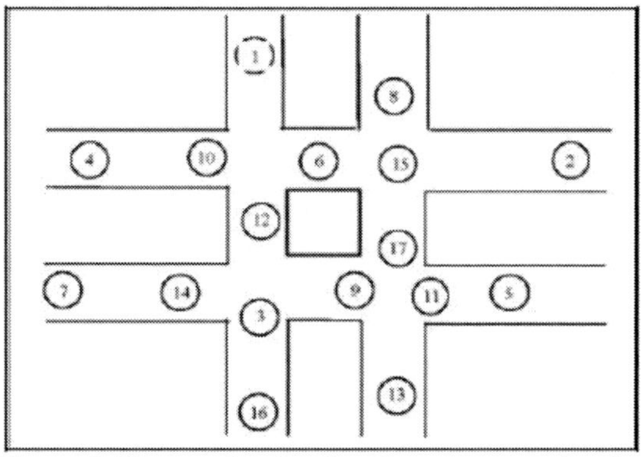

Figure-2. Topography showing the movement of nodes for Manhattan mobility model

DSR

The Dynamic Source Routing protocol is a simple and efficient routing protocol designed specifically for use in multi-hop wireless ad hoc networks of mobile nodes. Using DSR, the network is completely self-organizing and self-configuring, requiring no existing network infrastructure or administration. Network nodes cooperate to forward packets for each other to allow communication over multiple "hops" between nodes not directly within wireless transmission range of one other (V.Ramesh, et al., 2010). In the Figure-3, the DSR protocol is composed of two main mechanisms that work together to allow the discovery and maintenance of source routes in the ad hoc network

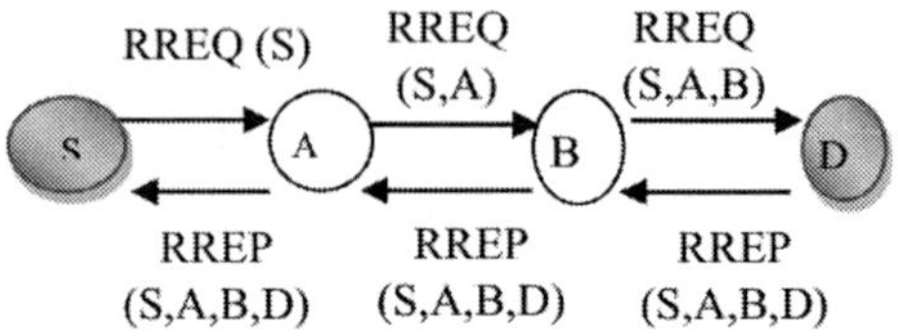

Figure-3. DSR protocol mechanism

The advantage of using DSR protocol are at intermediate nodes do not have to maintain it up to date information on the current routing packets skipped, because each packet always contains routing information in its header. Routing This type of process also removes also the route advertisement periodic and neighbor detection run by routing on other ad hoc networks. Compared with other on-demand routing protocol routing DSR has the best performance in terms of packet delivery ratio, Normalized Routing Load (on the package) and the average length of the path. Use of this would be optimal routing on a small number of nodes or less than 200 nodes.

Limitations DSR protocol is to build and depend on the route that unipath for each data transmission. Whenever there is a path that fails on the route, the DSR routing protocol needs to do a new route search process. This leads to high routing load and has a delay which is bad for the search process for a new route. For larger quantities would result in collisions between packets and cause increasing delay at the time of going to establish a new connection.

METHODOLOGY

Simulations was carried out by the Network Simulator NS 2.35. In the simulations used two different network sizes are: 500x500 m^2 and 1000x1000 m^2. In the Table-1, the following parameters in the simulation more:

Table-1. Simulation parameters

Parameter	**Value**
Mobile nodes	20,30,40,50
Simulation time	240,500 sec
Pause Time	20 sec
Number of connections	10
Transmission range	250m
Traffic size	CBR
Packet Size	512 byte
Maximum Packet	50
Maximum movement speed	20 m/s

Parameter	Value
Packet rate	1 p/s
Antenna Type	Omnidirectional
Network Size	$500x500m^2$, $1000x1000m^2$
Mobility model	Random Waypoint, Manhattan

We have used following two performance metrics :

Packet Delivery Ratio (PDR) - the ratio of total number of packet successfully received by the destination node to the number of packet generated by the source node. A high value of PDR indicates that most of the packets are being delivered and is a good indicator of the protocol performance.

$$PDR = (Pr/Ps) * 100 \quad (1)$$

Average End to End Delay - includes all possible delays caused by buffering during route discovery latency, queuing at the interface queue, and propagation and transfer times of data packets..

$$D = (Tr - Ts) \quad (2)$$

RESULTS

1. Comparison between the random waypoint and Manhattan with simulation time 240 sec.

Random Waypoint

Table-2. Random waypoint with area of 500x500 m^2

Node	PDR	Delay
20	98.69	52.84
30	98.14	50.88
40	97.18	124.91
50	98.23	18.66

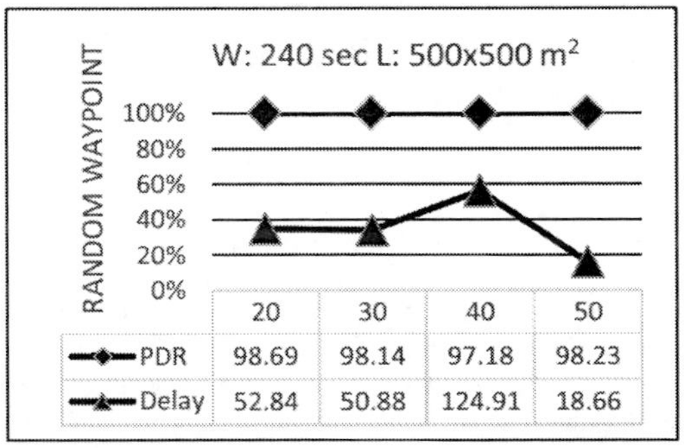

Figure-4. Random waypoint with simulation time 240 sec and area of 500x500 m^2

Table-3. Random waypoint with area of 1000x 1000 m^2

Node	PDR	Delay
20	90.83	2430.33
30	97.01	584.83
40	91.96	105.38
50	92.13	107.28

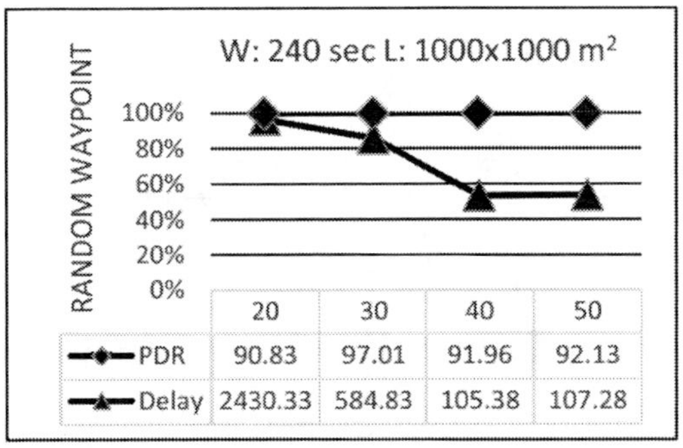

Figure-5. Random waypoint with simulation time 240 sec and area of 1000x1000 m^2

Manhattan

Table-4. Manhattan with area of 500x500 m^2

Node	PDR	Delay
20	99.53	282.13
30	99.95	9.45
40	100	11.49
50	99.90	9.23

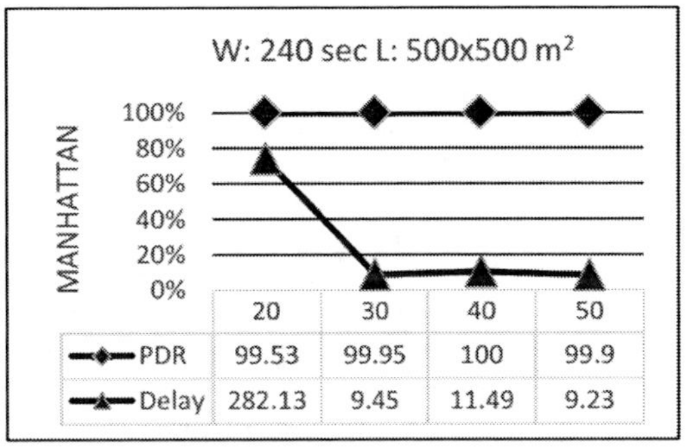

Figure-6. Manhattan with simulation time 240 sec and area of 500x500 m^2

Table-5. Manhattan with simulation time 240sec and area of 1000x1000 m^2.

Node	PDR	Delay
20	49.29	732.09
30	63.98	412.14
40	95.69	906.13
50	98.85	31.81

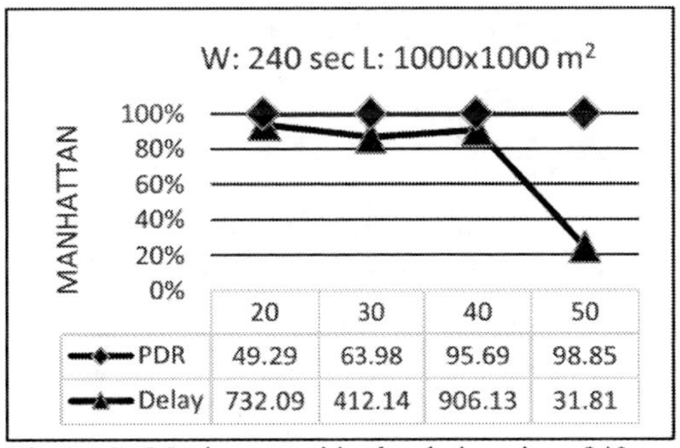

Figure-7. Manhattan with simulation time 240 sec and area of 1000x1000 m^2

2. Comparison between the random waypoint and Manhattan with simulation time 500 sec.
   Random Waypoint

Table-6. Random waypoint with area of 500x500 m^2

Node	PDR	Delay
20	98.89	53.70
30	99.31	57.43
40	98.73	25.03
50	99.58	117.33

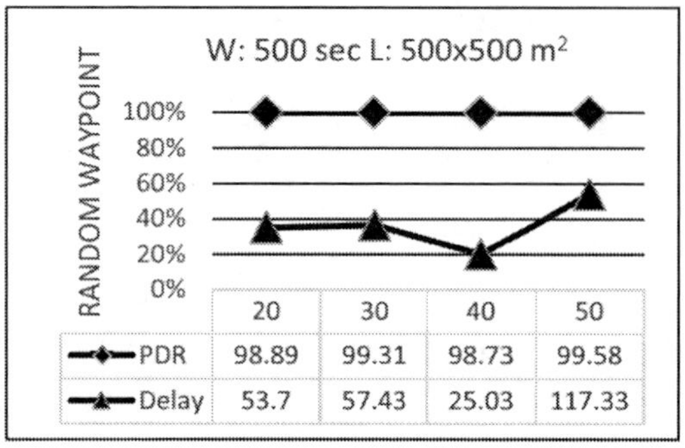

Figure-8. Random waypoint with simulation time 500 sec and area of 500x500 m^2

Table-7. Random waypoint with area of 1000x 1000 m^2

Node	PDR	Delay
20	94.98	2605.51
30	94.23	775.57
40	96.46	314.48
50	94.69	307.68

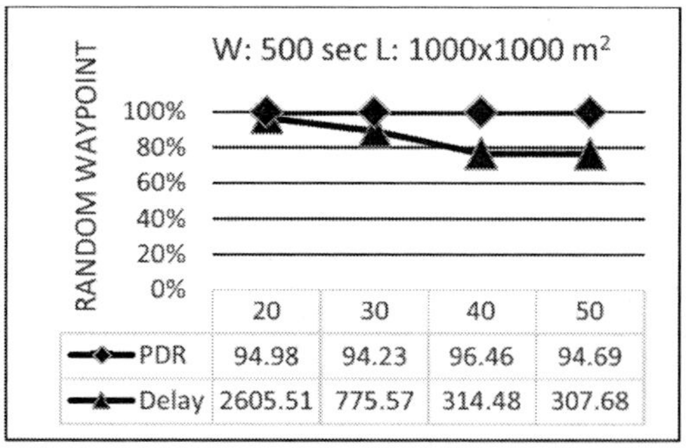

Figure-9. Random waypoint with simulation time 500 sec and area of 1000x1000 m^2

Manhattan

Table-8. Manhattan with area of 500x500 m^2

Node	PDR	Delay
20	100	12.10
30	99.91	11.99
40	99.93	8.25
50	99.39	19.15

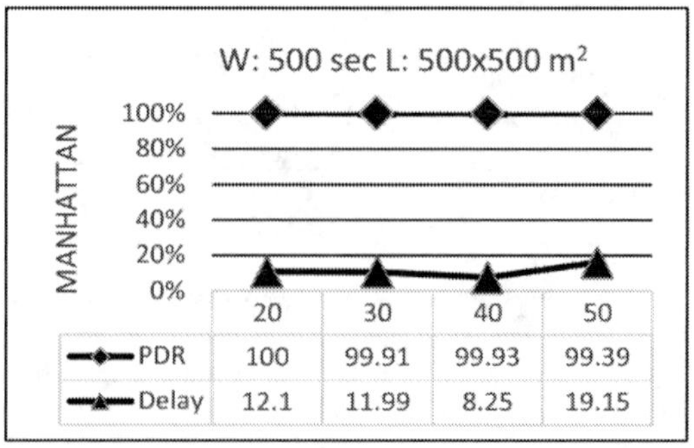

Figure-10. Manhattan with simulation time 500 sec and area of 500x500 m^2

Table-9. Manhattan with area of 1000x1000 m^2.

Node	PDR	Delay
20	77.74	1542.79
30	62.39	2466.67
40	91.96	1068.68
50	99.15	103.36

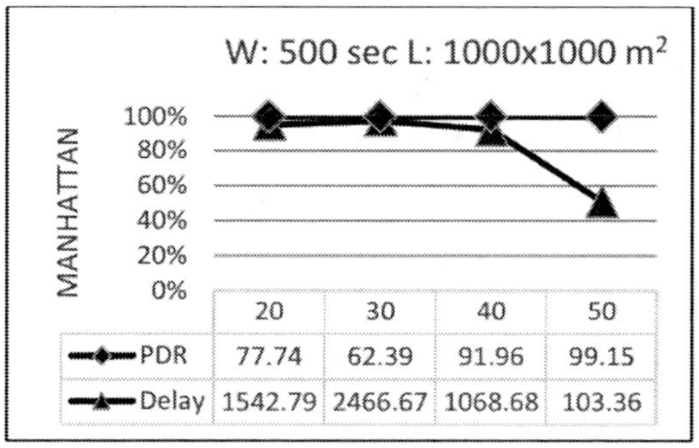

Figure-11. Manhattan with simulation time 500 sec and area of 1000x1000 m^2

DISCUSSIONS

By the time of 240 sec, random waypoint (Tables 2 and 3) and manhattan (Table 4 and 5) PDR greater mobility of manhattan is 99.84 compared with the random waypoint mobility in the area 500x500m². With a time of 500 sec, random waypoint (Table 6 and 7) and manhattan (Tables 8 and 9) PDR greater mobility of manhattan is 99.82 compared with the random waypoint mobility is on the area 500x500m² and the average delay is the smallest 12.87. The average delay is most likely in the area of mobility manhattan at 1000x1000 m² with a time of 500 sec that in 1295, 38.

This research by DSR experiences the lowest routing protocol overhead, on the count of higher average delays, particularly with Manhattan models, at higher node speeds. This protocol performs best with the random waypoint model.

CONCLUSIONS

In this paper we performed the simulation to evaluate the performance of DSR routing protocols on different performance metrics i.e. packet delivery ratio and end to end delay under the different mobility models with varying of mobile nodes number and network size.

DOI: 10.1051/matecconf/20165803011

From the different analysis of graph and simulations it can be conclude that performs of DSR in Manhattan mobility model is better than Random Waypoint. Under Manhattan mobility model, DSR experience the highest Packet Delivery Ratio and the lowest average end to end delay with the increase of mobile nodes number, simulation time, and network size.

REFERENCES

Bhavyesh Divecha, Ajith Abraham, Crina Grosan and Sugata Sanyal. 2007. Analysis of Dynamic Source Routing and Destination Sequenced Distance Vector Protocols for Different Mobility Models. in Proceedings of the First Asia International Conference on Modelling & Simulation (AMS).

F. Bai, A. Helmy. 2004. A Survey of Mobility Modeling and Analysis in Wireless Adhoc Networks. Book Chapter in the book on "Wireless Ad Hoc and Sensor Networks", Kluwer Academic Publishers.

F. Bai, N.Sadagopan, and A. Helmy. 2003. Important: a framework to systematically analyze the impact of mobility on performance of routing protocols for ad hoc networks. Proceedings of IEEE Information Communications Conference (INFOCOM 2003), San Francisco.

J. Broch, D. A. Maltz, D. B. Johnson, Y.-C. Hu, and J. Jetcheva. 1998. A performance comparison of multi-hop wireless ad hoc network routing protocols. Proceedings of the Fourth Annual ACM/IEEE International Conference on Mobile Computing and Networking(Mobicom98), ACM.

M. Sreerama Murty and M. Venkat Das. 2011. Performance Evaluation of MANET Routing Protocols using Reference Point Group Mobility and Random Waypoint Models. International Journal of Ad hoc, Sensor & Ubiquitous Computing (IJASUC), Vol. 2, No.1, pp. 33-42, March.

Rajesh Deshmukh, Asha Ambhaikar. 2010. Performance Evaluation of AODV and DSR with Reference to Network Size. International Journal of Computer Applications (0975 – 8887) Volume 11– No.8, December.

Sunil Kumar Singh, Rajesh Duvvuru, and Amit Bhattcharjee. 2013. Performance Evaluation Of Proactive, Reactive And Hybrid Routing Protocols With Mobility Model In Manets. International Journal of Research in Engineering and Technology (IJRET) Volume: 02 Issue: 08

V.Ramesh, Dr.P.Subbaiah, N.Koteswar Rao, N.Subhashini, Narayana.D. 2010. Performance Comparison and Analysis of Preemptive-DSR and TORA. International Journal of Ad hoc, Sensor & Ubiquitous Computing (IJASUC) Vol.1, No.4, December

MATEC Web of Conferences **58**, 03012 (2016) DOI: 10.1051/matecconf/20165803012
BISSTECH 2015

TERATOZOOSPERMIA CLASSIFICATION BASED ON THE SHAPE OF SPERM HEAD USING OTSU THRESHOLD AND DECISION TREE

I Gede Susrama Masdiyasa[1, 2], I Ketut Edi Purnama.[3], Mauridhy Heri Purnomo.[4]

[1] Electrical Engineering, Institut Teknologi Sepuluh November Surabaya
[2] University of Pembangunan Nasional "Veteran" East Java, Indonesia
[3] Dep. of Multimedia and Network Eng., Institut Teknologi Sepuluh November Surabaya
[4] Electrical Engineering, Institut Teknologi Sepuluh November Surabaya
Email: research.sperm@gmail.com

ABSTRACT

Teratozoospermia is one of the results of expert analysis of male infertility, by conducting lab tests microscopically to determine the morphology of spermatozoa, one of which is the normal and abnormal form of the head of spermatozoa. The laboratory test results are in the form of a complete image of spermatozoa. In this study, the shape of the head of spermatozoa was taken from a WHO standards book. The pictures taken had a fairly clear imaging and still had noise, thus to differentiate between the head of normal and abnormal spermatozoa, several processes need to be performed, which include: a pre-process or image adjusting, a threshold segmentation process using Otsu threshold method, and a classification process using a decision tree. Training and test data are presented in stages, from 5 to 20 data. Test results of using Otsu segmentation and a decision tree produced different errors in each level of training data, which were 70%, 75%, and 80% for training data of size 5x2, 10x2, and 20x2, respectively, with an average error of 75%. Thus, this study of using Otsu threshold segmentation and a Decision Tree can classify the form of the head of spermatozoa as abnormal or Normal

Keywords: Teratozoospermia, Classification, sperm Head, Otsu, Decision Tree

INTRODUCTION

Disorders in male sperm generally can affect fertility in men, since sperm is an important point to make fertilization. The most important thing for a conception is that the sperm must be in good condition and quality, since sperm cells must go through a long, full of obstacle journey to meet an ovum and finally fertilize it. One sperm abnormality that can be determined through sperm analysis is teratozoospermia (Jonge and Barrat, 2006). Teratozoospermia is a disorder in which many of the sperm have abnormal morphology (shape). Patients diagnosed with teratozoospermia have more than 30 percent sperm having abnormal shape (WHO Standard, 2010), while the normal ones have fewer than 30 percent. A normal sperm has a head and a tail, while an abnormal sperm may have two heads or two tails as in Figure 1. Normally, a good sperm must have an oval-shaped head, with a link on its middle part, and have a long tail. If the middle part is thickening, then it is said to be an abnormal spermatozoa. Now, a sperm morphology analysis is carried out by an expert using a microscope and counted manually, so there might be a difference reading between one expert and the others. A manual examination as explained before will require a lot of time and effort (Alias, et.al. 2011). Therefore, this study proposes a classification engine to distinguish the head of the normal and abnormal sperm

In (Abbiramy and Shanthi, 2010), the research conducted was just focused on the measurement of morphological parameters of spermatozoa and used it to perform the classification of normal or abnormal spermatozoa. The method used was preprocessing using a median filter, segmentation and feature extraction by using Sobel edge detection to detect the region of the head, neck and tail of spermatozoa to obtain statistical parameter morphology of spermatozoa, which would then be used to classify the morphology of normal or abnormal spermatozoa. In this classification process, there was still a manual calculation performed.

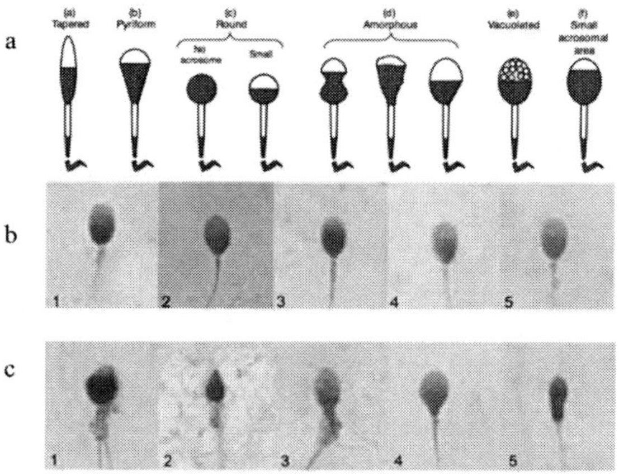

Figure-1. (a) Head Shape Abnormalities In Sperm morphology, (b) Normal sperm morphology, (c) Abnormal Sperm Morphology (Standard WHO, 2010)

Classification of normal or abnormal spermatozoa in humans can make use of some Fuzzy rules (Khachane, et.al, 2015). In his research, human spermatozoa segmentation based on the shape of the head, neck and tail of the sperm was conducted, using image processing techniques and classification only in the sperm head region. The digitized sperm image was preprocessed (by performing gray image conversion, Noise Removal and Binary Image Conversion), and then segmentation is applied to the Head, Neck, and tail of the sperm to obtain feature extraction using Region properties of Object. The proposed classification method was to use Fuzzy rules to categorize normal and abnormal sperm only based on the shape and area of the sperm head. This study did not mention the level of accuracy of the comparison between manual and computer analysis.

DOI: 10.1051/matecconf/20165803012

Another study of sperms was carried out by tracking the head and tail (Jun, et.al, 2013). In this study, the motion template method was used to detect and track the movement of sperm. The method was integrated with Kalman filter algorithm to maximize the intensity of the region to detect the location of the sperm tail. With the acquisition of the head and tail of the sperm, sperm movement and the amplitude of the measured pulse (flick) of the sperm tail can be investigated. To track the head of the sperm, several processes were performed: 1) silhouette 2) thresholding 3) erosion and dilation. The tracking of the tail was based on the sperm head and Sperm Tracking Region Of Interest (STROI) was used. However, the obtained location of the tail was often inaccurate so Kalman filter was used to acquire the exact position of the sperm tail. Results from this study were the tracking of the sperm head, sperm motility, and the amplitude of the sperm tail.

Research conducted by Qiaoliang Li and his team, made an application to detect and count the number of human spermatozoa in streaming video using OpenCV (Qiaoliang, et.al., 2012) The method used was combining Gaussian-modeling and morphological methods. Gauss-modeling algorithms were used to filter out non-target bright objects; whereas the morphological image processing methods were used to improve the video image quality of spermatozoa. Results from this study by merging several methods/algorithms had an average accuracy of 95% with the processing of streaming video in real-time from the test results of human spermatozoa. The downside of this study is the number of sperm obtained from the calculation is still less than the actual number. This is because the detection is based only on the sperm head. Sperm heads, which coincide (collide), should still be detected as two sperms, but in this study they were counted as one sperm.

Recent research conducted by Mohamed Elsayed, et al., is to create a computer-aided tool to analyze the motion of micro size spermatozoa with, or so-called Computer Assisted Sperm Analysis (CASA). Calculating the conventional motion characteristics of sperm makes CASA, studying the behavior of spermatozoa to provide a fluid flow, so a tracking and rheotaxis can be done to understand the behavior of spermatozoa (Elsayed, et.al. 2014)

In this research, an image analysis process of spermatozoa was performed, focusing on the shape of the head. From the shape of the sperm head, it was concluded whether the spermatozoa in the image was categorized as having normal or abnormal head of spermatozoa. The process began with a preprocessing to improve the sharpness of the image based on the level of contrast in the image. The process was continued by separating the head of spermatozoa with another portion of the image using Otsu threshold of segmentation, followed by an analysis of the region shape and the sperm head morphology. Results of feature extraction of spermatozoa head shape would be analyzed using decision tree classification engine.

The data used in this research was data from the World Health Organization (WHO Standards, 2010). The data was in the form of digital images; the sperm data used was only those that have been analyzed from the shape of the head, such as having a large oval head (macro shape), small oval head (micron form), flat head (lepto shape), two heads shape, and amorphous head shape as in Figure 1. The sperm image data was taken from WHO standards with the use of Otsu threshold, because the image data was clearly visible so it would be easier to perform segmentation. The end result of this study was to classify normal or abnormal sperm head, which will be referred to the analysis of WHO standard book as a comparison of accuracy of the result.

METHODOLOGY

Materials / Objects Spermatozoa

The object of research is the image of human spermatozoa in the book WHO Laboratory Manual For The Examination and Processing of Human Semen 2010 (Standard WHO, 2010). There were 20 normal spermatozoa head images and 20 abnormal ones. On those images, segmentation process was performed to classify spermatozoa as normal or abnormal ones.

System Design

Stages of the proposed method for identifying spermatozoa head shape, which will be used to determine abnormalities in sperm (teratozoospermia), with only normal and abnormal categories, are shown in figure 2.

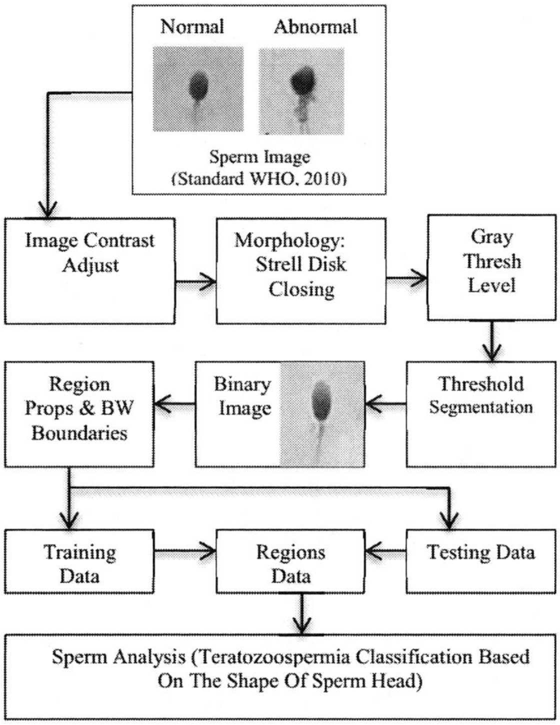

Figure-2. System Design of Teratozoospermia Analysis

Contrast settings

Spermatozoa images taken from WHO standards book have already a pretty good quality. Therefore, a brightness setting was carried out, by stretching the contrast. The contrast of an image is the distribution of light and dark pixels. A gray image (gray scale) with low contrast will appear too dark, too bright, or too gray. Contrast stretching is a useful technique to improve the contrast of images, especially those having low contrast. In contrast stretching, each pixel of an image U is transformed by using the function in equation (1):

$$o(i,j) = \frac{u(i,j) - c}{d-c}(L-1) \quad (1)$$

Where $o(i,j)$ and $u(i,j)$, respectively, were pixels before and after transformation in (i,j) coordinate; c and d, respectively, stated the maximum and minimum values of the pixels in input image, and L stated the maximum gray scale value. When the pixel value is less than 0, it will be made to 0, and when it is greater than (L-1), it will then be made to L-1. The results of contrast stretching setting from the original WHO standards image are shown in Figure 3. A2 for normal spermatozoa and B2 for abnormal spermatozoa.

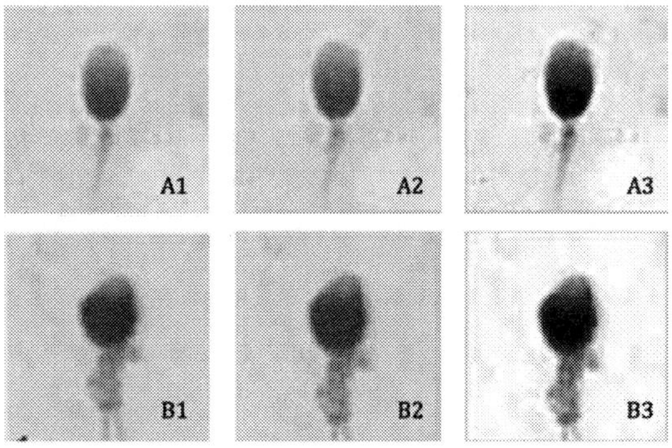

Figure-3. Results of WHO standards spermatozoa Segmentation

Morphology

Morphology is a branch of Biology that studies the form and structure of animals and plants. The same term in the context of mathematical morphology is as a tool for extracting image components that are useful in the representation and description of the shape of a region, such as boundaries, skeletons, and convex hull. Morphological techniques are also used for pre or post-processing, such as morphological filtering, thinning and pruning. Morphological operations use two input sets: binary image and structuring elements. Structuring elements (SE) is a generally small-size matrix. SE can be have a value of 1, 0 and don't care condition (usually left empty or filled with a cross sign). The basic operations in morphology processing are: dilation and erosion, as well as Opening and Closing.

This study used dilation and Closing basic operations. The Dilation operation was to increase the size of an object segment by adding a layer around the object. There are two ways to do this operation: One is by changing all background points which are the neighbors of the boundary point to an object point, or in other words, setting each neighbor point of an object point into that objects point. The second way is to change all points around the boundary point to be the object point. Mathematical formula for dilation process is in equation (2):

$$G(x,y) = f(x,y) + SE \quad (2)$$

Closing operation is a combination of dilation and erosion operations performed sequentially. First, the original image was dilated, and then an erosion process is applied to the result. This operation is used to close or eliminate small holes in an object segment, to combine adjacent objects and to smooth the limit of large objects without changing the object significantly. The Closing Math formula is as in equation (3):

$$\boldsymbol{f(x,y)\,.\,SE = (f(x,y) + SE) - SE} \quad (3)$$

The results of the closing operation are illustrated in Figure 3. A3 for the normal spermatozoa and B3 for abnormal ones.

Otsu Threshold

The purpose of Otsu method is to divide the histogram of the gray level image into two different areas automatically without requiring a user to enter a threshold value. The approach taken by Otsu method is by performing discriminant analysis, which is to determine a variable that can distinguish between two or more naturally risen groups. Discriminant analysis will maximize these variables in order to divide the foreground and background objects.

The formulation of Otsu method is as follows: Threshold values to be searched from a gray level image is denoted by k. K values range from 1to L, with L = 255. The probability value of each pixel on level-i can be formulated as:

$$Pi = \frac{n_i}{N} \quad (4)$$

Where n_i states the number of pixels on level-i and N denotes the total number of pixels in the image. The Zero^th cumulative moment, First cumulative moment, and total mean values can be expressed by the following formula:

$$\eta(k) = \frac{\sigma_B^2(k)}{\sigma_T^2} \quad (5)$$

Where

$$\sigma_B^2(k) = \frac{[\mu_T\omega(k) - \mu(k)]^2}{\omega(k)[1 - \omega(k)]} \quad (6)$$

And

$$\sigma_T^2 = \sum_{i=1}^{L}(i - \mu_T)^2 P_i \quad (7)$$

$$\mu_T = \mu(L) = \sum_{i-1}^{L} i\, P_i \quad (8)$$

$$\mu(k) = \sum_{i-1}^{k} i\, P_i \quad (9)$$

$$\omega(k) = \sum_{i-1}^{k} P_i \quad (10)$$

$$P_i = \frac{n_i}{N}, P_i \geq 0, \sum_{i=1}^{L} P_i = 1 \quad (11)$$

Region Props

Region properties (region props) method is a method used to measure a set of properties of each labeled region in a label matrix. A positive integer, which is an element of the matrix, corresponds to a specific region. The area and length of major and minor axes used are some of the properties generated by region props function. In a region props function, an object is represented as a region of a rectangular shape.

RESULTS AND DISCUSSION

From the analysis conducted, stretching the original image contrast (the results in Figure 3 carries out the pre-processing original image repair process. B2 and A2), the next process is to do the segmentation process with several stages of treatment include process Strell, gray thresh level and threshold Otsu method, the result is Figure 3. A3 and B3.

Before the process of identifying normal and abnormal head shape, carried out the training process of normal sperm head shape, taken from the original image of sperm in the WHO standard books. Image data captured by 20 for normal image data, from the image in search of its value regions prop

a b c

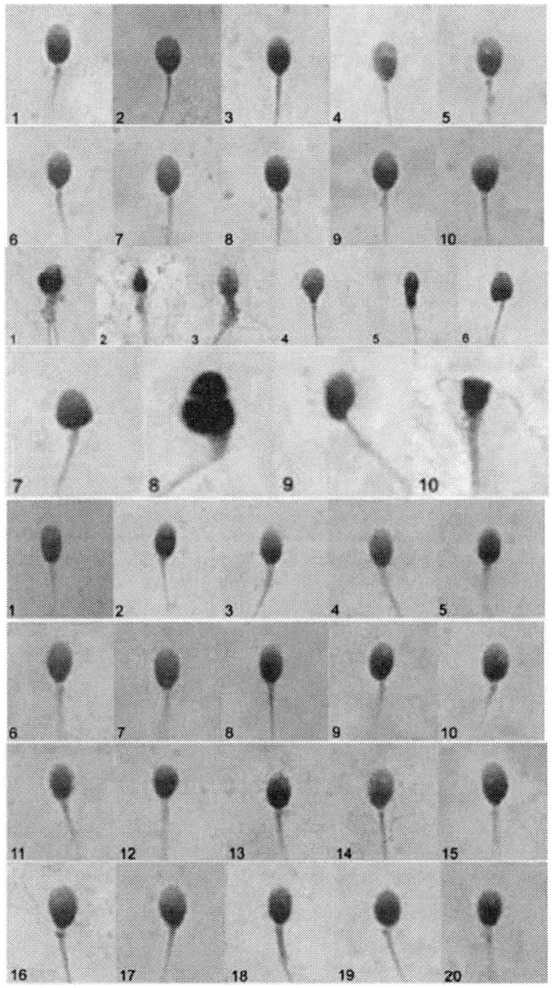

Figure-4. Data Image spermatozoa, (a) Example 10 Data Training Normal, (b) Example 10 Data Training Abnormal, and (c) Examples of Test Data

Table 1. Shows the results of the training image data of normal sperm. From this region props taken is Major Axis Length, Minor Axis Length: eccentricity, and Extent. This data is intended to compare the data of normal and abnormal data.

Table-1. Results of Training Data Region props

Data Training Normal Sperm			Data Training Abnormal Sperm		
Class	Extent	Eccentricity	Class	Extent	Eccentricity
1	0.9331	0.4482	2	0.9163	0.6304
1	0.9302	0.4228	2	0.9458	0.7078
1	0.9393	0.6894	2	0.8688	0.6919
1	0.9314	0.6397	2	0.9165	0.7299
1	0.9266	0.6639	2	0.9353	0.6981
1	0.9398	0.4677	2	0.9182	0.727
1	0.9244	0.698	2	0.9323	0.6949
1	0.9345	0.6483	2	0.8119	0.7349
1	0.9263	0.7013	2	0.9281	0.6663
1	0.9283	0.7591	2	0.923	0.756
1	0.9339	0.5531	2	0.923	0.7494
1	0.9073	0.8061	2	0.923	0.7139
1	0.9079	0.7668	2	0.923	0.8334
1	0.9248	0.751	2	0.923	0.7583
1	0.9328	0.7149	2	0.923	0.7718
1	0.9314	0.7352	2	0.923	0.7136
1	0.9282	0.7074	2	0.923	0.7122
1	0.9208	0.5938	2	0.923	0.7333
1	0.934	0.6853	2	0.923	0.6489
1	0.9238	0.7634	2	0.923	0.7766

The next process is to examine 20 normal spermatozoa and 20 abnormal spermatozoa images. Image data to be tested is analyzed and compared to existing training data in the data region. The result is normal and abnormal spermatozoa data, as shown in Table 2.

Table-2. Test Result of Normal and Abnormal Spermatozoa Data

Sperm	With Data Training = 5 x 2			With Data Training = 10 x 2			With Data Training = 20 x 2		
	Head Shape Standar WHO	Results of system test data	Status	Head Shape Standar WHO	Results of system test data	Status	Head Shape Standar WHO	Results of system test data	Status
1	Abn	Abn	Ok	Abn	Abn	Ok	Abn	Abn	Ok
2	Nor	Nor	Ok	Nor	Nor	Ok	Nor	Nor	Ok
3	Nor	Nor	Ok	Nor	Nor	Ok	Nor	Nor	Ok
4	Nor	Abn	No	Nor	Nor	Ok	Nor	Nor	Ok
5	Nor	Abn	No	Nor	Nor	Ok	Nor	Nor	Ok
6	Nor	Nor	Ok	Nor	Nor	Ok	Nor	Nor	Ok
7	Abn	Abn	Ok	Abn	Abn	Ok	Abn	Abn	Ok
8	Nor	Nor	Ok	Nor	Nor	Ok	Nor	Nor	Ok
9	Nor	Nor	Ok	Nor	Nor	Ok	Nor	Nor	Ok
10	Nor	Abn	No	Nor	Abn	No	Nor	Nor	Ok
11	Abn	Abn	Ok	Abn	Abn	Ok	Abn	Abn	Ok
12	Abn	Abn	Ok	Abn	Abn	Ok	Abn	Abn	Ok
13	Abn	Nor	No	Abn	Nor	No	Abn	Nor	No
14	Nor	Nor	Ok	Nor	Nor	Ok	Nor	Nor	Ok
15	Nor	Nor	Ok	Nor	Nor	Ok	Nor	Nor	Ok
16	Nor	Abn	No	Nor	Abn	No	Nor	Abn	No
17	Nor	Nor	Ok	Nor	Nor	Ok	Nor	Nor	Ok
18	Nor	Nor	Ok	Nor	Nor	Ok	Nor	Nor	Ok
19	Nor	Nor	Ok	Nor	Nor	Ok	Nor	Nor	Ok
20	Nor	Nor	Ok	Nor	Nor	Ok	Nor	Nor	Ok

Based on Table 2, the system has been tested on varied training data. The first test used 5 normal data and 5 abnormal data. Training result generated a decision tree classification engine which then tested and achieved an accuracy rate of 70%. Accuracy is obtained from the correctly classified data compared to the number of overall amount of test data

The second test was performed using 10 normal and 10 abnormal data as training. The result of the testing process gave an accuracy value of 75%. Meanwhile, the third test used 20 normal and 20 abnormal data and obtained an accuracy value of 80%. Testing data used in all three tests was the same set of data and training data in all three trials was another different set of data from testing data.

Average level of system accuracy of 75% of the three trials. The system accuracy is strongly influenced by the amount of training data and the type of training data. As more and more valid training data is used, it may increase the value of accuracy of the system since the basis of classification system used is a decision tree.

CONCLUSION

The proposed method in this study has been tested and has successfully identified spermatozoa based on the head shape with the highest accuracy value of 80%. The accuracy of this method is affected by two main factors, namely the classification engine and the amount of varied training data used, with an average accuracy of 75%. Accuracy value will increase if more training data is added in the process of classification. However, the value of accuracy is not proportional to the increase in the amount of training data. There was a 20% classification error occurred in this method due to the head shape that looks almost normal in the form of 2-dimensional (image), whereas in fact it has a flat shape in 3 dimensional. Thus, further development can be carried out using 3-dimensional analysis.

REFERENCES

C. D. Jonge and C. Barrat. 2006. The Cell: production, Maturation, Fertilization, Regenerasi. United States of America by Cambridge University Press, New York

L. Jiaqian, Kuo-Kun, D. Haiting and L. Yifan. 2014. Human Sperm Health Diagnosis with Principal Component Analysis and K-nearest Neighbor Algorithm: In Proc. International Conference on Medical Biometrics, pp. 108-113

L. Jun, L. Clement, L. Zhe and S. Yu. 2013. Quantitative Analysis of Locomotive Behavior of Human Sperm Head and Tail. IEEE Transactions On Biomedical Engineering, Vol. 60, No. 2, pp. 390-96

L. Qiaoliang, C. Xi, Z. Huisheng, Y. Li, C. Siping, W. Tianfu, L. Shumei, L. Xinyu, Z. Xiaofei, Z. Ruikai. 2012. Automatic Human Spermatozoa Detection in Microscopic Video Streams Based On OpenCV: In proc. 5th International Conference on BioMedical Engineering and Informatics (BMEI) IEEE, pp. 224-227.

M. Elsayed, T. El-Sherry, M. Abdelgawad, 2014. Development of Computer Assisted Sperm Analysis (CASA) Plugin for analyzing sperm in microfluidic environments using Image-J, In proc. 7th Cairo International Biomedical Engineering Conference (CIBEC), pp. 149-152.

M. F. Alias, N. A. Mat, S. A. Sulaiman and M. Mohamed. 2011. Sprague Dawley Rat Sperm Classification Using Hybrid Multilayered Perceptron Network, WSEAS Transaction on Information Science and Applications Issue 2 Volume 8, ISSN 1790-0832, pp. 109-118

M. Y. Khachane, R. R. Manza and R. J. Ramteke. 2015. Fuzzy Rule Based Classification of Human. International Conference on Electrical, Electronics, Signals, Communication and Optimization (EESCO): IEEE, pp. 1-5

V. S. Abbiramy and V. Shanthi. 2010. Spermatozoa Segmentation and Morphological Parameter Analysis Based Detection of Teratozoospermia. International Journal of Computer Applications. Foundation of Computer Science 3(7): 19–23.

Word Health Organization 2010. WHO Laboratory Manual For The Examination And Processing Of Human Semen Fifth Edition, WHO Library Cataloguing-In-Publication Data

THE APPLICATION OF EEMD AND NEURAL NETWORK BASED ON POLAK-RIBIÉRE CONJUGATE GRADIENT ALGORITHM FOR CRUDE OIL PRICES FORECASTING

M. Latif and Sri Herawati
University of Trunojoyo Madura, Indonesia
E-mail : latif.utm@gmail.com

ABSTRACT

Forecasting crude oil prices is very difficult to do because it has nonlinear and nonstationary characteristics. This research proposes a crude oil prices forecasting using a combination of EEMD and neural network. EEMD was used to decompose the price of crude oil into several IMFs and one residue. Before the training and testing was processed using FNN, EEMD output is normalized to fulfill network activation function. Data pattern of neural network was determined based on the results of normalization. The Learning method of neural network was based on Polak-Ribiére Conjugate Gradient algorithm. The output of neural networks on each component IMFs and the residue was aggregated using Adaline. The last process is denormalization of the Adaline output. Output of denormalization is the end result of the crude oil price forecasting. After forecasting results has been known, it then compared with the results of several neural networks learning algorithm. The result shows that the proposed method has better forecasting ability. This is indicated by the error value which was smaller than other forecasting algorithms for crude oil price forecasting.

Keywords: Crude Oil Price, Forecasting, EEMD, PCG, Adaline.

INTRODUCTION

The effect of crude oil prices have a large impact on the global economy. The prices will affect the financial markets, the price of fuel, and the price of goods and services (Haidar, et.al, 2008). Decline in crude oil prices will impact the budget deficit for oil exporting countries (Abosedra, Baghestani, 2004). The Fluctuations of crude oil market has a characteristics of nonlinear and nonstationary, so it is very difficult to forecast. The forecasting has important role in decision making such as: macro-economic policy (Xu and Ouenniche, 2012; Chang, 2012), financial stability (Azadeh et al.,2012; Naifar and Dohaiman, 2013), investment (Elder and Serletis, 2009), risk management and portfolio management (Reboredo and Castro, 2013). Additionally, forecasting aimed to reduce the impact of price fluctuations. The Research of forecasting crude oil prices which involves a combination of Ensemble Empirical Mode Decomposition (EEMD) and neural networks have been carried out by Herawati and Latif (2015). Resilient Backpropagation (Rprop) is used as an artificial neural learning. However, the algorithm Rprop takes a long time in the learning process. The research of crude oil prices has also carried out by Chatrath, et.al (2015), Priyadarshini (2015), Knetsch (2007).

This research proposes a crude oil prices forecasting using a combination of EEMD and neural network. The Learning method of neural network was based on Polak-Ribiére Conjugate Gradient (PCG). The PCG Algorithm is faster and has higher quality of separation and a lower cost in computational than others (Ghasemi-Fard, 2013). The Combinations of these methods have not been done in the case of crude oil price forecasting previously. The combination of EEMD and neural network based on PCG is intended to improve the weaknesses of the learning process in previous research.

LITERATURE REVIEW

Crude Oil Prices

World crude oil prices are determined based on the three benchmarks such as West Texas Intermediate (WTI), Brent and Dubai. WTI was used in the United States for industrial gas. Brent is the value of standardization in crude oil prices worldwide, especially used in Europe and OPEC market. Dubai is a product consisting of crude from Dubai, Oman or Abu Dhabi. Dubai is the main reference for Persian Gulf oil to be shipped to Asian markets. Based on data from the US Energy Information Administration stated that there were significant price increases for the price of WTI and Brent crude oil. In early 2003, the price of WTI is 32.95 US dollars per barrel, while the price of Brent is 31.18 US dollars per barrel. Price increases continue to occur until mid-2006, WTI through the price 74.41 US dollars per barrel, while the price of Brent reached 73.67 US dollars per barrel. The fluctuation of crude oil prices is influenced by many factors. According to Stevens (1995), demand and supply are also factors that affect the movement of crude oil prices. Fluctuations also very much influenced by the events of the past, present, and future irregular, such as weather, stock levels, the growth of GDP (Gross Domestic Product), political aspects and psychological expectations people (Yu et al, 2008). Based on the impact and the factors that affect the movement of crude oil price, the price forecasting is very necessary. Forecasting is used to reduce the impact of price fluctuations. Additionally, forecasting helps investors and individuals in making decisions related to the energy market.

Ensemble Empirical Mode Decomposition (EEMD)

EEMD is a data processing method of nonlinear and nonstationary (Wu and Huang, 2004). EEMD have a basic idea that any observed data is the combination of time series and the right noise. In fact, the average time series ensemble approach which is true if the data collected by the observation that a separate and different noise levels. Therefore, an additional step was taken with the addition of white noise to extract the correct signal in the data. White noise is a random signal having a frequency distributed evenly throughout the frequency range. The addition of white noise will fill the space evenly corresponding time frequency components that make up the signal with different scales. Components of different scales are automatically projected on a scale appropriate to the specified reference white noise. Procedure of crude oil prices decomposition using algorithms EEMD as follows:

a. Add a series of white noise into time series data.
b. The data is decomposed into some IMFs and one residue component.
c. Repeat steps a and b with different white noise for each time series data. The final result is obtained from the ensemble decomposition stating IMFs and residue appropriate.

Polak-Ribiére Conjugate Gradient (PCG)

Polak-Ribiére Conjugate Gradient is one of the learning neural network proposed by Polak and Ribiere. This algorithm searches the negative value of the gradient network since the first iteration and based on the direction conjugation. The algorithm of *Polak-Ribiére Conjugate Gradient* used is based on research Polak and Ribière (1969).

METHODOLOGY

The methodology in this research is shown in Figure 1.

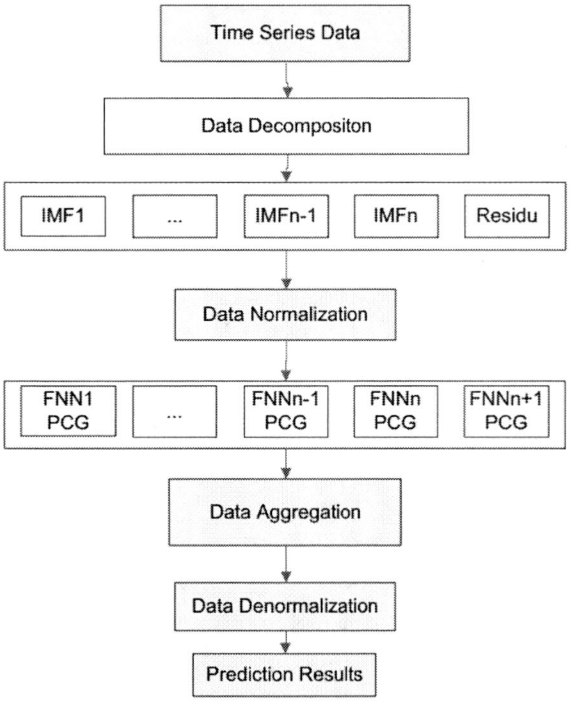

Figure-1. Methodology

Data Exploration

The research used monthly data time series of crude oil prices on the website of energy information administration (EIA) USA http://www.eia.doe.gov, that is West Texas Intermediate (WTI) and Brent. WTI data used were from January 1986 until January 2015 with a total of 349 data, while Brent data used were from May 1987 until January 2015 with a total of 333 data. Based on the two types of data, 85% data was used in the training process to construct forecasting models, while the remaining 15% data was used for testing the performance of the forecasting model. Mean Square Error (MSE) and Root Mean Squared Error (RMSE) was used to measure the accuracy of the forecasting performance. MSE and RMSE equation as shown in equation (1) and 2). A_t is the actual data at time t, where Ft is forecasting data at time t, and n is the number of data.

$$\mathrm{MSE} = \frac{1}{n}\sum_{t=1}^{n}(F_t - A_t)^2 \qquad (1)$$

$$\mathrm{RMSE} = \sqrt{\sum_{t=1}^{n}(F_t - A_t)^2} \qquad (2)$$

Data Decomposition

EEMD was used to decompose the international crude oil prices data into several IMFs and one residu. Number of ensemble members is set at 100 and the standard deviation of the white noise is set at 0.2. The decomposition process using parameter values-threshold-1 is 0.05, values-threshold-2 is 0.5, and the tolerance limit is 0.05.

Data Normalization

Data normalization was conducted to satisfy the requirements activation function neural network. The data was normalized using equation 3 (Kaastra and Boyd, 1996).

$$x' = TF_{min} + (TF_{max} - TF_{min}) \times \left(\frac{x - D_{min}}{D_{max} - D_{min}}\right) \qquad (3)$$

Where the variable x' is the result of data normalization, TF_{min} and TF_{max} is a minimum value and a maximum range of activation function, x is the actual time series data to be normalized, D_{min} and D_{max} is a minimum value and a maximum value of the actual time series data.

FNN with Polak-Ribiére Conjugate Gradient

Data pattern of neuron input, hidden and output of the network was determined based on the results of data normalization. Data are divided into two types, that is training and testing data. The training process will stop if the number of iterations reach the maximum value or the specified error value has been reached. Meanwhile, the testing process performed on the input data that has never been trained before. Testing data using the weights and biases of the results of the training FNN.

Data Aggregation

The testing results for all IMFs and residue was agregated using adaline. This process will produce the forecasting. The last process was denormalization of the Adaline output. Denormalization was used to restore data into real value.

RESULTS AND DISCUSSIONS

Result of decomposition using Ensemble Empirical Mode Decomposition algorithm is shown in Figure 2 for WTI data and Figure 3 for Brent data. Comparison of forecasting performance using PCG Learning with some artificial neural network learning algorithm is shown in Table 1 and Table 2, respectively for WTI and Brent.

Experiment begins with decomposing data using EEMD. Furthermore, the data of each IMF and the residue was normalized to fulfill activation function of FNN. Network architecture data patterns used is 6-10-1 (six input neurons, ten hidden neurons and one output neuron). FNN experiments performed iteration process as much as 10,000 times, learning rate value is 0.1, and error tolerance value is 0.0001. The final step is aggregating all the FNN output using adaline to generate forecasting. Then, determining the performance of the forecasting was done by finding the value of RMSE and MSE. After forecasting results using the PCG has been known, it was compared to the results of several neural network learning algorithm, such as Gradient Descent Backpropoagation (GDBP), Gradient Descent with Adaptive Learning Rate Backpropagation (GDALR), Levenberg - Marquardt Backpropagation (LMBP), Gradient Descent with Momentum Backpropagation (GDM), Gradient Descent with Momentum and Adaptive Learning Rate Backpropagation (GDMALR), Conjugate Gradient Backpropagation with Fletcher Reeves Updates (CGF), Powell-Beale Restarts (CGB), Batch training with weight and bias learning rules (B), Resilient Backpropagation (RPROP), Scaled Conjugate Gradient (SCG) and One Step Secant Algorithm (OSS).

Table 1 shows the best performance of WTI data gained from PCG learning algorithm with RMSE value is 0.04932 and the MSE is 0.00243. Meanwhile, the data Brent as shown in Table 2, showed that the best result was obtained with RMSE and MSE are respectively 0.04985 and 0.00249. In this research, RMSE and MSE is the main measure in assessing the performance of forecasting. Forecasting performance will be concluded to be good if it has a small error value. Forecasting performance using EEMD and neural networks with PCG learning algorithm is shown in Figure 4 and Figure 5, respectively for WTI and Brent. Both of these figures using testing data and showed a fairly accurate proximity to the actual crude oil prices data.

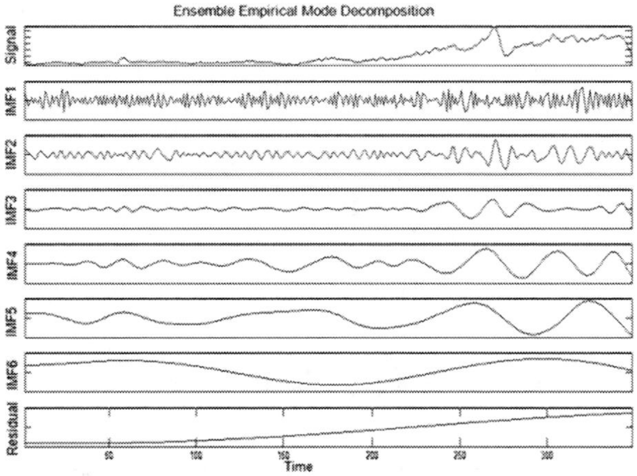

Figure-2. Decomposition Result of WTI Data

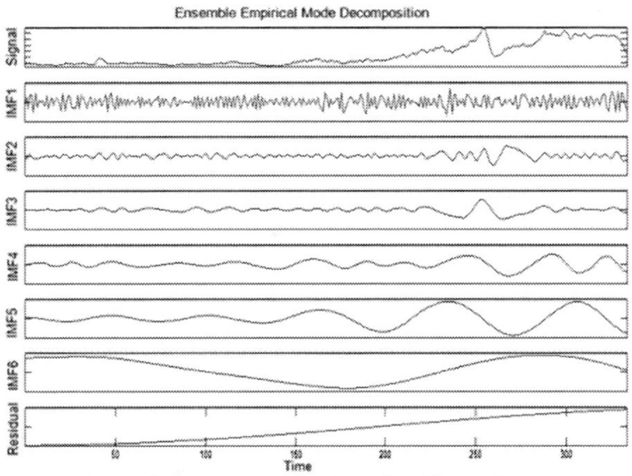

Figure-3. Decomposition Result of Brent Data

Table-1. Comparison of forecasting performance use WTI

Algorithm	RMSE	MSE
GDBP	0,05930	0,00350
GDALR	0,06890	0,00470
LMBP	0,05420	0,00290
GDM	0,05556	0,00309
GDMALR	0,05253	0,00276
CGF	0,05176	0,00268
PCG	**0,04932**	**0,00243**
CGB	0,05893	0,00347
B	0,05798	0,00336
RPROP	0,05410	0,00293
SCG	0,05137	0,00264
OSS	0,05051	0,00255

Table-2. Comparison of forecasting performance use Brent

Algorithm	RMSE	MSE
GDBP	0,05416	0,00293
GDALR	0,05306	0,00281
LMBP	0,05140	0,00264
GDM	0,05827	0,00339
GDMALR	0,05452	0,00297
CGF	0,05388	0,00290
PCG	**0,04985**	**0,00249**
CGB	0,05904	0,00349
B	0,06053	0,00366
RPROP	0,05191	0,00269
SCG	0,05637	0,00318
OSS	0,05330	0,00284

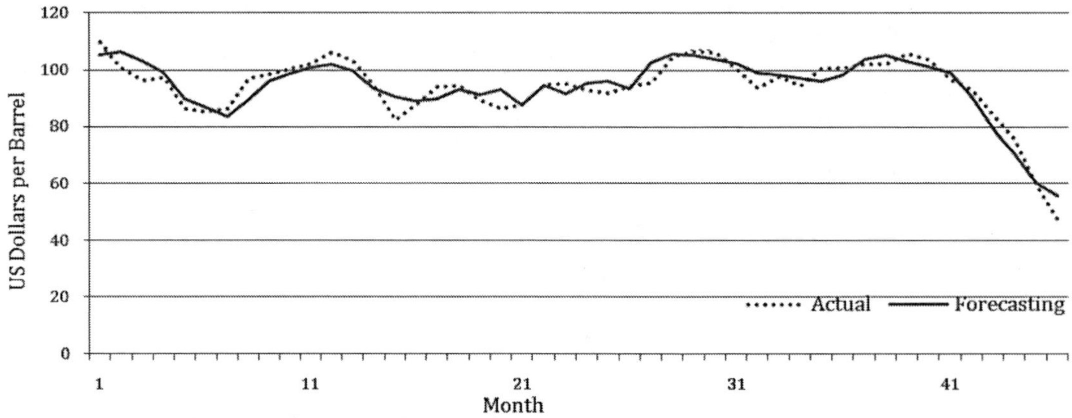

Figure-4. Comparison between forecasting results and actual data for WTI

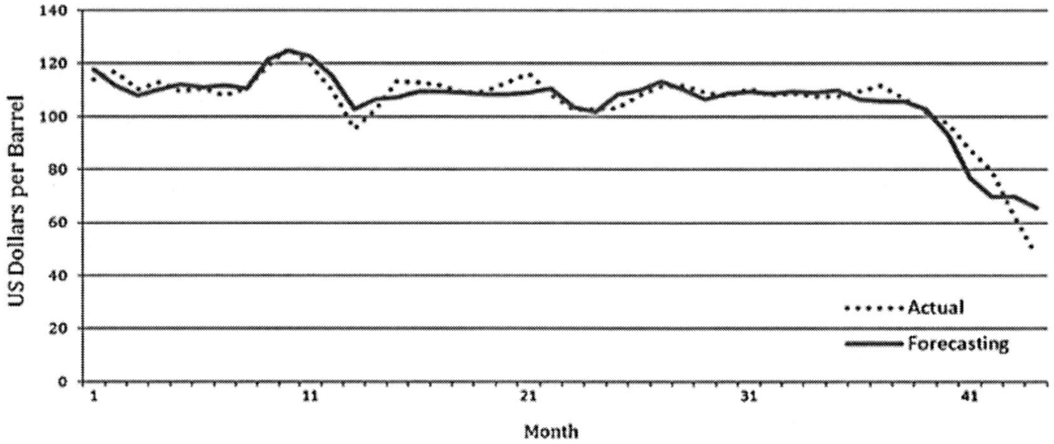

Figure-5. Comparison between forecasting results and actual data for Brent

CONCLUSIONS

According to the experiments, it can be concluded that the method of EEMD and neural network based on Polak-Ribiere Conjugate Gradient algorithm generates better RMSE and MSE than other artificial neural network learning algorithms. Forecasting result shows that the best performance is obtained with RMSE and MSE values of 0.04932 and 0.00243, respectively, for WTI data. Meanwhile, experiments conducted using data Brent generates RMSE and MSE of 0.04985 and 0.00249, respectively.

REFERENCES

Abosedra, I., Baghestani, H., 2004. On The Predictive Accuracy of Crude Oil Future Prices. *Energy policy.,* 32: 1389 - 1393.

Azadeh, A., Moghaddam, M., Khakzad, M., Ebrahimipour, V., 2012. A Flexible Neural Network-Fuzzy Mathematical Programming Algorithm for Improvement of Oil Price Estimation and Forecasting. *Computer & Industrial Engineering.,* 62: 421 – 430.

Chang, K.L., 2012. Volatility Regimes, Asymmetric Basic Effects and Forecasting Performance: An Empirical Investigation of the WTI Crude Oil Futures Market. *Energy Economics.,* 34, 294 -306

Chatrath, A., Miao, H., Ramchander, S. *and* Wang, T. *(*2015*),* The Forecasting Efficacy of Risk-Neutral Moments for Crude Oil Volatility. *J. Forecast.,* 34*:* 177–190.

Elder, J., Serletis, A., 2009. Oil price uncertainty. *Energy Economics.,* 31: 852 – 856.

Ghasemi-Fard, M., Ansari-Asl, K., Albera, L., Kachenoura, A., Senhadji, L., 2013. A low cost and reliable Polak-Ribiere conjugate gradient deflation ICA algorithm for real signals, in Electrical Engineering (ICEE) 21st Iranian Conference on. 1-5.

Haidar, I., Kulkarni, S., Pan, H., 2008. Forecasting Model for Crude Oil Prices Based on Artificial Neural Network. *Intelligent Sensors, Sensor Networks and Information Processing* (ISSNIP) *International Conference on,* 103-108.

Herawati, S., Latif, M., 2015. Forecasting Crude Oil Prices Using Ensemble Empirical Mode Decomposition and Resilient Backpropagation. *Seminar on Technology and Engineering* 1, 182 -186.

Kaastra, I., Boyd, M. 1996. Designing a Neural Network for Forecasting Financial and Economic Time Series, *Neurocomputing.,* 10: 215 – 236.

Knetsch, T. A. (2007), Forecasting the price of crude oil via convenience yield predictions. *J. Forecast.*, 26: 527–549.

Naifar, N., Dohaiman, M. S. A., 2013. Nonlinear Analysis among Crude Oil Prices, Stock Markets' Return and Macroeconomic Variables. *International Review of Economics and Finance.,* 27: 416 – 431.

Polak, B., Ribière, G., 1969. Note sur la convergence de méthodes de directions conjuguées. Rev. Fr. Inform. Rech. Oper. 16: 35–43.

Priyadarshini, E., 2015. A Comparative Analysis Of Prediction Using Artificial Neural Network And Auto Regressive Integrated Moving Average. *ARPN Journal of Engineering and Applied Sciences*., 10:3078 - 3081

Reboredo, J. C., Castro, M. A. R., 2013. A Wavelet Decomposition Approach to Crude Oil price and Exchange Rate Dependence. *Economic Modelling.,* 32: 42 – 57.

Wu, Z., Huang, N.E., 2004. Ensemble Empirical Mode Decomposition: a Noise Assisted Data Analysis Method. *Centre for Ocean-Land-Atmosphere Studies. Technical Report.*

Xu, B., Ouenniche, J., 2012. A Data Envelopment Analysis-based Framework for the Relative Performance Evaluation of Competing Crude Oil Prices' Volatility Forecasting Models. *Energy Economics.,* 34: 576 -583.

THE IMPLEMENTATION OF CACHING DATABASE TO REDUCE QUERY'S RESPONSE TIME

Arnaldo Marulitua Sinaga[1], Poppy Sibarani[2]
[1,2]Informatics Department, Faculty of Electronics and Informatics Engineering, Institut Teknologi Del, Indonesia
E-Mail: [1]aldo@del.ac.id

ABSTRACT

Performance is an imperative aspect of a web-based application. Response time is one of the most important performance parameters. Most of web-based applications are driven by database system. Reducing response time of its database system will improve the performance of a web-based application. Caching is one of techniques that can be implemented to reduce response time. The implementation of caching can be conducted in the application or database side. Caching is a technique to keep the executed query and hence the query will not be executed when called in the later execution. The result of an executed query will be stored in the cache therefore when that query is called, the database will retrieve the stored result without re-executing query. This mechanism will reduce the time to re-execute queries. However, this technique may be not suitable for all queries. This research is intended to investigate the effect of the implementation of caching into performance of a database. A series of experiments have been conducted by applying query cache. The response time of the application of query cache is compared with the response time of database without query. The results show that the implementation of query cache reduces response time significantly. However, this technique is more suitable for large database with many frequently asked queries. The number of clients accessing the database is found to be influencing the performance of caching. The more clients access the database, the more improvement is provided by query cache.

Keywords: database, caching, query, performance

INTRODUCTION

Many web-based applications are built using a multi-tier environment (Larson et al., 2004). Multi-tier client environment involves the use of client-tier, middle-tier and database-tier. Client-tier is the part that relates directly to the client, one of which is the browser. Through a browser, the client can connect with middle-tier. Middle-tier components relate directly to the database tier. Middle-tier contains business process that serves to process client requests that come through the client-tier to retrieve the data in the database-tier. Middle-tier itself can be divided into several types in use, such as web-based application or desktop-based. By the time the client is using a browser, an application that is used is a web-based application consisting of a web server and web client. Results from the web client request will be returned through the web server.

Each user who accesses the application will make a request through the browser, which is then forwarded to the web server. On the web server, the request is processed and routed to the database. Then, the application returns the response to the user via the browser. Web servers always retrieve data from the database to process each request that comes, regardless of whether a new request comes together with a previous request. The heavy workload of web servers can cause a problem into the network traffic (Qiong et al., 2002, Charles et al. 2008, Paweł and Krzysztof, 2011).

Caching is used to overcome the workload problem including Web Caching (Jesse et al., 2002), Web Databases Caching (Alexandros, 2010) and Database Caching (Qiong et al., 2002, Charles et al. 2008, Paweł and Krzysztof, 2011). This research investigates the affect the Query Cache into Database. Database Caching is an imperative method to overcome the problem in high-traffic web-based applications (Priya et al., 2011). With the implementation of Query Cache, is expected to reduce the response time in executing the query. On applying the Query Cache database, user queries are executed in cache on the local database or the database contained on the database server.

This research is intended to examine how the effect of the application of caching in the database of the performance, especially in query response time. From this study, it is expected to obtain the guidance of applying caching in database.

This research is conducted by using empirical approach. A series of experiments are held that is by applying the Query Cache in a subject application.

LITERATURE REVIEW

A. Performance Tuning Overview

Database performance tuning consist of three steps: Performance Planning, Instance Tuning and SQL Tuning. Performance tuning needs to be well planned involving all resources. Maintenance is important to decrease the problem in performance. Good management of all components of database in a system will reduce or avoid the problems during operation (Immanuel and Lance, 2014). Instance Tuning is involved when a database initialized that is to prevent bottlenecks. In the initial configuration of a system, resource allocation will be involved. The most important of instance tuning is to identify bottlenecks and make appropriate changes to eliminate the problem of the bottleneck. Tuning is usually done repeatedly, that is when the system starts to operate or while running. In general, tuning implement improvements in performance problems. In other words, tuning should be the part of the life cycle of an application system. Typically, tuning phase will be ignored until the database is generated. Tuning can be a reactive process, where most major bottleneck problems will be identified and repaired (Immanuel and Lance, 2014).

When the SQL statements executed on the database, query optimizer will determine the exact execution plan and factors relating to the object as well as the conditions specified in the query. The main purpose of query optimizer is to process the SQL statement and reduce the execution time. Throughout the process of execution, the query optimizer statistics gathered and used in reviewing the system to determine the best address data access and other considerations. We can reuse the (override) execution plan of the query optimizer by giving different input into the SQL statement.

B. Caching

Caching is a technique that can be used to improve the performance of a system. The cache can be interpreted as part of a computer's memory where information is stored so that the computer can find the information faster (http://www.merriam-webster.com/dictionary/cache, 2014). The stored information can be a result of the previous computation process or a duplication of the original information. There are some caching techniques that can be done to improve the performance of web-based applications that implement multi-tier concept. Caching can be performed on the database tier, the middle-tier and the client tier.

Caching technique in MySQL is named as Query Cache. The architecture of MySQL is shown in Figure 1.

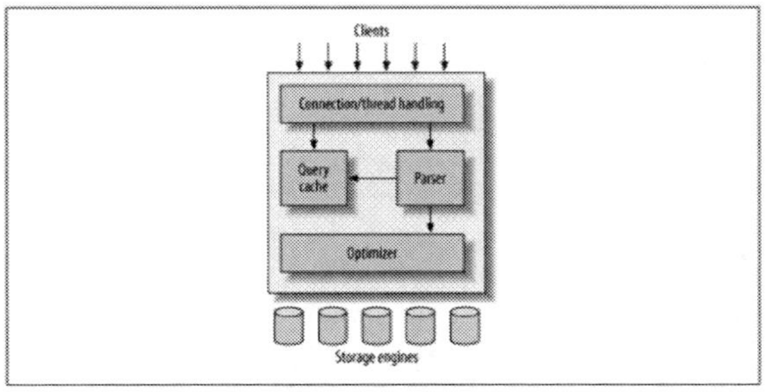

Figure-1. The Architecture of MySQL (Schwartz B. et al., 2012)

The upper layer of MySQL architecture contains one or more services that are needed by client/server (Schwartz B. et al., 2012), such as connection handling, authentication, security and so on. Important processes are held in the second layer, including code for query parsing, analysis, optimization, caching and all built-in function. The third layer contains storage engine that responsible for storing and retrieving data. The followings are steps of query execution (Schwartz B. et al., 2012):

1. Before parsing a query, MySQL checks whether Query Cache is active or not. If active, MySQL checks if the query has been stored in cache. MySQL the checks the privilege, if allowed MySQL returns Query Cache results to the client. When query is not in the cache, Query Optimization process will be activated to obtain the result of the query.
2. In Query Optimization process, MySQL parser divides a query into tokens and builds parse tree of those tokens. Parser ensures that the token is valid, and every part is in proper order.
3. Preprocessor checks related tables and columns and make sure the referenced name or alias is not ambiguous. Furthermore, the preprocessor checks privileges.
4. After the process of the parser and preprocessor, parse tree is ready to be converted into execution plans by the optimizer.

The query is executed on the Query Execution Engine in accordance with the execution plan that has been generated previously.

Query Cache is a technique that can be performed on systems with multi-tier concept. The purposes of applying the Query Cache is to increase performance and scalability of applications with distributed query processing (Qiong et al., 2002). At the time of Query Cache is placed on the middle-tier along with the application server, the Query Cache works like a database server. Query Cache can keep most of the existing data on the database server (Qiong et al., 2002). Thus, the server workload can be shared on the Query Cache. With the division of the workload of the database server, the response time to the user can be reduced, especially when the database server does the heavy work (Qiong et al., 2002).

To ensure the beneficial use of Query Cache, server operations have to be tested both with the cache on and off (http://dev.mysql.com/doc/refman/5.5/en/query-cache.html, 2014). The efficiency of Query Cache can be changed when the workload of the server is changed.

Query to be executed is compared with the correspondent query in Query Cache (http://dev.mysql.com/doc/refman/5.5/en/query-cache-operation.html, 2014). They have to be exactly the same. The value of variable Qcache_hits is incremented each time the query cache is being used. When a table has been changed/updated, all related queries in the cache will be deleted including all queries that have join operation with that deleted table.

MySQL stores query cache in the memory. The stored information in query is not just the result set, but also all the structure of the query including the relation between all involved tables with its query result and query text. Besides the housekeeping (base structure), query cache memory pool is provided in variable-sized block (Schwartz B. et al., 2012). When server is run, memory for query cache is initialized. Memory pool is initialized as an empty single block. When the server caches a result set, the size of allocated block in memory pool is set to query_cache_min_res_unit (Schwartz B. et al., 2012).

There are some constraints of Query Cache. Query Cache is not applicable to the following cases: query is a sub-query, the executed query is in a function, trigger or event, referring to the user or local stored program variables, referring to INFORMATION_SCHEMA, or PERFORMANCE_SCHEMA, referring to partition table, query uses temporary tables, query without table and user without any privilege to access the correspond tables (http://dev.mysql.com/doc/refman/5.5/en/query-cache-operation.html, 2014).

Query Cache store the text of the SELECT statement and the results of the statement are sent to the client (http://dev.mysql.com/doc/refman/5.5/en/query-cache-in-select.html, 2014). If the server receives the same statement/query as the previous statement, then the server will take the result of Query Cache and will not re-execute the statement. Query Cache is divided into sessions, so that a result set resulting from the query execution carried out by a client, can be reused as a response to another client (http://dev.mysql.com/doc/refman/5.5/en/query-cache-in-select.html, 2014).
Data stored in the cache is always the last data and trustworthy. All statements INSERT, UPDATE, DELETE or other modifications at a table will make the relevant data on the Query Cache eliminated (flushed) (http://dev.mysql.com/doc/refman/5.5/en/query-cache.html, 2014).

THE MATERIAL AND METHODOLOGY

This research is conducted by using empirical approach. A series of experiments are held that is by applying the Query Cache in a subject application. Del's library database system is used for the experiments. It is analyzed in order to design a caching application on the database. Data and running queries on the database will be reviewed if it is adequate for use in this study. Dummy data and the query will be added to improve the accuracy of the results.

A web-based application is used to simulate the application of the studied method. There are two types of environment being applied: standalone and client-server system. The experiments performed on a standalone system involve the database server and web server. Application on a web server executes queries in the database server. While the client-server system involves five clients that simultaneously execute queries in the database server. The response time of query execution with caching and without caching will be compared in both environments. The less response time the better method.

There are 17 queries involved in the experiments. Those queries use three tables in the database. Table 1 lists the used tables with their size. There are four phases (P1, P2, P3 and P4) of implementation that are intended to observe the influence of the size of the database. The phases are respectively involving the original tables, tables with 200% additional records of P1, tables with 300% additional records of P2, and tables with 400% additional records of P3.

Table-1. The size of the used tables

Table	Size (MB)			
	P1	P2	P3	P4
Books	1.57	4.72	16.25	76.13
Members	0.16	0.43	2.26	8.91
Book_issues	8.91	23.59	84.62	415.45

The selected queries are the frequently used queries in the library system. All queries uses SELECT operation and involve book_issues table. Queries can be classified as follows:

1. Q2 and Q7 are queries that involve a table with a condition in the WHERE clause.
2. Q1, Q8, Q10 Q11 Q14 and Q17 are queries that use a mathematical operation functions (COUNT and MAX).
3. Q5 and Q12 are queries that involve two tables and specific conditions in the WHERE clause.
4. Q15 is a query that involves two tables and two conditions on the WHERE clause.
5. Q3, Q4 and Q13 are queries that involve three tables with two conditions on the WHERE clause.
6. Q6 is a query that involves three tables and a condition in the WHERE clause.
7. Q9 and Q16 are queries that involve two tables and four conditions in the WHERE clause.

RESULTS AND DISCUSSIONS

In this section, the results of the experiments are presented. Table 2 presents the response time of query execution in the first two phases (P1 and P2) of standalone system.

Table-2. The result of P1 and P2 in standalone system

	P 1		P2	
Query	Non-Cache	Cache	Non-Cache	Cache
Q1	0.0743525	0.0010688	0.1889850	0.0010592
Q2	0.1146919	0.0037127	0.1088151	0.0020179
Q3	0.1090127	0.0013898	0.1257802	0.0016618
Q4	0.1462186	0.0028428	0.1043107	0.0018131
Q5	0.1391565	0.0022554	0.2379355	0.0111596
Q6	0.1246219	0.0016452	0.2268318	0.0025159
Q7	0.1020386	0.0014686	0.1680096	0.0019838
Q8	0.1099152	0.0011290	0.1914059	0.0010930
Q9	0.0705571	0.0012097	0.0971323	0.0011827
Q10	0.0900616	0.0011907	0.1091233	0.0010717
Q11	0.1212053	0.0011059	0.1911143	0.0010356
Q12	0.1332350	0.0012631	0.2111791	0.0013082
Q13	0.1217054	0.0013704	0.2193553	0.0017975
Q14	0.1300057	0.0063233	0.1944020	0.0012449
Q15	0.0853827	0.0016110	0.1256537	0.0019941
Q16	0.0746527	0.0011631	0.1303570	0.0014266
Q17	0.1011552	0.0011564	0.1737531	0.0010264

The response time of query execution in the phases P3 and P4 of standalone system are presented in Table 3.

The results in Table 2 and Table 3 show that response time of query execution for all queries with caching is lower than without caching. The improvement is significant with all types of queries. The bigger the involved data the higher improvement is resulted by applying caching.

Table-3. The result of P3 and P4 in standalone system

Query	P3		P4	
	Non-Cache	Cache	Non-Cache	Cache
Q1	0.5396976	0.0010229	2.2130160	0.0010281
Q2	0.6830338	0.0262220	3.5772230	0.1553261
Q3	0.4846241	0.0043907	1.8988990	0.0130687
Q4	0.7431542	0.0171590	4.1549310	0.1081332
Q5	0.7401709	0.0114675	3.0402140	0.0022376
Q6	0.6032453	0.0068385	2.2505520	0.0278870
Q7	0.5096860	0.0045215	2.2431590	0.0176957
Q8	0.6768925	0.0009779	2.3414040	0.0011835
Q9	0.4644715	0.0017498	1.8870290	0.0026999
Q10	0.4263281	0.0009145	1.9058490	0.0010663
Q11	0.6661291	0.0010140	2.4589780	0.0010838
Q12	1.1911524	0.0011866	2.9607830	0.0012625
Q13	0.7462513	0.0013488	2.4844530	0.0012071
Q14	0.7089699	0.0010861	2.4446860	0.0011466
Q15	0.4547726	0.0032481	1.9272760	0.0124869
Q16	0.4815077	0.0015633	1.9333240	0.0031702
Q17	0.5305893	0.0010612	2.1488340	0.0011355

Table 4 and Table 5 present the results for client-server system.

Table-4. The result of P1 and P2 in client-server system

Query	P 1		P2	
	Non-Cache	Cache	Non-Cache	Cache
Q1	0.0019607	0.0338124	0.0018267	0.0673393
Q2	0.0058976	0.0078376	0.0045933	0.0149182
Q3	0.0024565	0.0035589	0.0040213	0.0068385
Q4	0.0056106	0.0076440	0.0042841	0.0055331
Q5	0.0045032	0.0678604	0.0155843	0.1572880
Q6	0.0041579	0.0397551	0.0057334	0.0764448
Q7	0.0025960	0.0375549	0.0039457	0.0784872
Q8	0.0024780	0.0449281	0.0026435	0.0945814
Q9	0.0019519	0.0028731	0.0028062	0.0051634
Q10	0.0020485	0.0032194	0.0028013	0.0033805
Q11	0.0021062	0.0537042	0.0021254	0.1177131
Q12	0.0022890	0.0651954	0.0027685	0.1390019
Q13	0.0027707	0.0586838	0.0042931	0.1213539
Q14	0.0019491	0.0515476	0.0039212	0.0051362
Q15	0.0025961	0.0042301	0.0022024	0.1126135
Q16	0.0021842	0.0030673	0.0025272	0.0045962
Q17	0.0021624	0.0334407	0.0030330	0.0700180

Table-5. The result of P3 and P4 in client-server system

Query	P3		P4	
	Non-Cache	Cache	Non-Cache	Cache
Q1	0.0018960	0.6601156	0.0435279	4.1322056
Q2	0.0280391	0.6449867	0.3318795	4.4872204
Q3	0.0137948	0.0353426	0.0160832	1.6769764
Q4	0.0203815	0.5425361	0.1821219	4.9271134
Q5	0.0140861	0.8600350	0.0042396	5.2070280
Q6	0.0107668	0.6401707	0.0387153	7.2895811
Q7	0.0066079	0.6231424	0.0222229	5.2179774
Q8	0.0021964	0.7697019	0.0019169	4.8541472
Q9	0.0030772	0.0146861	0.0049721	1.9258970
Q10	0.0021125	0.0180916	0.0020366	1.9683380
Q11	0.0020450	0.8815631	0.0020068	5.6550891
Q12	0.0024103	1.0496134	0.0023406	6.2858812
Q13	0.0021318	0.9181136	0.0020022	5.9463682
Q14	0.0023554	0.5764136	0.0020882	5.6825592
Q15	0.0054374	0.0316479	0.0150830	1.5791501
Q16	0.0028682	0.0162668	0.0052960	1.9990924
Q17	0.0020640	0.7048124	0.0020359	3.7366625

The experiments in client-server system indicate the same results. The response time of query execution for all queries with caching is lower than without caching. The results with bigger data (P4) indicate that caching produce bigger saving. In this case, saving is the ratio of the response time of the system with caching and without caching. For example, saving for Q1 with P1 in client-server system is 94.20% ((1 - (0.0019607 / 0.0338124)) * 100). The average saving for all queries in both environment is presented in Table 6.

Table-6. The average saving

Environment	Average Saving (%)			
	P1	P2	P3	P4
Standalone	98.31	98.77	99.21	99.35
Client-server	68.26	71.93	92.89	99.06

Table 6 shows that the saving for all phases are mostly more than 90%, this indicates that the implementation of caching can reduce response time significantly. The highest improvement is achieved when it is applied into big size of data in client-server system

A statistical analysis is conducted to find out the difference of the two studied methods. The response time of each query in all phases is compared by using Wilcoxon Signed Test. This non-parametric test is chosen since the number of sample data is small (17). The p-values resulted from all comparison (all phases in both environment are less than 0.05. It indicates that the two methods are significantly different.

CONCLUSIONS AND SUGGESTIONS

The implementation of query cache into a database has been investigated experimentally in this research. The response time of query execution in database with caching is compared with database without caching. The two methods are implemented and compared in two environments, standalone and client-server system. The results show that the application of caching reduce the response time of query execution with big saving. The results also indicate that the application of caching in bigger data will produce bigger benefit. It has been analyzed statistically that the query caching can significantly improve the response time of a database.

However, the validity of this experiment needs to be enhanced in the next research. The application and data under tests with real operation need to be applied. It is also suggested to investigate the methods in more complex database with varies of query.

REFERENCES

Alexandros L., Qiong L. Jie X., and Wenwei X. 2010. Caching and Materialization for Web Databases. *Found. Trends databases* 2, 3 (March 2010), pp. 169-266.

Charles G., Amit M., Anastasia A., Bruce M., Todd M., Christopher O., and Anthony T. 2008. Scalable query result caching for web applications. *Proceedings Very Large Data Base* Endowment. 1, 1 (August 2008), pp. 550-561.

Immanuel C. and Lance A., "Oracle® Database Performance Tuning Guide 11gRelease 2 (11.2)", Oracle, 2014

Jesse A., Lawrence J., Xiang L., Jordan P., Zheng Z., Tie Z., Web caching for database applications with Oracle Web Cache, *Proceedings of the 2002 ACM SIGMOD international conference on Management of data,* June 03-06, 2002.

Larson, P.-A., Goldstein, J., Jingren Z. "MTCache: transparent mid-tier database caching in SQL server", *Data Engineering, 2004. Proceedings. 20th International Conference on,* pp. 177 – 188.

Mehmet A., Qiong L., Sailesh K., C. Mohan, Hamid P., Bruce G. L., Honguk W., Larry B., DBCache: database caching for web application servers, *Proceedings of the 2002 ACM SIGMOD international conference on Management of data*, June 03-06, 2002.

Paweł L. and Krzysztof S. 2010. Consistent caching of data objects in database driven websites. In *Proceedings of the 14th east European conference on Advances in databases and information systems* (ADBIS'10), Barbara Catania, Mirjana Ivanović, and Bernhard Thalheim (Eds.). Springer-Verlag, Berlin, Heidelberg, pp 363-377.

Priya G., Nickolai Z., Samuel M., "A Trigger-Based Middleware Cache for ORMs", Middleware 2011, Lecture Notes in Computer Science-Springer 7049, pp 329-349.

Qiong L., Sailesh K., C.Mohan, Honguk W., Hamid P., Bruce G. L., Jeffrey F. N., "Middle-tier database caching for e-Business", *Proceedings of the 2002 ACM SIGMOD international conference on Management of data*, Madison, WI, June, 2002.

Qiong Luo, Jerey F. Naughton, Rajasekar Krishnamurthy, Pei Cao, and Yunrui Li, "Active Query Caching for Database Web Servers", *International Workshop on the Web and Databases-WebDB 2000*, Lecture Notes in Computer Science-Springer 1997, pp. 92-104, 2001.

Schwartz B. et al.: "High Performance MySQL 3rd Edition", O'Reilly Media, United States of America, 2012

http://www.merriam-webster.com/dictionary/cache Merriam-Webster online dictionary. Accessed on26 April 2014

http://dev.mysql.com/doc/refman/5.5/en/query-cache.html. Accessed on 15 April 2014

http://dev.mysql.com/doc/refman/5.5/en/query-cache-operation.html. Accessed on 15 April 2014

http://dev.mysql.com/doc/refman/5.5/en/query-cache-in-select.html. Accessed on 15 April 2014

WEB-BASED TRAFFIC TICKET DATA PROCESSING INFORMATION SYSTEM

Faikul Umam[1], Rachmad Hidayat[2]
[1]Mechatronic Engineering Departement, Faculty of Engineering, University of Trunojoyo Madura Indonesia
[2]Industrial Engineering Department, Faculty of Engineering, University of Trunojoyo Madura Indonesia
E-mail : [1]faikul@trunojoyo.ac.id, [2]rachmad.trunojoyo@gmail.com

ABSTRACT

The purpose of the present study was to generate a WEB-based traffic ticket data processing information system to simplify the management of traffic offender data. This study used a structured analysis method, which is a technique to collect data and determine the facts in a systems design. A Traffic Ticket Data Processing Information System provides information concerning the amount of fines, articles violated, time and place of the trial, providing ease of information effectively and accurately. However, the characteristics of the information generated had not yet had a proper quality in terms of accuracy, completeness or timeliness of the information. As end-users, communities have not been satisfied with the information generated. The information system has not been functioning as a management information system for in the implementation stage there were still many obstacles and shortcomings so that the information generated could not be used for decision and policy making.

Keywords: traffic ticket data processing, ease of information, information timeliness, information quality

INTRODUCTION

Globalization and advances in science and technology have significant impacts on people's lives. A closer look at the impacts would indicate that a very practical and modern way of life almost occurs in all aspects of people's lives, no exception the legal aspect. The technological developments also have impacts on levels of crime and offenses committed by people. Law and society are two inseparable things since they are interrelated. Law enforcement on Road Traffic and Transportation covers the prosecution of violations and handling of Traffic Accidents. Prosecution of violations is carried out by on-road checks of motor vehicles. In the case of violations during the checks, prosecution may be taken by fast procedure and imposition of fines.

On-road checks of motor vehicles carried out by police officers and civil servant investigators LLAJ includes: (1) checks of Driving License, Certificate of Motor Vehicle Number, Certificate of Vehicle Test; (2) proof of passing compulsory vehicle test; (3) physical checks of motor vehicles; (4) load capacity and/or modes of transport; (5) Transportation Service Business License. Police officers have the authority for on-road checks of points 1 to 5. The civil servant investigators have the authority for checks of points 2 to 5. With regard to the authority for on-road checks of motor vehicles, police officers have the full authority, including that held by the civil servant investigators.

Authority for on-road checks of motor vehicles must have relevance to the previous governmental authority. Checks of motor vehicles shall have legal certainty. It should be based on official duties in the form of an Assignment Letter issued by the supervisor of the police officers or the supervisor of the civil servant investigators. Under Law No. 22 of 2009, traffic tickets is one form of quick examination procedure applied specifically to certain traffic violations. Actions categorized as violations are those that are visibly identified, do not need means of proof and do not need expert witnesses. Traffic ticket is a simplified police investigation report that specifies the identity of the offender, the type of offense, the location of the offense, the evidence requested, the time of trial and the data of the law enforcer along with the signature.

A ticket consists of 5 sheets of different colors. The red and blue sheets are for the offender, green for the court, white for the prosecutor and yellow for the police. The ticket serves as a summons to the District Court. It also serves as a cover note for the payment of fines to the Bank/Registrar. Another function is as a certificate of forfeiture of the seized proof (such as driver's license, vehicle registration certificate, motor vehicle). With regard to the settlement of a traffic ticket, there are two options that may be chosen by the people. One option is for those offenders who admit their guilt and are willing to pay a fine to the Bank and is represented by a designated police officer; the offenders are given a blue form. Another option is for those offenders who do not admit their guilt, those who are against the officers by refusing to sign the ticket and/or those offenders who commit offenses in the Traffic Order Zone.

In the first option, the police prosecute the offender using the blue form and suggest the offender to pay a fine to the Bank (with the maximum time limit of 5 days). After being validated by Bank's registration stamp, the police immediately restore the evidence. The traffic ticket files submitted cumulatively by the police to District Court are attended by a designated representative. In the second option, the police prosecute the offender using the red form. The day of trial must consider the provisions of the Court, which explain when and where the offender must attend the trial. If the offender is absent from the trial, the Police will make two summons and make an arrest for the third summons. Evidence shall be returned awaiting the

DOI: 10.1051/matecconf/20165803015

completion of the trial and after the offender pay the fine to the Registrar.

To assist the police and the people with regard to improving the settlement of traffic tickets, a WEB-based traffic ticket data information system is required, which can generate traffic ticket information quickly, at any time round the clock. This information system is aimed at reducing traffic violations committed by the people as well as increasing public awareness of safe, orderly and proper driving. The purposes of the present study were to improve efficiency and effectiveness of ticketed motor vehicle data management, to assist the police in addressing and reducing traffic offenses and to improve the order in driving.

In this study, the traffic ticket data information system used a Personal Home Page (PHP): Hypertext Preprocessor programming language. The PHP Preprocessor is a programming language that runs on a web server. PHP itself was created by the programmer of *Unix* and *Perl* named Ramsus Lerdorf in 1994. At first, Ramsus was trying to create a script in his website (Zajaczek J.E.W. et. al., 2006). PHP script itself is a programming language that runs on a web server, or often called a server side. Therefore, PHP can do anything that can be done by other CGI programs, such as processing any type of data. A server side itself means that three main programs, i.e., a web server (either IIS, Apache, or other web server that supports PHP), a PHP module and a web-browser, are required to run this program (Williams H.E., & Lane D., 2004). PHP can run on various operating systems, including Linux, Unix, Microsoft Windows, Mac OSX, RISC OS, SUN OS, and many more.

One of the considerations for choosing the PHP as the programming language of the server side is that there are many alternative operating systems and web servers that can be used. Outputs generated by PHP are not only limited to HTMLs, but also images, PDF files, as well as animated images using Libswf and Ming. Other outputs of text type may be XHTML and XML. Thus, a program with an interactive display can be created. The working system of PHP begins with a request coming from the web page by the browser. Based on the URL or address in the network the browser will find the address of the web server, identify the desired page and submit all information required by the web server (Davis M.E. and Phillips J.A., 2007).

When the page contains a PHP script, the web server will check the file type requested. If a PHP is requested, the web server will examine for a PHP script in the page. If a PHP script is not found in the page, the web server will be displayed in the browser. However, if the page contains a PHP script, the web server will continue it on the PHP module as the engine that translates and process the PHP script to be convertible to HTML codes and displayed to the user's browser (Valade J., 2007).

In building a sound and accurate application system, the system must also have a good data storage system. One good and widely used database storage system is a web-based application, such as MySQL. MySQL is selected as the database storage since it (1) has rapid access when used; (2) is multi-platform that can run on various operating systems; (3) can accommodate up to a terabyte of data; (4) is in accordance with the PHP programming. Additionally, its free-of-charge nature in its use requires no purchase [3]. Data is an essential part of modern programming so that all programming languages provide functions to access the database. The main standard database language is Structured Query Language (SQL). SQL is standardized as a language to create a database to store information into the database and retrieve information out of it (Ullman L.E., 2003).

METHODOLOGY

This study began with conducting systems analysis and preparing a use diagram and Sequence Diagram. Furthermore, it proceeded with the design of the data base and application of traffic ticket data processing information system. With PHP and MySQL, processing and retrieval of the required traffic ticket data can be done practically [7]. Access rights in the system are divided into: (1) user, in this case all users can access the added data; (2) admin is the user who has special access rights to data management and also data manipulation (insert, update, delete);and (3) hardware required is a personal computer (PC) or a notebook.

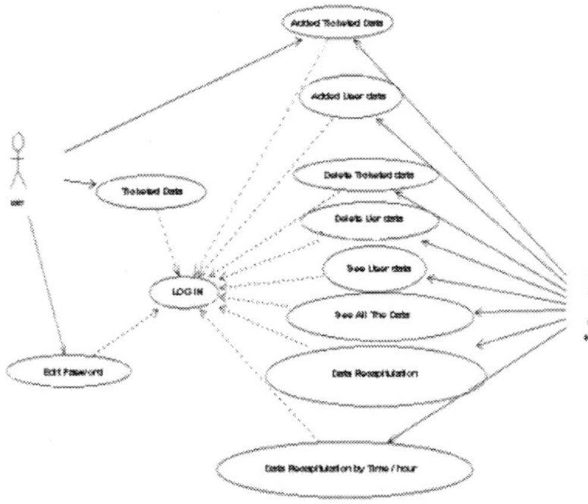

Figure-1. The Use Case Diagram

A Use Case diagram illustrates the functionality expected of a system. It represents an interaction between the actor and the system shown in Figure 1. A use case diagram presents the interaction between a use case and an actor, where the actor can be a person, equipment, or other systems that interact with the system being built. In the use case, there are actors who play roles in this application, the user and the administrator. This diagram illustrates the overall functions of the Ticketed Motor Vehicle Data Processing Information System application.

The user (in this case, or the Police District Command, locally known as the *Polres*) is required to login first to operate other activities, such as add data, view traffic ticket data and change password. The

Police Provincial Command admin can add, change and delete traffic ticket data. Additionally, it can view traffic ticket data and the summary of traffic ticket data and change the password. Prior to conducting any activity, the administrator must go through the login process first.

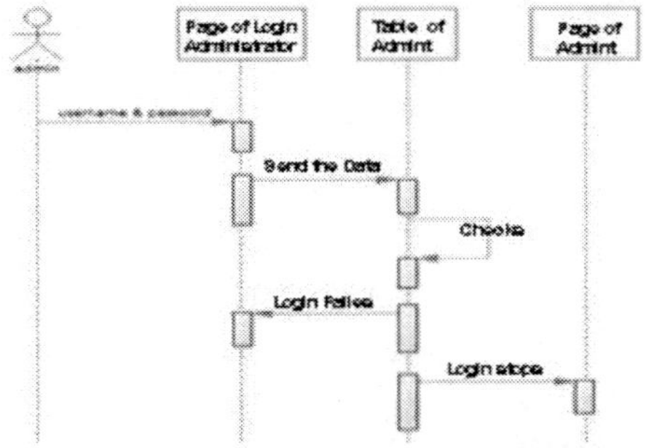

Figure-2. The SequenceDiagram of Admin Login

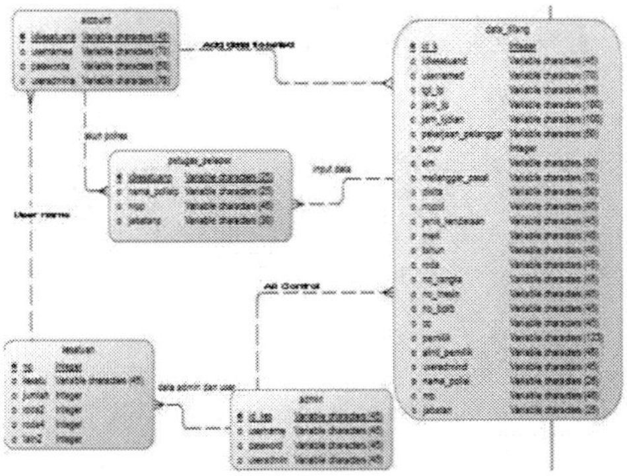

Figure-3. Conceptual Data Model

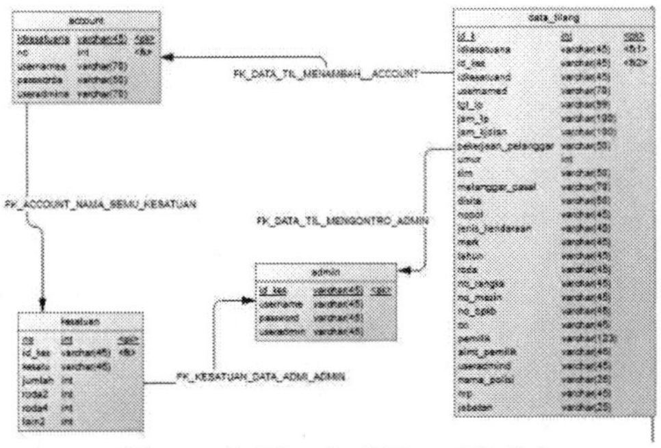

Figure-4. Physical Data Model

Inter-object interaction and communication is depicted in a Sequence Diagram shown in Figure 2. This diagram shows a series of messages exchanged by objects that perform a specific task or action. The objects are then sorted from left to right, with the actor initiating the interaction usually being placed the leftmost of the diagram.

In the design of the database, there are two kinds of model: (Figure 3) Conceptual Data Model (CDM) is a model based on the assumption that the real world consists of a collection of basic objects called entities and relationships between these entities. The CDM is usually represented in the form of an Entity Relationship Diagram. (Figure 4) Physical Data Model (PDM) is a model which uses a number of tables to describe the data as well as the relationship between these data. Each table has a column in which each column has a unique name.

RESULTS AND DISCUSSION

Traffic Ticket Data Processing Application

The two early steps to run this application are to enable the server module and then open the web browser by typing the web application's address in the address bar of the web browser. The first step to enter the Traffic Ticket Data Processing application is to enter the homepage of this application. Upon selection of the login menu, the login page will come into view. When the admin logs in, a homepage with admin menus will appear. Admin menus include "insert traffic ticket data", "change user and traffic ticket data", "browse all data", and "change admin data". The "Add Data" menu is used to add traffic ticket data and user data. Add Traffic Ticket Data Form is used to add traffic ticket data. The "Add Data Form" is shown in Figure 5.

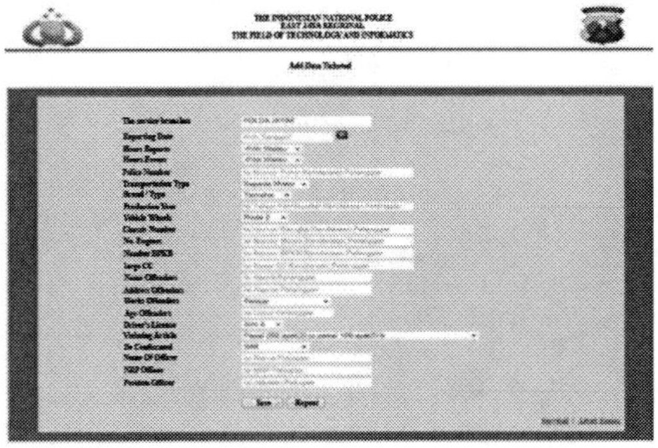

Figure-5. Add Traffic Ticket Data Form

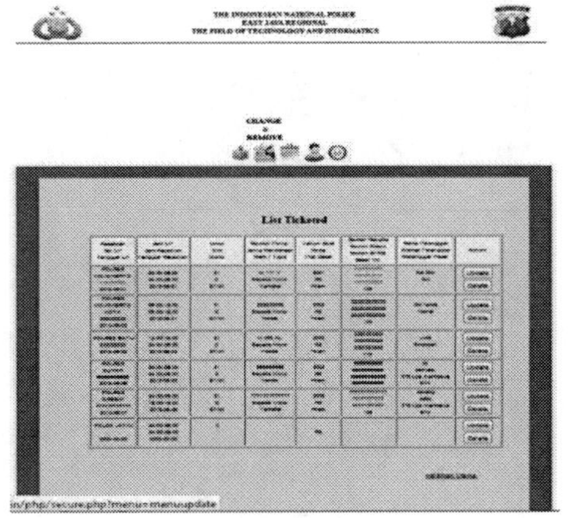

Figure-6. Page "Change Traffic Ticket Data"

Menus "Change Data" and "Delete Data". The page "Change Data" is shown in Figure 6. Menu

 DOI: 10.1051/matecconf/20165803015

"Change Account" of *Polres* is used to view and change profile data of *Polres*. Menu "All Data" is used to view all traffic ticket data, a summary of traffic ticket data by wheels, a summary of traffic ticket data by time, and a graph per *Polres*. Page "All Traffic Ticket Data" is used to view all traffic ticket data. Appearance of the traffic ticket data is shown in Figure 7.

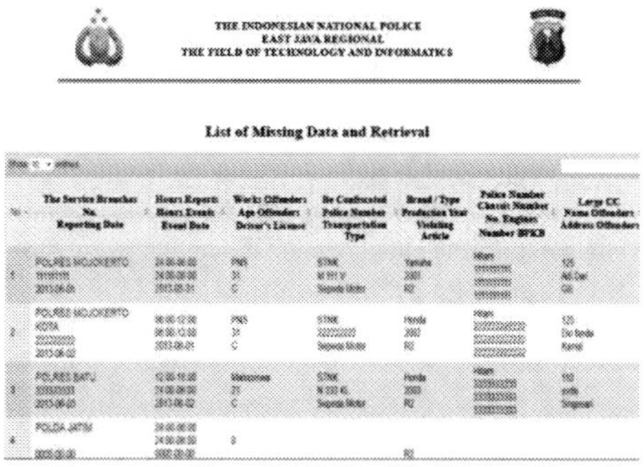

Figure-7. Page "All Traffic Ticket Data"

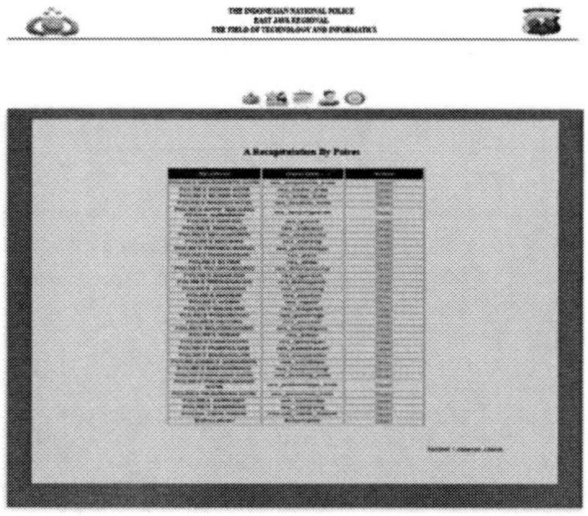

Figure-8. Page "View Data per Polres"

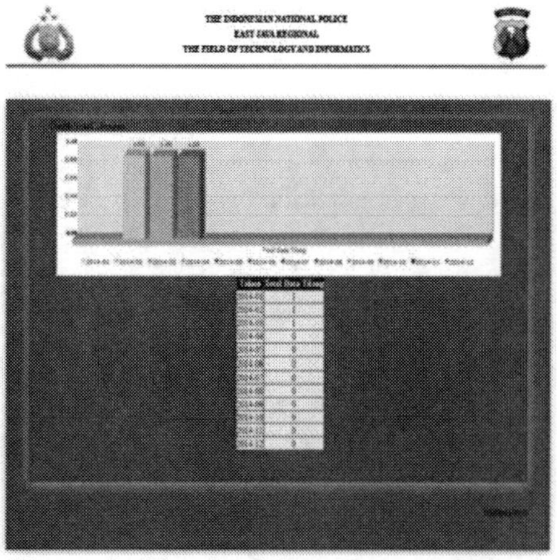

Figure-9. "Graph of Case Data per *Polres*"

Menu "Data per *Polres*" is used to view the summary of data per Polres. Menu "Summary of Data by Wheels" is used to view the summary of data by wheels per month. Menu "Summary of Data by Time" is used to view the summary of data by time or hours per month. Appearance of traffic ticket data is shown in Figure 8. Page "Graph of per User or *Polres*" is used to view the chart of data per *Polres*. "Graph of Case Data per *Polres*" is used to view the chart of case data per *Polres*. "Graph of All Case Data in Polda Jatim" is used to view the chart of all case data in East Java Police. The pages of the charts of data can be seen in Figure 9 and Figure 10.

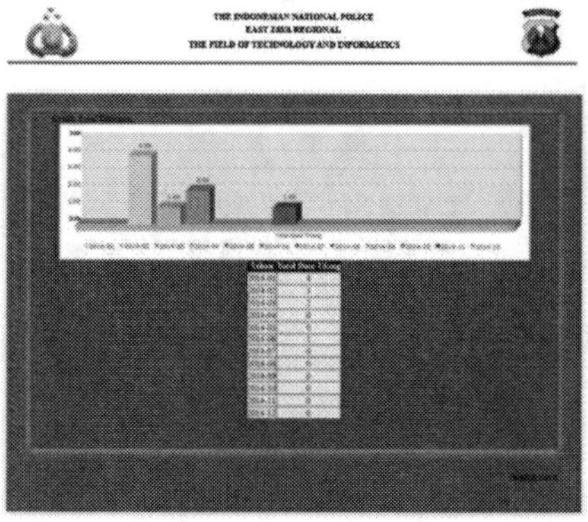

Figure-10. "Graph of All Case Data"

Traffic Ticket Data Processing Information Systems

Based on the study, the East Java Police have been implementing a Traffic Ticket Data Processing Information Systems, which could support managerial tasks and policy making. Availability and support of technological infrastructure, such as hardware, software, or network, to speeds up access to traffic ticket data processing Information Systems has not been adequate. An information system is a conceptual system and will run properly if the components, such as software, hardware, brain ware and network, functions properly to provide the correct decisions for its users (Kumar P.M., 2014).

Results of the study showed that the application of Traffic Ticket Data Processing Information Systems in each division has been being supported by software and hardware. Data and information between units/divisions have not been integrated since the intranet has not connected, which had negative impacts on the bureaucracy with regard to supplying the necessary data in a timely manner. Utilization of information technology within an information system provide an accelerated access and timeliness as well as the reduction of bureaucracy since it is capable of fast information generation with high accuracy (Iivari J., 2007), which has a positive impact on decision making.

In today's rapid technological developments, information systems support to a network constitutes a necessity to speed up acquisition of accurate and timely information fast. Information systems supported by an intranet network will have an impact on the high rate of data communication and provide advantages in a variety of transactions from a variety of different places that can cut the cost of the preparation and delivery of reports due to capability of online acquisition. Results showed that among the constraints and problems in the application of the traffic ticket data Processing Information Systems, in addition to the technological aspect, was the lack of understanding and awareness of human resources. More specifically, the assigned officers did not carry out their functions, duties and responsibilities properly at the operational level resulting in incomplete data needed for policy making.

Understanding and awareness of the functions and benefits of the information system will determine the quality of information produced by a system since the computer system has taken over human works in data processing; however, man has an important role as a user (Au N., and Cheng, T.C.E., 2012). (Al-Debei M.M., 2013) argue that the success of an information system is influenced by the perceived information quality and the perceived quality system which is a significant predictor for user satisfaction, while user satisfaction is also a significant predictor for the intended use and perceived individual impact.

Other components that play an important role in the handling and management of an information system are leadership and management. East Java Police's organizational structure consists of units or divisions with a considerable distance from *Polres* to another, requiring the structuring of the management information system to provide information for the benefit of the management for policy making. Presence of standard operating procedures and a clear organizational structure within each unit/division are needed to create a good collaborative relationship, which would in turn be able to create and maintain information systems useful and appropriate for decision making.

Results of the review of reports of Traffic Ticket Data Processing Information System showed that data inputting, data processing, data transformation and information (outputs) generation have not run as expected. The Format of Traffic Ticket Data Processing Information Systems has not been completed in full, as well as delivery of reports by each *Polres* was not on time, which adversely affected the quality of the information produced by the Traffic Ticket Data Processing Information Systems for management decision and policy making.

Quality information is measured by the accuracy of the information generated in the sense that it should be free from errors and not mislead those who receive it, as well as on time since delayed information would no longer useful and relevant information has benefits for recipients so that it can be useful for making a decision quickly and accurately in solving a problem (Choi Sue Young, et. al., 2010). Sophistication factor is no less important since a decision is a conscious and systematic effort to overcome the unfavorable situation in problem solving. Time orientation of a decision relates to expired information that will not support the decision-making process quickly and accurately.

Poor quality of information generated by a Traffic Ticket Data Processing Information Systems is inseparable from the handling and management of information systems, since the outputs of this system is information useful to stakeholders and decision makers in the East Java Police. Information systems that do not produce output but always receive input would put the input into a deep hole (Hevner A., and Chatterjee S., (2010).

Results also indicated that user's satisfaction with the Traffic Ticket Data Processing Information Systems was not significant. Satisfaction with the use of information technology can be measured from the aspect of behavior since behavior is one of the important aspects that relate directly to the user. Furthermore, interaction between user and computer devices is heavily influenced by perception, attitudes and affections inherent in human beings as users (Cheung C. M., and Lee M. K., 2008).

(Grispos G., et. al 2014) suggest that user behavior and a personal system are needed in the system development. This relates to the understanding and perspective of the system user since the perception of the personnel involved in the implementation of the system will take effect at the end of a system, whether the system is successful or not, acceptable or not, useful or not, if implemented. Furthermore, (Gil D.N., 2009) argues that the use of information technology by an individual, group, or organization is an important variable in the information system since its acceptance or rejection should be confirmed beforehand.

(Marchewka J.T., & Kostiwa K., 2014) define usefulness as a level in which an individual believes that the use of a particular subject will be able to improve his or her performance. Results of the present study also showed that information generated by the Traffic Ticket Data Processing Information Systems for management policy-making in the East Java Police has not been optimally utilized since there were still many obstacles and shortcomings in the implementation stage, leading to inability to support management policy making. A new information system has operational significance in the life of an organization if the system supports the decision-making process, which means that an information system must be capable of supporting resolution of problems faced by users of the system (Al-Mamary Y.H., et. al., 2013).

Management Information System constitute a tool to manage processes which provides information for managers to support operations (Al-Mamary Y.H., et. al., 2014). Additionally, management information system is also a formal method that provides accurate and timely information to management to facilitate decision making and to make the organization capable of performing the functions of planning, operating and controlling effectively (Peffers K., et. al., 2007).

CONCLUSION

Application of the Web-based Traffic Ticket Data Information System has not been running as expected due to: (a) with regard to the relationship between the technological aspects and the human aspect as user, the implementation of Web-based traffic ticket data information has not yet provided ease of use. The level of use of the Web-based traffic ticket data information system has not been maximized; (b) with regard to the relationship between the human aspect as user and the organizational aspects, the awareness and understanding of police officers were still low. The assigned officers did not carry out their tasks properly, resulting in sub-optimal implementation of the web-based traffic ticket data information system; (c) with regard to the relationship between the organizational aspects and technological aspects, the cooperation between units/divisions in terms of exchange of information has not been well-established due to the lack of a clear standard operating procedures.

Information generated by the Web-based Traffic Ticket Data Information System has not yet had the significant level of quality in terms of accuracy, completeness or timeliness of information. As end users, the public were not yet satisfied with the information generated by the implementation of the web-based traffic ticket data information system. The information systems generated could not function as an information management system since in the implementation stage there were still many obstacles and shortcomings, leading to inability of the information generated for use to support management decision and policy and making.

ACKNOLWEDGEMENT

We would like to thank all those who have helped this study, particularly the East Java Police Chief and the personnel directly related to the results of the study.

Reference

Al-Debei M.M., (2013). Developing and Implementing a Web Portal Success Model. *Jordan Journal of Business Administration*, *9*(1), 161-190.

Al-Mamary Y.H., Shamsuddin A., & Aziati N., (2013). The Impact of Management Information Systems adoption in Managerial Decision Making : A Review, *The International Scientific Journal of Management Information Systems*,. 8, (4), 010-017.

Au N., & Cheng, T. C. E. (2012). The formation of employee satisfaction with airline information systems. *Journal of Travel & Tourism Marketing*, *29*(4), 335-351.

Al-Mamary Y.H., Shamsuddin A., & Aziati N., (2014). The Role of Different Types of Information Systems in Business Organizations: A Review, *International Journal of Research,* 1 (7), 333-339.

Cheung C. M., & Lee M. K., (2008). The structure of Web-based information systems satisfaction: Testing of competing models. *Journal of the American Society for Information Science and Technology*, *59*(10), 1617-1630.

Choi Sue Young, Heeseok Lee, and Youngjin Yoo., (2010) "The Impact of Information Technology and Transactive Memory Systems on Knowledge Sharing, Application, and Team Performance: A Field Study." *MIS quarterly* 34.4: 855-870.

Davis M.E., & Phillips J.A., (2007). *Learning PHP & MySQL: Step-by-Step Guide to Creating Database-Driven Web Sites*. O'Reilly Media, Inc.

Gil D.N., (2009). Management Information Systems And Strategic Performance: The Role Of Top Team Composition, *International Journal of Information Management*, 29(2), 104-110.

Grispos G., Glisson W.B., Pardue J.H., and Dickson M., (2014). Identifying User Behavior from Residual Data in Cloud-based Synchronized Apps. *Conference on Information Systems Applied Research (CONISAR 2014)*, 6-9 November 2014, Baltimore Maryland, USA.

Hevner A., & Chatterjee S., (2010). *Design research in information systems: theory and practice* (Vol. 22). Springer Science & Business Media.

Iivari J., (2007). A paradigmatic analysis of information systems as a design science. *Scandinavian Journal of Information Systems*, *19*(2), 5.

Kumar P.M., (2014). Information Technology: Roles, Advantages and Disadvantages. *International Journal of Advanced Research in Computer Science and Software Engineering*, 4, (6), 98-109.

Marchewka J.T., & Kostiwa K., (2014). An application of the UTAUT model for understanding student perceptions using course management software. *Communications of the IIMA*, *7*(2), 10.

Peffers K., Tuunanen T., Rothenberger, M.A., & Chatterjee, S. (2007). A design science research methodology for information systems research. *Journal of management information systems*, *24*(3), 45-77.

Ullman L.E., (2003). *PHP and MySQL for dynamic web sites*. Peachpit Press.

Valade J. (2007). *PHP and MySQL for Dummies*. John Wiley & Sons.

Williams H.E., & Lane D., (2004). *Web database applications with PHP and MySQL*. O'Reilly Media, Inc.

Zajaczek J.E.W., Götz F., Kupka T., Behrends M., Haubitz B., Donnerstag F., & Becker H., (2006). eLearning in education and advanced training in neuroradiology: introduction of a web-based teaching and learning application. *Neuroradiology*, *48*(9), 640-646.

MATEC Web of Conferences **58**, 03016 (2016) DOI: 10.1051/matecconf/20165803016
BISSTECH 2015

THE DESIGN OF MOBILE APPLICATION FOR TEACHER AND PARENTS COMMUNICATION IN INDONESIAN SCHOOL

Sholeh Hadi Setyawan[1], Dhiani Tresna Absari[2], Susana Limanto[3*], Andre[4]

[1]Information System Department, University of Surabaya, Surabaya, Indonesia
[2]Information System Department, University of Surabaya, Surabaya, Indonesia
[3]Informatics Engineering Department, University of Surabaya, Surabaya, Indonesia
[4]Informatics Engineering Department, University of Surabaya, Surabaya, Indonesia
E-mail: susana@staff.ubaya.ac.id

ABSTRACT

One of the success factor to achieve education goals is the good communication between the student's parents and the school. Most of Indonesian schools use written communication in form of communicator book, renewed yearly when the student move to the new grade. All of the important information contained in the conversation will be lost if the book is missing, torn, damaged or replaced by the new book. The response time of the conversation is very slow, parent should wait the answer from the school until the student goes home. A mobile application is designed to replace the communicator book electronically. The application is hybrid, enables the parent to communicate easily to the class teacher, counselors, health services and also the school managements. The design has been reviewed dan feedback given from the users.

Keyword: Electronic communication, School communicator

INTRODUCTION

The good communication between parents and teachers plays important role to support the achievements of educational goals. Teachers need to daily inform the parents about the student's activities, achievements, homeworks, events and other official informations from the school. On the other hand, the parents also need to communicate to the school, either to the class teacher, the school counselor, the school administration or the principals about many aspects of the students. The regular bi-directional communication is required both to enhance academic achievements and students mental well-being (Thornton, 2011).

The media of the communication is usually a communicator book. The students should bring the book everyday. The teachers and the parents communicate by writing messages and informations on the book. The content of the book may vary for every school, but the main fields of information are the date of the message, subject, the message body, and the sign of both teacher and parent to verify that both party has already read the message or information.

Because of the nature of the book, some problems may arise. The book can be missing, left at home, torn or damaged. In that condition, the informations stored on the book will also be lost without any possibility to restore. Teachers or parents may have problem with writing by hand. The handwriting message may be difficult to read by other people. The result of this problem may lead to misunderstanding or miscommunication. The example of the written message is shown in Figure-1. The first column is the date of the message, followed by the message content, and the place for teacher and parent's signs.

The communication book is renewed yearly when the students enter the new level. All of the communication records stored on the old book are discarded. The parents usually should rewrite important messages to the new class teacher. On the other side, if the teacher need to send the same message to all of the parents in his class, he should write the same message on each student's book. Depends on the size of the class and the size of the message, this activity may consume the quality time that the teacher can use to do more important things to his class.

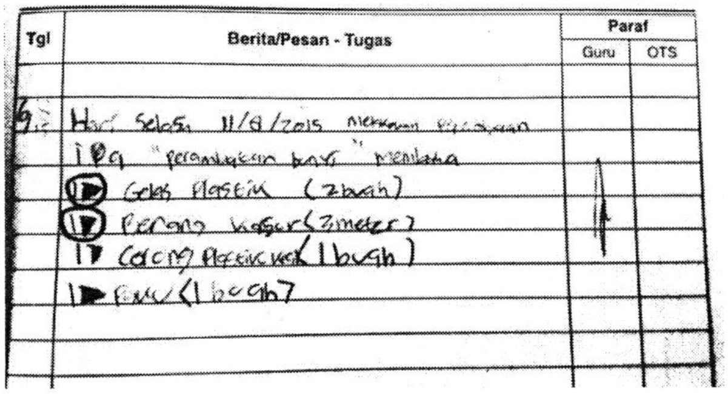

Figure-1. The content example of communication book

MOBILE APPLICATIONS IN EDUCATION

The rapid growth of smart phones and PC tablets usage drives the mobile application development. The mobile applications have the main different characteristics such as the less dependent of space and time, more personalization, and more integrated to other mobile application and services.

Among the many fields of the mobile application development, one of the emerging field is the education. Recent studies and developments of mobile application in education are educational game (Zhang and Loeb, 2013), mobile learning system (Petrova, 2007; Sung, et.al., 2005) and interactive learning table (Wilson, 2011). Studies show that mobile applications give better result than desktop applications in language learning (Jarvis and Achilleos, 2013), help to get better skills in math and writing (Aker, Ksoll and Lybbert, 2011).

Many mobile communication applications are available for free e.g. Whatsapp Messenger, Blackberry Messenger, Google Hangout, etc. and may replace the communication book. But because the applications are not integrated to the school database (teachers, students, classes, and parents records), the messages can not be furthermore processed by the school information system. Therefore the communication application can't contribute to the school data to be utilized for decision support system by the school management.

REQUIREMENT ANALYSIS

A requirement analysis should be performed before designing the system, to assure that the design meets the actual requirements of the system users. The analysis begin with a survey to the parents. A series of questions were asked to identify the problems to be solved, the requirements to be implemented and the expectation of the mobile application. The answers were processed and analyzed to generate the list of parents' requirements. The respondents were the parents of sampled play group, kindergarten, primary school and junior high school in Surabaya, Indonesia.

The other survey also conducted for the class teachers and the school managements, to identify the requirements of the school users. The answers of the questionnaires were processed and analyzed. Together with the parent requirements, a list of user requirements was written and become the main input of application design.

The common requirements of the mobile application are:

- The application runs on Android smart-phones
- The application should notify the user when there is a new message or reply. The notifications are also copied to the user's email.
- The user accounts and credentials are generated by system, using email address as the login name. The password is randomly generated, and sent to the user's email address. User can change the password after successfully log in to the system
- The user can search the messages by keywords, the date, or the destination
- The message can be marked as "urgent"
- "Calendar" view, so the messages can be grouped and searched by date

Parents need to communicate with the school for several reasons like :

- lack of information details from school letters, reports or announcements.
- To inform the student's condition if required, i.e when the students were in recovery condition after sick, when the student need medical attention from the teacher or schooll health services.

The parents requirements for the communication mobile application are :

- The parent may write messages not just to class teacher, but also to school management, school counselors, or school health services person.
- The parent can attach images taken from the mobile device
- The parent may have more than one students in the school. The conversation should be displayed separated for each student

The requirements identified from the teachers and the school management are:

- Teacher can write message to a single parent, multiple parents or all of the parents in his class in a single process
- There are 3 types of informations sent by the teacher: normal message, announcement, or letter from the school.
- Teacher can attach file to the message i.e. images, office documents, etc.
- The teacher can forward the message from the parent to another teacher or school management.
- The teacher can also access the system using web browser, especially to upload files from his computer.
- The teacher can forward a message from the parent to his colleague or to the school managements if needed
- The teacher can re-post old messages
- The teacher make carbon copy to other parents

THE SYSTEM DESIGN

After examining the requirements, the mobile application should be accessible from web browser, to facilitate the teachers to upload files from their computers. Therefore the application should be hybrid, developed using hybrid framework such as Phonegap (Shotts, 2014).

There are 13 entities designed for the application, as shown in Figure-2. and listed in Table-1.

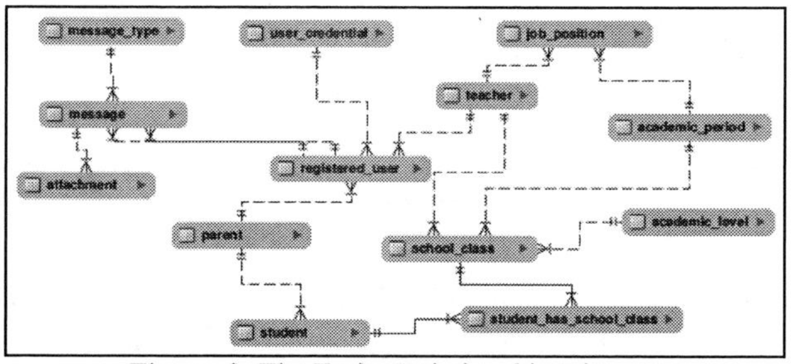

Figure-2. The Entity Relationship Diagram

DOI: 10.1051/matecconf/20165803016

Table-1. System entities

No.	*Entity Name*	*Description*
1.	message_type	Define the types of messages, i.e. announcement, conversation, notice, letter from school, etc.
2.	message	The messages between parents and schools
3.	message_attach ment	Define the multimedia attachement (if required) for each message
4.	registered_user	The central storage for usernames and passwords for each registered user
5.	user_credential	Define the access rights grouping for the users
6.	parent	
7.	student	
8.	teacher	
9.	academic_level	The level/grade of students
10.	academic_period	Yearly set up
11.	job_position	The job position of teachers during an academic_period
12.	school_class	The class opened for each academic period
13.	Student_has_sch ool_class	The assignment of each student for each shool class

The mobile application was designed for several users to handle processes as shown in Figure-3. The users of the applications are the system administrator, the school management, the health services person, the school counselors, the class teachers and the parents.

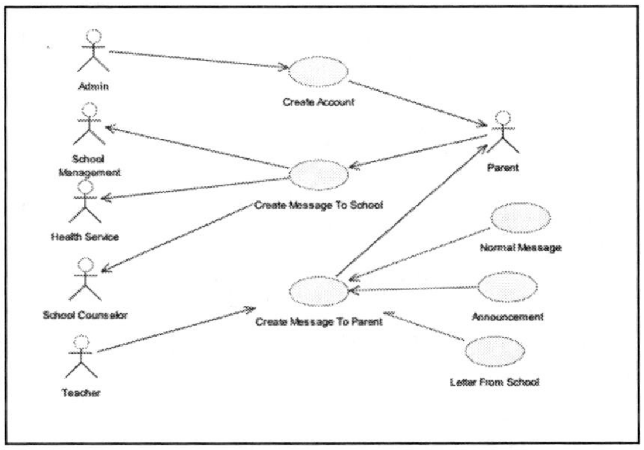

Figure-3. The use case diagram

The mobile application was designed to best suite the major smartphone screens around 4 – 7 inches, using Android Operating System User Interface standards (Lee, 2012). The design of the user interface includes :

1. Login screen: the first screen to show when user first accesses this application, as shown in Figure-4.

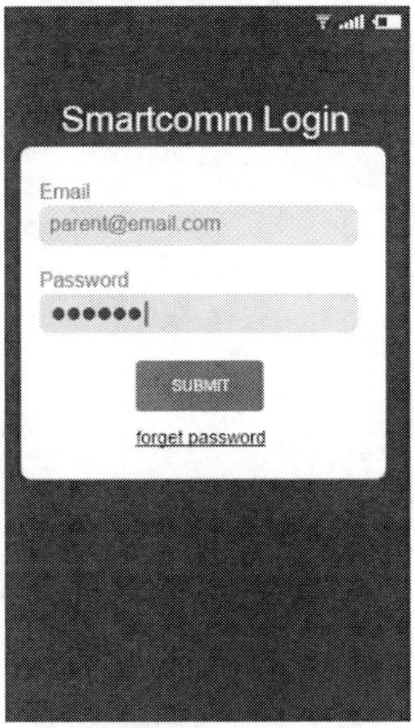

Figure-4. The login screen

2. Inbox : the list of incoming messages, as shown in Figure-5.

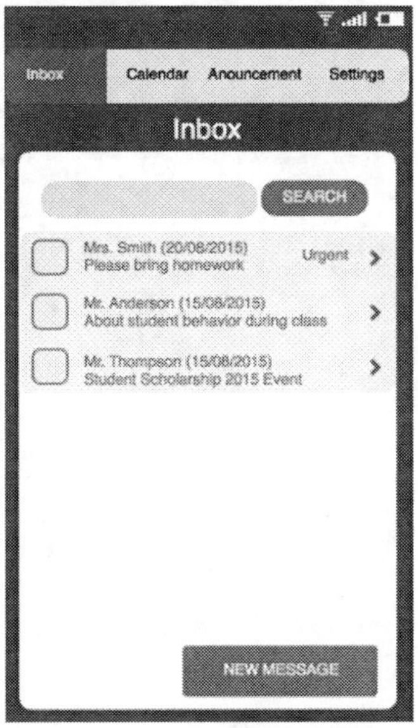

Figure-5. The inbox screen

3. Conversation screen : the time-line conversation messages between parent and the school, as shown in Figure-6.

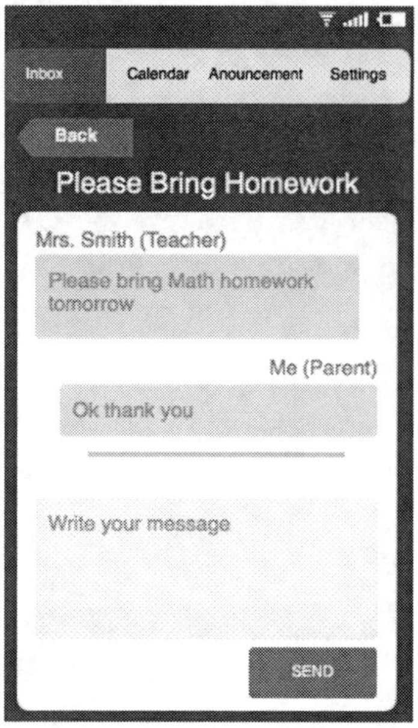

Figure-6. The Conversation Screen

4. Calendar view: to facilitate searching messages based on the date, as shown in Figure-7.

Figure-7. The Calendar View Screen

5. New message creation: user can choose the destination of the message, write the subject and the content, as shown in Figure-8.

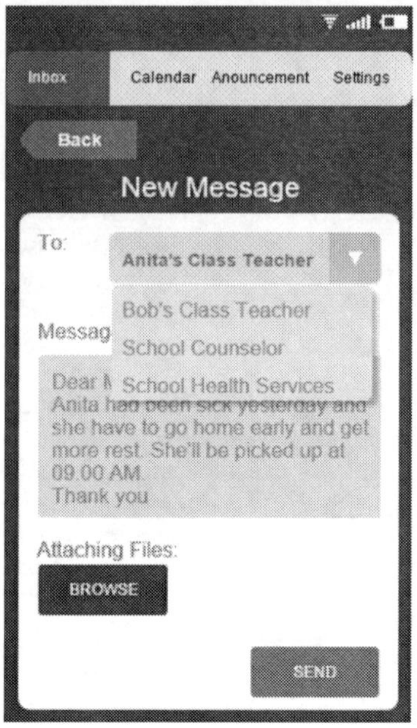

Figure-8. The New Message Screen

Conclusion

The system design validation was done by conducting focus group discussion involving the school managements, the class teachers, the school counselors, the school health services and the parents. The feedback from the respondents about the features and the interfaces has been implemented on this design, and reviewed by the related users.

References

Aker, J.C., Ksoll, C., and Lybbert, T.J., 2011. *Can Mobile Phones Improve Learning? Evidence from a Field Experiment in Niger*. Medford, Mass., Tufts University; Oxford, UK, University of Oxford; Davis, Calif., University of California, Davis. http://www.aae.wisc.edu/mwiedc/papers/2011/Aker_Jenny.pdf [Accessed September 2015].

Jarvis, H., and Achilleos, M., 2013, From Computer Assisted Language Learning (CALL) to Mobile, *The Electronic Journal for English as a Second Language*, 16(4), [online], Available at : <http://www.tesl-ej.org/wordpress/issues/volume16/ej64/ej64a2/> [Accessed August 2015]

Lee, W.M., 2012, *Beginning Android 4 Application Development*. [e-book] Indianapolis:John Wiley & Sons Inc. Available through : <blogs.wrox.com/files/2013/05/Chapter-8-9781118087299-2.pdf>

Petrova, K., 2007,. Mobile learning as a mobile business application, *International Journal of Innovation and Learning*, pp 1--13

Shelley R.T., 2011. *Supporting children's mental well-being in primary school: problem solving through communication and action*. PhD. Southern Cross University, NSW.

Shotts, K., 2014, *PhoneGap 3.x Mobile Application Development Hotshot*, Brimingham:Packt Publishing.

Sung, M., Gips, J., Eagle, N., Madan, A., Caneel, R., DeVaul, R., Bonsen, J., and Pentland, A., 2005, Mobile-IT Education (MIT. EDU): M-Learning Applications for Classroom Settings, *Journal of Computer Assisted Learning 21*, pp.229–237

Wilson II, G., 2011. *Experiences with the Mobile Interactive Learning Table.* Master of Science in Computer Science and Application. Virginia Polytechnic Institute and State University.

Zhang, E.Y., and Loeb, L., 2013. *Mobile Applications: Games that Transform Education* .[pdf], Hanover : Dartmouth Computer Science, Dartmouth College, Available at : <http://www.cs.dartmouth.edu/reports/TR2013-737.pdf > [Accessed September 2015]

A COMPARATIVE STUDY OF FINITE JOURNAL BEARING IN LAMINAR AND TURBULENT REGIMES USING CFD (COMPUTATIONAL FLUID DYNAMIC)

Susilowati[1,2] , Mohammad Tauviqirrahman[2], J. Jamari [2], Athanasius P. Bayuseno[2] and Muchammad[3]
[1]Universitas Pembangunan Nasional "Veteran" Jawa Timur, Surabaya-Indonesia
[2] Universittas Diponegoro, Semarang-Indonesia
[3] University of Twente, The Netherlands
E-mail: zuzisukasno@gmail.com

ABSTRACT

In this paper, an investigation of hydrodynamic performance for journal bearing lubricated by thin film is performed by three-dimensional CFD (computational fluid dynamic). Two regimes, i.e. laminar and turbulent regimes are of main interest. The sliding velocity was varied from 3000, 5000 and 10000. The numerical simulation shows that the velocity magnitude of the journal has a strong effect on the hydrodynamic pressure. In addition, it is found that an appropriate modelling of flow regime affects the predicted lubrication performance. From the overall analysis, both using laminar and turbulent regimes, the distribution of static hydrodynamic pressure shows a similar trend. It is also found that at the beginning of the contact, the static pressure is low, and then, it gradually increases to the highest point, and finally drops significantly.

Keywords: Computational fluid dynamic (CFD), journal bearing, lubrication, turbulent

INTRODUCTION

Recently, a thin film lubricated journal bearing has attracted much attention. Following the progress in numerical analysis and technology, many researchers began to use commercial computational fluid dynamics (CFD) program in their investigations. The main advantage of CFD code is that it uses the full Navier – Stokes equations and provides a solution to the flow problem based on finite volume method (Panday et al., 2012). Therefore, the CFD methods are applicable in very complex geometries.

Gertzos et al., (2008) worked on CFD method of lubricated journal bearing by Bingham lubricant. They solved Navier-Stokes equations by using the FLUENT package and compared the results of the developed 3-D CFD model with theoretical and experimental results of previous investigations, for both Newtonian and Bingham lubricants, and found to be in a very good agreement.

Stefani and Rebora (2009) worked on using Quasi-3D mass-energy-conserving method by extending a previous mass and energy-conserving algorithm. The proposed method produced temperature profiles in a good agreement with the experimental results, whenever consistent boundary conditions were induced. Taking into account the thermal deformations of the kinematic pair, it leads us to improve the reliability of the model.

Gururajan and Prakash (2002) investigated the effect of surface roughness in a narrow porous journal bearing. It was shown that for a smooth case, the results based on an approximate analysis could be safely used for thin-walled bearings of low permeability.

Peng et al., (2007) developed a new form of the Reynolds equation. The new Reynolds equation for the lubricating film was solved by using Finite Volume (FV) discretization. According to the numerical solution, the journal squeeze effect and the pad rotation effect on film force were discussed.

Basri and Gethin (1990) principally described a theoretical analysis of the thermal behavior of offset half, orthogonally displaced, three-lobe and four-lobe bearing geometries. Murthy et al., (1983) dealt with the analysis of a special hydrodynamic bearing incorporated in a precision cylindrical grinding machine.

For unsteady analysis, Krasowski (2011) studied laminar flow during periodic and unperiodic perturbations of bearing load. It was found that perturbations mostly occured during the starting and stopping of the machine. Reynolds Equation solution for unsteady laminar newtonian flow of lubricated oil was presented to enable initial opinion to hydrodynamic pressure and velocity distribution as a basic slide bearing operating parameter.

Sharma et al., (2011) studied on film lubrication performance to a certain degree. The finite element method was theoretically studied with respect to the performance of four- containers hydrostatic tapered bearing system.

Roy (2009) investigated thermo-hydrodynamic analysis of oil journal bearing and made a comparison at five different feeding locations of an axial grooved. Reynolds equation was solved simultaneously along with the energy equation and heat conduction equation in bush and shaft.

Based on the survey of literature, the analysis mostly neglects the cavitation phenomena which may be present in analysis. In addition, laminar regime is a particular interest. This paper is aimed at analyzing the journal bearing using laminar and turbulent approaches by employing CFD. This method is applicable by using cavitation model, hence it is more realistic to simulate the situation.

METHODOLOGY

This research uses laminar and turbulent regimes and makes a comparison between them to investigate

DOI: 10.1051/matecconf/20165804001

which best model to represent the phenomena. The sliding velocities used are 3000 rpm, 5000 rpm, and 10000 rpm. Operational pressure is set to 101325 Pa. Whereas, the ratio of L/D is 0.25. The 3-D simulation model is developed by using the CFD package FLUENT.

Mathematical model

The equation for mass conservation, or the continuity equation, can be writen as:

$$\frac{d\rho}{dt} + \nabla.(\rho\vec{\vartheta}) = 0 \qquad (1)$$

Eq.(1) is the general form of the mass conservation equation, and it is valid for both incompressible and compressible flows. Momentum conservation equations (the Navier Stokes equations) Conservation of momentum in an inertial (non-accelerating) reference frame is described by:

$$\frac{\partial}{\partial t}(\rho\vec{\vartheta}) + \nabla.(\rho.\vec{\vartheta}.\vec{\vartheta}) = -\nabla p + \nabla.(\bar{\bar{\tau}}) + \rho\vec{g} + \vec{F} \qquad (2)$$

Where $\rho\,\vec{\vartheta}$ and $\vec{F}$ are the gravitational and external body forces, respectively. The stress tensor τ is given by

$$\bar{\bar{\tau}} = \mu_a\left[(\nabla\vec{\vartheta} + (\nabla\vec{\vartheta})^T) - \frac{2}{3}.\nabla\right].\vec{\vartheta}I \qquad (3)$$

where the second term on the right hand side is the effect of volume dilation. All investigations that are referred in this work solve the fundamental Reynolds Eq. (4), or a modified form of (4) that takes into account the equations describing the non-Newtonian lubricant,

$$\nabla.\left(\frac{h^3}{12\mu}.\Delta\rho\right) = \nabla.(h\vec{U}) - V \qquad (4)$$

RESULTS

The simulation results are divided into main categories, i.e. laminar and turbulent analyses. This comparison would give which best model in predicting the characteristic of lubricated bearing.

Laminar Approach

Figures 1-6 show the static hydrodynamic pressure of the lubricant in the journal bearing by varying the rotational speed of 3000, 5000, and 10000 rpm for the laminar analysis. For the laminar approach for the rotational speed of 3000 rpm (Figures 1 and 2) from the static pressure, it can be concluded that zero pressure is present at inlet area. The score ranges around 640 to 8000 Pa. Whereas, for the rotational speed of 5000 rpm laminar (Figures 3 and 4), the result is almost the same with the rotational speed of 3000 rpm. The only difference is the lower score range of contour static. This also applies for the rotational speed of 10000 rpm (Figures 5 and 6). The result is the same with the rotational speed of 3000 rpm and 5000 rpm laminar.

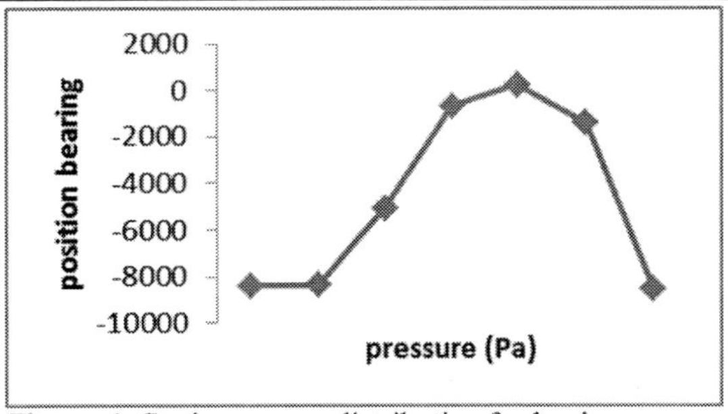

Figure-1. Static pressure distribution for laminar condition at 3000 rpm.

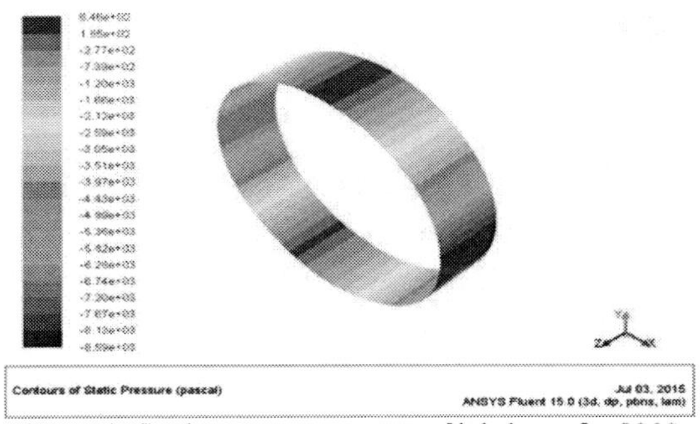

Figure-2. Static pressure contour of lubricant for 3000 rpm at laminar regime.

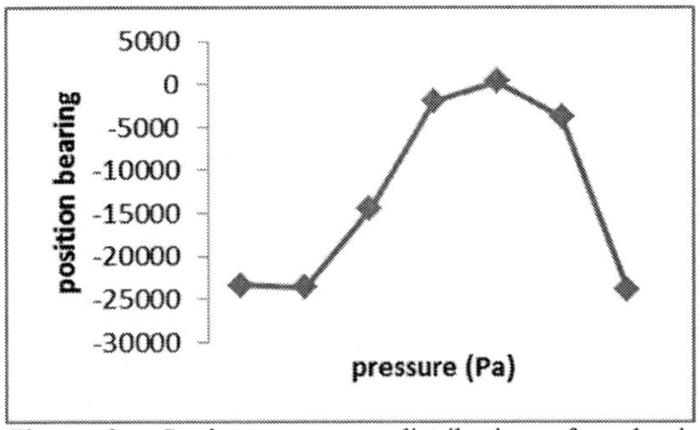

Figure-3. Static pressure distribution for laminar condition
at 5000 rpm.

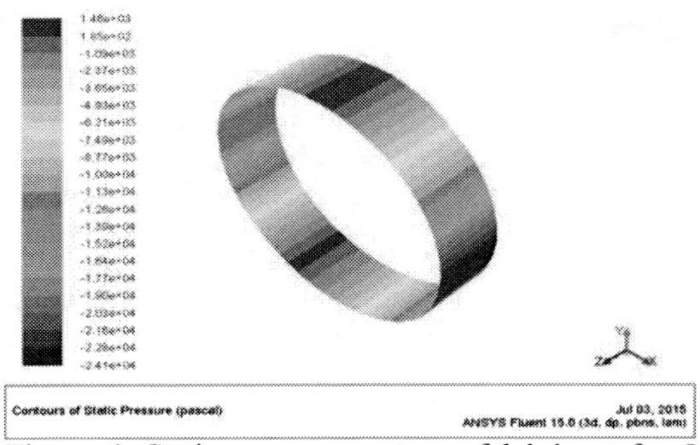

Figure-4. Static pressure contour of lubricant for 5000 rpm at laminar regime.

 DOI: 10.1051/matecconf/20165804001

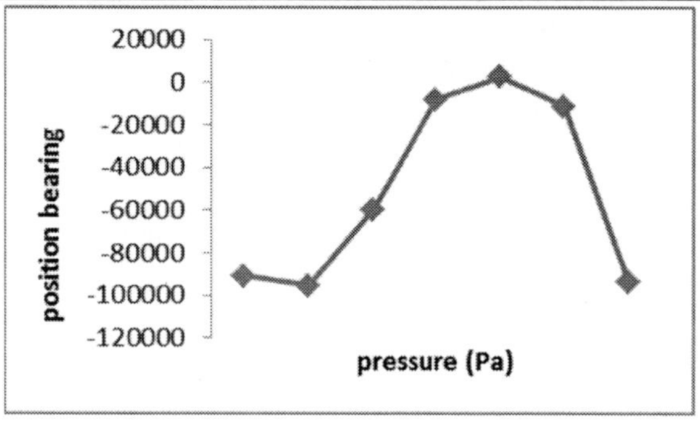

Figure-5. Static pressure distribution for laminar condition at 10000 rpm.

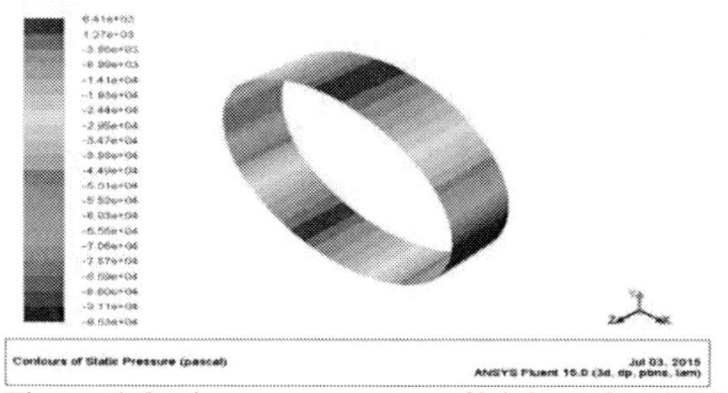

Figure-6. Static pressure contour of lubricant for 10000 rpm at laminar regime.

Turbulent Approach

Figures 7–12 depict the static pressure of the lubricant in the journal bearing by varying the rotational speed of 3000, 5000, and 10000 rpm for turbulent analysis. In the present study, *k*-epsilon model is used to represent the turbulence. It can be observed that for the rotational speed of 3000 rpm (see Figures 7 and 8), compared to the laminar pressure static, the analyzed score of pressure static using turbulent flow is lower. Most likely, this result comes from the presence of inertia. It is known that inertia is the most significant effect in contribution to make the lubricant very turbulent. Particularly from Figure 8, it can be concluded that the contour of the static pressure is more evenly distributed compared to the laminar flow.

The turbulent modelling approach for the rotational speed of 5000 rpm (Figures 9 and 10) shows that the pressure static is the same with the rotational speed of 3000 rpm turbulent. Similarly, the contours of pressure static are also more evenly distributed. Figures 11 and 12 indicate that the maximum pressure for the rotational speed of 10000 rpm turbulent is a little bit higher compared to the lower rotational speed. Whereas, particularly from Figure 12, the contours are more similar to the contour predicted by that lower rpm.

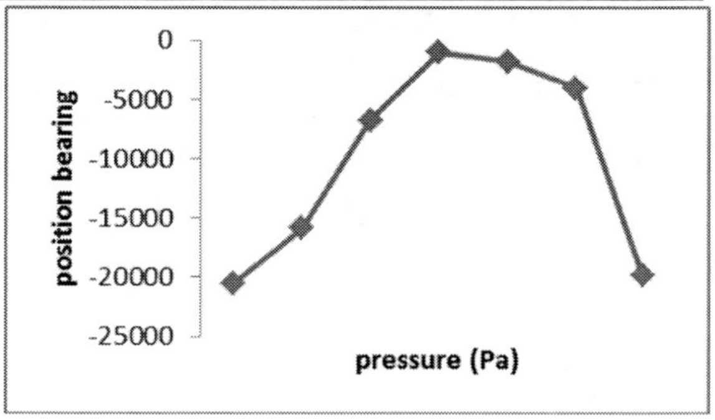

Figure-7. Static pressure distribution for turbulent condition at 3000 rpm.

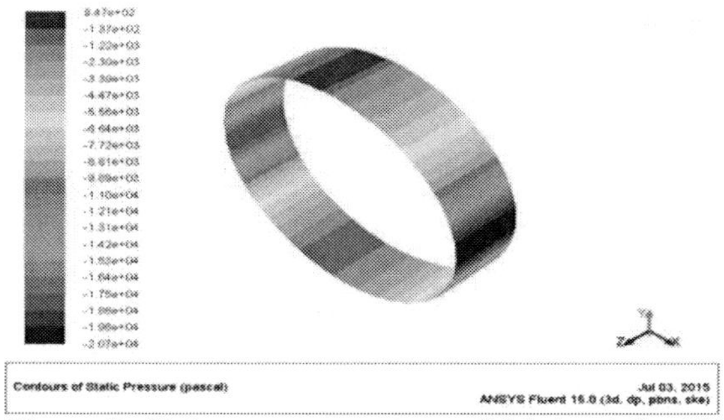

Figure-8. Static pressure contour of lubricant for 3000 rpm at turbulent regime.

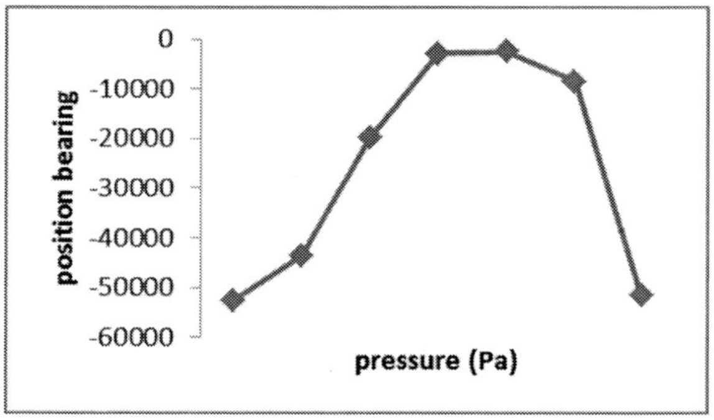

Figure-9. Static pressure distribution for turbulent condition at 5000 rpm.

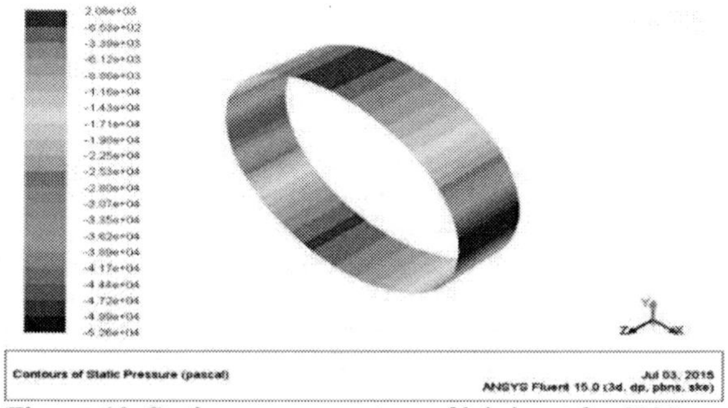

Figure-10. Static pressure contour of lubricant for 5000 rpm at turbulent regime.

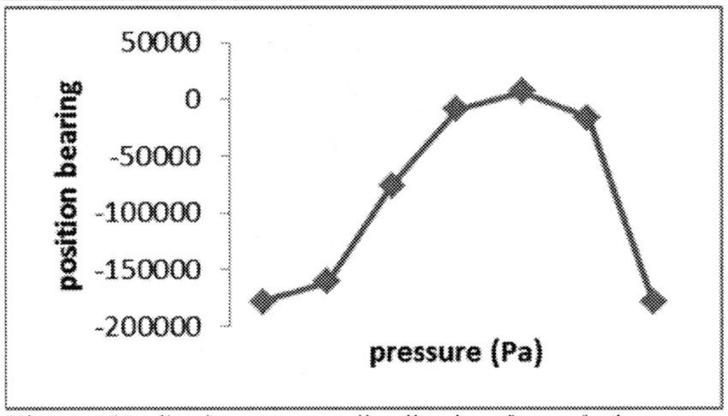

Figure-11. Static pressure distribution for turbulent condition at 10000 rpm.

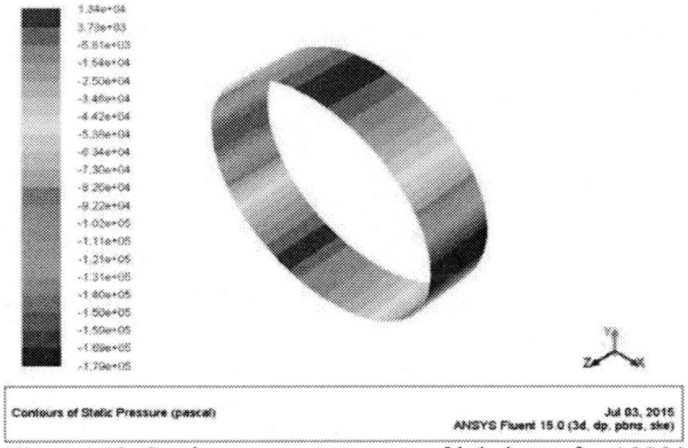

Figure-12. Static pressure contour of lubricant for 10000 rpm at turbulent regime.

DISCUSSIONS

As it is known, the journal bearing is one of the main components in the thermal fluid system. The load support can be generated by integrating the hydrodynamic pressure over the surface area. Further, this pressure generation can be analyzed by two different approaches, turbulent and laminar. In the real application, it is difficult to determine what appropriate model to use in analyzing the bearing characteristic. Therefore, the main contribution of this research is that by taking into account the cavitation effect, the turbulent as well as the laminar modelling is compared with respect to the hydrodynamic pressure.

From the overall analysis mentioned in the previous section, the distribution of static pressure shows the similar trend, which is: at the beginning, the pressure static is low, then, it gradually increases to the highest point, and finally drops significantly.

CONCLUSIONS

This paper presents the result of journal bearing simulation by using laminar and turbulent analyses. From the explanation, it can be concluded that both laminar and turbulent flows show the almost similar trend. It indicates that in the journal bearing, the turbulent effect can be ignored.

REFERENCES

Basri S. and Gethin D.T. 1990. A comparative study of the thermal behaviour of profile bore bearings.Tribology International, 23: 265-276.

Gertzos K.P., Nikolakopoulos P.G. and Papadopoulus C.A. 2008. CFD analysis Of journal bearing hydrodynamic lubrication by Bingham lubricant. Tribology International, 41: 1190-1204.

Gururajan K. and Prakash J. 2002. Roughness effects in a narrow porous journal bearing with arbitrary porous wall thickness. International Journal of Mechanical sciences, 44: 1003-1016.

Krasowski P. 2011. Velocity distribution in slide journal bearing gap by laminar unsteady lubrication. Journal of Kone Powertrain and Transport, 18(3): 177-184.

Murthy T.S.R., Balaramaiah Y. and Venkatesh V.C. 1983. An analysis of a special hydrodynamic bearing for machine tool spindles. Manufacturing Technology. 32(1): 319-325.

Panday K.M., Choudhury P.L. and Kumar N.P. 2012. Numerical Unsteady Analysis of Thin Film Lubricated Journal Bearing, 4(2) : 185-191.

Peng Y., Chen X., Zhang K. and Hou G. 2007. Numerical Research on water guide bearing of hydro-generator unit using finite volume method. Journal of Hydrodynamics, 19: 635-642.

Radfordand A. and Fitzgeorge D. 1977. The effect of journal lobing on the performance a hydrodynamic plain journal bearing wear, 45: 311-322.

Roy L. 2009. Steady state thermo-hydrodynamic analysis and its comparison at five different feeding locations of an axial grooved oil journal bearing is obtained theoretically, Tribology International, 42(8) : 187-1198.

Sharma S., Satish C., Phalle.,Vikas M. and Jain S.C. 2011. Performance analysis of a multirecess capillary compensated conical hydrostatic journal bearing. Tribology International , 44(5): 617-62.

Stefani F. and Rebora A. 2009. Steadily loadedjournal bearings:Quasi-3D mass-energy-conserving analysis. Tribology International, 42: 448-460.

DOMESTIC WASTEWATER TREATMENT MINIPLAN OF INSTITUTION USING A COMBINATION OF "CONETRAY CASCADE AERATOR" TECHNOLOGY AND BIOFILTER

Naniek Ratni Juliardi A.R*, Rudy Laksmono, Firra Rosariawari, Erwin Chandra
Environmental Engineering Department-University of Pembangunan Nasional "Veteran" Jawa Timur
Surabaya, East Java, Indonesia
E-mail: naniktlupn@yahoo.com

ABSTRACT

In Indonesia, the contamination from domestic wastewater is becoming the greatest pollutant and reaches up to 85% that goes into the body of water. This might worsen the river water quality. To anticipate this condition, there should be some way out, by making use of wastewater as a source of energy. One of those systems is by employing Cone-tray Cascade Aerator combined with Biofilter by using Hydrilla plant which can be used as an alternative of waste water treatment for both domestic and institutional scale. This research was aimed at determining the system capability of aeration combined with biofilter by using hydrilla plant to reduce or eliminate the pollutant load of BOD, COD, and TSS. This research was conducted in a continuous stream of water at 200L/hour. The result showed that the best system of Conetray Cascade Aerator combined with biofilter could reduce BOD as much as 69,11%, COD at 61,14%, and TSS at 65,45%.

Keywords: Domestic Wastewater, Aeration, Conetray Cascade Aerator, Biofilter, Hydrilla.

INTRODUCTION

Water is one of the main components to support the life of all living things. For human beings, water is one of the basic needs because they always need and use water. Therefore, the more population of the people the more water required. This fact makes the increase of pollution and the decrease of water quality due to the increase of the people's activity that might cause the danger of life's continuance of many kinds of aquatic biota. This might endanger human being's life. One method that might increase the water quality is by increasing the oxygen content in the water. Conetray Cascade Aerator is one instrument which is capable of dissolving oxygen into the water through air contact gravitationally by means of some steps.

Conetray Cascade Aerator is one type of gravity aerator, a kind of aeration that the way it works is based upon gravitation power. Meanwhile, the water that will be aerated will flow gravitationally due to the different height from one step to another. In every step in the water, it will occur some contact between BOD, COD, and TSS with oxygen so that there will occur oxidation reaction. The more steps the more perfect the oxidation reaction will be. Basically, the aerator consists of 4-6 steps, where every step has its height of 30 cm with its capacity about 0,01 m3/second per m2 to relieve rotation (turbulence) in order to increase the aeration efficiency. However, some hindrance often occurs on the edge of the instrument in every step. Moreover, the advantage of employing Conetray Cascade Aerator is that it does not need any treatment. Meanwhile, the weakness/disadvantage is that it needs a larger place to put on. Some factors that affect the process of Conetray Cascade Aerator are the height and the number of Cascade, the width of container and number of plants (Benefield,1990).

According to the research of Hartini E (2012), the use of one side of the cascade aerator can reduce 98.74% Mn, and the use of bubble aerator can reduce 76.47% Mn. COD parameter can also be lowered combination of aeration processes, adsorption, and filtration with a flow rate of 228.33 ml/min, the initial results of 323.55 mg/lt to58.95 mg/lt (Endahwati L. and Suprihatin, 2013)

Some results of domestic wastewater treatment with aeration and the addition of pseudomonas putida bacteria produce an 89.19% reduction in BOD with oxygen of 2 liters/min, time of 216 hours, and 10 cells/ml (Wibowo ME and Sajidan W, 2006). The process of oxidation pond aeration system - constructed wetland can also reduce COD (Xiaoou Wang et al, 2015).

Fate of the low arsenic concentrations during full-scale aeration and rapid filtration (JCJ Gude et al, 2015). According to John Colt and Eric Kroeger (2013), in general, the aeration process can also improve water quality. Moreover, when combined with vegetation treatment (Phragmites australis) in the scale of wetland, it can reduce COD (Henric Svensson et al, 2015). Also, Enrique J. La Motta(1996) said a pilot scale tray tower was built at The University of New Orleans urban waste management and research center to study the factors affecting carbon dioxide removal by plain aeration.

A feasibility study of iron-mediated aeration (IMA) was conducted by Yang Deng et al (2013) for treatment of secondary effluent, natural surface water, and simulated contaminated water, which achieved the following removal efficiencies: 82.2% of glyoxylic acid, 100% formaldehyde as an oxidation product of glyoxylic acid, 94% of Ca2þ and associated alkalinity, and 44% of chemical oxygen demand (COD).

According to Senlin Shao et al (2015), long sludge retention time (SRT) and continuous aeration are generally adopted in powdered activated carbon membrane bioreactor (PAC-MBR) process for micropolluted surface water treatment.The results showed that the intermittent aeration with proper aeration intervals barely reduced the removal efficiency of pollutants, due to the low DO consumption in the bioreactor and the low settling velocity of the sludge (mostly at the range of 0.04-0.08 m min^{-1}).

Domestic Waste Water

Originally, waste is classified into 3 (three) kinds: agricultural, industrial, and domestic waste. Agricultural waste is originated from pesticide or insecticide use. Plant waste, or more often is called industrial waste, is the waste resulted from the production process. Meanwhile domestic waste or it is often called household waste is the waste from households (Kusno P H,1986).

Working Principle of Conetray Cascade Aerator

Aeration is another term of gas transfer which is more specialized on the gas transfer of oxygen or process of oxygen addition into the water. Meanwhile, the success of an aeration process depends much on the temperature, oxygen saturation, and the characteristics and turbulence of the water. Meanwhile, the main function of aeration in water treatment is to dissolve the oxygen into the water to increase the content of dissolved oxygen in the water. Moreover, the suspended mixture of active sludge in bioreactor releases the content of dissolved gasses in the water and helps in water stirring. In an aeration process, there are some factors affecting the oxygen transfer, such as Temperature, Oxygen Saturation, Characteristics of Water, and Water Turbulence.

Moreover, in aeration process of conetray cascade aerator, the domestic waste water of institution is collected in one tank or container then let flowed through each of the 5 (five) traps.The water that has passed by each trap will pass through the bio-filter process available in hydrilla plant that has already been determined its number and weight. The water flowed will go to the installation provided for filtration (Benefield, 1990).

METHODOLOGY

Materials and Tools / Instruments

The materials used in this research covered:

- Domestic waste water of institution
- Hydrilla

Meanwhile, the tools required were:

- Conetray Cascade Aerator
- Valve
- Pump
- Container / Water Reservoir

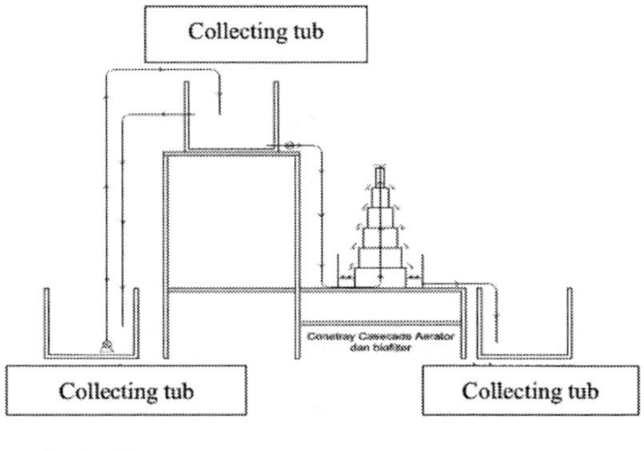

Figure-1. Tool/Instruments.

Variable

The variable used in this research was established variables and treatment variables.

a. The variables established were:
   - Rate of flow/discharge of waste water in the container of 200 liters/hour = 0.0033 m^3/Minute
   - The diameter of basin/container is 0.5 m
   - The number of Cascade: 5 traps
   - The Weight of Hydrilla

b. Treatment Variables
   - Total number of Plants: 0, 5, 10, 15, 20
   - Sampling Time: 5 minutes, 10 minutes, 15 minutes, 20 minutes, 25 minutes

Working Procedure

Domestic wastewater of institution available in the first water reservoir was pumped into the second reservoir then it was brought down by gravity by controlling the rate of flow from the faucets. The waste water will be picked up to the highest cascade, then it will drop by gravity to pass through each of trap or cascade. After passing through all cascades, the water would touch the hydrilla plant reaching to bio-filtration. Then water would come to the outlet to be collected in one reservoir. The analysis in this research was aimed at knowing the effectiveness of parameter decrease of BOD, COD and TSS in the domestic wastewater of institution by employing Conetray Cascade Aerator combined with biofilter by using Hydrilla.

RESULTS AND DISCUSSION

The variable employed in this research was dealing with the number of plants and single living/growing time. In 'without plant' process, it was required 5 plants of hydrilla, 10 plants of hydrilla, 15 plants of hydrilla, and 20 plants of hydrilla with some variations of living time: 5 minutes, 10 minutes, 15 minutes, 20 minutes, and 25 minutes.

Initial Characteristics of Domestic Wastewater.

The initial condition of domestic wastewater used in Conetray Cascade Aerator process was combined with the biofilter that had a meaning that the analysis was conducted before the process was begun. This was meant to obtain some initial characteristics description of the material to be processed in Conetray Cascade Aerator which would be combined with a biofilter. The analysis result of initial characteristics could be seen in table 1 below.

Table-1. Initial Characteristics of Domestic Wastewater of Institution.

Number	Analysis Result		
	COD mg/L	TSS mg/L	BOD mg/L
1	176	440	137
2	376	468	342
3	324	313	267

(Sources: Analysis Result of Laboratory, 2015)

Table 1 above showed that BOD, COD, and TSS had not been appropriate with the criteria of quality standards of domestic wastewater (East Java Governor's Regulation No.72, 2013).

In this research, the variable employed was ratio number of plants and time controlling. Some processes are without any plants, 5 plants of Hydrilla, 10 plants, 15 plants, and 20 plants with time variations of 5 minutes, 10 minutes, 15 minutes, 20 minutes and 25 minutes. The domestic wastewater that the institution used in Conetray Cascade Aerator process at the initial condition had the meaning that the analysis was conducted before the process was started in order to produce the initial description of material characteristics to be processed in Conetray Cascade Aerator. Based on table 1 above, it could be seen that the ratio of COD, BOD and TSS was still appropriate with the criteria from the quality standard of domestic waste water (the decision of State Minister of Environment, no. 112, 2003).

Having some test of oxygen transfer or aeration been conducted by employing Conetray Cascade Aerator tool combined with biofilter using hydrilla plants, it could be seen that the result of aeration showed there was some fluctuation in the decrease of BOD, COD, TSS in domestic wastewater.

The Effect of Hydrilla Quantity and Living Time in BOD Elimination.

Variable Variation of Hydrilla Quantity and Living Time could be seen in Table 2 below:

Table-2. Elimination Percentage of BOD

Time (minute)	Elimination Percentage of BOD (%)				
	Without Hydrilla	5 Hydrillas	10 Hydrillas	15 Hydrillas	20 Hydrilas
5	25,32	51,16	2,92	28,36	11,98
10	37,88	54,96	13,45	40,64	24,71
15	44,31	57,08	14,32	47,07	33,33
20	47,37	58,68	26,61	56,43	42,69
25	47,51	66,21	40,64	69,11	67,41

(Sources: Analysis Result of Laboratory)

Based on Table 2 on 'The Effect of Hydrilla Quantity and Living Time in BOD Elimination', it could be described in the following chart/graph:

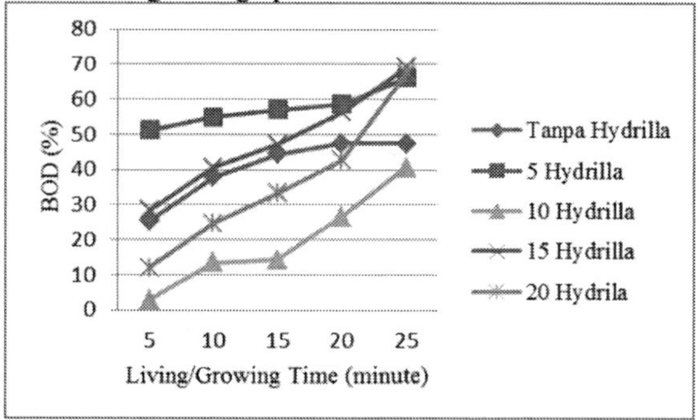

Figure-2. The Relationship between Elimination of BOD (%), Living/Growing Time (minute) and the Number of Hydrilla.

The parameter used in figure-2 was a process of conetray cascade aerator combined with a biofilter. Based on our observation, this is the best score of BOD. In accordance with the effect of hydrilla quantity to the decrease of BOD, it was found that the best variable was that of employing 20 plants of hydrilla with its living time of 25 minutes which resulted in its BOD elimination reaching up to 67,41%.

BOD could decrease in accordance with the number of cascade and duration (time length) of aeration process. This was due to the fact that the longer the aeration time required, the more oxygen to be transferred. This might cause the decomposition increase from organism population to grow better. Moreover, Unus Suriawiria (2008) said that biofilter was a group of microbes, like bacteria and fungus, which was capable of decomposing some organic and inorganic substances available in the wastewater by employing some certain plants.

The Effect of Hydrilla Amount and Living/Growing Time to COD Elimination.

Table-3. Elimination Percentage of COD

Time (minute)	Elimination Percentage of BOD (%)				
	Without Hydrilla	5 Hydrillas	10 Hydrillas	15 Hydrillas	20 Hydrilas
5	21,45	43,27	4,25	30,58	12,11
10	34,54	47,63	6,11	36,71	14,64
15	38,9	52	8,24	44,94	29,29
20	38,9	52,27	16,48	51,06	38,85
25	43,27	60,72	22,61	59,31	61,14

(Sources: Analysis Result of Laboratory)

Based on Table 3 above, it could be summarized that the Effect of Hydrilla Amount and Living/Growing Time to COD Elimination could be described in the following chart/graph:

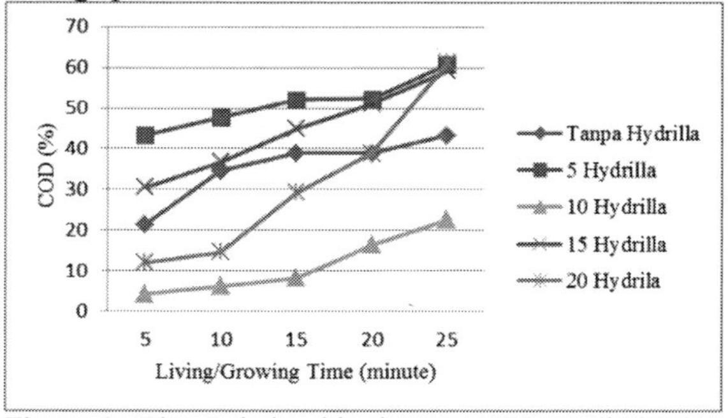

Figure-3. The Relationship between COD Elimination (%), Living/Growing Time (minute) and the Amount of Hydrilla.

The parameter used in figure-3 was a process of conetray cascade aerator combined with a biofilter. Based on our observation, this is the best score of COD. In accordance with the effect of hydrilla quantity to the decrease of COD, it was found that the best variable was that of employing 20 plants of hydrilla with its living time of 25 minutes which resulted in its COD elimination reaching up to 61,14%. COD elimination could increase as the other variable (living time) and the amount of plants (cascades) did. Meanwhile, the amount of oxygen required as an aeration effect had been fulfilled to oxidize some organic substances. Moreover, Unussuriawiria (2008) said that biofilter, like BOD, was one of the ways/solutions of disposal processings by employing plants which had some capability (like bacteria and fungus) of decomposing some organic and inorganic substances/compounds.

The Effect of Hydrilla Amount and Living/Growing Time to TSS Elimination

Table-4. The Effect of Hydrilla Amount and Living/Growing Time to TSS Elimination

Time (minute)	Elimination Percentage of BOD (%)				
	Without Hydrilla	5 Hydrillas	10 Hydrillas	15 Hydrillas	20 Hydrilas
5	15,22	57,72	21,36	31,41	8,22
10	18,18	60	23,93	32,47	8,75
15	19,31	61,81	25,64	35,04	9
20	19,77	64,31	28,21	36,53	9,43

25	57,04	65,45	29,91	38,88	10,57

(Sources: Result Analysis of Laboratory)

Based on Table 4 above, it could be summarized that the Effect of Hydrilla Amount and Living/Growing Time to TSS Elimination could be described in the following chart/graph:

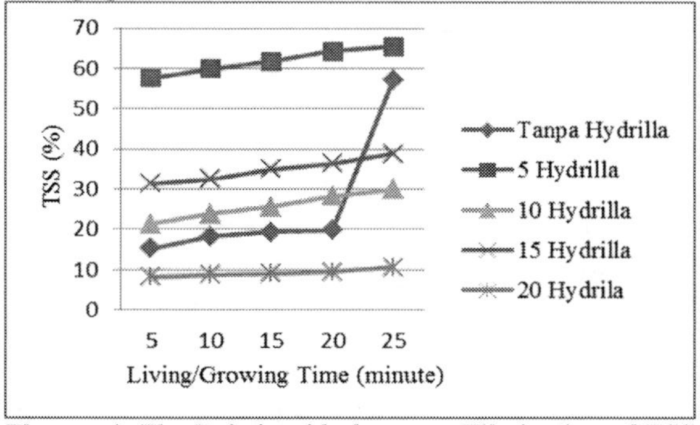

Figure- 4. The Relationship between Elimination of TSS (%) and Living/Growing Time (minute) to the Amount of Hydrilla.

Figure-4 showed that the best variable was that of employing 5 plants of hydrilla with a living time of 25 minutes for TSS elimination. As Unus Suriawiria (1993) said that a group of microbes, like bacteria and fungus growing/living symbiotically around plants, both plants living in soil and aquatic habitat could grow/live dependently to those plants. These groups of microbes are usually called 'rhizosphere microbes'. Therefore, the higher the number of available plants, the higher the TSS will be.

CONCLUSION

Referring to the research result and discussion above, it could be concluded as follows: A combination of conetray cascade aerator and a biofilter has been created, where the best system could reduce BOD as much as 69.11%, COD 61.14%, and TSS 65.45% on debit of 200L/hour.

ACKNOWLEDGMENTS

Acknowledgments are addressed to the Directorate of Research and Community Service, the Directorate General of Higher Education, Ministry of Education and Culture, that have funded the research in 2015, Seeded research university No. SPP/002/Lit/ LPPM/II/2015.

REFERENCES

Benefield, Larry D. 1990. "Process Chemistry For Water and Wastewater Treatment", Prentice Hall Inc., New Jersey.

Enrique J. La Motta et al. 1996. Corrosion Control of drinking water Using Tray Aerator, Jurnal Environmental Engineering, 122: 640-648.

East Java Governor's Regulation No.72, 2013.

Endahwati L dan Suprihatin. 2013. A combination of aeration processes, adsorption and filtration in wastewater treatment fishery, the scientific journal Environmental Engineering Vol 1 No 2.

Hartini E. And Sajidan W.2012. Effectiveness and Bubble Aerator Cascade aerator in lowering levels of Manganese Water Well Drilling, Journal of Public Health, Pack 8 (1): 44-52.

Henric Svensson, B€orje Ekstam, Marcia Marques, William Hogland. 2015. Removal Of Organic Pollutants from oak leachate in pilot-scale wetland systems
: How efficient are aeration and vegetation treatments?, Water Research 84: 120-125.

John Colt and Eric Kroeger. 2013. Impact aeration and alkalinity on the water quality and product quality of transported tilapia – A simulation Study, Aquacultural Engineering 55: 46-58.

J.C.J.Gude, Rietveld, L.C., van Halem, D. 2015. Fate of low arsenic concentrations during full-scale aeration and rapid filtration, Water Research,doi 10.1016/j, watres 2015.10.034.

Kusno P H.1986.Environmental Health, Public Health Faculty, University of Indonesia.

Muhammad Eko Wibowo Romayannto et al. 2006. Domestic waste water treatment with aeration and addition of Pseudomonas putida, Bioteknologi 3 (2) :42-49.

Metcalf and Eddy. 2003. "Wastewater Engineering Treatment and Reuse", 4th Edition, Mc.Graw – Hill Inc., Singapore.

Reynolds T.D. and Richards P.A., 1996. "Unit Operations and Processes in Environmental Engineering", 2nd Edition, Boston, PWS: Publishing Company.

Sadatomi Michio. 2011. "Micro Bubble Generation Rate and Bubble Dissolution Rate into Water by a Simple Multi Fluid Mixer with Orifice and Porous Tube", Graduate School of Science and Technology, Kumamoto University. Japan.

Said N.I. 2005. "The removal method of Iron and Manganese in Drinking Water In Domestic",Water Journal Indonesia (JAI), 1(5): 239-250.

Sutrisno. 1982. ""Water Supply Technology, BinaAksara. Jakarta.

Suriawira Unus. 2008. "Microbiology water and the foundations of processing emissions are biologically",Alumni, Bandung.

Suriawiria Unus. 1993. Water Microbiology, first printing , second edition, ITB, Bandung.

Senlin Shao, Fangshu Qu, Heng Liang*, Kai Li, Huarong Yu, Haiqing Chang, Guibai Li.2015. Powdered activated carbon - membrane bioreactor operated under intermittent aeration and short sludge retention times for micro-polluted surface water treatment, International Biodeterioration & Biodegradation 102:81-88.

The Decision of State Minister of Environment no. 112, 2003.

Xiaoou Wang, Yimei Tian, Xinhua Zhao, Sen Peng, Qing Wu, Lijian Yan. 2015. Effects of aeration position on organics, nitrogen and phosphorus removal in combined oxidation pond-constructed wetland systems, Bioresource Technology 198: 7-15.

Yang Deng, James D. Englehardt ,Samer Abdul-Aziz, Tristan Bataille,Josenrique Cueto, Omar De Leon, Mary E. Wright, Piero Gardinali,Aarthi Narayanan, Jose Polar, Shibata Tomoyuki .2013. Ambient iron-mediated aeration (IMA) for water reuse,Water Research 4 7 : 8 5 0 -8 5 8.

HEAT REMOVAL USING THE HOLLOW ROOF IN THE JAVANESE HOUSE

Mohammad Pranoto S
Architecture Department, University of Pembangunan Nasional "Veteran" Jawa Timur, Surabaya, East Java, Indonesia.
E-mail: mprants@yahoo.co.id

Abstract
Ventilation plays a vital role in securing optimum air quality and thermal comfort in buildings. In fulfilling this need, it touches on a wide range of building design and ventilation performance. Early attempts to minimize ventilation energy loss frequently resulted in poor indoor quality and comfort problems. This paper explains about natural air circulation, indoor and outdoor air temperature and also air velocity on the hollow roof in Javanese house, which has a specific shape of geometry and how far that roof can reduce the heat gain in the building. This method using computational fluid dynamics based program have become available to predict airflow and pollutant transport throughout a space. These approximate the space by the grid system containing many control volumes or elements to which numerical equations of flow are applied. And, the result of this research, the cavity roof in Javanese house is able to add indoor air rate velocity, with the average value of 0.5 m/s, and giving the indoor majority cooling effect average value of 0.55 ° C.

Keywords: heat removal, hollow roof, Javanese house

INTRODUCTION

This paper discusses the applicability of the openings in the roof of the house based on traditional Javanese house. According to Lippsmeier (1994), proper placement of the building against the sun, wind, shape and construction as well as the selection of appropriate materials, the indoor temperature can be lowered a few degrees without mechanical equipment. These three elements are influencing the amount of solar radiation global / global irradiance (W/m2).

In addition, there are two elements that have an influence on the amount of the heat resistance: the type of roof material and the design of the roof space, with and without ventilation. The energy-saving building is designed to consume less energy, but still functional, such as by the use of non-conventional energy and energy conservation (Wong, 2007). According to Lechner (2009), there are several factors that will affect the air exchange process that occurs naturally in a room or building. These factors are wind speed and direction outside the building, the temperature, and humidity inside and outside the building, and ventilation holes specification (inlet and outlet position, dimensions, shape and feature supports). Many factors are interrelated and support in creating air exchange either in a room or building. The selection of roof materials and insulation systems in certain climatic regions will also decrease the amount of radiant thermal in buildings which mostly come from the roof, like a hot humid climate such as Surabaya (Lau, et al. 2008)

According to Santosa (1994) and which was then confirmed by Wonorahardjo (2000), in the hot humid climate region that has high humidity causes to be a specific problem including high temperature. Overheating, where outside of the building above the comfort temperature, it is avoided by the human. Passive cooling efforts that have been made, are used to reduce the heat load in the building.

To achieve these conditions is by reducing the impact of unfavorable climatic conditions. There are some things that must be addressed for it. Is the acceleration of wind flow into the building will be able to remove and reduce the accumulation of heat that occurs?

The warm-humid climate is climate controlled with a difficult character. The main problem in the warm humid climate that the inside air temperature of the building is almost equal to the outside air temperature of the building, where it is caused by high solar radiation. According to Santosa (1994), climate character, especially in Indonesia, can be summed up as follows: The thermal comfort ranges in Indonesia is between 25.4°C until 28.9°C . The comfort condition occurs at night on 11.00 p.m – 05.00 a.m., beyond human active hours.

Natural ventilation is used to purge the building of hot air and cool the building fabric. The methods may include window opening, wind towers (including the hollow roof), solar chimneys and atria. In each case, the system is designed to take advantage of prevailing driving forces. Air movement is also an important aspect of ventilation cooling since it offsets increases in air temperature while maintaining comfort, provided that the increased air velocities are under occupant control. On solar radiation and radiation section, the global irradiance value can be of high value at all the time. Some things that can be concluded is the total radiation on a horizontal surface reaches a maximum at midday.

Solar gain occurs as a result of the absorption of infra-red radiation by surfaces exposed to sunlight. The air temperature of these surfaces can rise considerably, resulting in the heat transfer to the surrounding air stream. The Southern surface will receive the maximum radiation in December while the Northern surface will receive the maximum radiation in June. The Eastern and the Western surface will receive a high radiation throughout the year. Wind speed ranges between 1 m/s until 3m/s. The wind climate is potential for passive cooling in buildings. According to the driving force theory by Liddament (1996) which was then confirmed by Lechner (2009), natural ventilation can be regulated by the wind thermally. Natural ventilation plan is affected by the position of the openings and value of the pressure distribution. The pressure distribution is affected by the elements:

Firstly, the stack of pressure means stack effect is built as a result of the difference between the air temperature and water density, between inside and outside of the building (Fig.1). It produces an imbalance in the pressure gradient of inside and outside of air masses which is the result in an outside vertical pressure which is the result in a vertical pressure difference.When the indoor temperature is greater than that of the outdoors, the air passes through the openings at the lowest level of the building and throughout the openings at the highest level.The second one is the pressure of the wind. That means that the wind enters the building and induces a positive pressure on the windward face and negative pressure in the opposite direction in the area throughout of the sides of buildings (Fig.2). Third, mixing between the use of wind and stack pressure, this system is planned to attempt a merger, that the effects of the wind and stack action complement and supported each other.The passive stack and design of atria (space of void) is the solution to the combination of two elements. Alternatively, the pressure distribution can be modified through the mature plan inlet or in the availability of a shelter belt in order to reduce wind effects.

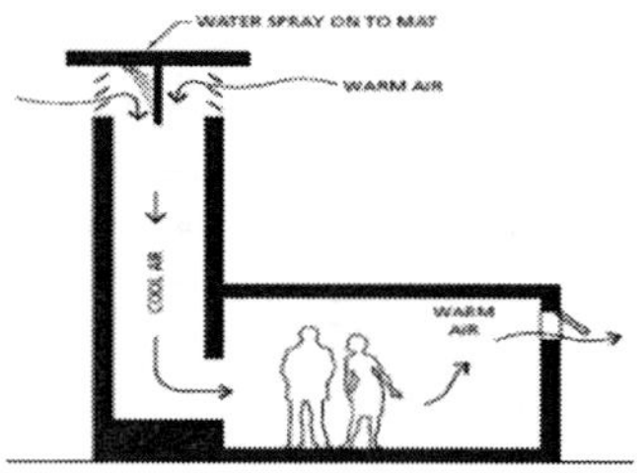

Figure - 1. Cool tower mechanism (Lechner, 2009)

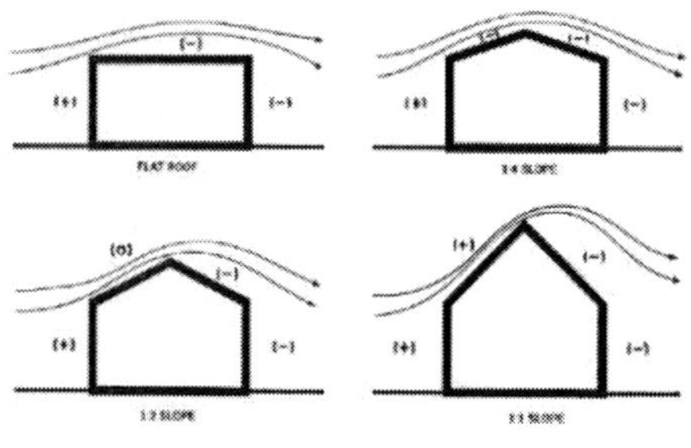

Figure - 2. Pressure of Leeward (-) and Windward (+) (Lechner, 2009)

Ecological roof and the cooling strategy are part of the Javanese's hollow roof home as a method of providing cooling with not always associated with air flow system. The Advantages of this system is a thermal satisfaction that affects the psychological condition in the house. It is also associated with research on thermal performance on roofs with a slope angle of 30° made of insulated concrete roof tiles in Klang Valley, Malaysia (Zakaria, et al. 2008). Evaporative cooling occurs when air coming into the building across the surface moist or via sprays or wet materials in the area of the opening. The heat is removed from the house by air by the transfer and absorption system. The aim of this study is to reveal and explain the phenomenon of air flow that goes through the hollow roof of Javanese house so as to transfer and remove heat in the house and the air temperature becomes lower than the previous

METHODOLOGY

As neither the field measurements nor the wind studies provided truly definitive information on the ventilation performance of The Javanese House, an effort to identify other analysis techniques were undertaken. One potential promising technique was computational fluid dynamics (CFD) based program, as was done in this study with CFX-5 (Fig 3). According to Koenigsberger et.al (1973), The movement of air past the skin surface accelerates heat dissipation in two ways: 1). Increasing convective heat loss 2). Accelerating evaporation. The physiological cooling indicator is the wind speed. Wind speed indicators are necessary for physiological cooling based on the relationship three climate variables, namely, the wind speed, the air temperature and the air humidity (Satwiko, 2004)

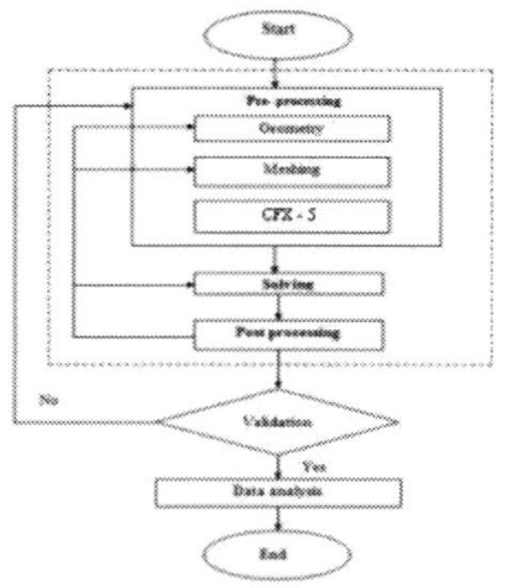

Figure - 3. Flow Chart for CFX-5 (CFD-Based) Simulation Process

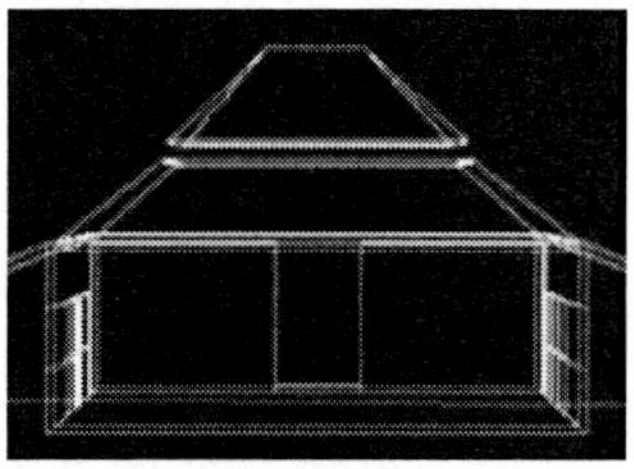

Figure – 4. The model of Javanese House in 3D (from behind)

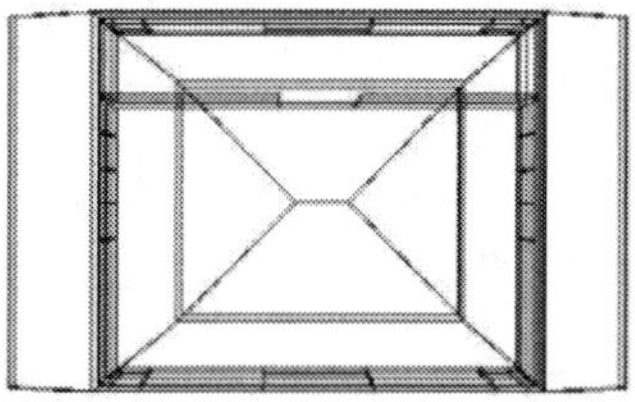

Figure - 5. The model of Javanese House in 3D (from above)

To measure air temperatures experienced in this house, the wind pressure generated by hollow roof depends on the particular position of the opening on the façade, the exact geometry of the building, wind speed, orientation of buildings in the direction of the wind, and the shading effects of the nearby buildings (Fig 6).

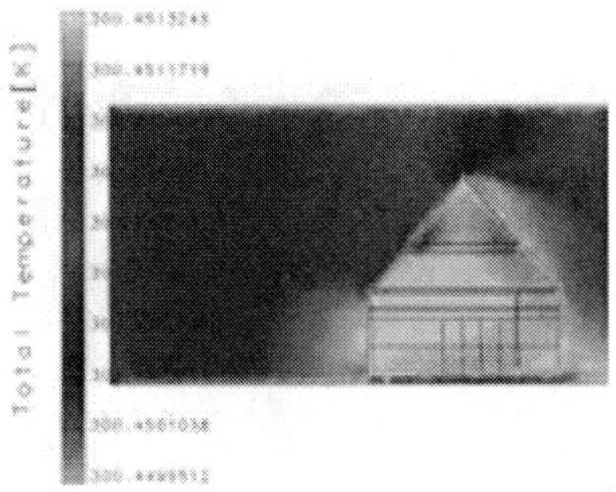

Figure - 6. The air temperature of Javanese House with hollow roof in 2D View

Although the pressure may vary from point to point on the façade, the model will make use of a single averaged value for the pressure at all points on each facade (Fig. 8). Roof hollow opening on the same facade as it will have the same wind flow generated by the external wind pressure. Wind pressure to physiological cooling required when the air humidity reaches 60% (Aynsley, 1990)

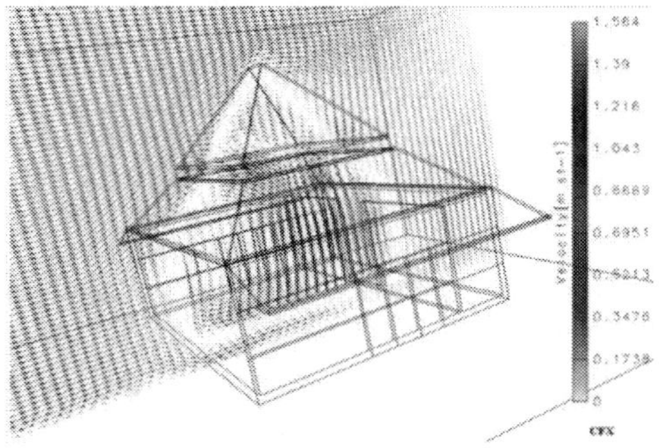

Figure - 7. The pattern and velocity of wind of Javanese House with the hollow roof in 3D view

Free wind speed is generally measured at a certain reference height above ground level. The relationship between the wind speed at the height of the roof and the reference altitude is determined by the shape of the atmospheric boundary layer (Fig. 7). It is important to realize that the effects of a particularly hollow roof to reduce heat gain caused by radiation outside the roof

The concept mentioned above is able to eliminate the internal heat load due to increased air temperature inside the house. As indicated previously, it is affected by the opening of the effectiveness of the total permeability of the home as determined by the total area of the cavity opening. Airflow is driven by the pressure field is fundamental to the whole flow process. Pressure is maintained by a combination of driven air or forced convection and by buoyancy forces or natural convection. The Roof covering material in coating can help in protecting the roof from ultraviolet rays and reduce the heat from infrared light while saving the range of energy between 20% until 50% (Poonia. S.J. et al. 2011)

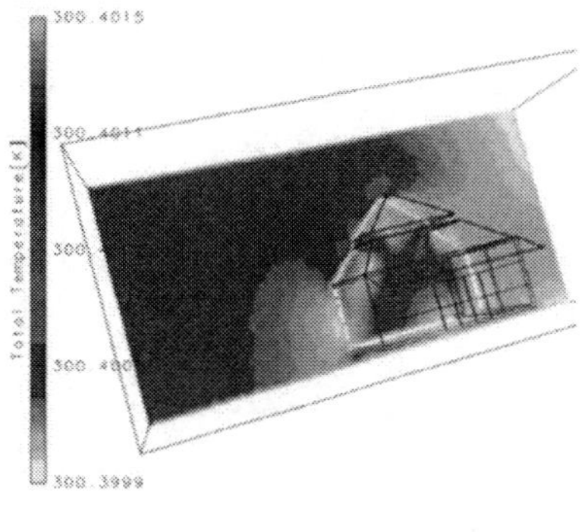

Figure - 8. The air temperature of Javanese House with hollow roof in 3D View

RESULTS

On the development trend of the performance of the housing at this time, it is rather difficult to reform the traditional buildings in order to obtain optimal natural ventilation. The planning design, the performance of the air in the contemporary home is: Firstly if the outdoor airspeed is low, it brings the movement of hot air, which flows upwards (out of the building) through the ventilation mesh (vertical movement).. And the second, if air moves by applying cross the ventilation system, it moves from the opening area on wall to the other side (horizontal movement) (E, Prianto, at.al 2000)

A hollow roof model is proposed which uses a good balance between empirical data and theoretical equations. The model shows the layout of the building, permeability openings, openings of different height above ground level, wind speed and direction, the temperature difference between indoor and outdoor air as well as the effect of the surrounding area where the building is located. Building using roof ventilation openings are based on the performance data of the building with reference to climate data in Surabaya. Wide openings allowed for the effects of ventilation stacks are openings with a percentage of 5% of the total roof area (Fig. 9).

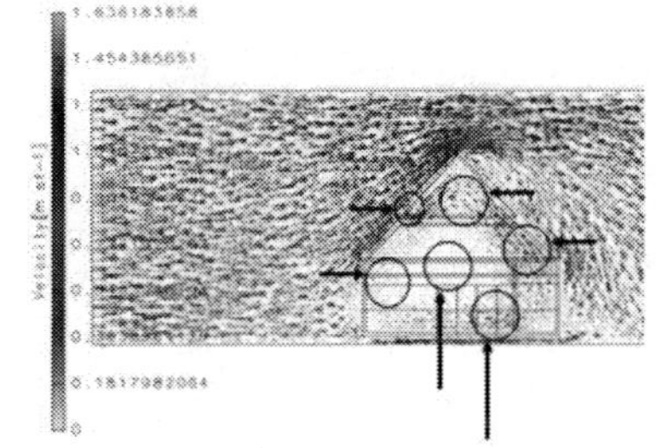

Figure - 9. The pattern and velocity of wind of Javanese House with hollow roof in 2D view

In the case study, the complex interactions between heat and airflow models are illustrated by investigating the combined effect of roof vents and the dimensions of the cavity. The effectiveness of opening roof area has reduced the heat gradually with the addition of more velocity of the wind.The data of climate is regarded from the data of speed of the wind, the temperature of the air and the direction of the wind of Surabaya City. The average of data amount obtained in each type are as follows, namely: Data of average air temperature of hottest month in 2009 until 2014 was in May that was an equal to 27.3°C, data of the average speed of wind on hottest of month on these years, was 2.3 Knot (1.18 m/s).Direction of wind data on the hottest of month in average in these years, was 180°

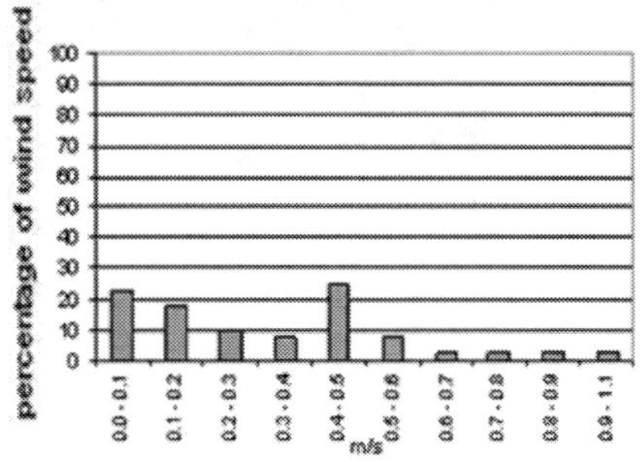

Figure - 10. The value of velocity of wind of Javanese House with hollow roof

The effect of internal heat gain and solar radiation affecting indoors condition through the opening of the roof and other elements of the building envelope is to raise the indoor air temperature (Szokolay, 1980). Effect stacks opposed to relying on thermal power, set the density differences (caused by differences in air temperature) between indoor and outdoor air. This can occur through the opening roof of the cavity: warmer and lighter indoor air will flow from the top and the air will flow through the cracks at the bottom. The principle is the same as in wind generation. The dominant wind speed that occurred when roof vent openings opened by percentage of 5% is between 0.4 m/s until 0.5 m/s (Fig.10).

Special provisions can be made for it in the form of ventilation shafts. The higher the shaft, the greater the cross-sectional area and the larger the temperature difference; the greater the motive power, the more volume of air to be moved. That is the phenomenon of the condition as described in the table below (Table-1).

Table -1: Condition of Indoor space of roof of Javanese House

Condition	Wind speed (m/s)	Air Temp. (°C)	Opening percentage (%)	Dimension of opening (m2)
Existing	0	27.75	0	0
With cavity roof	0.5 m/s	27.45	5	7

According to Koenigsberger (1977) which was then confirm by Moore (1993) and Allard (1998), the indoor wind movement are able to occur due to the differences in wind dynamics and the temperature of the air. Movement of the wind occurs due to differences in air pressure between the two sides, which the natural wind is going to move from high air pressure to low air pressure (Santamouris and Asimakopoulus, 1996). In the hollow roof in the Javanese house, the wind pressures on house surfaces, which is the direction of the wind (windward) will give positive air pressure (+) in that building, while on the edge, back and top of the building, it is going to give the negative air pressure(-).

CONCLUSION

The provision of air flow and the removal of heat by mass transfer are the first two tasks of ventilation. Where a relatively small volume of air exchange is adequate, the problem is easily manageable. The passive cooling systems described here include not only the well-known traditional techniques mentioned above but also more sophisticated modern techniques (with the CFD-based program) and this technique should be considered as an aid to thinking and should be used intelligently, not only as a mechanical tool. Without cross-ventilation, there will be no appreciable air movement. When the inlet opening is large, the velocity of entering air will be low, but the volume of air flow will be large. Given an elevational area, the highest air velocity within the room is obtained with a small inlet and large outlet opening. As much as possible, passive cooling uses natural forces, energies, and heat sinks. The result and analysis on the condition of the Javanese house with the hollow roof can turn out to lower the air temperature inside the building. The hollow roof in the Javanese house (with 5% of percentage of the opening area of the roof), can provide the dominant speed of wind changes on that house, from its state of original (existing condition) at 0 m/s to 0.5 m/s, and can remove the heat or in the other hand is providing effect of cooling of 0.55 °C , reducing the average of temperature of the air from the value of 27.7°C, to the value of 27.2 °C.

REFERENCES

Lippsmeier. G. 1980. Tropenbau Building In The Tropics. Van Nostrand Reinhold. NY.

Satwiko. P. 2004. Solar-Wind Generated Roof Ventilation System (S iVATAS) for Humid Climate, The International Journal of Ventilation. Vol.3. No.3. December 2004.(ISSN 1473-3315)

Santosa. M. 1994. Geometrical Plan Design and Roof Construction As Main Aspects Of The Housing Living Comfort Level, Research Report, Research Institutions, ITS Surabaya, pp 7-13.

Givoni. B. 1976. Man, Climate and Architecture, Applied Science Publisher Ltd, Essex, UK

Poonia. S. Jethoo. AS. Poonia, MP. 2011. A Short Review On Energy Conservation In Buildings Using Roof Coating Materials For Hot An Dry Climates". Universal Journal of Environmental Research and Technology. Vol.1.Issue 3: 247-252.

Wong. N.H. 2007. A Study of Effectiveness of Passive

Climate Control In The Naturally Ventilated Residential Buildings In Singapore. The Journal of Environment and Buildings 42. 2007. 1395-1405. Department of Building. School of Design and Environment. National University of Singapore. The Republic of Singapore 117566. Singapore

Lau. A.K.K. Elias. S. and Lim. C.H. 2008. The Thermal Performance Evaluation of Roofing Systems and Materials in Malaysia Residential Development, In: Proceeding of 9th SENVAR and 2nd ISESEE Humanity and Technology Shah Alam. Malaysia. 387-395

Aynsley. R and Spruill. M. 1990. The Thermal Comfort Models For Outdoor Thermal Comfort In Warm Humid Climates and Probabilities of Low Wind Speeds, Journal of Wind Engineering and The Industrial Aerodynamics 36 1990. pp.481-488

Zakaria. N.Z. Peter. W. and Ahmad. R. 2008. Thermal and Energy Evaluation of Roofing and Ceiling Insulation For The Residential Building in Tropical Climate. In : Proceeding of 9th SENVAR and 2nd ISESEE Humanity Technology. Shah Alam. Malaysia, 577

Prianto. E. Bonneaud.F. Depecker. P. Peneau.P.J. 2000. Tropical Humid Architecture in The Natural Ventilation Efficient Point of View. The Reference of Traditional Traditional Architecture in Indonesia, The International Journal of Architecture Science, Vol.1. No. 2. P.80-95.

Moore. F. 1993. Environmental Control System: Heating, Cooling, Lighting, McGraw - Hill Inc, USA

Santamouris. M. and Asimakopoulus. D. 1996. Passive Cooling of Building. James and James Science Publishers. London.

Allard. F. 1998. The Natural Ventilated Building Design Handbook. James and James Science Publisher. London. pp 32-33.

Szokolay. S.V. 1980. Environmental Science Handbook For Architects and Builders, The Construction Press, Lancaster, pp 381 -383.

AEA Technology plc. 2002. Introduction to CFX-5, CFX International, AEA Technology plc, 8.19 Harwell Didcoot, Oxfordshire OX11 0RA, United Kingdom.

Koenigsberger. O. Ingersol. T.G. Mayhew. A and Szokolay. S.V. 1973. Manual Of Tropical Housing and Buildings. Part 1: The Climatic_Design. Bombay: Orient Longman

Climate Datas of Surabaya City from 2009 - 2014, Geophysics and Meteorological Board (BMG) Juanda Climate Station

Wonorahardjo. S. 2000. A Study Of Roof Ventilation Types and Their Influence On Indoor Thermal Comfort. SENVAR 2000. Edt: Santosa. M . The Laboratory Of Architectural Science and Technology. Department Of Architecture. Faculty Of Civil Engineering and Planning. ITS Surabaya

IMPLEMENTING CLEANER PRODUCTION AS AN ENVIRONMENTAL MANAGEMENT EFFORTS IN SMALL INDUSTRIES OF CASSAVA CHIPS

Erina Rahmadyanti[1] and Andre Dwijanto W. [2]
[1]Department of Civil Engineering, Faculty of Engineering, State University of Surabaya, Indonesia
[2]Departement of Management, Faculty of Economic, State University of Surabaya, Indonesia
E-mail: erinarahmadyanti@unesa.ac.id

ABSTRACT

Small and Medium Enterprises (SMEs) is one of the major driving factors for Indonesian economy, especially in food processing industries. The cassava-based industry is one type of food and beverage industry with chips as its major product. The limitations of knowledge caused their activities to only aim at pursuing economic benefits and ignoring the environmental balance. The most appropriate preventive method used, according to the characteristics of SMEs in Indonesia, is Cleaner Production. This study aims to reduce the risk of environmental pollution caused by the waste production of small chips industries by implementing cleaner production. The method used in this study is quick scanning by analyzing mass balance, energy, and utilities that aim to find an inefficient process to minimize losses. Implementation of cleaner production may include good housekeeping, reducing, and reusing. Based on the assessment of alternative eligibility criteria, the equipment modifications are the main factor in implementing cleaner production that drives the profits by providing efficiency of cutting as much as 80 percent and optimizes the profits into 57.62 kg in a month or 691.44 kg in a year. If the price of cassava chips is IDR 40,000 in a kg, then it would save IDR 27,657,600 in a year.

Keywords: small industries, cassava chips, cleaner production.

INTRODUCTION

Indonesia's current economic structure is primarily focused on agriculture and industries which extract and harvest natural resources. There are only limited industries which focused on products with added value. In addition to this, there is a development gap between western and eastern parts of Indonesia. Masterplan for Acceleration and Expansion of Indonesia Economic Development 2011-2025 (MP3EI) is aimed at accelerating and expanding the economic development in Indonesia as a starting point towards making the nation more equitable. As a country consisting of thousands of islands, the Indonesian archipelago has a unique combination of economic with specific major islands or regions having its own strategic future role in achieving Indonesia vision. By taking into consideration these potentials and strategic roles of each major island, there are six corridors that have been identified as follows: Sumatra economic corridors, Java economic corridors, Kalimantan economic corridors, Sulawesi economic corridors, Bali-Nusa Tenggara economic corridors, and Papua-Kepulauan Maluku economic corridors (Coordinating Ministry For Economic Affairs, 2011).

The prime purpose of MP3EI is to enable Indonesia to become a developed and prosperous country with a national GDP of around USD 4-4.5 Trillion in 2025 and becoming the 9th largest economies in the world. To achieve this goal, approximately 82 percent or equivalent to USD 3.5 Trillion will be targeted as a contribution to GDP from economic corridors. By implementing MP3EI, Indonesia's overall GDP is expected to grow more rapidly and broader, both for areas within the six corridors and for areas outside the corridors. By applying MP3EI, the annual national GDP growth will be approximately 12.7 percent nationally with regional growth within the corridor at 12.9 percent. Growth in the areas outside of the corridors would also increase by 12.1 percent as a result of the spillover effects of economic development within the corridors areas (Coordinating Ministry For Economic Affairs, 2011).

Driver for national industry and services provision are main themes of the Java economic development corridor. East Java province as part of the Java corridor should contribute the success of the economic development. Although Indonesia's fundamentals are strong, achieving a high level of growth will not happen automatically. A number of challenges must be overcome in order to realize the sustainable development for a successful Indonesia. To support this, a special strategy will be implemented to further develop Java corridor industries that support the conservation of water and the environment.

One of the main focus of Java corridor development will be on food and beverage industry. Food and beverage industry in Indonesia, including in East Java province is dominated by thousands of small and medium enterprises (SMEs). SMEs play an important role in the production, innovation, and employment creation. SMEs are enterprises classified as having fewer than 50 employees; make up 80-90 percent of all industrial establishments in most countries (Behnsen et al. 2002).

The role of SMEs has also been actualized in the past to the present crisis. During times of economic crisis up to now, SMEs are able to stand as a major driving factor for the Indonesian economy. Especially when the crisis investment and government spending is very limited, then the role of SMEs as a form of the people's economy is very large. Furthermore, SMEs contributes for about 56.7% and non-oil exports contribute more about 15%. However, in the range of entrepreneurship, SMEs still donate 99% and has a 99.6% share in employment (Indonesians statistic, 2011). Although SMEs constitute the backbone of an economy, they also account for the bulk of human resources and polluting activities of the environment (Sohair et al, 2007). In the other hands, The Asian Development Bank (ADB) notes that most of the pollution in the Asian countries comes from SMEs, and is especially concentrated in its megacities (ADB, 2000). Environmental problems that generally occur in the food and beverage industries are (Ramjeawon, 2000): 1.) high water consumption, especially problematic in water-poor or water-stressed countries; 2.) the generation of liquid effluent with high organic content; and 3.) the generation of large quantities of sludge and solid wastes.

Their environmental impact may not be great, but their cumulative impact is substantial. It is caused by several factors including lack of access to resources allowing for investments in pollution control, low level of technology, lack of space, unavailability of trained personnel, and the unwillingness of management to invest in environmental protection (Ramjeawon, 2004). In 1987, the United Nations published a document that challenged the economy orthodoxy of the relationship between development and environmental degradation. Called *Our Common Future*, it emphasized the concept of 'sustainable development'. In theory, sustainable development simply means meeting the needs of the present generation, without compromising the needs of future generations. The true challenge of sustainable development lies in putting the theory into practice.

In order to compete with larger enterprises in the market, SMEs should perceive environment-oriented management, regard as cost intensive that hampers innovation and development. Competition for capital and new investment opportunities is driving the environmental performance of SMEs in the 21 st century. Achieving this objective requires financial performance and a commitment to innovation and best practice. Also, sustainability-driven companies achieve their business goals by integrating economic, environmental and social growth opportunities in a pro-active, cost-effective and responsible manner today, so they will outpace their competitors and be tomorrow's winners (Abou-Elela, 2002 and Sohair et al, 2007). To achieve such goals, the implementation of pollution prevention measures, cleaner production, and low-cost treatment technologies are required (Abou-Elela, 2002).

SMEs limitations require an easy method as a preventive action to protect the environment. Several advantages obtained when implementing various preventive action are (Fresner et al, 2010): 1) reducing waste and emission generally means using smaller quantities of materials and energy, which has the potential for economic savings; 2.) reduction of waste and emissions usually triggers an innovative process in the company because of the intensive focus on the analysis production processes; 3) risks regarding environmental liability and disposal are reduced to a minimum; 4.) reduction of waste and emissions means moving towards sustainable economic development.

The most appropriate preventive method is used to the characteristics of SMEs in Indonesia is Cleaner Production. The SMEs should adopt cleaner production methods compared to the traditional end of pipe approaches to pollution control. Cleaner production offers an opportunity for win-win solutions as SMEs reduce their operational costs and environmental and other liabilities by using less energy, water and materials, handling chemicals and their waste safely and generating less waste and pollutants (Van Berkel, 2011). Cleaner production is commonly achieved through a diversity of practice, including good housekeeping, process and/or equipment modification, input material substitution, on-site reuse and recycling and change in product and/or technologies (Van Berkel, 2007a,b). Cleaner Production techniques are dynamic, and although industries have improved their environmental performance over the last two decades, there is continuing room for improvement.

Cleaner Production has been widely applied in various industries of cassava, but the majority in cassava starch. On the other hand, there are some industries that also produce derivative products of cassava besides flour including cassava chips. Previous studies related to the implementation of cleaner production in cassava chips industry is still limited, especially in small industries. Therefore, the cleaner production method may still be developed in this industry, especially in small industries of cassava chips which vary process depending on the origin of the production areas.

Cassava is the fourth most important source of calories in the tropics after rice, wheat, and maize (Okunade dan Adekalu, 2013). Cassava production in Indonesia makes up about 30 percent of the production of Asia and the Pacific, and Indonesia is the fourth largest cassava exporter in the world. The total amount of production is almost the same as Thailand's, the world largest cassava exporter, however, Indonesia's cassava export is less than 10 percent of Thailand's (based on FAOSTAT, http://faostat.fao.org in Sugino and Maerowani, 2009). Production can be observed in all over the country, but the major production area is located in the islands of Sumatra and Java). East Java is one of the provinces with the largest area of cassava production in Indonesia which produced 20 percent of the national output (Indonesia statistics, 2006). Implementation of cleaner production in cassava-based small medium industries in East Java province will increase the competitiveness of the products. This will have an impact on increasing the contribution of the province of East Java in strengthening the national economy.

Cassava contains cyanide which limits utilization in the fresh unprocessed form. The roots need to be processed to reduce the cyanide content to safe levels before consumption. Once harvested, cassava tubers begin to deteriorate and can not be stored for more than a few days. They are bulky with about 70 percent moisture content which makes transportation of the tubers to urban markets difficult and expensive. Thus, there is a need for rapid processing of the tubers (before spoilage sets in) into various products with increased shelf life. These shelf-stable products are easier to transport and market and cost less to transport and contain less cyanide (Udoro, Gbadamosi, and Taiwo, 2008). Cassava-based small and medium industries in East Java produces three major products namely chips, crackers, and flour.

Cassava chips are unfermented, dry products of cassava. Roots are chipped into smaller sizes for fast drying that also help the process of detoxification. Cassava can be dried naturally by the sun or artificially in an oven (FIIRO, 2005; Irinkoyen et al, 2008). In some countries, chips are utilized in the production of flour and starch.

Traditionally, cassava is processed into dry whole roots that have undesirable color, irregular shapes and are often contaminated with dust. Production of chips by SMEs traditionally includes the steps of peeling, washing, slicing, seasoning, frying, oil draining, and sorting. The process of washing and seasoning results in the production of wastewater. Most cassava processing industry discharges significant quantities of wastewater whose major component is cyanide to rivers, lakes, sewage canals, agricultural fields and the environments which eventually flows back to streams or downstream surface water locations (Oliviera et al., 2001 in Okunade and Adekalu, 2013). This will lead to contributing significantly to environmental pollution and depletion of water resources as they contaminate surface water, particularly during dry seasons. Currently, there is no specific method of disposal or treating the cyanide-laden wastewater resulting from cassava. In addition to waste water, production processes produce waste in the form of chips cassava outer shell and the remaining pieces of the slicing process.The resulting wastes are often left in mounds which generate a highly offensive odor and are rather unsightly.

One of the strategic measures to reduce cyanide content of processed cassava is to improve processing methods used for conversion of roots storable cassava product such as chips and flour. A number of processing methods or a combination of them (peeling, slicing, retting, fermentation, boiling, drying, roasting, pounding, milling etc) are available that attempt to remove poisonous principle as much as possible (Amsalu and Esubalew, 2011). Cassava contains the potentially toxic compound like cyanogenic glycosides. These compounds can cause acute cyanide poisoning and death in humans and animals when consumed in sufficient quantities (White et al, 1998 in Amsalu and Esubalew, 2011). This study aims to reduce the risk of environmental pollution caused by the waste production of small chips industries by implementing cleaner production.

METHODOLOGY

A waste audit procedure is a systematic tool used to identify the opportunities of cleaner production. The information from a waste audit can be starting point for investigating pollution issues at any facility. Such an assessment of waste generation as well as raw material and energy consumption can highlight areas for potential intervention and provide a baseline for comparing subsequent increases or decreases in specific waste stream Based on the UNEP's Audit and Reduction Manual for Industrial Emissions and Waste (1991) and other relevant literature (UNEP, 1991), the methodology to be used in this study to identify waste streams and energy usage were developed and implemented. The method used in this study is quick scanning by analyzing mass balance, energy, and utilities that aim to find an inefficient process to minimize losses.

Data for material and energy balances often can be taken from accounts (for a year), but have to be weighed and measured for more detailed balances. This makes this step time consuming and sometimes difficult in practice if process documentation is not in place and has to be developed for the cleaner production project (Fresner, 2010).

All the record of production for a 12-month period were used for the audit. All the available information resources (raw material purchase records, product quantities, water usage and wastewater discharge data, etc), as well as the checklist and personal interviews, were used to gather all the available data for the mass balances. A data collection from the mass balance are translated into a waste reduction plan.

RESULTS

The process outputs from each of cassava chips unit operation were quantified in the following figure:

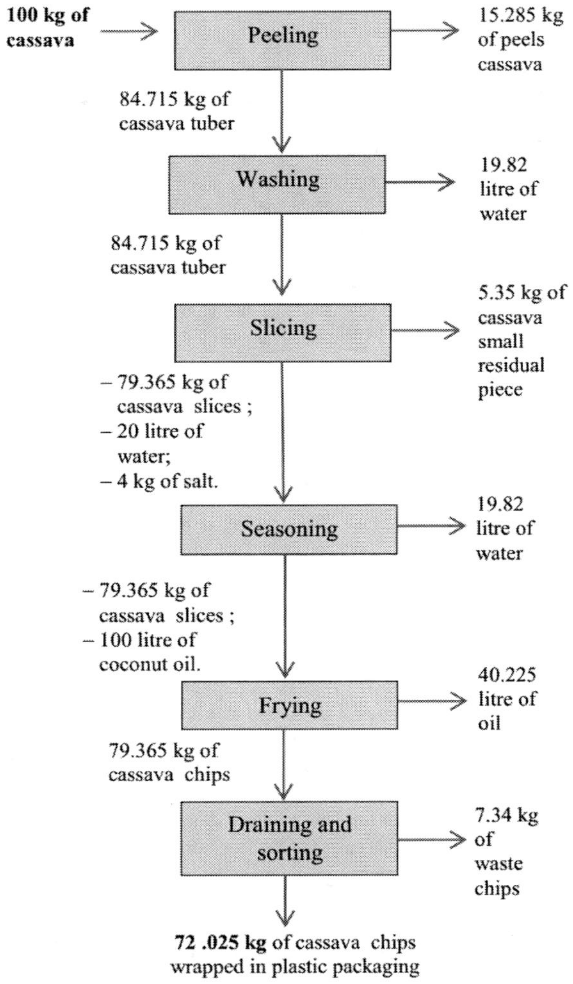

Figure-1. Mass balance at every stage of the production of cassava chips.

Peeling

This peeling process is manually done using the knife. For every 100 kg of cassava, production process generates waste as much as 15.285 kg of cassava peel. Peeled cassava is then soaked in a tub of water for washing.

Washing

The washing process is gradually done until the whole amount of 84.715 kg of cassavas is completely cleaned. Gradual immersion process requires 10 pieces sized 25-litre bucket filled water. During this process, it wastes 19.82 liters of water.

Slicing

The whole cassava which amount 84.715 kg cut using the manual cutter designed in circular thin. The slicing process produces small residual pieces as much as 5.35 kg.

Seasoning

The soaking process of 79.365 kg cassava slices in shaped thin circle into 20 liters water which has been given saline 4 kg and whiting 2 kg takes 1 hour to make the salt seep into slices and whiting is able to reduce the water content in cassava. After 1 hour, it is drained and sliced using an ordinary filter before fried. In this process, the waste water soaking is 18.326 liters.

Frying

Frying is done twice by using 50 kg of cooking oil. 79.365 kg cassavas are gradually fried until cooked. The frying process is done in two stages. The first stage uses 25 kg cooking oil and takes 15 minutes before subsequently drained by the filter. The second stage uses 50 kg cooking oil which has been heated for 10 minutes before used to fry the chips. The remaining oil from the frying process is 40.225 kg and would be a waste the third use. These processes use a frying pan with the diameter of 50 cm heating energy source of Liquid Petroleum Gas (LPG) are placed side by side.

Draining and sorting

79.365 kg of cassava that has been fried subsequently drained of its oil use newspaper until losing the heat then sorting the chips and packing it using 1 mm polypropylene plastic. The sorting process aims to select chips that may be damaged by frying. This last process produces 72.025 kg cassava chips.

DISCUSSIONS

Waste Identification

Wastes produced by this cassava chips industry are cassava peel, residual water from washing and soaking, pieces of cassava from the slicing process, remaining cooking oil, and pieces of cassava chips from frying.

Cassava peel waste produced is 15.285 kg per day, but this waste is not discharged into the environment because it is taken by local farmers using as cattle feed. The wastewater from washing or soaking along with the remaining of seasoning are directly discharged into a small drainage channel away without any processing. This liquid waste is composed of 19.82 liters of water containing dissolved cassava starch and 18.326 liters of water marinade which contains salts and whiting. Cassava pieces that can not be sliced into chips are dried using the sun for 7 days to be stored in sacks and powdered. The flour is used to make cakes for own consumption.

Cooking oils used in the frying process will be the waste after 3 times used, because, after the third use, the cooking oil will change the color of cassava chips, decrease the quality of taste, and make the products become not good for consumption. The waste of cooking oil is collected around 80.45 kg per week into four plastic containers of 25 liters.

In the final production process, the pieces of damaged or burnt chips because of the frying process showed the efficiency of the production process. This can be minimized by using good equipment that has the ability to cut into same thickness, yet it results in the same size and same cooking time.

Opportunities in implementing cleaner production

Implementing cleaner production in cassava chips industry prevents or minimizes environmental impacts through the product life cycles from raw material to be a product until final disposal. The strategy applied in cassava chips industry are strategies to look at the process and the final product.

The strategy carried out by observing the damage prevention process in the form of raw materials, minimizing energy use, eliminating the use of raw materials that are poisonous and dangerous as well as reducing levels of toxins contained within the emissions and waste before leaving the process. Strategy on the final product includes reducing environmental impact throughout the product cycle from the manufacture of the product until final disposal.

Feasibility assessment

Technical feasibility

Technical feasibility assessment is done by weighting towards the implementation of cleaner production. Some cleaner production alternatives offered by the condition of technical support. Technical analysis of these alternatives are:

Good-housekeeping

Good-housekeeping implemented aims to keep the environment from potentially contaminated in the various stages of the production process. Water spills on residual leaching or soaking cassava seasoning is one of the stages that could potentially contaminate the environment. At the production area, the majority of SMEs typically have clay floor which used in displacement washing and soaking location adjacent seasoning with a floor made of cement layer so it is easy to clean. Washing and soaking in the production process are done at the same hygiene which does not meet any safety standards because the clay floor spilled water may become slippery. The technical aspect for maintaining cleanliness is important to note as well as owners and workers require awareness.

Improving operation procedures

This effort is minimizing the impact on the effectiveness of production time. Production can be done every day with each operation process timing appropriately. This improvement can also be done by creating a Standard Operation Procedure (SOP) in implementing production process. Standard is the basis for labor in doing his job.

Equipment modifications

Modification of the equipment is required to minimize the residual pieces of cassava slicing process. In the process of production, every 100 kg of cassava produced 5.35 kg of small pieces per day. If production is done every day, in a month it will produce 160.5 kg of waste. It means the waste produced during the one-month production process is about 150% of the raw materials a day.

At this present, the waste is dried using sunlight just above the bamboo to be made into flour and use to make various cake for own consumption. The waste is quite a lot because of the unstandardized equipment for cutting process. In this process, the cutting blades are assembled or can be said to be conventional. By using the conventional tools, the waste produced also affect the next process of soaking and frying. Only because of the cutting tool used cannot produce pieces with the same thickness, this condition will get worse if the process is done by five different workers.

The different thickness of Cassava slice will affect the absorption of seasoning and the cooking time. Therefore, as expected that after the frying process, the chips will be found into a different level of cooked, such as sufficiently cook, evenly cooked, and charred. Overall, cutlery affects the quality of cassava chip products. Failure to redesign the equipment will potentially reduce the profits earned by the owner of SMEs.

Reuse leftover cooking oil

In the production of cassava chips, the owners do not reuse cooking oil for frying. It because after being used for three times the cooking oil will affect the color of cassava chips though the color of the remaining cooking oil is still clear. The owner of SMEs agreed to maintain the quality of the products to give satisfaction to the consumers. The waste of cooking oil is 80.45 kg per week and in a month it becomes 321.8 kg. This leftover cooking oil is purchased by others for the production of liquid soap. Waste cooking oil is purchased at a price of IDR 50,000.00/plastic containers.

Economic feasibility

Assessment is based on the economic feasibility of the profits and payback periods. Earnings calculation is done by comparing the earning before and after implementing cleaner production. The SMEs incomes through cassava chips production prior to implementing cleaner production is about IDR 150,000,000.00 in a year. Production cost in a year is IDR 68,800,000.00 so the profits is IDR 81,200,000 in a year.

Good-housekeeping

Good housekeeping is applied to the spill or spills of water in the washing process and soaking seasoning, divert the process that is originally located above the clay floor into place with cement floor which is only 2 meters away from the pan. Although the spilled water does not affect the profits but it causes unhygienic environment.

Improving operation procedures

Improving operation procedures for the production process has no effect on the economic or financial savings but on the contribution of efficiency which gains by the workers in the production process of cassava chips.

Equipment modification

Equipment modifications done is by introducing a slicer machine to replace conventional mower. It aims to minimize the wasted pieces on every production process. Slicer machine has a capacity of 50 kg per hour by using an electrical operating power of 250 watts. This means that the process of slicing the raw material for a total of 100 kg per day only takes 2 hours.

The production process for one month would require as much as 15 kWh of electricity supply. If electricity tariff (TDL) for households in 2015 amounted to IDR 1321.00, within one month (30 days of production) the costs required for the machine slicer is IDR 1.9815. A very low cost compared to the loss resulted by conventional cutting knives. Slicer machine is assumed to be able to minimize the waste disposed of by 80 percent as part of a cutter that uses stainless steel materials are designed to be able to set its thickness so it cuts the same thickness optimally. This slicer machine has a payback period 0.083 years or about 1 month then the cost by using this technology can be returned. The use of this equipment for the SMEs also does not require power adjustment because power is managed by 1.300 kWh.

Reuse leftover cooking oil

Purchasing leftover cooking oil by the liquid soap manufacturer gives its own profits for the owner of SMEs because the leftover cooking oil from the production process of cassava chips is purchased for about IDR 200,000 or about IDR 800,000 in a month.

Environmental feasibility

Selection of cleaner production alternatives based on the environment consideration namely through the introduction of machine slicer, good housekeeping, and reuse of leftover cooking oil. There are alternatives related to the environmental advantages of cleaner production: 1) the efficiency of water; 2) the efficiency of electricity use; 3) sanitation and hygiene company and workers; and 4) capacity of the environment use.

Implementing cleaner production

Some cleaner production alternatives offered can be implemented in accordance with the results of the feasibility assessment. Alternative production that showed the highest score is the modification of equipment. This alternative is easily implemented by SMEs.

CONCLUSIONS

Based on the previous discussions, it can be concluded that the SMEs of cassava chips are not familiar with cleaner productions. Implementation of cleaner production may include good housekeeping, reducing, and reusing. From the assessment of alternative eligibility criteria, the equipment modifications are the main factor in implementing cleaner production that drives the profits by providing efficiency of cutting as much as 80 percent and optimizes the profits into 57.62 kg in a month or 691.44 kg in a year. If the price of cassava chips is IDR 40,000 in each kg, then it would save IDR 27,657,600 in a year.

REFERENCES

Amsalu N and Esubalew G. 2011. Soaking and drying cassava roots reduced cyanogenic potential of three cassava varieties at Jimma, Southwest Ethiopia. World Journal of Agricultural Sciences 7(4):439-443.

Asian Development Bank. 2000. Industry and the Environment. Chapter 6 in Asian Environmental Outlook 2001. ADB Report.

Behnsen, A., Hilebrecht, K., Kurzingar E., and Kranert, M. 2002. Waste minimization in small and medium enterprises. International conference of appropriate environmental and solid waste management and technologies for developing countries. Turkey, Istanbul: ISWA.

Coordinating Ministry For Economic Affairs. 2011. Masterplan for Acceleration and Expansion of Indonesia Economic Development.

FAO. 2000. The world cassava economy: facts, trends, and outlook. Food and Agriculture Organizations of the United Nations. FAO report, pp 7-15.

Fresner J. 2010. The theory of inventive problem solving (TRIZ) as option generation tool within cleaner production projects. Journal of cleaner production 18 (2010) 128-136.

Okunade D.A., and Adekalu K.O. 2013. Physico-chemical analysis of contaminated water resources due to cassava wastewater effluent disposal. European international journal of science and technology 2(6), pp.75-84.

Ramjeawon T. 2004. A case study of cleaner production opprtunities in small and medium enterprises on the island of Maritius. Electronic Green Journal 1(20):1-14.

Sohair I, Abou –Elela, Fayza A.Nasr, Saber A. El-Shafai. 2008. Wastewater management in small-and medium-size enterprises: case studies. Environmentalist 28: 289-296.

Sugino T and Mayrowani H. 2009. The determinants of cassava productivity and price under the farmer's collaboration with the emerging cassava processor: A case study in East Lampung, Indonesia. Journal of Development and Agricultural Economics Vol 1(5), pp.114-120.

Van Berkel R. 2007a. Cleaner Production and Eco-Efficiency. In Marinova D, Annandale D, and Phillimore J (eds). The international handbook of environmental technology management. Edward Elgar Publications, Cheltenham, UK, pp 67-93.

Van Berkel R. 2007b. Cleaner Production and Eco-Efficiency in Australia small firms. International Journal Environmental Tehnology Manage 7(5-6):672-693.

Van Berkel R. 2011. Evaluation of the global implementatio of the UNIDO-UNEP National Cleaner Production Centres (NCPC) Programme. Clean Technology Environmental Policy 13:161-175.

Udoro E.O., Gbadamosi O.A., and Taiwo A.K. 2008. Studies on the production and utilization of dried cassava chips as human food. International Union of Food Science and Technology.

UNEP, 1991. United Nations Industrial Development Organization's Audit and Reduction Manual for Industrial Emissions and Wastes.

White, W.L.B., D.I.Arias-Garson, J.M. McMahon and R.T. Sayre. 1998. Cyanogenesis in cassava: the role of hydroxy nitrile lyase in root cyanide production. Plant Physiology 116:1219-1225.

MOMENT CONTRIBUTION CAPACITY OF TENDON PRESTRESSED PARTIAL ON CONCRETE BEAM-COLUMN JOINT INTERIOR ACCORDING TO PROVISIONS ACI 318-2008 CHAPTER 21.5.2.5(c) DUE TO CYCLIC LATERAL LOADS

Made Dharma Astawa[1], IGusti Putu Raka[2], Tavio[3]
[1] Civil Engineering Departement, University of Pembangunan Nasional"Veteran", Surabaya-East-Java- Indonesia and Doctoral Student of Civil Engineering (structure), Faculty of Civil Engineering and Planning-ITS Surabaya-Indonesia.
[2,3] Profesor at Department of Civil Engineering, Faculty of Civil Engineering and Planning-ITS, Surabaya.
E-mail: mdastawa@upnjatim.ac.id, masdawa@yahoo.com

ABSTRACT

This research designed a partial prestressed concrete beam-column with reinforced concrete interior joint, using square columns of 400/400 mm, reinforcement 6 D16 + 4D13, section beam 250/400 mm, tensile reinforcement 5 D13, compression reinforcement 3 D13 + 2 strand tendon D12,7 mm , and joint without plastic hinge, then tested in laboratory with lateral cyclic loads on peak column, static axial load 1120 kN on the centre column, to get the tendon capacity to assume positive and negative bending moments due to lateral load, according to provisions of ACI 318-2008 part 21.5.2.5 (c). Test results showed that the moment tendon contribution on beam section, in the tensile area, the positive and negative moment both on the left side or the right side column are all qualified (<25 %). In compression area, the positive moments on left and right columns are not qualified at all (> 25 %). As for the negative moment, either left or right side column are all qualified (< 25 %). Ductility, compression, and tensile ductility on drift ratio 3.50% are all qualified (>4.0). Although the contribution of positive moment capacity tendon in compression areas does not qualify, in overall, the reliability and ductility of the structure qualify.

Keywords: Interior beam-column joint, partial prestressed, tendon moment contribution, ductility.

INTRODUCTION

In Storey Frame Building Structure design, especially in areas of Strong Earthquake Region, shall be designed with earthquake resistance (*SNI 03-1726-2012*). This research aimed to design a model element beam-column joint structure capable of ductile behavior during the earthquake lateral work on the building frame structure (*Purwono R et al., 2005*). The Beam-column joint is composed of the elements of *column reinforced concrete - partially prestressed concrete beams*, that monolithicly connects without special plastic hinge design (*Blakeley & Park., 1971*). In another reference, a comparative study on exterior RCC beam-column joint is subjected to monotonic loading (*S.V. Chaudhari et al., 2014*).

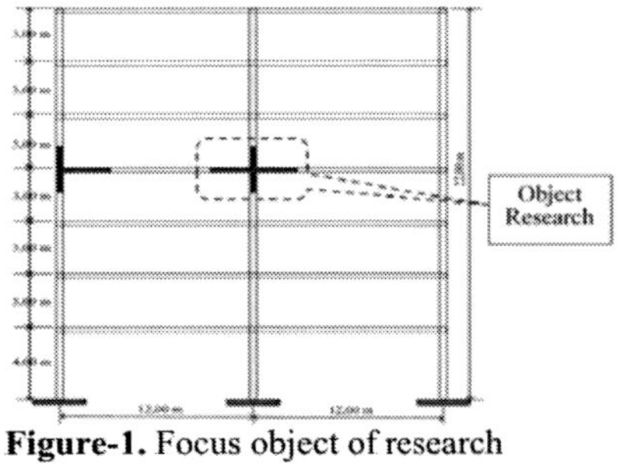

Figure-1. Focus object of research

The selection of the model 's structure, especially on long-span beam structure would be more economical than full Prestressed Concrete structures in Building Structures. Due to the structure analysis of prestressed full strength, the tendon is considered to work with full withstand bending loads, regardless of the share of the burden of reinforcing steel bending. The steel is considered as installed practical reinforcement so that the design would have less economical structure (*Lin., 2002*). An illustration of Beam-Column Joint allocation to be studied is shown in the picture of building's Framework above.

Significance of Research

Referring to the design of Beam - Column Joint models with elements of Partial Prestressed Concrete Beam-Reinforced Concrete Column Interior, forming a partnership between the strand tendon and rebars in bearing of bending load and lateral load is based on references as follows: acceptance criteria of the moment frames based on structural testing and commentary (*Ronald Klemencic et al., 2005*), guide for testing reinforced concrete structural elements under slowly applied simulated seismic loads (*Sergio M Alcocer et al., 2013*), proposed revisions to 1997 NEHRP recommended provisions for seismic regulations for precast concrete structures Part-2, and seismic force resisting system (*Hawkins & Gosh., 2000*). The prestressing steel shall not contribute to more than one-quarter of the positive or negative flexural strength at the critical section in a plastic hinge region and shall be anchored at or beyond the exterior face of the joint. (*ACI 318, 2008*).

This research aimed to calculate the contribution of tendon moment capacity in total moment beams capacity at the Joint Bearing and the level of ductility Joint Structure Interior. From this study, it was expected to give a result in a more economical of design structure than the full Prestressed Concrete Beams structure, especially for buildings that need to widen the span of the rooms (in concept) (*Made D Astawa et al., 2013*).

Specimen Beam-Column Joint Interior

There is one piece of Model Structure Beam-Column Joint Interior, with full-scale design, which is using Partial Prestressed Concrete Beams elements-Reinforced Concrete Columns, by referring to the

DOI: 10.1051/matecconf/20165804005

procedure for calculation of concrete structures for building construction (*SNI 03, 2013*).

METHODOLOGY

First, we made the one piece specimen of Beam-Column Joint with Partial Prestressed Concrete Beams element-Column Reinforced Concrete, specimen forming of mold, and then the concrete was cast. After 28 days, the initial force prestressed of beams (stressing) was given. Within an interval of 24 hours, the specimens were tested with static axial loads and lateral cyclic loads on the peak of the column.

Design of Load

Lateral load plan is cyclic load, the magnitude of the lateral cyclic loading work is controlled by the Drift ratio ranging from 0.00% to 3.50%, according to NEHRP 1997 or ACI 318-2008, ACI 374.1-05 and 374.2R-13, that is categorized as Cyclic load (pseudo-dynamic). Determining Drift achievement ratio of 3.50% is enough to evaluate the qualified ductile structure. Axial load static vertical column has a stability of the structure by 1120 kN.

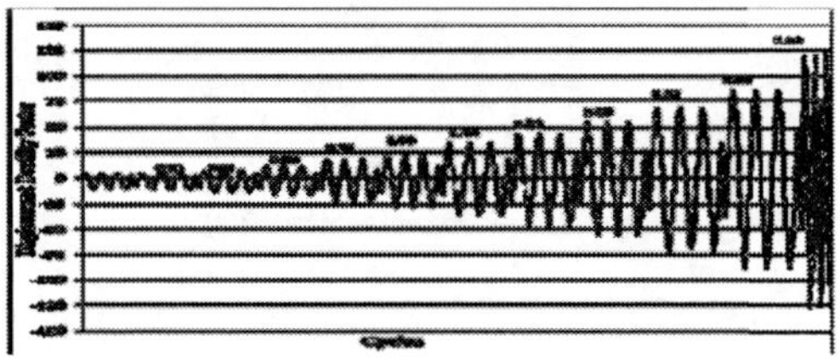

Figure-2. Cyclic Load Diagram (Hawkins & Gosh, 2000)

Test Set-up Spesimen

The test specimens were conducted using the actuator with a loading capacity of 1000 kN for lateral load and 2000 kN for vertical loading.

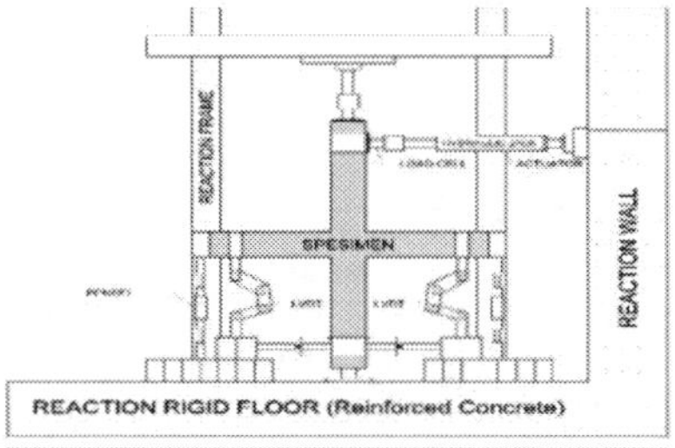

Figure-3. Models test set-up Specimens

Design of Partial Prestressed Beams elements

The dimension of the beam section was made to 250/400 mm, concrete cover is made of 35mm thick with the following Materials Properties:

Top Reinforced 5D13 with As = 663.7 mm^2

Bottom Reinforced 3D13 with As=398.2 mm^2

Using 1 tendons and 2 strands, with diameter (D) = 12.7 mm, sectional area Aps = 2(98.71) = 197.42 mm^2, and fps=1030Mpa. In order to meet the requirements of under-reinforced, then:

$$\omega_p = \frac{A_{ps} \cdot f_{ps}}{b.d.f'_c} \leq 0,3 \qquad \text{...................(1)}$$

Effective high d =400–35–Ø8-½D_{13}= 350,5mm

$$\omega_p = \frac{197,42\ (1030)}{250\ (350,5)40} = 0,06 < 0,3 \ldots (\text{OK})$$

Comparing the tendon section area and concrete section area according to ACI and UBC, then:

$$\frac{197,42}{250\ .400} = 0,002 < 0,007 \ldots (\text{OK})$$

The effective high requirements of concrete compressive stress block are taken: $a = 0,235d \leq 0,2h\ s/d\ 0,25\ h$

a=0.235(350.5) = 82.4 mm ≤ 0.25(400); 82.4<100…(OK).

Spacing stirrup partially prestressed beams (ParkThompson, 1980) is between 1" to 7" (25.4 mm–78 mm), and chosen ∅ 8 - 75 mm, so it is in accordance to the provisions.

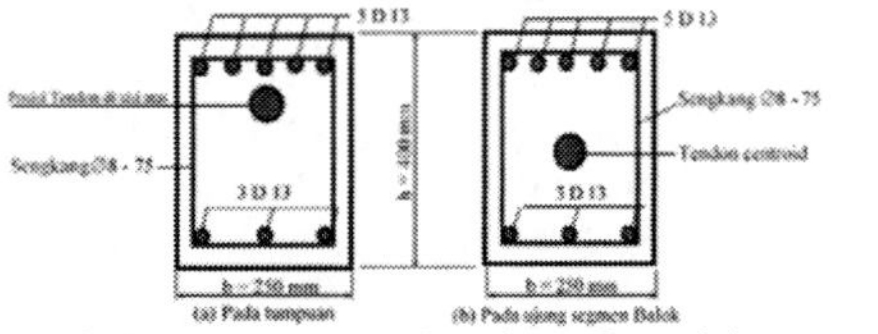

Figure-4. Concrete Beam Section of Partial prestressed

Flexural beam load capacity

Nominal Momen Mn = T(d-a/2) = (663,7.400) (350,5-½.82,4)10^{-6} =82,12 kNm

Tensile capacity of prestressed tendons is:

$f_n = f_{ps} \cdot A_{ps}$ = 1030 (197.42). 10^{-3} = 203.34 kN

With Partial Prestressing Ratio (PPR):

$$PPR = \frac{A_{ps} \cdot f_{ps}(d_p - \frac{a}{2})}{A_{ps} \cdot f_{ps}(d_p - \frac{a}{2}) + A_s \cdot f_y(d_s - \frac{a}{2})} \qquad \text{........................... (2)}$$

Should meet: $0 < PPR < 1$

PPR = 0.56, then 0 < 0.56 < 1 ... (OK)

Effective high beam is computed from Tendons:

d_p=400-35-8-25-25-50/2=282 mm

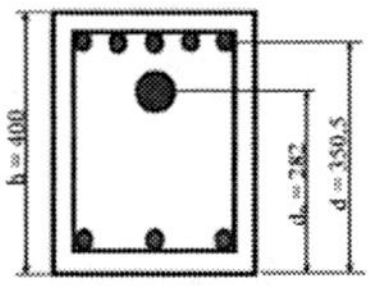

Figure-5. The effective high beam

Controls global reinforcing index, (ω), is the ratio of strand tendons and reinforcement used for calculation of moments of a strong structural component:

$$\omega = \omega_p + \omega_s - \omega'_s \leq 0,36\beta_1, \qquad \text{........................... (3)}$$

After the calculated: β_1=0.77 ; ω_p=0.06; ω_s=0.076; ω_s'=0.045, making: $\omega = 0,06 + 0,076 - 0,045 = 0,091$

$\omega < 0,36\beta_1 \rightarrow 0,091 < 0,36.0,77 \rightarrow 0,091 < 0,277 \ldots (\text{OK})$

Horizontal Actuator capacity = 1000 kN, effective 80 % = 0.8 (1000) = 800 kN. In the design of load capacity of the structure, all of the conditions of the structure specimens were taken into account in the elastic state, so the structure has not cracked.

Actuator moment due to lateral forces P:

800 kN (1.00) . m = 800 kNm, this became the primary moments. The members' stiffness:

After the calculation: $k_{1,3} = \frac{3\,EI_1}{L_1} = 4,0$; $k_{2,4} = \frac{3.EI_2}{L_2} = 6,4$

With moment distribution factors:

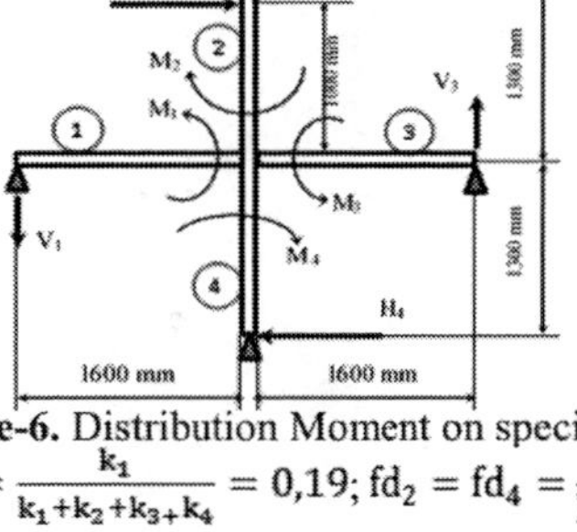

Figure-6. Distribution Moment on specimens

$$fd_1 = fd_3 = \frac{k_1}{k_1 + k_2 + k_3 + k_4} = 0,19; fd_2 = fd_4 = \frac{k_2}{\sum k} = 0,31$$

$M_1 = M_3 = 0.19\ (800) = 152$ kNm ; $M_2 = M_4 = 0.31\ (800) = 248$ kNm, Momen of Reinforcement:
M_{n1}=As.fy(d-a/2)=663,7.400(365-82,4/2)10^{-6} =85,96 kNm
Mn of Tendon Prestressed:
$X = a/\beta_1 = 82.4/0.77 = 107$ mm; e = 282-107 = 175 mm
M_{n2} = F (e) = 203.34 (175).10^{-3} = 35.58 kNm

Design of elements Reinforced Concrete Columns

The dimension of column sectional is 400/400 mm.

a. Column reinforcing and stirrup design

According to SNI 03-2847-2002 (part 23.4.3), column longitudinal reinforcing ratio is taken between $0.01 \le \rho_g \le 0.06$. In this design, it was taken as $\rho_{g(min)} = 0{,}01$, so it still qualifies.

Reinforcing section area As = ρ_g. Ag = 0.01 (400.400) = 1600 mm^2, used 6 D_{16} + 4D_{13}= 1837.3 mm^2.

The sectional area of stirrups was taken from value equation according to the capture SNI 03-2847-2002 (section 23.4.4) for a square-shaped stirrup as follows:

$$A_{sh} = 0{,}09\left(\frac{s.h_c.f_c'}{f_{yh}}\right) \quad (4)$$

The outer clearance between stirrups is 400 - 2(40) = 320 mm, while stirrups range from center to center towards high column=320-10=310 mm, so that: h_c = 310 mm, and stirrups range is determined as 50 mm (for single stirrups). The sectional area of horizontal stirrups, after the calculation, was: A_{sh} =139.50 mm^2; while sectional area one foot:

$A_{sh}1$ = 139.5 / 2 = 69.75 < area 1Ø10 = 78.50 mm^2…(OK)

Planning of shear stirrups in the joint

According to the provisions of SNI 03-2847-2002 part 23.4.4):

$$A_{sh} = 0{,}3\left(\frac{s.h_c.f_c'}{f_{yh}}\right)\left[\left(\frac{A_g}{A_{ch}}\right) - 1\right] \quad (5)$$

Or $A_{sh} = 0{,}09\left(\frac{s.h_c.f_c'}{f_{yh}}\right)$ Stirrups of Ø10, cover of concrete = 40 mm

A_{ch}=(320)(320)=102400 mm^2; h_c=20–2(0.5.10)=310 mm.

Stirrups space S, was taken as 50mm

After the calculation:

$M_n = M_{n1}$+25 % (M_{n2}) = 85.96 +0.25(35.58) = 94.86 kNm < 152 kNm (OK)A_{sh} = 261.56 mm^2; or A_{sh} = 139 mm^2, used a great value; $1\emptyset 10 \rightarrow As = 78{,}5\ mm^2$. With the amount of stirrups: 261.56/78.5 = 3.33, it takes 4 stirrups, but since S = 50 mm and an empty of high space beam=400-2(35)-2(8)-2(13) =288mm, we used the amount of stirrups = 288/50 = 5.76 rounded as 6 pieces with A_{sh} = 471 mm^2.

The calculation of shear strength of joint is:

$$V_{col} = \frac{2T_b.Z_b + V_b.h_c}{l_c} \quad (6)$$

$$V_{jh} = V_{col}\left(\frac{l_c}{Z_b} - 1\right) - V_b\left(\frac{h_c}{Z_b}\right) \quad (7)$$

V_b = 0.4. 0.25. 24(1.6) + 192/1.6 = 123.84 kN

After the calculation: V_{col} = 68 kN; V_{jh} = 274.30 kN.

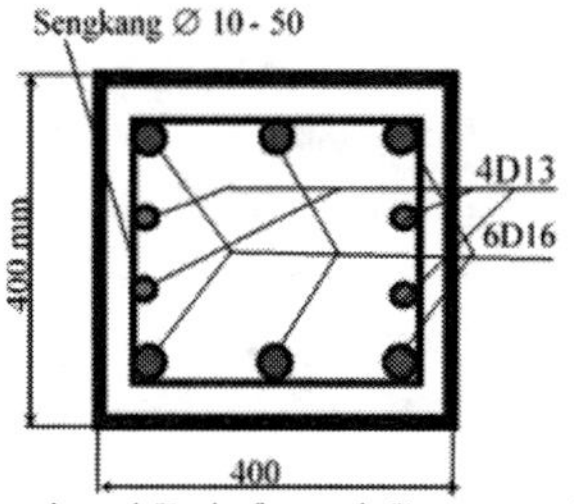

Figure-7. Sectional Reinforced Concrete Columns

Fabrication and assembling specimen

Cutting and bistaat reinforcement in such a way was adapted to the design of beam-column joint specimens Interior. Furthermore, assembling was done according to the shape of the specimen Interior. The results of the model specimen assemblies are as illustrated in the following figure:

Table-1. Recapitulation specification Beam-Column Joint specimens

Type of Structure	Element of Structure (mm)	Reinforcement	Stirrups	Tendon quantity	Specimen amount
Interior Beam-Column Joint	Beam 250/400	Pull Reinforced 5D_{13}	Ø8 - 75	1(2 strand)	1
		Press Reinforced 3 D_{13}			
	Column 40/40	6 D_{16} + 4D_{13}	Ø10 -50	-	

LVDT, Wire-gauge, and Starin-gauge

Linear Variable Displacement Transducer (LVDT), Wire-Gauge (WG) and Strain-Gauge (SG) are used to measure the amount of strain due to deformation (dift) on the specimen when the cyclic lateral load works. LVDT and WG were installed on the outside of the columns and beams while SG was installed on the fiber outer parts.

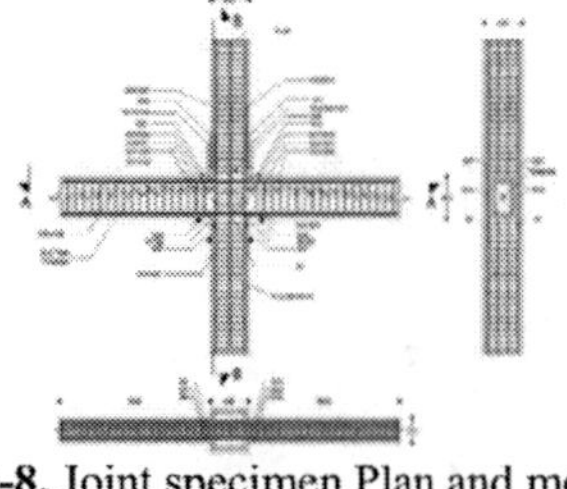

Figure-8. Joint specimen Plan and mounting location Strain Gauge

Concrete beams and columns inside of the Joint were pulled and pressed while SG for reinforcement and Strand was installed in each reinforcement beams and columns and Strand on the side of the joint.

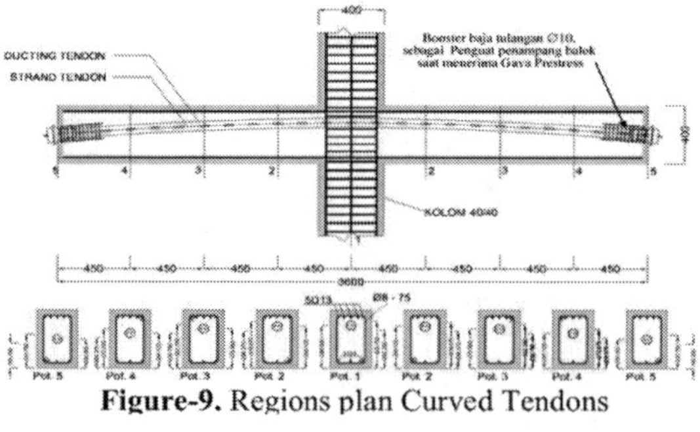

Figure-9. Regions plan Curved Tendons

Scope of Analysis

Moment contribution of Tendon Analysis

Moment tendon contribution at the moment capacity of the beam is according to the provisions ACI 318-2008 part 21.5.2.5(c) which were ≤ 25 % from total capacity, both positive and negative moments.

Structure Ductility

Ductility analysis of structure was taken from the maximum lateral deformation which is proportional to the initial deformation structure.

RESULTS

Hysteretic curve test results

Test results used hysteretic curve data, with several sensors were installed at critical points, among others: Linear Variable Displacement Transducer (LVDT), Wire-gauge (WG) and Starain-gauge (SG). Each outcome data, at any point, will be presented sequentially in the form of graphs.

Hysteretic curve Wire-gauge (WG) was mounted on one point, top of the column, because the maximum lateral deformation will occur at the top of the column at the location of the working force by means of Cyclic Lateral Horizontal Actuator. Results of Wire-gauge recordings are perfect for maximum strain deformation capability of these tools, up to 500 mm, so the deformation of the wire specimen is not high enough to break up. The Curves of Hysteritic LVDT and WG on top of the column as a representative of other points are as follows:

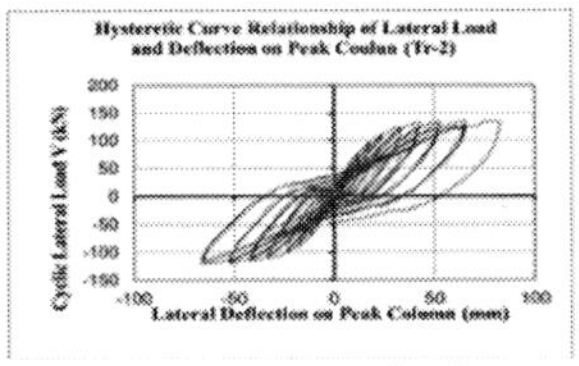

Figure-10. Hysteretic Curve LVDT on Peak Column

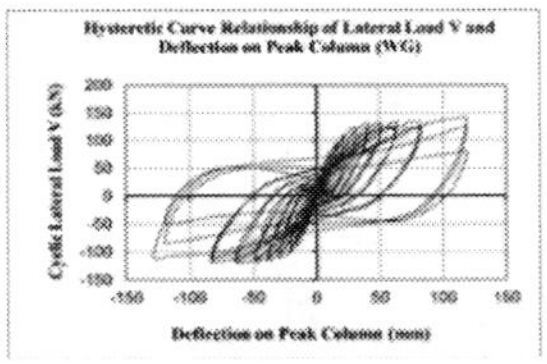

Figure-11. Hysteretic Curve WG on Peak Column

Hysteretic curves Strain-gauge (SG) beam are represented by the SG-6, SG-7 and SG-8, while for reinforcement Pull the right side of the column were represented by SG-19, SG-20 and SG-21. As for the press and the left side of the column, reinforcement is represented by SG-11 and SG-12, on the right side column is of the SG-24 and SG-25. SG-24 is dead, but that only represented the SG-25. Strain Gauge Strand Tendons is attached to the left column SG-31 and SG-32, while right-hand column is of the SG-33 and SG-34, but only SG-31 that lives, so it only represented the SG-31 alone. SG Picture Hysteritic curve will be presented just as the following. Keep in mind that the conditions recorded by the strain-gauge can not be as perfect as that of recorded by the LVDT.

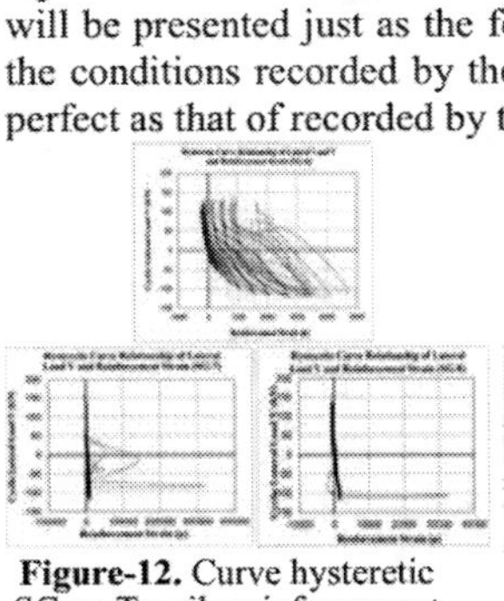

Figure-12. Curve hysteretic SG on Tensile reinforcement the left side column

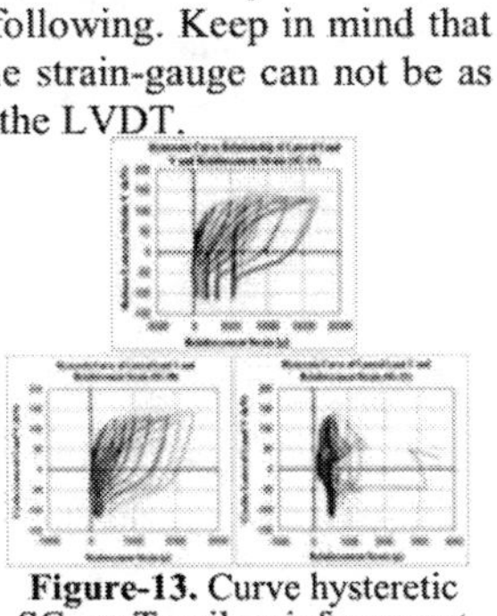

Figure-13. Curve hysteretic SG on Tensile reinforcement the right side column

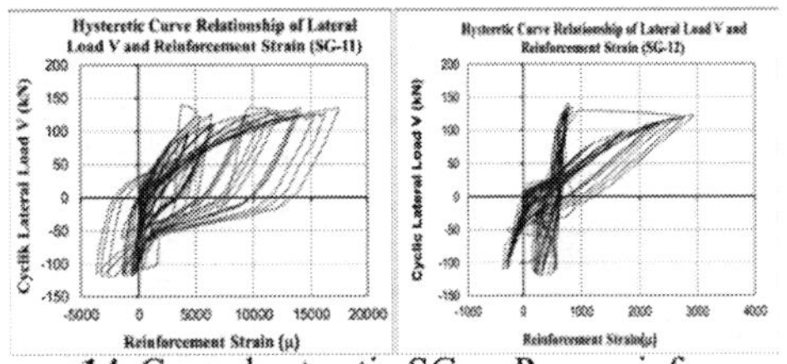

Figure-14. Curve hysteretic SG on Press reinforcement on the left side column

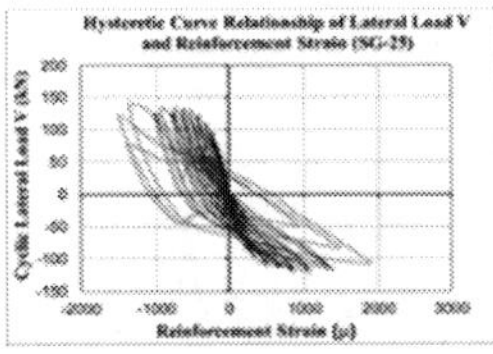

Figure-15. Curve hysteretic SG on Press reinforcement on right side column

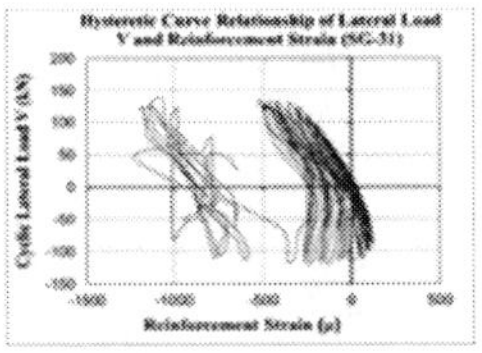

Figure-16. Curve hysteretic Strand SG on left and right side columns

Test result of lateral cyclic loads and drift ratio

Lateral Cyclic loading results of a mock earthquake load (Pseudo-Dynamic) and magnitude of each load cycle can be read by the data logger, which can be seen from Thrust (press) and Pull force. The results of testing at each stage of deformation to inelastic stage at Story Drift 5.0% are arranged in a table as follows:

Table-2. Lateral Load V Data and Deflection at peak Column

No.	Press Load V	δ Tr2	Story Drift	No.	Pull Load V	δ Tr2
	(kN)	(mm)	(%)		(kN)	(mm)
0	0	0	0.000	0	0	0
64	38.00	4.78	0.200	77	-33.10	-4.84
90	36.40	4.80	0.200	103	-32.40	-4.86
116	36.40	4.78	0.200	129	-32.10	-4.80
155	47.60	6.04	0.250	170	-38.30	-5.94
185	44.60	6.00	0.250	200	-40.30	-6.12
215	44.00	5.92	0.250	230	-40.00	-6.02
256	61.80	8.32	0.350	270	-58.20	-8.42
284	59.20	8.36	0.350	298	-55.90	-8.28
312	58.50	8.40	0.350	326	-55.60	-8.38
351	77.70	11.96	0.500	366	-75.40	-11.92
380	75.70	12.06	0.500	394	-72.80	-11.90
408	70.50	12.00	0.500	422	-70.50	-12.08
448	95.60	17.84	0.750	462	-94.60	-18.20
476	93.30	17.90	0.750	490	-92.00	-17.86
504	90.30	17.86	0.750	518	-90.60	-17.88
544	108.50	23.86	1.000	558	-106.20	-23.80
572	105.90	23.90	1.000	586	-103.20	-23.86
600	103.90	23.90	1.000	614	-101.90	-23.84
641	121.10	33.58	1.400	657	-118.10	-33.54
673	114.50	33.42	1.400	689	-115.50	-33.58
705	112.50	33.44	1.400	721	-113.50	-33.48
749	125.40	42.76	1.750	765	-124.70	-41.94
781	122.10	41.70	1.750	797	-121.80	-41.78
813	120.40	41.66	1.750	829	-120.40	-42.80
857	131.40	52.66	2.200	873	-130.00	-52.50
889	128.70	52.54	2.200	905	-127.10	-52.28
921	125.70	52.42	2.200	937	-125.40	-52.42
966	138.00	65.60	2.750	984	-133.70	-65.56
1002	134.00	65.50	2.750	1020	-131.70	-65.72
1038	130.70	65.54	2.750	1056	-130.40	-65.56
1086	138.30	82.78	3.500	1104	-137.60	-83.39
1122	133.70	83.29	3.500	1140	-134.30	-83.39
1158	130.70	83.29	3.500	1176	-131.00	-83.39
1201	142.30	119.28	5.000	1218	-135.70	-128.78
1217	131.40	119.78	5.000	1244	-129.40	-120.28
1233	114.80	119.48	5.000	1270	-99.30	-120.18
Load max.	142.30	119.28	Beban Maks:		-137.60	-83.39

Momen capacity of Strand Tendon

The following is an overview of Contributions Moment of Strand Tendons in accordance with ACI-318-2008 part 21.5.2.5(c). Force and moment of the strand tendons: Beam using two Strand Tendons, strain gauge attached life is SG-31 (see figure 16).
From the reading of this strain gauge:
Pull load = 119.1 kN; Press load = 126.7 kN.
Calculation of Moment Capacity Strand Tendons:

$$Mn^+ = \left(dp^+ - \frac{a^+}{2}\right)$$

After the calculation: Mn^+=28.68 kNm
For 2 strand = 2 (28.68) = 57.36 kNm

$$Mn^- = T^-\left(dp^- - \frac{a^-}{2}\right)$$

$$a^- = \frac{T^-}{0,85fc'.b} = \frac{126,7.10^3}{0,85.40.250} = 14,90 \text{ mm}$$

After the calculation: Mn^- =9.57 kNm
For 2 Strand = 2(9.57) = 19.14 kNm
Force and moment of reinforcement:
At pull reinforcement:
On left side column: Strain gauge SG-6, 7 and 8 (see figure 12) the result read is: Pull load 119.1kN, 117.5 kN, 119.1kN. Read result Press load: 139.3 kN, 137 kN, 133.7 kN. By taking the average of the three SGs:
T^+ = 1/3(119.1+117.5+119.1) = 118.57 kN
T^- = 1/3(139.3+137+133.7) = 136.67 kN.
There are 5 Pull reinforcements:
Total T^+ = 5(118.57) = 592.85 kN
Total T^- = 5(136.67) = 683.35 kN
On right side column: Strain gauge SG-19, 20 and 21 (see figure 13) the result read is: Pull load 118.8 kN, 117.5 kN, and 116.1kN. The result read at Press load is: 134 kN, 137kN, and 139.3 kN. By taking the average of the three SGs:
T^+ = 1/3(118.8+117.5+116.1) = 117.47 kN
T^- = 1/3(134+137+139.3) = 136.77 kN.
There are 5 Pull reinforcements:
Total T^+ = 5(117.47) = 587.35 kN
Total T^- = 5(136.77) = 683.85 kN
Calculation of Moment Capacity Pull Reinforcing:

$$Mn^+ = T^+(d^+ - \frac{a^+}{2})$$

On left side column:

$$a^+ = \frac{T^+}{0,85fc'.b} = \frac{592,85.10^3}{0,85.40.250} = 69,74 \text{ mm}$$

After the calculation: Mn^+ =187.12 kNm.

$$Mn^- = T^-\left(d^- - \frac{a^-}{2}\right)$$

$d^- = d^+ = 350,5$ mm ; a- = 80.40 mm; Mn^- =212.04 kNm.
On right side column:

$$Mn^+ = T^+(d^+ - \frac{a^+}{2})$$

a^+ =69.10 mm; Mn^+ =185.57 kNm.

$$Mn^- = T^-\left(d^- - \frac{a^-}{2}\right)$$

$d^- = d^+ = 350,5$ mm; a^- = 80.45 mm; Mn^- =212.18 kNm.
At press reinforcement:
On left side column: Strain gauge SG-11 and 12 (see figure 14), the result read is: Pull load 119.1kN, 111.2 kN, Press load: 134.3 kN and 123.1 kN
The average of the two SGs are:
T^+ = ½(119.1 + 111.2) = 115.15 kN
T^- = ½(134.3 + 123.1) = 128.7 kN
There are 3Pull reinforcements:
Total T^+ = 3(115.15) = 345.45 kN
Total T^- = 3(128.7) = 386.1 kN
On right side column: Starin gauge SG-25 (see figure 15), the result is: Pull load = 116.5 kN, Press load = 122.4 kN
There are 3 Press reinforcements:
Total T^+ = 3(116,5) = 349,5 kN
Total T^- = 3(122,4) = 367,2 kN
The calculation Capacity of Moment Reinforcement Press:

$$Mn^+ = T^+(d^+ - \frac{a^+}{2})$$

On left side column:
After the calculation: a^+ =40.64 mm; Mn^+ = 114.06 kNm.

$$Mn^- = T^-\left(d^- - \frac{a^-}{2}\right)$$

$d^- = d^+ = 350,5$ mm; a^-=45.42 mm; Mn^-=125.56 kNm.
On right side column:

$$Mn^+ = T^+(d^+ - \frac{a^+}{2})$$

a^+ =41.12 mm; Mn^+ = 115.31 kNm.

$$Mn^- = T^-\left(d^- - \frac{a^-}{2}\right)$$

$d^- = d^+ = 350,5$ mm ; a^- =43.20 mm; Mn^- =120.77 kNm
The moment capacity of Structure**:**
Moment Capacity is a combination of Moment Capacity Strand Tendons with Reinforcement Moment, either a positive or negative Moment.

The Compilation of Strand Tendons Moment with Pull Reinforcing Moment is:

a. On left side Column:
Mn^+ = 57.36 + 187,12 = 244.48 kNm
Mn^- = 19.14 + 212,04 = 231.18 kNm.

b. On right side column:
Mn^+ = 57.36 + 185.57= 242.93 kNm
Mn^- = 19.14 + 212,18 = 231.32 kNm.

The Compilation Strand Tendons Moment with Compression Reinforcing Moment is:

a. On left side column:
Mn^+ = 57.36 + 114.06 = 171.42 kNm
Mn^- = 19.14 + 126.56 = 245.70 kNm

b. On right side column:
Mn^+ = 57.36 + 115.31 = 172.67 kNm
Mn^- = 19.14 + 120.77 = 139.91 kNm

Contributions Moment Capacity Strand Tendons to Moment capacity Structure

In the pull reinforcement beam

On left side column:

Positive Moment:

Mn^+ Strand Tendon = 57.36 kNm, with total Capacity Mn^+ = 244.48 kNm
The Percentage of Strand Tendon contribution:

$$Mn^+ = \frac{57,36}{244,48} X\ 100\% = 23.46\% < 25\% \text{ (OK)}$$

Negative Moment:

Mn^- Strand Tendon = 19.14 kNm, with total Capacity Mn^- = 231,18 kNm.
The Percentage of Strand Tendon contribution:

$$= \frac{19,14}{231,32} X100\% = 8,27\ \% < 25\ \%\text{(OK)}$$

On right side column:

Positive Moment:

Mn^+ Strand Tendon = 57.36kNm, total Capacity Mn^+ = 242.93 kNm

The percentage of Strand Tendon contribution:

$Mn^+ = \frac{57,36}{242,93} X\ 100\%$= 23.61% < 25% (OK)

Negative Moment:

Mn^- Strand Tendon = 19.14 kNm, with total Capacity Mn^- = 231.32 kNm.

The Percentage of Strand Tendon contribution:

$= \frac{19,14}{231,32} X100\% = 8,27\ \% < 25\ \%(OK)$

In Compression Reinforcement beam:

On left side column:

Positive Momen:

Mn^+ Strand Tendon = 57.36 Nm, with total Capacity Mn^+ = 171.42 kNm

The Percentage of Strand Tendon contribution:

$= \frac{57,36}{171,42} X\ 100\% = 33,46\ \% > 25\ \%\ (Not\ OK)$

Negative Moment:

Mn^- Strand Tendon = 19.14 kNm, with total Capacity Mn^- = 231.32 kNm.

The Percentage of Strand Tendon contribution:

On right side column:

$= \frac{19,14}{231,32} X100\% = 8,27\ \% < 25\ \%(OK)$

On right side column:

Positive Moment:

Mn^+ Strand Tendon = 57.36 Nm, with total Capacity Mn^+ = 172.67 kNm

The Percentage of Strand Tendon contribution:

$= \frac{57,36}{171,42} X\ 100\% = 33,46\ \% > 25\ \%\ (Not\ OK)$

Negative Moment:

Mn^- Strand Tendon = 19.14 kNm, with total Capacity Mn^- = 139.91 kNm.

The Percentage of Strand Tendon contribution:

$Mn^+ = \frac{19,14}{139,91} X100\% = 13,68\% < 25\ \%\ ...\ (OK)$

Structure Ductility:

In Story Drift 3.50 % of the first cycle, it yielded boundary condition, while working press force cyclic: δy=82.78mm. The stable condition on Drift Ratio 0.75% in first cycle, δi = 17.84 mm.

Level of Ductility $\mu = \delta y/\delta i \geq 4.0$

μ = 82.78/17.84 = 4.64 > 4.0.........(OK)

While working pull force cyclic: δy = 83.39 mm

The stable condition on Drift Ratio 0.75% in first cycle, δi = 18.20 mm.

Level of Ductility $\mu = \delta y/\delta i \geq 4.0$

μ = 83.39/18.20 = 4.58>4.0.........(OK)

DISCUSSIONS

Maximum compressive lateral cyclic loading occurs at Drift Ratio of 5.0% in cycle 1 = 142.30 kN, δ_{max} = 119.28 mm, while the stable condition at Drift Ratio 0.75% in the first Cycle, δi = 17.84 mm. The maximum tensile force was happened on Drift Ratio 3.50% cycle-1 = 137.60 kN with δ_{max} = 83.39 mm, while the strain in stable condition at the drift ratio 0.75% in first Cycle, δi = 18.20 mm. The compression maximum load on the drift ratio to 3.50% cycle-1 = 138.30 kN. The analysis will be performed on condition drift ratio of 3.50 %, because if the drift ratio is 3.50%, the structural conditions are already qualified in accordance, therefore the frame structure of the building meets the structural reliability (NEHRP, 1997, ACI 318-2008, ACI 374.1- 05 and 374.2R-13).

From the analysis of the contribution of tendon moment capacity to the total capacity moment of beam, for area tensile, the positive moments in the left side column = 23.46% <25% (OK), the negative moments = 8.28% <25% (OK), the positive moments on the right hand side column = 23.61% <25% (OK), and the negative moments = 8.27% <25% (OK). In the compression area, the positive moments in the left side column = 33.46%> 25% (not OK), the negative moments = 8.27% <25% (OK), while on the right hand side column, the positive moment = 33.22%> 25% (not OK) and the negative moment = 13.68% <25% (OK).

The results of ductility analysis is that drift ratio is 3.50%, compression ductility μ = 4.64 > 4.0 (OK), and tensile ductility μ = 4.58> 4.0 (OK)

The case of cracks on the column was only hairline cracks with two lines cross. Story Drift 5.0% on the third cycle was inelastic conditions; reinforcing steel in the section beam of everything to break down. For Strand and reinforcing steel on top of the beam section through the drift ratio of 5.0%, it had no break. Here is a figure of the pattern of cracks on the beams and columns:

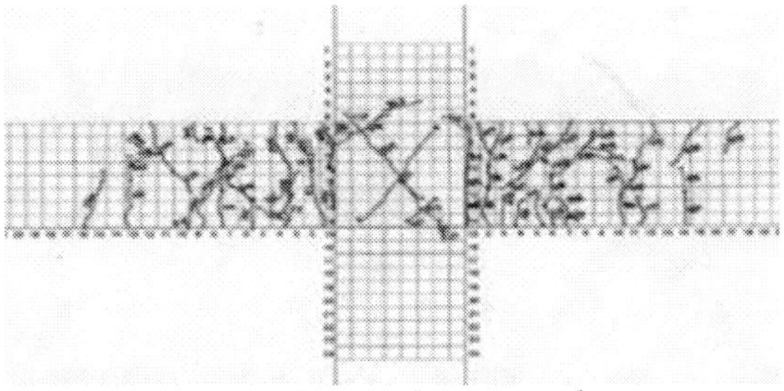

Figure-17. Fracture Pattern specimen

CONCLUSIONS

From the analysis test results of this model specimen, we can draw the following conclusions.

On the tensile region, the positive moment and negative moment both on the left side or the right side column are all qualified.

In the compression area, all positive moment on the left and right columns are not qualified, while, for the negative moment, the left or right-hand side column are all qualified.

Compression and tensile ductility on drift ratio of 3.50% is all qualified.

Although the contribution of positive moment capacity tendon in compression areas does not qualify, in overall, the reliability and ductility of the structure qualify.

ACKNOWLEDGEMENTS

In the writing and design of this research, we got a lot of help from various parties, especially Promoter Prof. Dr. Ir. I.G.P. Raka, DEA & Prof. Tavio, ST, MS, Ph.D., I would like to show my gratitude. Thank yous also go to LPPM-ITS for financial support through the

 DOI: 10.1051/matecconf/20165804005

decentralization Ditlitabmas Research Programme, Directorate General of Higher DP2M Kemdikbud, PT. Wijaya Karya Beton and PT. Freyssinet Indonesia, as well as to all those who have helped the success of this research.

REFERENCES

American Concrete Institut(ACI 318M-08)., 2008 ***Building Code Requirements for Structural Concrete and Commentary***. First Printing June 2008, 335 pp

Blackeley Roger W. G, Park Robert,. 1971 ***Seismic Resistance of Prestressed Concrete Beam-Column Assemblies***. ACI Journal. 217 pp

Barbara Chang, et al., 2014. ***Seismic Performance of Beam-Column Subassemblies with Hight Strength Steel Renforcement***. ACI Struktural Journal, 1330-1332 pp.

Dharma Astawa Made., 2006. ***Concrete Structure 1***, Teaching Modules, Grant Competition A_1, Civil Engineering Department UPN "Veteran" East Java, Surabaya, ISBN : 978-979-1005-21-0.

Dharma Astawa Made., 2007. ***Earthquake Engineering***, Teaching Modules, Civil Engineering Departement UPN "Veteran" East Java, Surabaya, ISBN:98-979-1005-20-3.

El-Sheikh Magdy T, et al., 1999. ***Seismic Behavior and Design of Unbonded Post-Tensioned Precast Concrete Frame.*** PCI Journal 512 pp.

Hawkins Neil M &Ghosh S. K ,, 2000. ***Proposed Revision to 1997 NEHRP Recommended Provisions for Seismic Regulations for Precast Concrete Structures Part 2 Seismic Force Resisting Systems.*** PCI Journal,35-42 pp.

I Gusti Putu Raka., Tavio., Made Dharma Astawa., 2014. ***State-of-the-Art Report on Partially-Prestressed Concrete Earthquake-Resistant Building Structures for Highly-Seismic Region.*** Elsevier Prosedia Engineering 00 (2014), 3-11 pp.

Liang., Parra., 2004. ***Seismic Behavior of Reinforced Concrete Column-Steel Beam Subassemblies and Frame Systems.*** Journal of Structural Engineering @ ASCE, February 2004, 311-314 pp.

Lin T. Y, Burns Ned. H, Indrawan Daniel, 1996 ***"Desain Struktur Beton Prategang Jilid 1"***, Erlangga Jakarta.

Lin T. Y, Burn Ned. H, Mediana., 2000. ***Desain Struktur Beton Prategang Jilid 2.*** Erlangga Jakarta.

Made D Astawa., Tavio., I G P Raka., 2013. ***Ductile Structure Framework of Earthquake Resistant of Highrise Building on Exterior Beam-Column Joint with the Partial Prestressed Concrete Beam-Column Reinforced Concrete.*** Elsevier Prosedia Engineering 54 (2013), 419-427 pp.

Nakano K, Tanabe K, Machida S, & wada S., 2001. ***Damage Controlled Seismic Design by Precast-Prestressed Concrete Structure with Mild-Press-Joint, Part 1***, Basic Consept of Design, AIJ Summeries of Technical Papers of Annual Meeting, Japan, 1-5 pp.

Naaman Antoine. E., 1982. ***Prestressed Concrete Analysis and Design***. McGrow-Hill Book Company, New York, San Francisco,Auckland, Bogota, Hamburg, Johannesburg, London, Madrid, atc.

Naaman Antoine E, Siriaksorn A., 1978. ***Analysis and Design of Partially Prestressed to Statisfy Serviceability Criteria.*** A Study Report by a Research Fellowship Award from tht Prestressed Concrete Institute and by the University of Illionis, Chicago, 12-17 pp.

Naaman Antoine E., 1985. ***Partially Prestressed Concrete (Review and Recommendation)***. special report, PCI. Journal, 112-116 pp.

Neil M. Hawkins., 1978. ***State-of-the-Art Report on Seismic Resistance of Prestressed and Precast Concrete Structures.*** PCI Journal/January-February 1978, 42-53 pp.

Paulay T, Priestley M. J. N., 1992. ***Seismic Design of Reinforced Concrete and Masonry Buildings.*** John Wiley & Sons, New York, Chichester, Brisbane, Toronto, Singapore.

Raka I Gusti Putu., 1987. ***Speech ITS Anniversary Scientific in order to XXVII.*** ITS, November 1987.

Raka I Gusti Putu., 1993. ***Ductility Sectional Round Hollow Prestressed Piles Compaction Centrifugal results***. Research Report, Civil Engineering Departement, ITS Surabaya., 3-8 pp

Salah Khalfallah., Dahbia Guerdouh., 2014. ***Tension stiffening approach in concrete of tensioned members***. International Journal of Advanced Structural Engineering 2014, 6:2, 2-6 pp.

Said M. Allam et al., 2013. ***Evaluation of tension stiffening effect on the crack width calculation of flexural RC members.*** Alexandria EngineeringJournal, January 2013, 164-172 pp.

Sergio M. Alcocer et al., 2013. ***Guide for Testing Reinforced Concrete Structural Elements under Slowly Applied Simulated Seismic Loads.*** Reported by ACI Committee 374.2R-13, 8-14 pp.

Shiohara Hitoshi., 2012. ***Reinforced Concrete Beam-Column Joints : an Overlooked Failure Mechanism and its Influence on the Future Seismic Design.*** Symposium on Earthquake and Resistance Building in Jakarta on Earthquake, Jakarta 17 Desember 2012, 5-7 pp.

SNI 03-2847., 2013. ***Procedure for Calcula-tion of Concrete Structures for Buildings,*** Concrete Laboratory ITS Surabaya.

SNI 03-1726-2012, ***Structure Earthquake Resistant Buildings.*** Badan Standardisasi Nasional (BSN), Jakarta.

S.V.Chaudhari., K.A.Mukane., M.A.Chakrabarti., 2014. ***Comparative study on Exterior RCC Beam Column Joint Subjected to Monotonic Loading.*** International Journal of

Computer Applications (0975 – 8887) Volume 102– No.3, September 2014, 35-48 pp.

Tavio., Benny Kusuma., Prio Suprobo., 2012. ***Experimental Behavior of Concrete Columns Confined by Welded Wire Fabric as Transverse Reinforcement under Axial Compression.*** ACI Structural Journal, May-June 2012, 2-4 pp

Thompson Kevin J., Park Robert., 1980. ***Ductility of Prestressed and Partially Prestressed Concrete Beam Section***. PCI Journal, March-April 1980, 219-223 pp.

UBC., 1997. ***Uniform Building Code.*** Chapter 16, DIV. IV-1997.

TEMPERATURE APPROACH OPTIMIZATION IN THE DOUBLE PIPE HEAT EXCHANGER WITH GROOVE

Putu Wijaya Sunu, Daud Simon Anakottapary, Wayan G. Santika
Department of Mechanical Engineering, Bali State Polytechnic, Badung-Bali, Indonesia.
E-mail: wijayasunu@pnb.ac.id

ABSTRACT

Heat transfer in double pipe heat exchanger with circumference-rectangular grooves has been investigated experimentally. The volume flowrate of cold and hot water were varied to determine its influence on the approach temperature of the outlet terminals. In this experimental design, the grooves were incised in annular room that is placed on the outside surface of the inner pipe. The shell diameter is 38.1 mm and tube diameter 19.4 mm with 1 m length, which is made of aluminum. The flow pattern of the two fluids in the heat exchanger is a parallel flow. The working fluid is water with volume flow rate of 27.1, 23.8 and 19.8 l/minute. The temperature of water on the inlet terminals are 50±1^{0}C for hot stream and 30±1^{0}C for cold stream. Temperature measurements conducted on each terminal of the inlet and outlet heat exchanger. The results showed that the grooves induced the approach temperature. The change of the approach temperature from the grooves compared to that of without grooves decreased by 37.9%. This phenomenon indicates an increase in heat transfer process and performance of the heat exchanger. Groove improves the heat surface area of the inner pipe, increasing the momentum transfer and in the other hand, reducing the weight of heat exchangers itself.

Keywords: Approach temperature, groove, heat exchanger, optimization

INTRODUCTION

A heat exchanger is a heat transfer device that exchange heat between two or more process fluids at different temperature. Heat exchangers have widespread applications in industrial such as industry process, petroleum transportation, power, heat recovery, RHVAC and many domestic applications. One of the simply heat exchanger is double pipe heat exchanger (DPHE). Their usual application is for small duties and they are suitable for high pressure and temperature. Another advantage is the design of DPHE serviceable easily than the other heat exchanger and can be connected in series or parallel. Heat exchangers have to fulfill some requirement such as effective temperature difference. The effective temperature driving force is a measure of the potential for heat transfer that exists at the design conditions. Effective temperature difference is defined by the log mean temperature difference (LMTD). Other simply method to measure thermal performance of heat exchanger is using temperature approach. Temperature approach referred to difference of temperature between hot and cold outlet of heat exchanger.

The need for lightweight, space-saving, and economical heat exchangers has driven the development of compact surfaces. Many engineering techniques have been developed for enhancing the rate of heat transfer through the wall surface of heat exchanger. The enhanced of heat transfer accompanied by pressure drop and drag increase. One of the flow surface applications to increase heat transfer in turbulence flow is groove. Using grooves is well known as passive flow control. The passive flow control as concerned wide consideration as it need no auxiliary power, increase surface area, induced pressure drop, simplicity and hence give more benefits (Sunu *et al.*, 2015). Changing the surface of pipe with circumference rectangular groove can be used to control flow separation. The groove provided as vortex generator and increased the boundary layer momentum.

Over the years, numerous researchers have carried out experimental studies on flow characteristics through internally grooved pipe. Sunu (2015) investigated increased pressure drop on internal wall surface in pipe with rectangular groove. Sunu *et al* (2015) revealed the effect of groove numbers on pressure drop in pipe flows. Investigation about pressure drop characteristics in a periodically grooved was conducted by Adachi *et al* (2009). The investigation on pipe bend with rough surface produced 4 % drag reduction (Nakao, 1995). Lee and Jang (2005) studied v-shaped groove over curved body. Choi and Orchard (1997) reveal about turbulence management using passive device for momentum transfer. Ding *et al* (2009) present a correlation to predict frictional pressure drop of R410A-oil mixture inside internal spiral grooved microfin tube. Alam and Yong Kim (2012) conducted mixing analysis in microchannel with rectangular grooves. A number of studies have been carried out to optimize the enhancement the heat transfer of rough surface geometry. Investigation of turbulent channel flow conducted by Katoh *et al* (2000), reveal the relationship between drag and thermal enhancement by rough surface. Aroonrat *et al* (2013) investigated the relationship of heat transfer and flow characteristics in horizontal internally grooved tube made by stainless steel. The result showed the thermal enhancement factor is 1.4 to 2.2 for grooved tube. Graham *et al* (1999) studied of pressure drop and heat transfer in copper tube with axially grooved. Lorenz *et al* (1995) investigated the distribution of the heat transfer coefficient in a channel with periodic transverse grooves.

In the several paper describe above, most studies focus on heat transfer, flow characteristics and the both relationship inside/outside grooved tube. The studies of heat transfer through grooved pipe in a room between tube and shell has received little attention. Consequently, the objective of this present study is to optimize the temperature approach of double pipe heat exchanger with groove placed in annulus room.

METHODOLOGY

The flow pattern of the two fluids in a general heat exchanger can vary depending on the heat exchanger design. One of the flow patterns is a parallel flow, which is the hot fluid and cold fluid move in the same direction. This investigation used parallel flow in the flow direction. The hot side inlet temperature is in contact with the cold side inlet temperature, the two fluid progresses through the heat exchanger and at the other terminal point the cold side outlet temperature is in contact with the hot side outlet temperature. In a parallel flow heat exchanger, the approach temperatures are the difference between the outlet temperature of one stream and the outlet temperature of the other stream

The experiment was performed using the experimental apparatus shown schematically in Fig.1. The experiment installation consisted two centrifugal pump which was stabilized by electric stabilizer to circulate hot and cold water through a loop. The control valve used to vary flow rate through the test section and the water tank to hold water from the system.

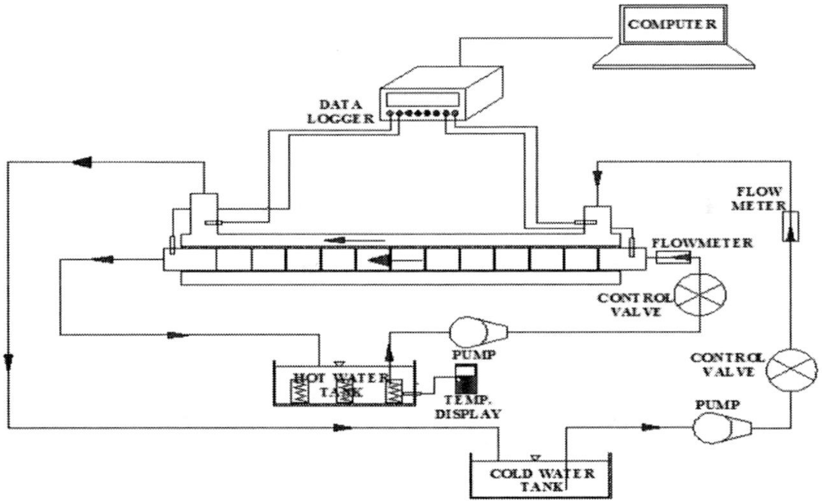

Figure-1. Experiment set-up

The hot water temperature was 50±1^{0}C and the cold water was 30±1^{0}C. The volume flow rate was varied by control valve set some distance from test section as 27.1, 23.8 and 19.8 l/minute respectively. The tube test outer diameter is 19.4 mm which has circumference rectangular groove 40 x 0.4 mm. While, the shell test section has 38.1 mm outer diameter. All of the tube made of aluminum pipes.

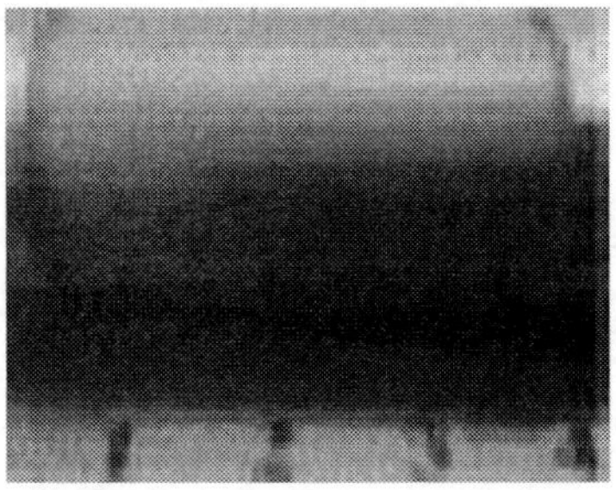

Figure-2. Groove geometry

Temperatures were measured by thermocouples placed in the inlet and outlet terminal of hot and cold water from the test section. A data logger digitalized signal from thermocouples before being recorded in a computer memory. The approach temperature (AT) was determined using the data point from hot and cold terminals of 750 s.

The percentage of approach temperature (% AT) from the grooves was compared to that of a smooth pipe (without grooves) and estimated by using the equation bellow:

$$\%AT = \left[\frac{(AT)_{smooth} - (AT)_{groove}}{(AT)_{smooth}} \right] x100\% \qquad \text{Eq.1}$$

RESULTS AND DISCUSSIONS

Preliminary experiments were done on the double pipe heat exchanger without groove (smooth pipe) to establish the time series temperature and approach temperature. This data was then used as a comparison to the data from the double pipe heat exchanger with groove. The approach temperature data in DPHE are shown in fig. 3 and 4.

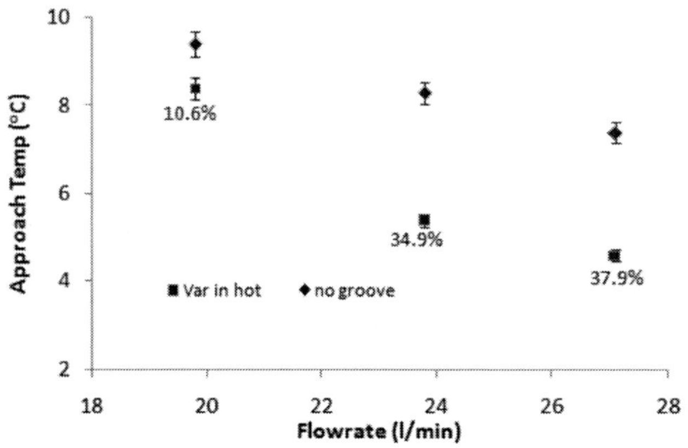

Figure-3. Average approach temperature for hot water flowrate variation.

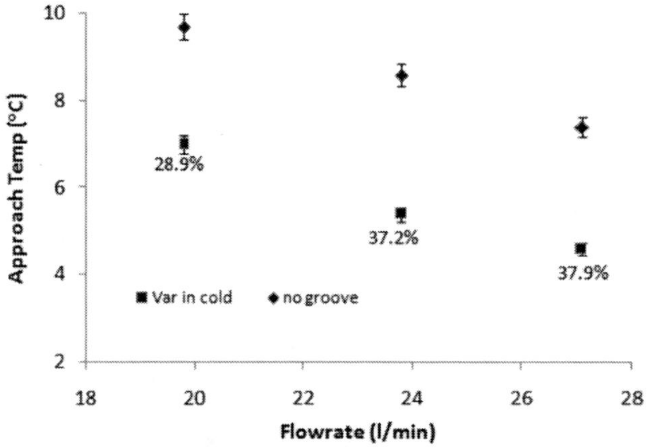

Figure-4. Average approach temperature for cold water flowrate variation.

Fig. 3 shows the average value of approach temperature at various hot water flowrate with constant cold-water flowrate. While figure 4 shows the average value of approach temperature at various cold-water flowrate with constant hot water flowrate. The rectangular pointed denotes the average approach temperature in smooth pipe DPHE. The percentage values on the plot show percentage approach temperature estimated using equation 1. For variation in hot water supply 19.8, 23.8, 27.1 l/min, the approach temperature data of grooved DPHE indicates reduction of about 10.6%, 34.9%, and 37.9% respectively. For variation in cold water supply 19.8, 23.8, 27.1 l/min, the approach temperature data of grooved DPHE indicates reduction of about 28.9%, 37.2%, and 37.9% respectively. This phenomenon indicates the increases of heat transfer process between two fluids.

It was observed that approach temperature decreased with increasing flow rate in all case. The approach temperature obtained from the circumference-rectangular grooved was lower than that those from the smooth pipe. This occurred because the grooved pipe provided the enhancement of heat transfer by increasing recirculation region of the flow. Groove produced swirl effect on the fluid near to the pipe surface in the annulus room which lacerating the boundary layer. The fluid thermal boundary layer became thinner and the increasing fluid mixing process increased the rate of heat transfer which represented by the rate of approach temperature.

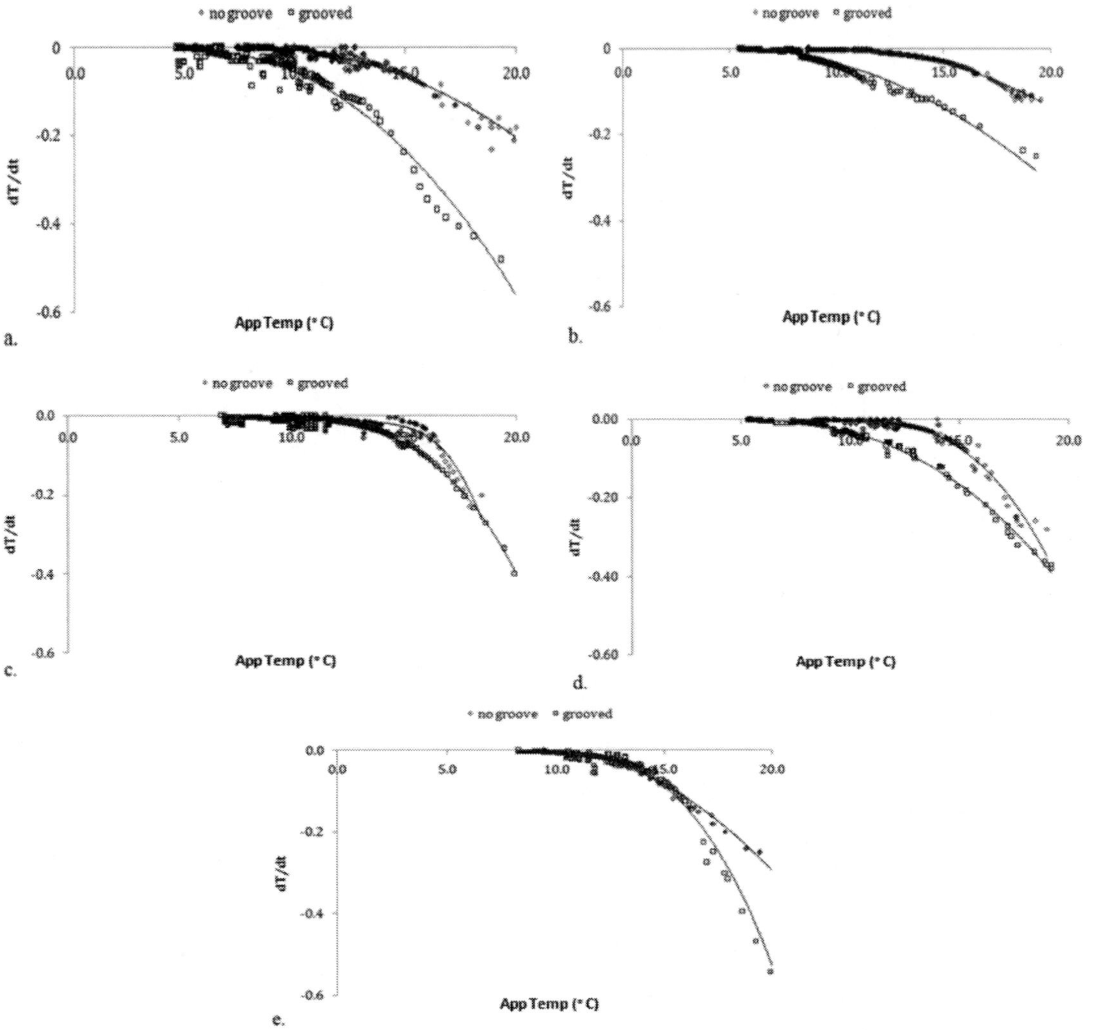

Figure-5. The rate of approach temperature (a) Q_{hot}= Q_{cold}= 27.1 l/min; (b) Q_{hot}= 27.1 l/min, Q_{cold}= 23.8 l/min; (c) Q_{hot}= 27.1 l/min, Q_{cold}= 19.8 l/min; (d) Q_{hot}= 23.8 l/min, Q_{cold}= 27.1 l/min; (e) Q_{hot}= 19.8 l/min, Q_{cold}= 27.1 l/min

The rate of approach temperature as a function of approach temperature is presented in figure 5. Marked with circle indicates the condition of smooth pipe (without groove) and marked with rectangular point represent with grooved pipe. Negative sign on ordinate axis, indicating the occurrence of a decrease in approach temperature and indicates greater of heat transfer process. For dT/dt grooved pipe, the rate of approach temperature for each approach temperature is more negative then that the smooth pipe. The greater the flow rate, the more negative the value dT/dt will be. Increased flowrate is accompanied by an increase in Re. Greater Re will increase the randomness of fluid motion, resulting in higher momentum convection and energy transfer through interchange of fluid particles.

CONCLUSIONS

Heat transfer in double pipe heat exchanger with circumference-rectangular grooves was investigated using experimental method. The results showed that the grooves induced the approach temperature. Groove improves the heat surface area of the inner pipe (tube side), increasing the momentum transfer, produced swirl effect on the fluid near to the pipe surface in annulus room. The change of the approach temperature from the grooves compared to that of without grooves decreased by 37.9%, which indicates an increase in heat transfer process and performance of the heat exchanger.

REFERENCES

Adachi Takahiro, Yamato Tashiro, Hirofumi Arima, Yasuyuki Ikegami. 2009. Pressure drop characteristics of flow in a symmetric channel with periodically expanded grooves. Chemical Engineering Science 64: 593-597

Alam Afroz, Kwang-Yong Kim. 2012. Analysis of mixing in a curved microchannel with rectangular grooves. Chemical Engineering Journal 181–182: 708–716.

Aroonrat, K., Jumpholkul, C., Leelaprachakul, R., Dalkilic, A.S., Mahian, O., Wongwises, S. 2013. Heat transfer and single-phase flow in internally grooved tube. International Communication in Heat and Mass Transfer 42: 62-68.

Choi, K.S and Orchard D.M. 1997. Turbulence Management Using Riblets for Heat and Momentum Transfer. Experimental Thermal and Fluid Science 15: 109-124.

Ding G., Haitao Hu, Huang X., Bin Deng, Yifeng Gao. 2009. Experimental investigation and correlation of two-phase frictional pressure drop of R410A–oil mixture flow boiling in a 5 mm microfin tube. International Journal of Refrigeration 32: 150–161.

Graham D., J.C. Chato, T.A. Newell. 1999. Heat transfer and pressure drop during condensation of refrigerant 134a in an axially grooved tube, International Journal of Heat and Mass Transfer 42: 1935–1944.

Katoh K., Choi, K.S.,Azuma, T. 2000. Heat-transfer enhancement and pressure loss by surface roughness in turbulent channel flows. International Journal of Heat and Mass Transfer, 43: 4009-4017.

Lee, S.J. and Jang, Y.G. 2005.Control of flow around a naca 0012 airfoil with a micro-riblet film. Journal of Fluid and Structure 20: 659-672.

Lorenz S., Mukomilow D., Leiner W. 1995. Distribution of the heat transfer coefficient in a channel with periodic transverse grooves. Experimental Thermal and Fluid Science 11: 234-242

Nakao, S.I. 1995. Effects of riblet bends on pipe flows. Applied Scientific Research 54: 237-247.

Sunu P.W. 2015. The characteristics of increased pressure drop in pipes with grooved. *Adv .Studies Theor. Phys.* 9(2): 57-61. http://dx.doi.org/10.12988/astp.2015.412152

Sunu P.W., ING Wardana, A.A.Sonief, Nurkholis Hamidi. 2015. The effect of wall groove numbers on pressure drop in pipe flows. Int. J. Fluid Mech. Resch., 42(2): 119 - 130

MATEC Web of Conferences **58**, 04007 (2016) DOI: 10.1051/matecconf/20165804007
BISSTECH 2015

ANALYSIS OF THE CONTACT AREA OF SMOOTH AND ROUGH SURFACES IN CONTACT WITH SPHERE INDENTER USING FINITE ELEMENT METHOD

Kartini[1, 2,*], E. Saputra[2], R. Ismail[2], J. Jamari[2], A.P. Bayuseno[2]

[1)]University of Pembangunan Nasional "Veteran" Jawa Timur, Surabaya, East Java, Indonesia
[2)]Department of Mechanical Engineering, University of Diponegoro,
Semarang, central Java, Indonesia.
E-mail: kartiniwachman@gmail.com

ABSTRACT

This paper analyzes the contact area of the contact between a deformable rough surface (smooth and rough) and a hard smooth sphere indenter using finite element method. A method was introduced to generate a three dimensional rough surfaces using Computer Aided Design (CAD) software. The rough surface model was developed based on the surface measurement data, while the smooth surface model was generated from the CAD software. Contact area and contact deformation were analyzed. Results showed that the contact area between rough surface versus sphere and smooth surface versus sphere is different.

Keywords: rough surface; finite element simulation; contact area; contact pressure.

INTRODUCTION

In micro scale, the real surface topography of mechanical component that has smooth appearing is rough surface. It is the component of surface texture due to the production process, excluding waviness and deviation of form (van Beek, 2012). Machining process in component manufacturing will affect the surface quality, because the final process of machining is determined by surface roughness. Therefore, a variety of machining process will generate different surface roughness.

The rough surface can be described as the collection of hill and valley, or called as "asperity" (Jamari, 2006). Therefore, the real contact interaction in engineering surfaces is the contact interaction between rough and rough surfaces. In other words, the real contact interaction in the engineering surface is asperity against asperity. In this case, topographical changes that occurred was actually at the level of asperity. Based on this condition, the behavior of interaction of two contact surfaces can be analyzed at the asperity level.

To simplify this, a lot of research on the interaction between the asperity contacts was made with simple model. It is contact interaction of sphere against sphere or sphere against flat, where the sphere is assumed as one asperity. Hertz (1882) proposed a theory of static contact model in elastic condition, where the asperity was modeled as sphere and cylinder geometries. Chang et al. (2008) generated static contact model in elastic-plastic condition analytically where asperity was modeled in sphere geometry. Zhao et al. (2005) also created static contact model in elastic-plastic condition where asperity was modeled in sphere geometry. Kogut and Etsion (2002) and Jackson and Green (2005) also analyze it in elastic-plastic condition. Later, Jamari and Schipper (2006) modeled the asperity in ellipsoid geometry.

Analysis of the real rough surface is negligible because of the difficulty to build a model of real rough surface and also to simplify it in order to save the time calculation (Peyrou et al., 2007). Indeed, the surface roughness plays an important role in influencing the friction, wear, and lubrication (Lee and Ren, 1996). Therefore, the rough surface model to approach the real surface and in order to get results closer to the actual topography changes is needed where the real condition of contact interaction on rough surfaces can be seen from the contact area.

The purpose of this research is to study the contact area on smooth and rough deformable surfaces in contact with a hard smooth sphere indenter using finite element analysis.

MODELING OF STATIC CONTACT

Commonly, the first step was to create rough surface based on mathematical calculations in the form of point clouds with extension (*.xyz) randomly. These point clouds were then sent to the CAD software to be converted into three-dimensional (3D) model. By using CAD software, it was changed to IGES extension (*.iges) or other extensions which were compatible with FE software. The static contact simulation on rough surface was performed using ABAQUS commercial software. These processes will be explained in more detail.

Rough surface model

The rough surface model could be directly created in finite element (FE) software as had been done by Thompson. The rough surface by utilizing the nodal or elements had been formed first, then the height of the nodal needed to be changed randomly (Thompson, 2007). That method had disadvantage of being very sharp peak asperity, whereas the surface with a sharp peak asperity should be avoided since it would bring extreme stress. In this research, the rough surface that was created would decrease these weaknesses by using alternative method. The method was a combination of mathematical, CAD and FE software.

The rough surface in this study was based on a mathematical equation and constructed using mathematical software. Gaussian equations were used to create a surface with random roughness. Figure 1 shows the code used in mathematical software to generate rough surface randomly by Gaussian height distribution (GHD) equation. The rough surface can also be obtained from

VEECO 3D images data which are then exported to the mathematical software for conversion into a form of point clouds with extension (*.xyz). The GHD equation will produce the rough surface with vary asperity heights randomly. The GHD equation that applied into a mathematical software consists of several variables. *N* is the number of points in one surface, *rL* is the length of the surface area, *h* is height of the root means square (rms) and *clx* is the correlation between points in a surface distance (pixel).

```
N=200;
rL=1;
h=0.05;
clx=0.05;
x = linspace(-rL,rL,N);
y = linspace(-rL,rL,N);
[X,Y] = meshgrid(x,y);
Z = h.*randn(N,N);
F = exp(-((X.^2+Y.^2)/(clx^2/2)));
f                                              =
2/sqrt(pi)*rL/N/clx*ifft2(fft2(Z).*fft2(F));
surf(X,Y,f,'FaceColor','interp',...
'EdgeColor','none',...
'FaceLighting','phong')
axis equaloff
```

Figure-1. The code of Gaussian height distribution (GHD) for rough surface.

The implementation of *N*, *rL*, *h* and *clx* with different values would resulting a different rough surface. And also, the same value of *N*, *rL*, *h* and *clx* would generate different rough surface every time, for the same rms value. As an illustration, the variation of *h* is presented in Table 1 and and its generating surfaces as are shown in Fig. 2 to Fig. 5.

Table-1. Variation of *N*, *rL*, *h* and *clx*.

Variation	N	rL [mm]	h [mm]	clx [mm]
Var1	200	1	0.0025	0.05
Var2	200	1	0.01	0.05
Var3	200	1	0.05	0.05
Var4	200	1	0.5	0.05

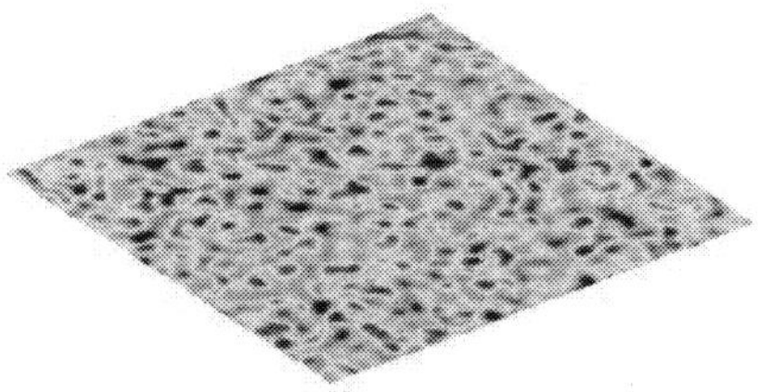

Figure-2. The rough surface with Var1 (*N*=200; *rL*=1; *h*=0.0025; and *clx*=0.05).

Figure-3. The rough surface with Var2 (*N*=200; *rL*=1; *h*=0.01; and *clx*=0.05).

Figure-4. The rough surface with Var3 (*N*=200; *rL*=1; *h*=0.05; and *clx*=0.05).

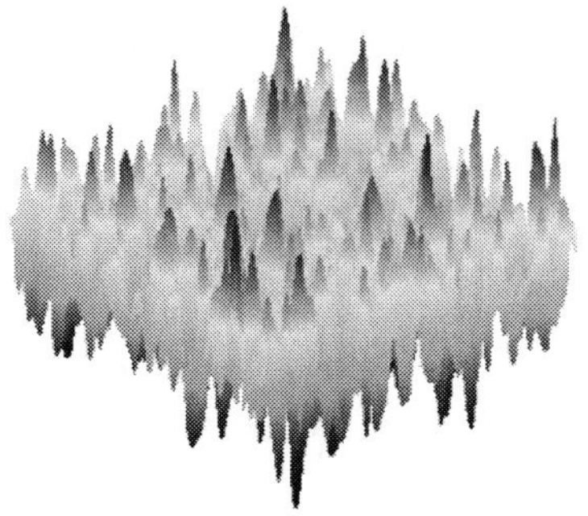

Figure-5. The rough surface with Var4 (*N*=200; *rL*=1; *h*=0.5; and *clx*=0.05).

Figure 2 to Figure 5 show that value of *N*, *rL*, *h* and *clx* give different rough surfaces. As can be seen, the *h* variable plays an important role in this equation. The *h* variable represents *Rt* in the surface roughness standard parameter. *Rt* is the distance between the highest asperity and the lowest valley measured over the entire assessment length or usually called as peak-to-valley height (van Beek, 2012).

```
N=200;
rL=2;
h=0.5;
clx=0.05;
x = linspace(-rL,rL,N);
y = linspace(-rL,rL,N);
[X,Y] = meshgrid(x,y);
f = h.*cos(5.*X).*cos(5.*Y)
surf(X,Y,f,'FaceColor','interp',...
'EdgeColor','none',...
'FaceLighting','phong')
axis equal off
```

Figure-6. The code of Gaussian height distribution (GHD) for sinusoidal surface.

Figure-7. The sinusoidal surface with (*N*=200; *rL*=2; *h*=0.05; and *clx*=0.05).

Another example, the GHD surfaces are presented in sinusoidal surfaces. Figure 6 shows the code for generating the GHD sinusoidal surface. With the variation of *h*, different surfaces can be obtained, see Fig.

7 and Fig. 8. It can be seen from these figures that the topography of these surfaces are well-regulated. Therefore, this surface will not be used in the present study due to its simplicity.

Figure-8. The sinusoidal surface with (N=200; rL=2; h=0.05; and clx=0.05)

3D rough surface model

After generating the roughness (2D), next is developing 3D rough surface model. The rough surface point clouds (Fig. 9) was then analyzed to form a 3D surface using CAD software and then converted to form solid 3D (Fig. 10). The 3D surface model was then converted into (*.iges or *.sat) extension. The IGES (initial graphics exchange specification) is a common form of the extension that can be used in any CAD or FEA software. Then, the 3D model was exported to the FEM software.

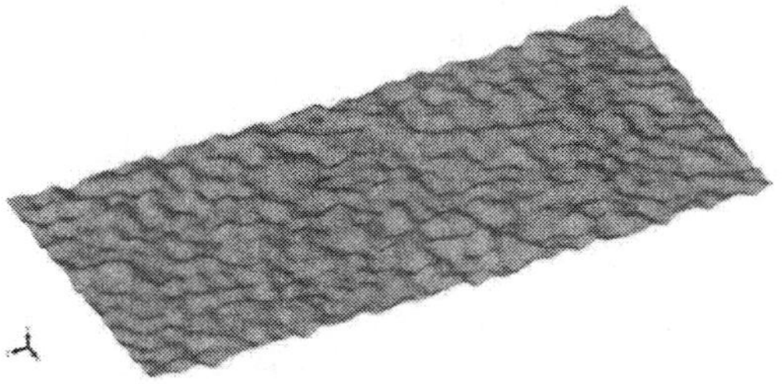

Figure-9. Point clouds of rough surface.

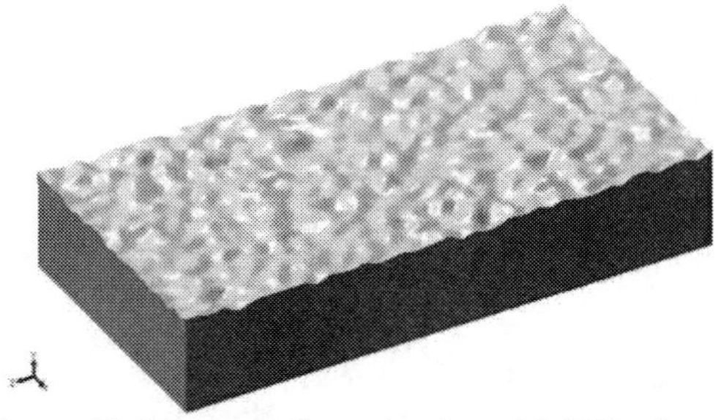

Figure-10. The three dimensional model (3D) of rough surface.

Modeling the contact

Analysis performed in this study was a static contact. A commercial finite element analysis software ABAQUS was employed. The steps for finite element analysis were creating of parts (create geometry), property (define the material properties), assembly, step (specify the type of analysis and simulation stages), interaction (define a contact interaction), load (specify loads and boundary conditions), mesh (discretize geometry into several elements), job (ordered the software to perform iterations) and visualization (see the results of the simulation). The 3D finite element model with the generated mesh of the rough surface is presented in Fig. 11.

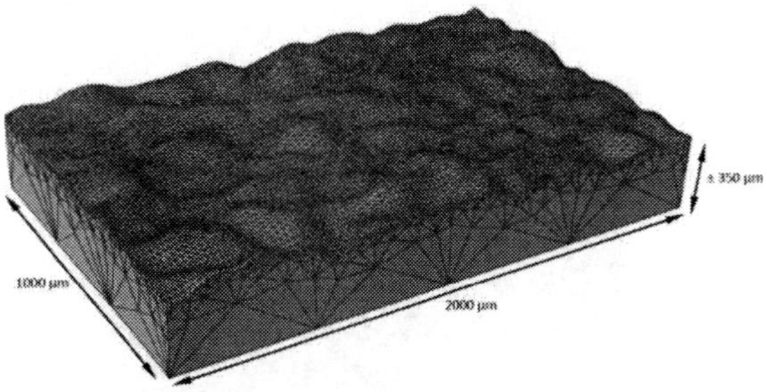

Figure-11. The 3D model with mesh for static contact simulation

Figure 12 shows the finite element contact model of the contact between a rough surface and a smooth sphere. For the case of the contact between the smooth surface and the smooth sphere, the rough surface is replaced by the smooth surface. The rigid sphere indenter was loaded with a normal force F of 5N toward to the deformable rough surface in -z direction. All nodal at the bottom of the deformable rough surfaces were fixed. The elastic-plastic contact analysis was performed for the elastic-perfectly plastic material property.

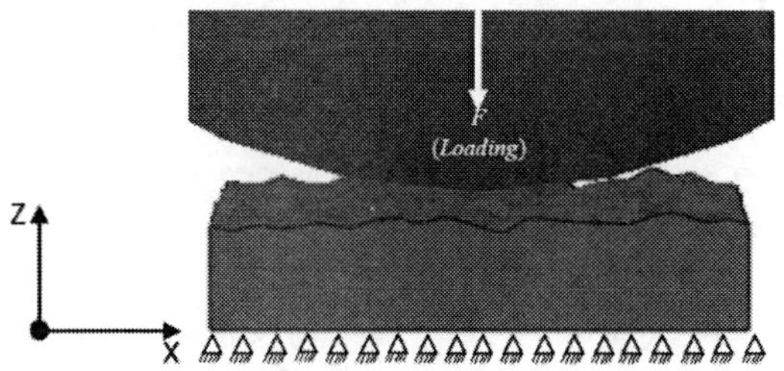

Figure-12. FEM model of the contact between a rough surface and a smooth sphere.

RESULTS AND DISCUSSION

Validation

Results of the developed finite element contact model was validated with the experimental result of Jamari et al., 2006. In the experiment of Jamari et al., 2006, a static contact between a steel ball and an aluminum rough surface was performed. Diameter of the steel sphere indenter is 10 mm with the rms roughness value of about 0.05 µm. This roughness is relatively smooth compare to the rms roughness of the aluminum of about 4 µm. The applied normal load was 1 Newton. The same contact conditions were simulated in the present study. The material properties of the contact pair is presented in Table 2.

Table-2. Material properties of steel and aluminum, Jamari et al., 2006.

Components	E [GPa]	σy [GPa]	ν [-]
Steel sphere indenter	210	2.67	0.3
Aluminum flat	75.2	0.086	0.34

DOI: 10.1051/matecconf/20165804007

Figures 13(a) and (b) show the results of the experiment of Jamari et al., 2006 and the present finite element analysis, respectively. Here, the data of the surface roughness was taken from the measured data of Jamari et al., 2006. On the qualitative comparison, the present model shows a good agreement with the experiment.

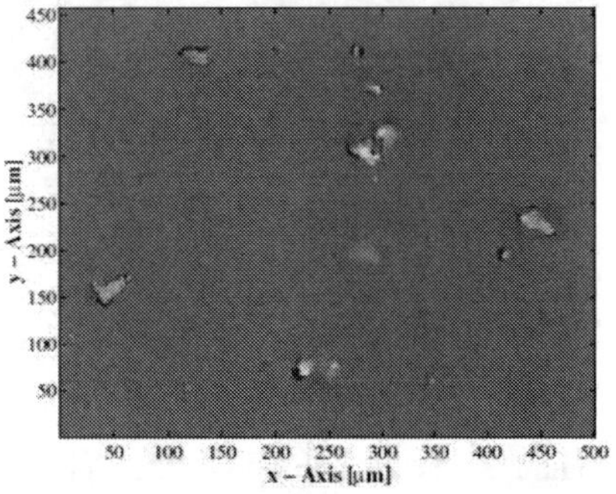

(a)

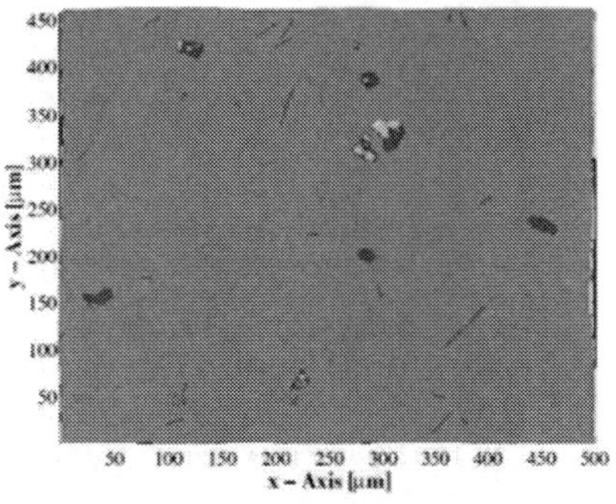

(b)

Figure-13. Contact are of a) the experimental result of Jamari et al., 2006, and b) the finite element analysis of present study.

Smooth surface and rough surface contacts

Figure 14 shows the results of the contact area for the contact of the hard smooth sphere against the deformable rough surface (a) and for the hard smooth sphere against the deformable smooth surface (b) for the same load.

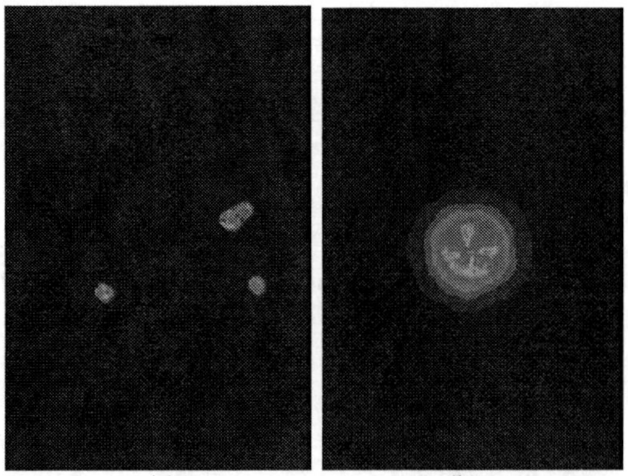

(a) (b)

Figure-14. Contact area of a) rough surface, b) smooth surface.

The total contact area for the rough surface is smaller than the total contact area for the smooth surface. The difference of the contact area is caused by the difference deformation mode. In the rough surface there are many asperities deform plastically while in the smooth surface the asperity deform elastically.

CONCLUSION

Analysis of the contact area of the contact between a deformable rough surface (smooth and rough) and a hard smooth sphere indenter using finite element method have been performed. The 3D rough surface model was developed based on the surface measurement data and the CAD software. Results showed that the developed rough surface finite element contact model agree well with the experimental result. For the contact area simulation it was found that the contact of rough surface versus sphere is smaller than the contact of smooth surface versus sphere.

REFERENCES

Chang, W. R., I. Etsion and D. B. Bogy; "An elastic-plastic model for the contact of rough surfaces," Journal of Tribology, **109**, 257 – 263 (1987).

Hertz, H.; "Uber die beruhrung fester elastische korper und uber die harte (On the contact of rigid elastic solids and on hardness)," Verhandlungen des Vereins zur Beforderung des Gewerbefleisses, Leipzig, Nov (1882) (For English translation see Miscellaneous Papers by H. Hertz, Eds Jones and Schott, Macmillan, London, 1896).

Jackson, R. L. and I. Green; "A finite element study of elasto-plastic hemispherical contact against a rigid flat," Journal of Tribology, **127**, 343 – 354 (2005).

Jamari, J. and D. J. Schipper; "An elastic-plastic contact model of ellipsoid bodies," Tribology Letters, **21**, 262 – 271 (2006).

Jamari, J., Running-in of rolling contacts, PhD Thesis, University of Twente, Enschede, The Netherlands (2006).

Kogut, L. and I. Etsion; "Elastic-plastic contact analysis of a sphere and a rigid flat," Journal of Applied Mechanics, **69**, 657 – 662 (2002).

Lee, S. I. C. and N. Ren; "Behavior of elastic-plastic rough surface contacts as affected by surface topography, load, and material hardness," Tribology Transactions, **39**, 67 – 74 (1996).

Peyrou, D., F. Pennec, H. Achkar, P. Pons and R. Plana; "Effect of contact force between rough surfaces on real contact area and electrical contact resistance," published in MEMSWAVE Workshop, Barcelona, Spain (2007).

Thompson, M. K., A multi-scale iterative approach for finite element modeling of thermal contact resistence, PhD Thesis, Massachusetts Institute of Technology, USA (2007).

van Beek, A., Advanced engineering design: Lifetime performance and reliability, Delft University of Technology, The Netherlands (2012).

Zhao, Y., D. M. Maietta and L. Chang; "An asperity microcontact model incorporating the transition from elastic deformation to fully plastic flow," Journal of Tribology, **122**, 86 – 93 (2000).

AUTHOR INDEX

AUTHOR INDEX

EDP Sciences – Web of Conferences
17, Avenue du Hoggar
Parc d'Activité de Courtabouf
BP 112
F-91944 Les Ulis Cedex A
France

ISBN 978-1-5108-2464-5

CURRAN ASSOCIATES INC.
proceedings.com